Quantitative Methods in the Humanities and Social Sciences

More information about this series at http://www.springer.com/series/11748

Maurizio Forte · Stefano Campana
Editors

Digital Methods and Remote Sensing in Archaeology

Archaeology in the Age of Sensing

Springer

Editors
Maurizio Forte
Department of Classical Studies, Art, Art History and Visual Studies
Duke University
Durham, NC
USA

Stefano Campana
Ancient Topography, Head of LAP&T Lab, Department of History and Cultural Heritage
University of Siena
Siena
Italy

ISSN 2199-0956 ISSN 2199-0964 (electronic)
Quantitative Methods in the Humanities and Social Sciences
ISBN 978-3-319-40656-5 ISBN 978-3-319-40658-9 (eBook)
DOI 10.1007/978-3-319-40658-9

Library of Congress Control Number: 2016948800

Printed on acid-free paper

This Springer imprint is published by Springer Nature
The registered company is Springer International Publishing AG
The registered company address is: Gewerbestrasse 11, 6330 Cham, Switzerland

Preface

The 1990s will probably be remembered in the history archaeology as the age of GIS. At that time, the introduction of digital technology in archaeological research was in its infancy. Software and hardware had only a limited capacity to integrate the range and complexity of information involved in the archaeological process. In the following decade, however, the archaeological community became gradually aware of the need for a consistency of approach across the whole framework of archaeology, while rapid advances in software and hardware made it possible to envisage a significant renewal of the whole or large parts of the archaeological process. This was the age of the Digital Revolution.

At the same time, remote sensing gained an increasing relevance and application within archaeology and throughout the scientific community. Up to this stage, the definition of remote sensing had focused on the analysis of data collected by sensors that were not in physical contact with the objects under investigation, using cameras, scanners, radar systems, etc., operating from spaceborne or airborne platforms. Now, a wider characterization began to take hold, treating remote sensing as ***any*** nondestructive approach to viewing the buried and nominally invisible evidence of past activity. Spaceborne and airborne sensors (now supplemented by laser scanning) became joined by ground-based geophysical instruments and undersea remote sensing, as well as—for *some* archaeologists at least—by other noninvasive techniques such as surface collection or field-walking survey. Within this broader interpretation, ***any*** method that enables observation of the evidence on or beneath the surface of the earth, without impacting on the surviving stratigraphy, can legitimately be included within the ambit of remote sensing. This and other impulses have also resulted in a rapid growth in multidisciplinary working within and around archaeology and related cultural studies.

From the methodological point of view, the most important change over the past few years has been the burgeoning capacity of archaeologists and cultural historians to collect—relatively easily and quickly—massive 3D datasets at the landscape, local, site, and object scale. Initially, archaeologists did not know exactly how to manage this vast array of 3D information. They readily grasped the idea of its huge potential but did not see how to exploit it. The all pervading presence of the third dimension prompted the need for new perceptions of archaeological features and processes at an intellectual level, in terms of "3D thinking"—or better 4D thinking considered that as archaeologists, we cannot avoid dealing with the chronological dimension—and at a procedural level, challenging long-established approaches to archaeological documentation and therefore to the interpretation process as a whole.

Now, in the early years of the present decade, we feel that we are ready—or at least ***nearly*** ready—to embrace these new methods of recording, interpreting, conceptualizing, and communicating archaeological data and relationships across the passage of time. Technological, cultural, and epistemological advances are enticing us to encompass new and completely different perspectives based on immersive, interactive 3D and 4D environments for managing archaeological data at both the scientific and interpretative levels.

Everybody, in the next few years, will have the opportunity to blend the physical world with a sensory-rich "virtual" world where archaeologists can naturally and intuitively manipulate, navigate, and remotely share interpretations and case studies. Our understanding of archaeology will be taken to a new level, enhancing our capacity to develop interpretations and to present them to fellow specialists and to the general public as simulated scenarios in 4D. Rapid developments in ICT, including hardware and software for immersive environments, will even allow us to communicate and interact with one another through further cultural experiences such as sound, smell, and tactile interfaces. The transformation of the traditional remote sensing in "something else" defines new borders for this research field and suggests a new methodological approach. "Polysensing" rather than "remote sensing" can better define this revolutionary approach. It is quite interesting to notice that archaeology plays as primary actor in this revolution because of its multidisciplinary character and mission.

Welcome in the Age of Sensing!

Durham, NC, USA Maurizio Forte
Siena, Italy Stefano Campana

Acknowledgements

This book is not a proceeding of the conference "The Age of Sensing," but without the conference, we could not have this book. The Age of Sensing has been an extraordinary event and an inspirational opportunity for scholars and students for discussing cutting-edge research projects and applications.

As with most collaborative projects, there are many characters and groups whose assistance has proven crucial to the production of this volume.

First of all, we are sincerely grateful to a good many friends and colleagues who made a long journey to join the meeting at Duke University, Durham, North Carolina, and who made the publication that developed from it possible.

Particular thanks are offered of course to the University of Siena, the Department of History and Cultural Heritage and the Laboratory of Landscape Archaeology and Remote Sensing for the support provided for the organization of the conference and the Summer School in Siena and Vulci.

Also of great importance to the organization of the conference has been the role played from Melissa Huber, Ph.D. candidate at that time at Duke University who managed greatly the general secretariat making our work much easier before, during, and after the conference. In this respect, we are also particularly appreciative to the assistance of a number of Duke students, throughout the symposium to its sooth running.

The conference was supported by the Trinity College of Art & Science, the Department of Classical Studies, the Department of Art, Art History and Visual Studies, the Trent Foundation grant, Institutional sponsors were ICIP-ICOMOS, NASA JPL and UNESCO (Cultural Sector).

A special thanks to Vanda Gianpaoli and Roberto Rabitti for their very generous and outstanding contribution to the social activity of the conference "The Age of Sensing"

Durham, NC, USA — Maurizio Forte
Cambridge, UK; Siena, Italy — Stefano Campana

Contents

About the Editors

Maurizio Forte is William and Sue Gross professor of classical studies art, art history, and visual studies at Duke University. He is also the founder of the DIG@Lab (for a digital knowledge of the past) and Director of the Graduate Program in Classical Studies at Duke. In 2006–2011 was in the advisory board of the UNESCO Remote Sensing Archaeology—Open Initiative. His main research topics are as follows: digital archaeology, classical archaeology, and neuroarchaeology. He was a professor of World Heritage at the University of California, Merced (School of Social Sciences, Humanities and Arts), and director of the Virtual Heritage Lab. He was the chief of Research at CNR (Italian National Research Council) of "Virtual Heritage: integrated digital technologies for knowledge and communication of cultural heritage through virtual reality systems," senior scientist at CNR's Institute for Technologies Applied to the Cultural Heritage (ITABC), and professor of "Virtual Environments for Cultural Heritage" in the "Master of Science in Communication Technology-Enhanced Communication for Cultural Heritage" at the University of Lugano. He has coordinated archaeological fieldwork and research projects in Italy as well as Ethiopia, Egypt, Syria, Kazakhstan, Peru, China, Oman, India, Honduras, Turkey, the USA, and Mexico. Since 2010, he is the director of the 3D Digging Project at Çatalhöyük and since 2014 he is the Director of the fieldwork and digital project Vulci 3000 (Vulci, Italy). He published pioneering books on digital archaeology, virtual archaeology and cyber archaeology.

Stefano Campana is currently faculty member of the University of Siena (Italy), in the Department of History and Cultural Heritage, where he has engaged in teaching and research as senior lecturer in ancient topography. He is specializing in landscape archaeology, remote sensing, GIS, and archaeological methodology for purposes of research, recording, and conservation. His work is focused on the understanding of past landscapes from prehistory to the present day. The principal context for his work has been Tuscany, but he has also participated in and led research work in the UK, Spain, Turkey, Palestine, Iraq and Asia. He has been very active in the international sphere and has established a sound reputation for innovative research. In 2011, he was proposed and admitted as a fellow of the Society of

Antiquaries of London (FSA), and in 2012, he was invited to be a member of the General Management Board of HIST, the Governing Board of the International Centre on Space Technologies for Natural and Cultural Heritage, under the auspices of UNESCO and the Chinese Academy of Sciences. From 2014 to 2016 he was Senior Marie Curie Research Fellow at the University of Cambridge (UK), Faculty of Classics. There, he initiated a totally new project under the title 'emptyscapes', aimed at stimulating change in the way archaeologists in the Mediterranean world study the archaeology of landscapes, moving from an essentially site-based approach to a truly landscape-scale perspective.

Introduction

The roots of this book lie in the 5th International Conference on Remote Sensing in Archaeology, the Age of Sensing which took place from October 13 to October 15, 2014, at Duke University within the "From Space to Place initiative."

The initiative started in Beijing on 2004 where took place the first conference organized from the Chinese Academy of Sciences, and in the years, this experience has been developed—independently—from the editors of the present book organizing conferences, workshops, and summer schools in Italy, at Roma and Tuscany (2006), India at Tiruchirappalli (2009), China at Beijing (2012), USA at Berkeley California (2012), France at Marseille (2013), Italy at Siena and Vulci (2014), and of course back to the USA at Durham (2014).[1]

From space to place, initiative could be considered within the framework of current orthodox scientific environment, anarchic, amorphous, and self-referential. To be fair after more than 10 years, it is even difficult to find any other definition than "initiative" to describe it. It is not at all a national or an international association, a scientific society, or whatever. We would probably define it as an ongoing forum aimed to bring around the world every one, sometimes two or three years yet depending from various reasons, the discussion on remote sensing in archaeology, intended from the broader point of view. Priority is given to places where it is more unusual to have the opportunity to put large number of outstanding scientists together or experts belonging to different environments (scientific as well geographic). Moreover, privileged areas are those under threat due to fast and unrestrained economic development, population growth, global warming, environmental pollutions, war, terrorism, and so forth.

After the first experiences, we thought that the publication conference proceedings were unsuitable to deploy the actual meaning of the seminar. Therefore, from the 5th edition of the conference, we decided to change policy (to be fair, we never had any) editing a monograph volume on the main topic of the conference. Therefore, this book is by no means just as a selection of best papers of the 5th

[1]Campana S., Forte M 2006 and Campana S., Forte M., Liuzza C., 2010.

conference rather than the meeting has been useful to identify a group of authors dealing with themes that would be developed according to the purpose of the present book.

The book opens providing an overview of a leading-edge technology of data recording impacting substantially the last years. The range choice was definitively very wide, lot of technologies had a major influence on archaeology in the last decades, but we decided to focus on laser scanning, terrestrial, and airborne. Indeed, laser scanning from the one hand played a very important role within archaeology, raising new attention on third dimension and form the other hand airborne Lidar provided for the first time a power full tool aimed to explore the 'black hole' of landscape archaeology: forested areas. Part "Data Collection and Technology" is organized into two contributions: The first one written by Nicola Lercari (University of California, Merced) provides a rather comprehensive critical overview of terrestrial laser scanning (TLS) in the age of sensing. The second paper written by Rachel Opitz (University of Arkansas) delivers an extended synopsis of the state of the art of the application of airborne laser scanning in archaeology with particular regard to undercanopy case history.

Part "Image and Digital Processing" deals with image and digital processing in relation to visual representation and methodological strategies and sites' analyses. The first manuscript written by Stacy Curry, Roy Stine, Linda Stine, Jerry Nave, Richard Burt, and Jacob Turner (University of North Carolina) deploys a case study on TLS and ground-penetrating radar (GPR) imaging spatial integration conducted near the third line action at the battle of Guilford Courthouse (American Revolutionary War, March 15, 1781), located at the Guilford Courthouse National Military Park, Greensboro USA-NC. The TLS dataset demonstrated the possibility to discern the concave surface found in the dense overgrown and obstructed wooded area integrating the subsurface feature seen in the GPR data before entering a heavily wooded area. The next paper by Till Sonnemann, Eduardo Herrera Malatesta, and Corinne Hofman (Leiden University) deals with image processing aimed to identify evidence from Unmanned Aerial System (UAS) photogrammetric surveying. The cultural as well the environmental context is tremendously challenging: precolonial settlements in northern Hispaniola (the Dominican Republic). The result is very interesting proving great potential for fast and precise recording of archaeological sites in difficult terrain providing a fast, detailed, and affordable opportunity to monitor changes to the landscape, caused by agriculture, new development, illegal looting, and so forth.

Part "Landscape Representation and Scales" is a quite dense one, facing within landscape studies, theoretical, methodological, and practical issues. The main topics are focused on multiscale landscape study, visibility and emptiness, landscape representation and reconstruction, and accuracy and visual analysis.

Frank Vermeulen (University of Ghent) in his contribution provides an excellent example of critical thinking developing an integrated multiscale and multisurvey approach to the study of the now-rural but formerly urban historical Mediterranean landscapes.

Stefano Campana (University of Siena, Italy) delivers a paper aimed to contextualize and clarify the state of the art of landscape survey focusing on the massive gap in quality and intensity of the research (substantially increased in the last decades) between the analysis of past rural landscapes and past cityscapes; this paper presents a case study aimed to reconcile city and countryside.

Dave Cowley focuses on an analytical review of the causes of gaps, uncertainties, and absences within the archaeological record mainly based on Central and northern Europe but which can be extended to a much broader area.

Heather Richards-Rissetto with her colleagues, Shona Sanford-Long, and Jack Kirby-Miller bring the reader to Copan (Honduras) presenting an stimulating project. Research design and goals instead of being focused on the typical computer-based visualization showing buildings and monuments surrounded by a mass of emptiness in a lunar landscape is aimed developing 3D visualization tool and workflow that have value for examining changes in land use, environment and settlement displaying 3D synchronic patterns visualization as well 4D landscape transformation across time. Back to Europe (Iberia), Edward Triplett uses spatial technologies and particularly the combination of the volumetric and GIS viewshed analysis methods. The case study reveals how between the twelfth and fourteenth centuries, frontier institutions (Muslim and Christian) controlled territory valued landscape visibility as a measurement of security and surveillance, while also acknowledging how vision affected architecture-scale decisions at a military- monastic complex on the frontier.

The contribution Kasper Hanus (Adam Mickiewicz University in Poznań) and Emilia Smagur (University of Sydney), is focused on Cambodia where up to recent time archaeologists have made a substantial progress in the research on the medieval landscape of the urban complex of Angkor neglecting regional survey. This paper presents the large-scale reconnaissance survey based on satellite imagery, which has been implemented to fill the gap in the understanding of the past landscapes in the region. Finally, John K. Millhauser (North Carolina State University) and Christopher T. Morehart (Arizona State University) argue on how Imaging and spatial analysis technologies can revolutionize archaeological methods and archaeologists' perceptions of space. More specifically the use of spatial data in the past recalls the importance of human experience in the representation and description of the empirical world.

Part "Simulation, Visualization and Computing" presents an overview of 3D archaeology, collaborative research, computing, modeling, and supercomputing. Maurizio Forte (Duke University) discusses theory and practice of cyberarchaeology at the intersection of digital embodiment, 3D polysensing environments, and neuroscientific perspectives. The introduction of mass application of virtual reality in research, education, and entertainment is changing completely the human approach to cultural transmission and the reconstruction of the past.

Bill Seaman (Duke University) seeks in his paper to define a new holistic approach to cyberarchaeology including new forms of multimodal sensor hardware to work in conjunction with current sensor systems. Here, we are in the sphere of polisensing systems—parallel multimodal sensing over time—enable the creation of a form of media object that can be given additional metadata and can be explored via state-of-the-art search algorithms, new metamarkup methodologies, and virtual visualization.

Nicoló Dell'Unto (University of Lund—SE) discusses how the development and use of three-dimensional geographic information system (3DGIS) are affecting the way archaeologists retrieve and analyze material detected in the field in support of more accurate archaeological interpretations.

Devin A. White (University of Tennessee) delivers an introduction to the world of high-performance computing, focusing on the present and the future of archaeological supercomputing, using several ongoing projects across a broad swath of the discipline as examples of where we are now and signposts for where we are heading, concluding with some thoughts on the art of the possible, given current and emerging technological trends. One of the goals of this paper is that the reader will come away feeling less intimidated by the idea of using supercomputing to solve archaeological problems and knowing that they can and should take full advantage of the computing power available today as well as help drive how the systems of tomorrow are designed.

Part "Interpretation and Discussion" takes on a central issue of the archaeological debate: how to approach data analysis, interpretation, archiving, and data sharing in large-scale scenarios. Fred Limp (University of Arkansas) presents a comprehensive essay on measurement and the analytical process that characterized our field. Indeed, oversimplifying the process sequence before high-density survey and measurement (HDSM) was observe, interpret/abstract, measure, record, and analyze. HDSM breaks us out of this process in that it pushes us toward a recursive and reflexive engagement with the data, in which we observe, record, measure, analyze, and abstract/interpret repeatedly and in various orders. The growth in the use of HDSM methods is paralleled by increasing applications of computer-based visualization. Effective use of both requires attention to a scholarly digital ecosystem that addresses the archive and reuse of these digital objects and includes strategies to reuse these digital objects in other scholarly representations along with the tools for citation and other aspects of scholarly discourse.

Jakob Kainz (University of Vienna) presents an approach combining archaeological excavation with geophysical prospection. This is achieved by a combination of magnetometry, magnetic susceptibility, ground-penetrating radar (GPR), and pXRF measurements, on archaeological features before and during excavation. The aim is addressed to establish the full archaeological potential of the various prospection methods as these measurements can help corroborate excavation results as well as providing further archaeological data that cannot be seen by the excavator's eye.

Willem Vletter (University of Vienna) and Sandra Schloen (University of Chicago) provide an original contribution aimed to validate chronological interpretation of airborne laser scanning (ALS) data in reconstructing historical road and path networks in forested areas. The chronological model makes use of both the Harris Matrix Composer (HMC) and the Online Cultural and Historical Research Environment (OCHRE), developed at the University of Chicago.

Part "Cultural Resource Management: Communication and Society" confronts a broad subject characterized by several declinations: cultural resource management (CRM), public archaeology, theory and practice of digital archaeological communication, museums and sensing, and social outcome.

Eva Pietroni (Italian National Research Council) argues that despite the exciting perspectives opened in education and in terms of social and economic growth, research in the domain of virtual museums has not reached a sufficient level of maturity, such as cinema or game sectors. There is still a disconnection between the research, that develop tools with little interest in their wide application, and the industry, that build ten-year plans addressing the market. Given the need for enhancing the emotional and cognitive impact of virtual museums, some criteria and good practices are discussed, exemplified through concrete case studies, among which the *Tiber Valley Virtual Museum*, dealing with engaging storytelling, embodiment, and novel solutions in the interaction design and in the integration of media.

Riccardo Olivito (Scuola Normale Superiore), Emanuele Taccola (University of Pisa), and Niccolò Albertini (Scuola Normale Superiore) in their contribution provide a critical view of virtual immersive environments delivering through the case study of the agora of Segesta, an excellent example on how this technology can play a key role for the archaeological practice allowing the visualization and analysis in real time of different types of data and the interaction with them.

Durham, NC, USA — Maurizio Forte
Cambridge, UK; Siena, Italy — Stefano Campana

Part I
Data Collection and Technology

Terrestrial Laser Scanning in the Age of Sensing

Nicola Lercari

Abstract For more than a decade, Terrestrial Laser Scanning (TLS) has been a primary remote sensing technique for disciplines related to archaeology, architecture, built heritage, earth science, metrology, and land survey. The increasing precision, range, and survey speed of TLS make this technology even more viable for large-scale data capturing in the Age of Sensing. This chapter reviews the state of the art of Terrestrial Laser Scanning in 2015 with the aim to assess its applications in a context of lower data capturing costs for alternative technologies, such as new commodity sensors, Image-based 3D Modeling, Unmanned Aerial Systems (UAS), optical 3D scanning, and Airborne Laser Scanning. More specifically, TLS still maintains a fundamental role in the documentation and interpretation of archaeological contexts at intrasite scale: (i) Terrestrial Laser Scanning delivers high-fidelity data of surfaces and structures of buildings as well as ultra-precise measurements of the morphology of stratigraphic layers; (ii) research in remote sensing proved that TLS point clouds can be successfully interpolated with data recorded with other instruments and techniques, such as magnetometry, Ground Penetrating Radar, Unmanned Aerial Vehicles, Image-Based Modeling, in order to generate hybrid documentation and new knowledge on natural and cultural heritage sites. Inevitably, the current advancements in TLS bring new questions. For example, how can micro-differences only visible in the point clouds change the analysis and interpretation of layers and buildings? How to improve the monitoring and conservation of a site via automated analysis of TLS data? How to enhance the mapping process of built-heritage using data segmentation or semi-automatic feature extraction of TLS point clouds? This chapter proposes a new approach to TLS based on multi-modal capture workflows, semi-automated post processing, online archiving, and online visualization and management of point clouds with the aim to open new horizons for digital archaeology, architectural survey, and heritage conservation.

N. Lercari (✉)
School of Social Sciences, Humanities, and Arts,
University of California Merced, Merced, USA
e-mail: nlercari@ucmerced.edu

M. Forte and S. Campana (eds.), *Digital Methods and Remote Sensing in Archaeology*, Quantitative Methods in the Humanities and Social Sciences,
DOI 10.1007/978-3-319-40658-9_1

Overview of New Data Capture, Processing, and Visualization Systems in Relation to Terrestrial Laser Scanning

The second half of the 2010s witnesses a deep transformation in the domain of data recording, data processing, and visualization. A number of cutting-edge remote sensing technologies and methods are now production-ready tools that can be deployed in the fields, do the job, and challenge established survey technologies, such as Terrestrial Laser Scanning. The Age of Sensing is characterized by the rapid diffusion of cost-effective and incredibly versatile technologies such as computer vision-based 3D scanners, inexpensive cameras and sensors, mobile or web apps for real-time processing, and interactive platforms for data sharing in the cloud. Commodity sensors, such as accelerometers, three-axis gyroscopes, proximity sensors, ambient light sensors, and Global Positioning System (GPS) receivers are becoming ubiquitous in smart phones, cameras, household electronics, cars, and wearables.

As of the beginning of 2015, new generations of sensors are ready to go mainstream while their manufacturers openly express the ambition to transform the way people interact with the real world through their digital devices.

Great examples of the new era of commodity sensors are: (i) revolutionary optical 3D scanning solutions, such as the Structure Sensor, now available for smart phones and tablets users to be employed in the digitization of objects, interior environments, and artifacts; (ii) low-cost motion tracking technologies, such as the Intel RealSense, embedded in new tablets and laptops that promise to change the way users interact with computers (Intel RealSense 2015); (iii) commodity thermal imaging sensors, such as the Seek Thermal XR camera, which enable smart phones to detect infra-red light and record thermal information opening new possibilities for basic spectral analysis for the masses (Seek Thermal 2015).

The effects of the mass diffusion of sensing technologies on the society at large are yet to be assessed. What is already clear is that new, low-priced, and increasingly powerful tools for data capture, processing, and visualization have started to transform the field of remote sensing and its applications.

This new scenario opens research opportunities linked to the development of novel methods, bringing scholars to experiment hybrid techniques and workflows that integrates more established tools, such as TLS, with cutting-edge technologies often developed by small, start-up companies, research centers, or universities.

The following sections of this chapter will analyze in detail the transformational shift described above, especially in regards to Terrestrial Laser Scanning. The aim is to ponder new advancements in the fields of data recording, processing, and simulation and discuss whether TLS still matters today.

Alternative 3D Capture Systems

In the Age of Sensing, TLS is no longer the only viable solutions to survey heritage sites, buildings, and archaeological excavations in 3D.

Image-based 3D modeling techniques, also known as Structure from Motion (SfM), have long proved viable for the documentation of heritage (Pollefeys et al. 2001; Remondino and Menna 2008), stratigraphic layers in archaeological excavation (Doneus and Neubauer 2005a, b; Forte et al. 2012), and artifacts (Kersten and Lindstaedt 2012).

What is remarkable is that one can now digitize an entire indoor environment in real-time using commodity 3D data capture systems based on depth cameras technologies or structured light devices. The effects of Microsoft Kinect sensor have been largely documented (Zhang 2012); especially in regards to data capture accuracy (Khoshelham 2011), and mapping of indoor environments (Khoshelham and Elberink 2012). The performance of low-cost 3D scanning devices has also been assessed in relation to their employment in the cultural heritage domain (Guidi et al. 2007).

In 2015, it is now possible to 3D capture, process, and virtually reconstruct both the built environment and objects in real-time using sensors, such as Microsoft Kinect or Structure Sensor by Occipital (Structure Sensor 2015) combined with mobile devices (Raluca Popescu and Lungu 2014). A Structure Sensor records colored triangular mesh of its surrounding space or objects—located within 2 or 3 meters from the device—in a matter of seconds. It uses an iPad, or smartphone, to process the captured data, render its geometry, and align multiple point of views in real-time (Fig. 1).

The possibility to capture, process, and instantaneously visualize the 3D scans on a mobile device implies that the survey of built heritage or archaeological sites can potentially be verified on the go. Differently than TLS, this capability makes data post-processing inexpensive and fast.

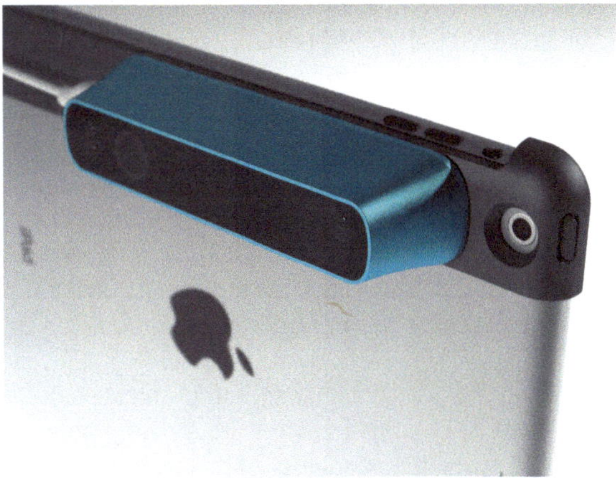

Fig. 1 Structure Sensor uses iPad for real-time data processing—courtesy of Occipital

A foreseeable effect of this new technology is that dense data capture becomes now available to anybody who owns a tablet or smartphone and is willing to spend few hundred additional U.S. dollars to purchase a Structure Sensor. There is no doubt that this capability will open new horizons for community-based heritage preservation performed by cultural associations, volunteers, students, and local communities. More broadly, one can envision that heritage diagnostics of the built environment or the digital documentation of archaeological remains could be immediately discussed on site, few instants after the survey is completed.

A discourse on alternative 3D capture systems need to go beyond a cost-benefit analysis of purchase price, survey time, and ease of use. Thus, this chapter needs to assess whether the new optical scanning solutions also challenge TLS in regards to data fidelity. One can now record, align, and process in real-time very precise colored point clouds of the interior of a building or the shape of complex objects using a DPI8 scanner developed by DotProduct (DPI8 2015). This hand-held 3D scanner is operated via the operating system Android and relies on a low-cost tablet PC for processing data in real-time. DPI8 delivers fairly accurate measurements within a range of 0.6–5 m when used with optimal ambient conditions. In February 2015, the author of this chapter had the opportunity to test a DotProduct scanner for a test survey of the interior of a warehouse located at Fort Mason Center, in San Francisco, during the *REAL 2015* conference (REAL 2015). Such preliminary testing showed that a DPI8 optical scanner is able to deliver precise data when scanning the interior of a building, which has been evenly lit. Undoubtedly, further testing on DPI8 is needed to call this portable 3D capture system a mature technology for heritage documentation. Given a price tag of few thousands of U.S. dollars, it is relevant to mention that the data fidelity of this optical scanner is acceptable if compared to a TLS unit, such as a FARO Focus^{3D} X330, which costs about ten times more (FARO Focus^{3D} X330 2015). No doubt, DPI8 already presents the characteristics needed to become a leading technology in the domain of artifacts digitization and documentation of interiors of buildings.

The current revolution of data capture platforms is not solely related to indoor surveys and artifacts scanning. New tools for landscape surveying and built environment 3D mapping are now available. Such new systems combine lightweight Unmanned Aerial Vehicles (UAVs), uncalibrated cameras, and Image-based 3D Modeling software, posing major challenges to the viability of TLS for what concerns intersite documentation or landscape surveying.

In 2015, advanced 3D mapping standalone software, such as Pix4D, allows scholars, architects, and heritage practitioners to perform accurate 3D mapping of entire sites and landscapes (Pix4D 2015). Other cloud-based UAS platforms, such as DroneDeploy (DroneDeploy 2015) provide archaeologists, land surveyors, and geoscientists, with new effective tools for 3D mapping cultural landscapes and natural environments simply using Android or iPad devices to manage mission planning, data capturing, and server-based data processing. Currently, the most widespread technique for the 3D documentation of archaeological heritage is the standalone Image-based 3D modeling software Agisoft Photoscan Pro (Photoscan 2015). In the Age of Sensing, the popularity of this technology is so widespread that Photoscan is becoming a standardized method for intrasite and intersite documentation. Photoscan provides archaeologists and conservators with an incredibly efficient workflow that reduces the cost and time of single context data recording, while enhances on-site data-driven discussion and interpretation (Forte et al. 2015, pp. 45–46) (Fig. 2).

The viability of standalone and cloud-based UAS platforms for 3D documentation in archaeology—specifically Photoscan Pro and DroneDeploy—were positively tested in the summer 2015 at the archaeological sites of Çatalhöyük and Boncuklu Höyük, in Turkey. In the field season 2015, a DJI Phantom 3 Pro multirotor copter equipped with a 4K RGB camera and DroneDeploy server-based mission planning was employed to conduct several missions for indoor survey inside the permanent shelters (Lercari and Lingle 2016), as well as for outdoor 3D mapping survey (Forte et al. 2016). Such UAS operations were aimed to enhance the 3D survey of Çatalhöyük buildings for conservation and monitoring purpose. UAS data capture was also employed to 3D map the landscape of Çatalhöyük and its environs with the goal to provide further understanding of the site's relationship with other Neolithic settlements in the Konya plain, such as Boncuklu Höyük.

The above mentioned survey methods open new horizons for heritage conservation and documentation in a time of decreasing funding for archaeological excavation or cultural heritage preservation. Thus, micro UAS platforms challenge commercial photogrammetry or airborne LiDAR services in relation to intersite surveys. Their capability to render the morphology and multispectral properties of heritage sites and landscapes with high accuracy and in a cost-effective way, allows the new multi-sensor data capture systems to also challenge laser scanning in regards to intrasite documentation.

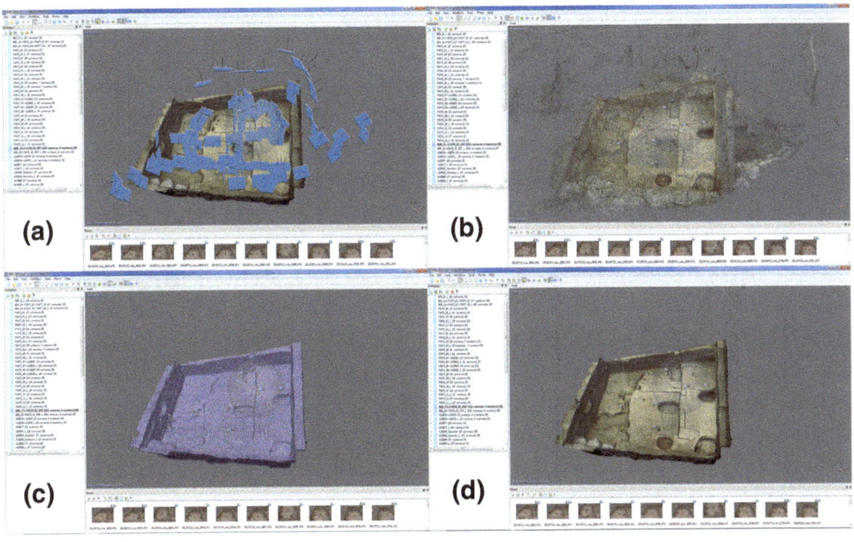

Fig. 2 Processing of 3D data captured at the UNESCO site of Çatalhöyük, Building 89 in Agisoft Photoscan showing. **a** Camera positions and ground control points. **b** Georeferenced dense cloud. **c** Edited triangular mesh in *Wireframe mode*. **d** Optimized triangular mesh in *Shaded mode*

The Historic Buildings and Monuments Commission for England—also known as Historic England or English Heritage—provides surveyors with thorough guidelines with the aim to help identify the best application scenarios for Airborne LiDAR, TLS, or other 3D capture methods in relation to different deliverables and specific precision and accuracy goals (Crutchley and Crow 2009).

Currently available technologies give surveyors the advantage to cut the duration of the survey process from data capture to final delivery by one order of magnitude. For instance, one can now fly an affordable thermal camera, such as a FLIR Tau 2, and a compact RGB camera, such as a mirror-less Sony RX100, mounted on a consumer multi-rotor UAW manufactured by DJI (DJI 2015) or 3D Robotics (3DRobotics 2015) for few thousands of U.S. dollars (FLIR&DJI 2015).

The current major shift in 3D mapping and 3D modeling is due to proven computer vision technologies based on Structure from Motion (SFM) and Dense Stereo Matching (DSM) algorithms (Verhoeven 2011; Verhoeven et al. 2012; De Reu et al. 2013; De Reu et al. 2014). SFM and DSM proved to be reliable technologies that can be used to process large datasets of aerial photographs captured by uncalibrated digital cameras mounted on lightweight aircrafts flying GPS waypoint missions.

Nonetheless, the main disadvantage of the new 3D capture technologies is that the new 3D scanners mostly rely on depth cameras or electro-optical sensors that still do not work outdoors, or at night, or underperform in scenarios where the subject is overexposed or not evenly lit. Thus, the quality and accuracy of the new 3D digitizers highly depend on environmental conditions such as the temperature, illumination, and reflectivity of the area of interest. One needs to notice that such constraints may be overcome by future technological development, but currently represent a strong drawback to the adoption of the new 3D capture technologies in many professional fields and academic disciplines. One also needs to underline that some of the above mentioned limitations might apply to traditional laser-based data capturing tools. For instance, digital archaeological work at the UNESCO site of Çatalhöyük, in Turkey, proved that the documentation of stratigraphic layers may be very complex or not feasible when an high-accuracy optical laser scanner (e.g. Minolta Vivid 910) was employed in the field to document the stratigraphy of a complex midden sequence (Forte et al. 2015, pp. 43–44). When compared to optical technologies, time-of-flight and phase comparison laser scanners are less affected by adverse lighting conditions; the accuracy and precision of such scanners can decrease in heavily lit scenarios, unless such equipment is specifically manufactured for long-range and outdoor usage. More broadly, one also needs to mention that extremely hot or cold temperatures can affect the majority of data capture sensors. Extreme environmental conditions may become an issue for surveyors. For example, the author of this chapter has often experienced TLS equipment warnings and shutdowns while scanning archaeological heritage inside the permanent shelters of Çatalhöyük where air temperature may be above 45° C in a hot summer afternoon.

In terms of survey range, the new commodity 3D scanners offer very limited options when compared with time-of-flight or phase comparison TLS technologies. Optical and TLS structured light data capture systems have very limited survey range—usually from 0.5 m to maximum of few meters from the sensor—and present a number of constraints that make them not very feasible for large sites or whole-building surveys (Fig. 3).

Moreover, mass consumers are not very interested in expensive or complicated calibrations operations or data fidelity. These propensities are reflected in the way the new commodity data capture tools are designed and built. The new 3D digitizers are rarely rugged enough to perform well outdoors or in the fields and do not support custom color and sensor calibration.

A comprehensive cost-benefit analysis of the alternative technologies and methods discussed in the previous pages goes beyond the scope of this chapter, but will need to be examined in future publications.

User-Oriented Data Processing and Open-Source Software

In the Age of Sensing, data processing is also more effective, faster, user-friendly, and occasionally freely available. For instance, the end-to-end 3D platform developed by Matterport allows users to perform the following with great ease Matterport (2015): (i) to scan and upload 3D data via a Matterport Pro 3D camera, an optical solution for data capturing, or simply via any mass-market mobile devices equipped with a Matterport 3D capture app; (ii) to automatically process the captured data in the cloud

Fig. 3 Indoor usage of the Structure Sensor to record artworks—courtesy of Occipital

using Matterport Cloud Service; (iii) to enable anybody to interact with and share the processed data using a web browser or mobile app (Matterport). The functionalities of the Matterport platform make this tool a comprehensive and very easy to use system able to 3D map the world, display content on virtual reality headsets, such as Oculus Rift (Oculus Rift 2016), HTC Vive, Samsung Gear VR (Samsung Gear VR 2016), or the Web, while enable mass mobile technologies to become 3D capture systems.

Sequoia is a multiplatform, standalone software that allows to easily reconstruct the surface of large point clouds made of several billions of points. Sequoia is able to convert huge data sets of laser scanning data and particle data to triangular mesh geometry in few minutes (Sequoia 2015). The start-up company Thinkbox Software developed Sequoia's architecture to handle massive amounts of laser scanning data through a progressive processing workflow. The result of this approach is that Sequoia is able to visualize the final result of the processing even before all the data is loaded. Moreover, this software is able to handle huge data sets that can be larger than the actual memory available in the computer where the processing is performed. Sequoia also allows users to perform operations such as smoothing, decimation, color and texture projection on mesh (Thinkbox 2015).

One of the exceptional aspects of Sequoia is that this application makes large point clouds processing accessible and easy to handle even for non-experts in TLS data processing. In fact, Thinkbox Software developed this application for architecture, engineering and construction markets with the goal to directly compete with more established data processing platforms such as the 3D authoring tools developed by 3D Systems (3D Systems 2015).

In regards to 3D Systems' products, one needs to spend few words on Geomagic Design X, formerly known as Rapidform XOR. In the Age of Sensing, Geomagic Design X is one of the most advanced point cloud processing software capable of combining the parametric approach of Computer Aided Design (CAD) software with advanced 3D data scan processing capability. Nonetheless this tool is part of a specialized software platform primarily created for reverse engineering and manufacturing projects, Geomagic Design X is user-friendly and presents a number of functions able to automatically extract features and components directly from the point clouds. The applications of Geomagic Design X for the documentation and mapping of sites and the drawing of artifacts are endless; one can employ Geomagic Design X for point cloud to CAD operations. This allows surveyors to generate accurate maps of entire sites or sections of walls and facades starting from TLS survey data. One can also use Geomagic Design X for authoring precise 2D drawing of artifacts and other material culture objects that were previously scanned (Geomagic Design X 2015). The main downsides of the commercial software referenced above are: (i) the high cost for acquiring the license of these proprietary platforms; (ii) the ongoing cost for maintaining them; (iii) the closed-source code; (iv) commercial strategies non-quite friendly to educational institutions.

In the Age of Sensing, viable alternative solutions to the above-mentioned software are available free of charge. MeshLab is an free software application for mesh and point cloud data editing that is incredible popular among scholars, educators, cultural institutions, and private firms involved in the digital documentation of heritage sites and 3D data processing (Cignoni et al. 2008).

The widespread diffusion of MeshLab is due to the powerful tools and filters it provides to its users (Fig. 4) (MeshLab 2016). This software is distributed under GNU General Public License. MeshLab is the product of the invaluable dedication and cutting-edge research of the Visual Computing Laboratory at CNR-ISTI research center. What is remarkable about MeshLab, is that it is developed by a team of scholars committed both to develop free software for cultural heritage as well as to advance virtual heritage research (Callieri et al. 2011; Dellepiane et al. 2012; Siotto et al. 2014).

CloudCompare is a multiplatform open-source solution for 3D point cloud editing that can be also employed to process triangular mesh (Girardeau-Montaut 2011; CloudCompare 2015). This software was initially created in 2004 in the division for Research and Development of the public utility company Électricité de France (R&D E.D.F. TP 2011). In 2009, CloudCompare was released as free software under GNU General Public License. CloudCompare architecture exploits octree structure techniques to visualize and handle large point cloud data sets (Chien and Aggarwal 1986). This application offers a large variety of cloud processing algorithms spanning mesh-cloud comparison, registration, resampling, color and picture projections, and interactive or automatic segmentation. CloudCompare is especially relevant for evaluation and comparison of 3D Data (Scollar and Giradeau-Montaut 2012; Rajendra et al. 2014) (Fig. 5).

Viable workflows for data capture and processing rely on: (i) transparency of the data acquisition process, (ii) use of open file formats (e.g. Wavefront .obj or Polygon

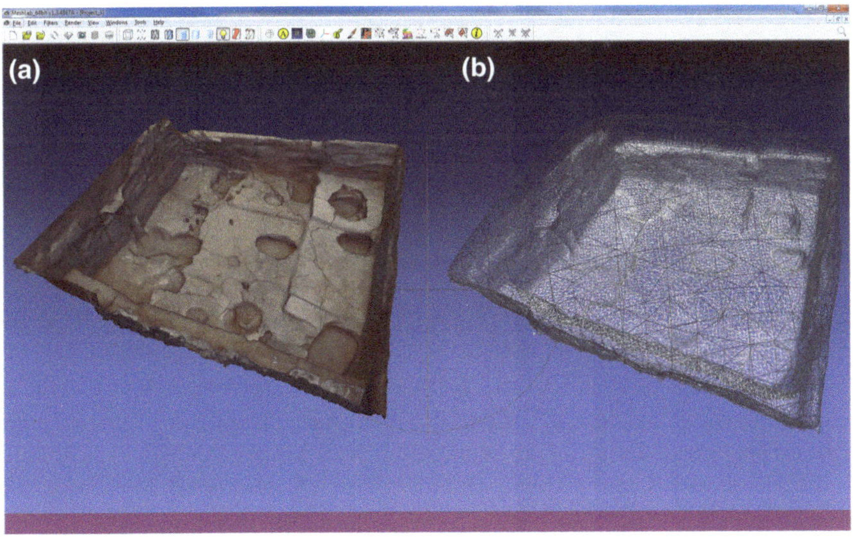

Fig. 4 Triangular Mesh of Çatalhöyük Building 89 in MeshLab. **a** Flat mode view with lighting. **b** Wireframe mode view showing poisson surface reconstruction

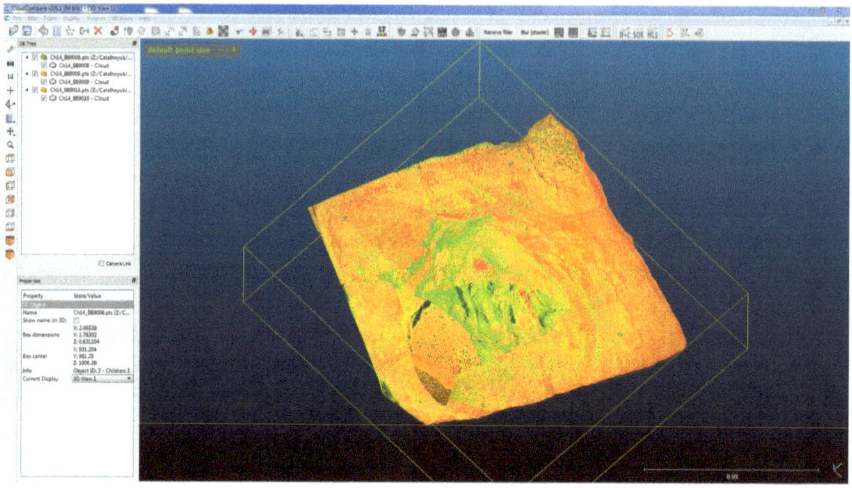

Fig. 5 Çatalhöyük Feature 3484 point cloud analysis and comparison in CloudCompare

File Format .ply) or manufacturer-independent file formats (e.g. ASTM E57 .e57), (iii) delivery of data that can be processed and visualized with open source or free software, (iv) open access to the end results (Lercari 2010). These four factors determine the sustainability of a workflow or technology over time and apply to both TLS and new tools available in the Age of Sensing.

3D Web Visualization and Cloud Services

To further advance this discussion on data recording and processing methods in the Age of Sensing, one has to mention that the increasing diffusion of high-speed networks—such as *next generation* wired connections able to transfer data at 10 or 100 Gb/s or Long Term Evolution (LTE)-A mobile connections able to download data at 1 Gb/s—create new opportunities to process 3D data in the cloud or to render complex 3D scenes directly over the Internet.

Web-based 3D reconstruction services have been utilized for years (Vergauwen and Van Gool 2006), but the availability, effectiveness, and versatility of the cloud services now available for 3D data processing have greatly expanded since the 2010s. In addition, the wide diffusion of open web 3D standards such as X3D (X3D 2015) and WebGL (WebGL 2015) and open-source frameworks, such as X3DOM (Behr et al. 2009; X3DOM 2015), has made it possible to visualize 3D data natively on a web browser, without the need to install additional plug-ins. These new scenarios are enabled by empowered web browsers (e.g. Mozilla Firefox 38.0 or Google Chrome 50.0) that are able to directly access the graphics card's acceleration capabilities to perform online, real-time rendering of 3D content (Evans et al. 2014).

Previous work demonstrates the potential of a complete workflow from the field to the 3D web. TLS data were captured at a natural heritage site, then processed, and finally simplified to be suitable for web visualization using X3D, WebGL, and X3DOM standards (X3DOM 2015; Silvestre et al. 2013).

New scholarship shows the potential of 3D visualization of cultural heritage data on the web using WebGL and SpiderGL (Callieri et al. 2015); as well as custom systems, such as 3DHOP, designed to optimize the online visualization of 3D cultural objects (Potenziani et al. 2014). While cloud computing tools have been used for years in the visualization of 3D cultural data in online virtual environments (Lercari et al. 2011), cloud platforms for 3D data processing and visualization are relatively new.

In recent years, big corporations in the field of remote sensing and 3D authoring software (e.g. Leica, FARO, and Autodesk) have engaged in the development of new cloud-based systems able to process, visualize, mark-up, and share point clouds and triangular mesh over the Internet. Autodesk Recap 360 or Recap 360 Ultimate (Autodesk Recap 2015), FARO SCENE WebShare Cloud (SCENE WebShare Cloud 2016), Hexagon Imagery Programme (HxIP 2015), and Leica CloudPro (CloudPro 2015) are good examples of new commercial cloud platforms created for online 3D data processing, visualization, and sharing of TLS or ALS data.

These new commercial cloud processing and interactive visualization systems enable surveyors, clients, and collaborators, to remotely access and share survey data on buildings, landscapes, and even entire sites. Moreover, these cloud platforms make it possible for stakeholders to work together to create and share mark-ups and interpretations of the TLS post-processed data.

As of 2015, many different models are available for 3D processing and visualization in the cloud. Web-based cloud services, such as Autodesk Recap 360, allow users to process and visualize both TLS and IMB 3D content using their Internet browser (Fig. 6). In addition, the hybrid standalone and cloud-based software Autodesk Recap Ultimate provides further options for TLS automatic data registration and processing. The cost of Autodesk cloud services is U.S. dollars 500/year per user for Recap 360 and U.S. dollars 2000/year per user for Recap 360 Ultimate (Recap 2015).

FARO Technologies also offers a Platform as a Service (PaaS) cloud-based hosting solution that promises to revolutionize access to TLS data online. In fact, FARO SCENE WebShare offers to its users incredibly easy to use tools aimed at data processing, managing, and sharing 3D data directly in the cloud. SCENE WebShare offers different levels of subscriptions that target Small Enterprise (€1.490/year for 100 GB of storage or 1000 scans), Medium Enterprise (€2.950/year for 200 GB of storage or 2.000 scans), and Large Enterprise (€7.750/year for 500 GB of storage or 5.000 scans) (SCENE WebShare Cloud 2016).

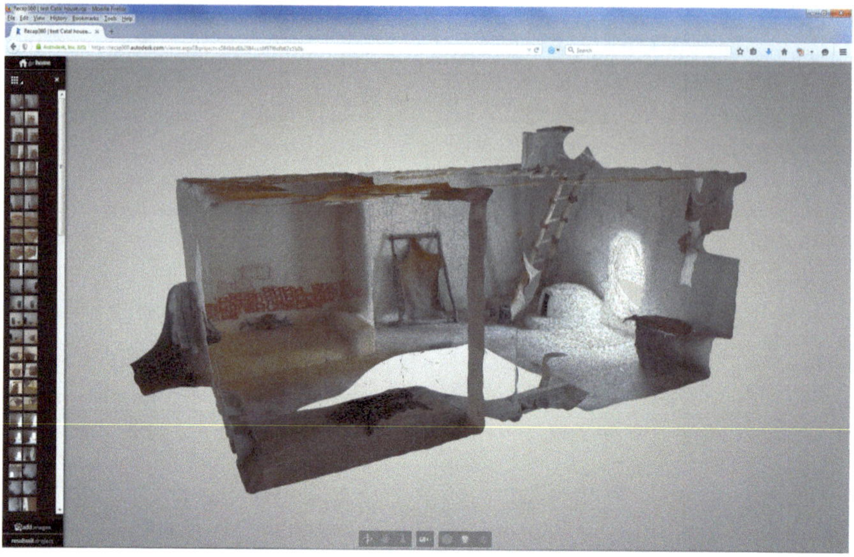

Fig. 6 Image-based 3D model of Çatalhöyük experimental house generated in Autodesk Recap 360

Hexagon Imagery Programme (HxIP) is a cloud platform that promise to take ALS data from the sky to the cloud. HxIP provides on-demand access—via third party software such as ESRI ArcGIS, Valtus, or Hexagon Power Portfolio—to high quality aerial imagery captured by Airborne LiDAR systems manufactured by Leica Geosystems.

It is clear that cloud technologies open new horizons for TLS and ALS in the Age of Sensing. Specifically, cloud-based systems give heritage surveyors the opportunity to show the collected point clouds to other stakeholders who are not physically on-site. A positive consequence of these new cloud services is that a more collaborative and inclusive interpretation of TLS data can now be performed via web browser, without the need to purchase additional software licenses. The downside of TLS data processing in the cloud is related to expensive subscription plans that could make commercial cloud solutions not feasible for heritage projects with limited budget. Whilst open access 3D repositories have been available for several years (Koller et al. 2009; Guidazzoli et al. 2012), open-source cloud solutions, such as HP Helion Eucaliptus (Helion Eucaliptus 2015), are still a rare exception (Nurmi et al. 2009).

To overcome this issue, the expansion of TLS data processing in the cloud would need more support from the research community; national and international institutions for heritage preservation should also provide local communities, local heritage institutions, and educational or not for profit organizations with new opportunities for making cultural data processing and sharing freely available in the cloud.

State of the Art of Terrestrial Laser Scanning in 2015

Since the early 2000s, Terrestrial Laser Scanning has led the process of digital documentation in fields such as architecture, earth science, and landscape and built heritage survey, providing scholars and professionals with incredibly precise and reliable tools for metric measurement and 3D data capture (Mills and Andrews 2011).

Concurrently, archaeologists have extensively tested TLS techniques in combination with photogrammetric methods to document archaeological heritage and entire monuments (Neubauer et al. 2005; Pescarin and Pietroni 2007), or archaeological landscape (Forte et al. 2005; Francovich and Campana 2005). Most importantly, archaeologists proved TLS viable to conduct digital documentation in single-context archaeological excavations where each stratigraphic unit's surface needs to be recorded with geometrical precision and centimeter-level accuracy (Doneus and Neubauer 2005a). Nonetheless TLS could be time consuming, employing a laser scanner in the excavation proved viable and time-effective and allowed. Using TLS, digital archaeologists may manage to save considerable amounts of time in the recording of the morphology and texture of stratigraphic surfaces, walls, and sections, when compared to traditional contact measurements tools (e.g. measuring tape) or other non-contact measurement tools (e.g. total station) (Forte et al. 2012). Seminal archaeological work also proved that the interpretation of the stratigraphy of an excavation can be enhanced by the integration of TLS data and photogrammetric data in a Geographic Information System (GIS) able to render the geometrical, topographical, and stratigraphical characteristics of a site (Doneus and Neubauer 2005b).

In the Age of Sensing, the increasing precision, range, and survey speed of TLS make this technology even more viable for large-scale data capture of buildings and heritage sites; this is especially true when TLS is combined with photogrammetric tools (Andrews et al. 2009). For example, archaeological workflows that integrate reflexive methods and employs 3D technologies in combination with GIS and tablet-based digital drawings proved viable in the digital documentation and interpretation at-the-trowel-edge of Çatalhöyük's archaeological heritage (Berggren et al. 2015).

TLS data processing used to be a bottleneck in 3D survey workflows because it implied lengthy and costly manual procedures for point cloud filtering, registration, editing, segmentation, and surface reconstruction.

State of the art 3D authoring and processing tools (e.g. 3D Reshaper developed by Hexagon Metrology) currently enable faster and more efficient point cloud segmentation and processing (3D Reshaper 2015). Processing functions, such as point cloud automatic separation and cleaning, best geometrical shapes extraction, semi-automatic feature extraction (Fig. 7), and point cloud to CAD are now common in many TLS processing platforms. The availability of new, semi-automatic or automatic processing tools helps to reduce cost and TLS data processing time, making this 3D survey technology more feasible in the Age of Sensing.

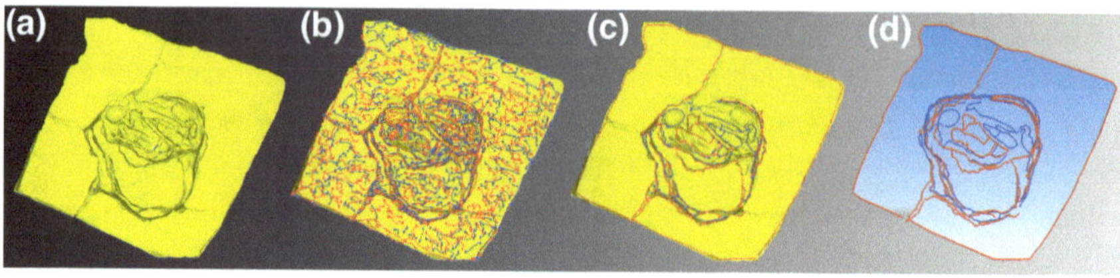

Fig. 7 Semi-automatic feature extraction in 3D Reshaper. **a** Triangular mesh of Skeleton 30928 excavated in Building 89 at Çatalhöyük. **b** Automatic convex and concave lines detection. **c** Semi-automatic lines filtering. **d** Semi-automatic measured drawing of Skeleton 30928

Therefore, new advancements in TLS make it possible to affirm that laser-based metric survey methods still maintain a fundamental role in the 3D documentation and three-dimensional interpretation of archaeological sites at intrasite scale (Forte et al. 2015, pp. 46–48). Thus, the aim of the following section of the chapter is twofold: the first being the discussion of the advantages and disadvantages of TLS when employed for the documentation of heritage; the other being the discussion of the state of the art of TLS through a number of explanatory case studies that best resemble the current advancements in conservation, analysis, interpretation, and visualization of cultural and natural heritage using this metric survey technology.

Pros and Cons of Terrestrial Laser Scanning

Programs developed by the U.S. Department of the Interior, such as the Historic American Buildings Survey (HABS), the Historic American Engineering Record (HAER), and the Historic American Landscapes Survey (HALS), are in charge of developing guidelines and protocols for the documentation of built heritage, historic engineering materials, and cultural landscapes. Among HABS/HAER/HALS (H^3) standards, specific guidelines examine the pros and cons of TLS (Lavoie and Lockett, n.d.). These recommendations underline that TLS offers advantages over other types of heritage documentation in terms of versatility, accuracy, long range, vertical reach, and survey speed. Conversely, H^3 guidelines identify cons in TLS workflows in regards to problems with occlusion between architectural elements, low accuracy in façade details (e.g. molding features), inability to pass through vegetation and built structures, as well as issues with long-term permanence of the digital formats of the point clouds for archival purposes. Cultural preservation guidelines, available in the United Kingdom, also provide thorough recommendations on the usage of TLS for metric survey that can help surveyors better understand advantages and limits of this technology and its applications in the heritage domain (Bryan et al. 2009). Other countries may have developed their own protocols and guidelines, but the above-mentioned sources could provide valuable guidance for TLS surveying to an international audience.

Following the recommendations and standards provided by H^3 and by the Historic Buildings and Monuments Commission for England, the comparison of data capture technologies discussed in these pages needs to analyze the following factors: (a) data acquisition and operation expense; (b) data processing time; (c) survey speed; (d) versatility in relation to environmental conditions; (e) portability; (f) range accuracy; (g) survey range; (h) noise; (i) positioning and/or georeferencing; (j) semi-automated or automated functions for 3D data post-processing, (k) usability; (l) online data curation and sharing.

The main cons related to the usage of TLS in heritage preservation or archaeological fieldwork are related to items (a) and (b). After more than fifteen years since the first terrestrial laser scanning technology was patented (Kacyra et al. 1999), the cost of a non-contact terrestrial laser measurement unit (a) is still very high. In 2015, a time-of-flight or phase comparison TLS unit is still priced between fifty to one hundred and fifty thousand U.S. dollars. More specifically, in January 2015 a FARO Focus^{3D} X330—a top-notch phase comparison TLS that features a long range from 0.6 to 330 m in indoor or outdoor environments, distance accuracy up to ±2 mm at 10–25 m, measurement speed up to 976,000 points per second, noise reduction of 50 %, and a built-in RGB coaxial camera able to deliver up to 70 Megapixel of colors (FARO Focus^{3D} X330 2015)—was priced at fifty eight thousand U.S. dollars.

Annual calibration, warranty extensions, software and equipment updates entails additional maintenance costs to the operation of a TLS unit. For instance, a 3-year Standard Warranty Plan for a FARO Focus^{3D} S120 unit, including annual laser scanner certification/calibration, parts, labor, and return shipping charges, costs a little more than ten thousand U.S. dollars (FARO Warranty department 2015). Thus, this discussion of the state of the art of TLS needs to emphasize that acquisition and operation cost still make this survey technology inaccessible for many cultural institutions, universities, and other stakeholders involved in heritage preservation.

The time needed for data processing (b) may also be another drawback for the adoption of TLS workflows. A typical acquisition and processing workflow entails long and costly operations that may include: (1) survey measurements/scanning on site (Fig. 8); (2) scan registration/georeferencing; (3) deliverables generation, such as point cloud/unrefined mesh, rendered images/2D or 3D drawings, and animations or decimated/edited mesh; (4) analysis; (5) conclusions (Mills and Andrews 2011, pp. 11–15).

The amount of time needed to complete steps (1)–(5) is proportional to the number of scans and to the complexity and extension of the case study. One also needs to bear in mind that highly specialized skills and dedicated software are needed to perform such tasks.

The pros of terrestrial laser scanning technologies over alternative systems are largely related to items (c), (d), (f), (g), (h), and (i).

In the Age of Sensing, cutting-edge time-of-flight and phase comparison terrestrial laser scanners manufactured by FARO, Leica, Riegl, and Trimble features a vast set of advanced functions and sensors. For instance, some of the new TLS units offer incredibly high survey speed (c) that makes these instruments able to record up to 976,000 points/second (FARO Focus^{3D} X330) or even 1 million

Fig. 8 FARO Focus ^{3D} S120 scanning Neolithic buildings in the North Area of Çatalhöyük's East Mound

points/second (e.g. Leica ScanStation P40 and Trimble TX8) (Leica ScanStation 2015); they are very versatile in relation to most environmental conditions (d) (Leica ScanStation P30 and Leica ScanStation P40 operate from 0 to 50 °C), performing optimally in bright outdoor settings, shaded indoor environments, and even in the complete darkness of caves and mines; they are also adaptable in terms of portability (e) (e.g. FARO Focus^{3D} X130 and X330 weigh only 5 kg); they present improved range accuracy (f) up to ±1.2 + 10 ppm over 120–270 m (e.g. Leica ScanStation P40), 2 mm over 120 m (e.g. Trimble TX8), or ±2 mm at 10 m and 25 m (e.g. FARO Focus^{3D} X330); (g) they offer extremely large survey range (g) up to 2000 m (Riegl VZ-2000) or ultra-large range up to 6000 m (e.g. Riegl VZ-6000); they feature enhanced noise reduction algorithms (h) (all the TLS units listed in this page grant noise reduction spanning 20–50 %); they are equipped with integrated GPS receivers (e.g. FARO Focus^{3D} X130 and X330, Riegl VZ-2000 and Riegl VZ-6000), GNSS positioning (e.g. Riegl VZ-2000 and Riegl VZ-6000), compass (e.g. Riegl VZ-2000, Riegl VZ-6000, and all FARO Focus^{3D} X330), altimeter (e.g. FARO Focus^{3D} SX130 and X330), and dual axis compensator or inclination sensor (e.g. all the TLS units listed above) to improve scans positioning and georeferencing (i).

Pondering item (j) and (k), one can infer that the new alternative 3D capture systems present more automated data processing features and seem more usable than TLS units. This is due to the fact that the new 3D digitizers integrate optical, mobile, and cloud technologies. The new optical data capture solutions are often designed and developed by dynamic start-up companies whose goal is to explore alternative ways of process, visualize, and disseminate 3D data, making these new tools incredibly adaptable to new data capture and processing workflows.

In the Age of Sensing, TLS platforms have also improved in terms of semi-automated or automated processing functions (j). For example, SCENE 5.4 (SCENE 2015)—the TLS operating software licensed by FARO Technologies—is able to perform automatic, target-less registration of point clouds using information obtained by the sensors embedded in the scanner, such as GPS, compass, and altimeter, or overlapping scan data (Fig. 9). The point cloud data processing suite Leica Cyclone (Cyclone 2015) offers surveyors the possibility to easily perform features and coordinates extraction through its Cyclone Virtual Surveyor function. In addition, Leica Cyclone-MODEL is able to process TLS point clouds and automatically generate objects for export into CAD systems for additional measured drawing operations. 2D or 3D CAD drawings often represent the most suitable option to manage digital documentation derived from large datasets of TLS point clouds (Christofori and Bierwagen 2013).

In regard to usability (k), one needs to report that TLS tools have also become more user-oriented in the Age of Sensing. For instance, FARO Technologies—one of the leading manufacturers of laser scanners as of 2015—has simplified significantly the operation of its devices. The FARO Focus^{3D} S120, X130, or X330 laser scanners feature touch-panels and smartphone-like user interfaces developed in Adobe Flash, giving users the possibility to use tablet PCs equipped with Wi-Fi connection and a flash-enabled web browser to operate the scanners. This option allows users to preview and download the results of their work one a larger and brighter screen while

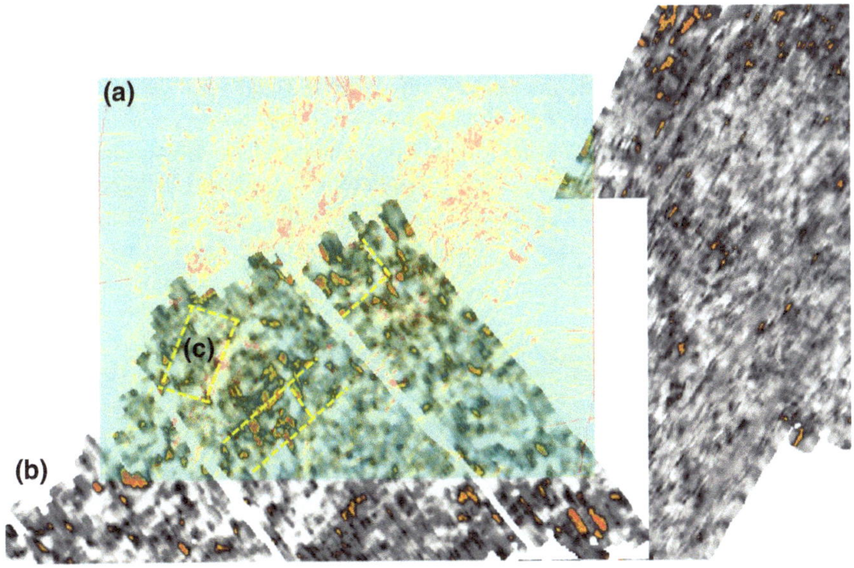

Fig. 9 Comparison of **a** top view and **b** cloud to cloud target-less automatic registration of point clouds of Çatalhöyük's GDA and TPC areas in FARO SCENE 5.4

still in the field, at a cost of few hundreds of U.S. dollars to purchase a standard tablet device.

Online data management and sharing (l), is vastly supported by the alternative 3D capture and processing systems through mobile, 3D web, and cloud technologies. In the last few years, online data management and sharing have also become more and more available in traditional TLS platforms.

In the Age of Sensing, point cloud datasets can be easily managed or viewed online without the need of installing costly and complex, supplementary software. By utilizing free 3D viewers available as web apps (e.g. Autodesk Recap), plug-ins (e.g. Leica TruView for Microsoft Internet Explorer) or standalone clients (e.g. Faro WebShare 2Go) it is now possible to view, measure and mark up point clouds directly over the Internet or local area connection.

The employment of SCENE WebShare Server (SCENE WebShare Server 2016) in combination with WebShare 2Go (SCENE WebShare 2Go 2016) or Cyclone-SERVER (Cyclone SERVER 2016) and Leica TruView (TruView 2015) offer even greater functionalities to FARO and Leica customers. These client-server technologies allow surveyors to install a robust application on their own server, enabling them to publish and manage point cloud data on the Internet or local area network. Other parties can interact with the recorded 3D survey data using FARO SCENE WebShare 2Go or Leica TruView plugins to directly access the point clouds from their web browsers. Affordable cloud service subscriptions are related to WebShare Server or Cyclone-SERVER. The option to curate 3D data in-house is particularly relevant to heritage and cultural institutions that want to protect the copyright or public access to their content while avoiding potential controversies related to uploading their data onto third party repositories.

The above list of facts and functions that belong to some of the terrestrial laser scanners available in the Age of Sensing is far from being comprehensive. The goal of this section is to provide the reader with arguments and facts able to show that TLS still presents competitive advantages over other non-contact 3D capture methods based on computer vision and depth cameras. This is especially true in regards to intra-site archaeological survey, outdoor survey, and documentation of vast areas and entire buildings. Thus, this chapter reiterates that current TLS features, such as extremely high accuracy, user-oriented interfaces, range accuracy, and interactive data curation, shows that Terrestrial Laser Scanning is far from disappearing in 2015.

Case Studies

The goal of this section is to illustrate six explanatory case studies that define new directions in remote sensing applied to cultural heritage for purposes such as conservation, analysis, interpretation, processing, and visualization.

Some of the following examples will describe the integration of TLS with other remote sensing technologies to obtain hybrid data capture workflows and data fusion. Others will illustrate new, semi-automated or automated ways to interpolate and interpret Terrestrial Laser Scanning data to produce drawings, maps and, more in general, new knowledge in a fast and effective way.

Finally, other case studies will discuss the state of the art of 3D web visualization of TLS data that rely on versatile technologies and new ways of data curation over the Internet.

Case Study 1

Drones in archaeology

Authors	Neil Smith, Luca Passoni, Said al-Said, Mohamed al-Farhan, and Thomas E. Levy
Year	2013
Methods	Integrated data capture, processing, and dissemination in archaeology
Data acquisition and processing	Unmanned aerial vehicles, terrestrial laser scanning, image-based modeling
Site	Dedan, al-Ula Valley, Saudi Arabia

In the last decade, small Unmanned Aerial Vehicles have been employed in archaeology to create 3D models of built structures, orthophotos and digital elevation models (Lambers et al. 2007; Guidi et al. 2008).

UAVs has been instrumental in the 3D mapping of Nabatean remains at the excavation of ancient Dedan in the al-Ula Valley in Saudi Arabia (Smith et al. 2014). Low-altitude photographs taken from multi-rotor copters equipped with GPS and barometric sensors were integrated with image-based 3D models and laser scanning data (captured with a FARO Focus^{3D} laser scanner) to produce and test new ways for rapid documentation of cultural heritage (Smith et al. 2014). This integrated approach proved feasible to document building features as well as entire built structures, such as the Lihyanite "lion tombs', specifically monumental tombs carved in the city's cliff faces using a construction style comparable to the one employed in the Nabatean capital of Petra. The combination of different survey methods allowed a team of scholars from UC San Diego and King Saud University to exploit the advantages of the three survey technologies that were utilized. Specifically, this project represents the state of the art in data capture because it was able to integrate: (i) reach and versatility typical of drone survey to access occluded and vertical features; (ii) easy-of-use, high survey speed, and short post-processing time typical of Structure from Motion; (iii) high resolution and precision measurements typical of terrestrial laser scanners.

Case Study 2

Surface and subsurface multimodal data capture at Çatalhöyük

Authors	Nicola Lercari, Maurizio Forte, Stefano Campana, Gianfranco Morelli, Gianluca Catanzariti, Krishopher Strutt, Ashley Lingle
Year	2011–2015

(continued)

(continued)

Authors	Nicola Lercari, Maurizio Forte, Stefano Campana, Gianfranco Morelli, Gianluca Catanzariti, Krishopher Strutt, Ashley Lingle
Methods	TLS for intrasite spatial analysis; integration of multiple technologies for above/below surface archaeological survey
Data acquisition and processing	Hybrid data capture bridging together terrestrial laser scanning, magnetometry, GPR, and Image-based 3D modeling
Site	Çatalhöyük, UNESCO World Heritage site, Turkey

Çatalhöyük is a Neolithic proto-city and a UNESCO site located in the Konya plane in central Anatolia, Turkey. Terrestrial laser scanning underpins the digital recording process of Çatalhöyük East Mound at intrasite level. In 2011, terrestrial laser scanners, such as Trimble FX (phase comparison) and FARO Focus^{3D} S120 (phase comparison), proved successful in the documentation of stratigraphic layers of a Neolithic house, specifically B.89 (Forte et al. 2012). To comply with UNESCO site management guidelines, Çatalhöyük requires thorough and systematical survey of the archaeological remains for conservation purpose. In 2012, TLS started being employed for scanning entire areas (North Area and South Area); this type of survey continued in 2013 also including the TPC area where archaeologists study the late Neolithic phases of the site. In 2014 all the currently excavated areas were documented via TLS (Forte et al. 2015, pp. 46–48). In 2015, a DJI Phantom 3 Pro multirotor copter equipped with a 4 K RGB camera and Drone Deploy server-based mission planning was also employed to enhance the survey speed of the areas inside the permanent shelters (Lercari and Lingle 2016) as well as for outdoor 3D mapping survey (Forte et al. 2016). This case study represents the state of the art of TLS because it fosters the integration of multimodal data capturing techniques for intrasite survey that combine TLS, IBM, and UAS platforms with other sub-surface survey techniques. More precisely, in 2012 TLS was employed to measure the morphology of small quadrants of the East Mound landscape located south and north of the North Area. This data were subsequently interpolated with magnetometry and GPR prospections elaborated by the University of Siena and the University of Southampton (Campana et al. 2013) to produce new knowledge on the sections of Çatalhöyük that have not been excavated.

Case Study 3

Informing historical preservation with the use of non-destructive diagnostic techniques

Authors	Michael Hess, Dominique Meyer, Aliya Hoff, Dominique Rissolo, Luis Leira Guillermo, and Falko Kuester
Year	2014
Methods	Non-destructive methods for cultural heritage diagnostics and new processing techniques for immersive visualization in C.A.V.E. systems

(continued)

(continued)

Authors	Michael Hess, Dominique Meyer, Aliya Hoff, Dominique Rissolo, Luis Leira Guillermo, and Falko Kuester
Data acquisition and processing	Terrestrial laser scanning, stereo panoramas, high-resolution imagery, aerial photography, thermal imaging, immersive visualization
Site	16th-century Church of Boca Iglesia, Ecab, Quintana Roo, Mexico

A team of experts affiliated with the Center of Interdisciplinary Science for Art, Architecture and Archaeology (CISA3) at UC San Diego surveyed the ruins of an early church and curate's house at the site of Ecab, located in a remote area in the tip of the Yucatan peninsula, in Mexico. The site of Ecab witnesses the first interaction between the Spanish and Maya communities in the early 16th century (Hess et al. 2014). The goal of the project was to deploy an array of non-destructive technologies such as terrestrial laser scanning, stereo panoramas, high-resolution imagery, aerial photography, and thermal imaging to digitally document the site. Survey data were provided to Mexico's National Institute of Anthropology and History (INAH) to develop a site conservation plan (Hess et al. 2014). In a very short time span of only two days, a FARO Focus^{3D} scanner was used to measure more that 800 million points of the built structures at Ecab and conduct visual and structural diagnostics off-site and serve as a digital scaffolding for other digital data collected on site. Multi-rotor UAVs were also employed to perform low-altitude photography to be used for structure from motion 3D reconstruction in combination with high-resolution imaging recorded with a stereo photography rig (CAVEcam) to improve color depth in the visualization of the ruins in immersive virtual environments (Smith et al. 2013). Previous work on monitoring ancient buildings using thermal imaging (Grinzato et al. 2002) inspired the UCSD team to employ Infrared thermography (FLIR A615) at Boca Iglesia with the goal to create thermal imaging mosaics able to document surface and subsurface data on the ruins that are not visible on site.

Case Study 4

Combined use of ground-based systems for cultural heritage conservation monitoring

Authors	Antonio Montuori, Guido Luzi, Salvatore Stramondo, Giuseppe Casula, Christian Bignami, E. Bonali, Maria Giovanna Bianchi, Michele Crosetto
Year	2014
Methods	Multi-technique approach for cultural heritage monitoring and restoration based on the integration of GBSAR, RAR and TLS sensors
Data acquisition and processing	Ground-based Synthetic Aperture Radar (GBSAR), GB Real Aperture Radar (RAR), and Terrestrial Laser Scanner
Site	Church of Sant'Agostino, Cosenza, Italy

New advances in cultural heritage monitoring entail new surveying methods able to provide dynamic and sustainable response to climate change and natural disasters. In addition to widely used techniques such as manual and topographic measurements, 3D mapping, GPS surveys, and multispectral imaging, this case study proposes the adoption of new radar techniques, usually employed to monitor earthquakes, avalanches, and flash floods, in combination with laser scanners (Montuori et al. 2014). More specifically, conservation work at the Church of Sant'Agostino, Cosenza, in Southern Italy developed a new monitoring approach that integrates ground-based systems, such as Ground-Based Synthetic Aperture Radar (GBSAR) and GB Real Aperture Radar (RAR), with TLS. A GB Radar System is a powerful tool with interferometric capabilities for both topographic deformation and structural vibration monitoring and measurements (Casula et al. 2009). Structural vibration can be measured via interferometric techniques that capture the position of an object comparing the electromagnetic waves it reflects at different time (Montuori et al. 2014). The radar systems employed in this project produced maps of topographic deformation via interferometric processing—with an accuracy of cm/year—as well as displacement time series of vibrating structures, with a precision of tens of microns (Luzi et al. 2012). This case study illustrates the state of the art in TLS because proposes a completely new approach to heritage conservation and monitoring that integrates TLS and radar technology in a new, feasible way that opens new perspectives for risk assessment related to heritage and natural and structural hazards.

Case Study 5

Automatic extraction of façade details of heritage building using TLS

Authors	Kenza Ait el Kadi, Driss Tahiri, Elisabeth Simonetto, Imane Sebari, and Hakim Boulaassal
Year	2014
Methods	Geometric and radiometric heritage survey, automatic point cloud segmentation and features extraction
Data acquisition and processing	Terrestrial laser scanning, custom methods for data segmentation using Delaunay triangulation and alpha-shape algorithm
Site	Casablanca old Medina, Morocco

The restoration of historic buildings in the old Medina district in Casablanca, Morocco pushed local authorities to require the production of a high number of CAD-based measured drawings of façades and built structures for leading the conservation planning. A mixed team from the Hassan Institute in Rabat, Morocco the School of Land Surveyors in Le Mans, France and the University of Science and Technologies, FST, Morocco was involved in the project. Scholars from the three institutions completed the task using new methods that involve TLS and automated systems for point cloud segmentation. This project represents the state of the art of point cloud post-processing because it proposes a new approach that exploits both geometric and radiometric information contained in a colored point clouds (Aitelkadi et al. 2014). RGB values, reflectance, and position of scanned points are used to perform automatic segmentation operations able to extract the façade of buildings from the rest of the point clouds with the aim to automatically generate 3D CAD drawings. The methodology used in this project proposes a four-pronged approach that includes: (a) geometric processing; (b) radiometric processing; (c) noise reduction; (d) component and contour detection (Aitelkadi et al. 2014). The automatic processing methods implemented in this project are able to filter the data resulting from segmentation of point clouds through Delaunay triangulation. Moreover, an alpha-shape algorithm is employed to detect the contour of details of the façade with the aim to categorize and separate the interior and exterior boundaries of various features in the façade. This work also presents thorough evaluations of the proposed automatic façade detail extraction methods that validate the feasibility of this technique in comparison to other manual or semi-automatic approaches.

Case Study 6

The Visionary Cross Project—3D Scanning and web dissemination of 3D content

Authors	Chiara Leoni, Marco Callieri, Matteo Dellepiane, Daniel Paul O'Donnell, Roberto Rosselli Del Turco, Roberto Scopigno
Year	2012
Methods	3D digitization of artifacts via TLS, 3D Web visualization based on HTML5, JavaScript, XML, WebGL, SpiderGL and Nexus library
Data acquisition and processing	Triangulation terrestrial laser scanning, high-res photography, point cloud, mesh editing, and color projection in MeshLab
Site	7–8th-century Ruthwell Stone Cross, Ruthwell Church, Dumfriesshire, Scotland

This case study discusses methods and techniques used to digitize the 7–8th century monumental stone artwork known as the Ruthwell Stone Cross, located in the Ruthwell Church, Dumfriesshire in Scotland, and to create its interactive 3D visualization on the Web. This initiative was developed as a collaboration between the Visionary Cross Project (Visionary Cross 2015), ISTI/CNR, the University of Pisa in Italy, and the University of Lethbridge in Alberta, Canada. The main goal of the project was to develop a web-based digital edition of the *Dream of the Rood*, one of the earliest Christian poems written in Old English, whose text is carved in runes on the Ruthwell Stone Cross (Leoni et al. 2015). A Minolta Vivid 910 triangulation laser scanner and digital cameras have been employed to capture in great details the geometry of the artwork, runes, as well as the true color of their surfaces. The 3D scanning data were later post-processed using MeshLab software (Cignoni et al. 2008); advanced tools for color data processing where also employed to apply the true color of the Cross to the recorded point clouds in MeshLab (Ranzuglia et al. 2012). A multimedia presentation of the scanned content was then prepared to make the recorded data available to the general public of the Web. Specifically, a critical edition of the *Dream of the Rood* that combines digitized images of the 10th century text *Vercelli book*, which contains a more recent version of the poem, was curated online. This case study represents the state of the art in TLS and 3D content online data curation because it proposes new ways to display laser scanning content on the Web using cutting-edge technologies and new standards such as HTML5, WebGL, Spider GL. Additional compliance with standards used in the digital humanities is granted by XML support for data curation and presentation (Leoni et al. 2015).

Conclusions or Why Terrestrial Laser Scanning Matters in the Age of Sensing

Since the early 2000s, Terrestrial Laser Scanning has been a primary remote sensing technique for disciplines related to archaeology, architecture, built heritage, earth science, metrology, and land survey. The increasing precision, range, and survey speed of TLS make this technology even more viable for large-scale data capturing in the Age of Sensing.

This chapter discussed novel methods and techniques for data capture, processing, and visualization of archaeological, cultural, and natural heritage data. The goal was to illustrate the state of the art of Terrestrial Laser Scanning as of 2015 as well as to analyze advantages and disadvantages of the usage of laser-based metric survey techniques for heritage conservation, analysis, interpretation, processing, and visualization. At the same time, this contribution wanted to verify the feasibility of TLS as a data capture technology in relation to alternative digital documentation methods such as new commodity sensors, Image-based 3D Modeling, UAS, optical 3D scanning, and Airborne Laser Scanning.

Therefore, Section "Overview of New Data Capture, Processing, and Visualization Systems in relation to Terrestrial Laser Scanning" discussed alternative data capture and processing tools, illustrating the current advancements in computer vision-based 3D capture systems, processing software, and visualization platforms on the Web and in the cloud. The analysis conducted in this section confirmed that the employment of new optical techniques for 3D data capture is not yet a feasible alternative to TLS because the new platforms still do not match the accuracy of laser scanner units nor can be used in outdoor surveys and scientific documentation of heritage sites.

Section "State of the Art of Terrestrial Laser Scanning in 2015" explored new horizons for digital archaeology and heritage conservation projects that need accurate, fast, and large-scale survey technologies. More specifically, Section "Pros and Cons of Terrestrial Laser Scanning" referred to standards and guidelines for historic preservation and heritage conservation to illustrate a series of principles and technological requirements that still define terrestrial laser scanning as an indispensable asset for the monitoring and survey of heritage sites and material culture.

Thus, this section discussed the state of the art of Terrestrial Laser Scanning in relation to disciplines such as architecture, earth science, and landscape and built heritage survey, with particular focus on single context archaeology.

The six case studies discussed in Section "Case Studies" shed light on the significant changes that involve TLS in the Age of Sensing. The proposed examples underlined current trends related to TLS: laser-based metric survey techniques are increasingly integrated with other survey methods (e.g Airborne LiDAR, Image-based 3D Modeling, UAS, and commodity sensors) with the goal to produce data fusion-based documentation able to provide data redundancy and reliability of the final deliverables. These new hybrid forms of data capture define multimodal workflows for heritage documentation that had already proved viable in relation to digital documentation of archaeological heritage and built environments (Forte et al. 2012; Campana et al. 2013) and digital data curation in museums and historic parks (Lercari et al. 2013, 2014) (Fig. 10).

The ultimate aim of this chapter was to ponder the significance of Terrestrial Laser Scanning and respond to the following question: does TLS still matter in the Age of Sensing?

The discussions and reflections brought forward by this chapter strived to provide a response to such dilemma. For instance, Section "State of the Art of Terrestrial Laser Scanning in 2015" thoroughly discussed new advancements in TLS, making it possible to affirm that laser-based metric survey methods still maintain a fundamental role in the 3D documentation and three-dimensional interpretation of archaeological sites at intrasite scale (Forte et al. 2015, pp. 46–48). The discussed examples showed that TLS still presents competitive advantages over other non-contact 3D capture methods based on computer vision and depth cameras. This is especially true in regards to intra-site archaeological survey, outdoor survey, and documentation of vast areas and entire buildings. Thus, this chapter has reiterated that cuttently available TLS features (e.g. extremely high accuracy, user-oriented interfaces, range accuracy, and interactive data curation) strongly underline that Terrestrial Laser Scanning still matters in 2015.

What this chapter makes evident is that the increasing precision, range, and survey speed of the most current TLS units, make this technology even more viable for large-scale data capture of buildings and heritage sites. This is especially true when TLS data are integrated and fused with information collected using other remote sensing techniques such as Image-based 3D Modeling, UAS, magnetometry, GPR, radar, and Airborn Laser Scanning.

To conclude this reflection on the significance of Terrestrial Laser Scanning in the Age of Sensing, one can affirm that laser-based metric survey techniques still play a fundamental role in highly specialized fields such as the digital survey of built environment, the digitization of artifacts, and the three-dimensional documentation of archaeological heritage.

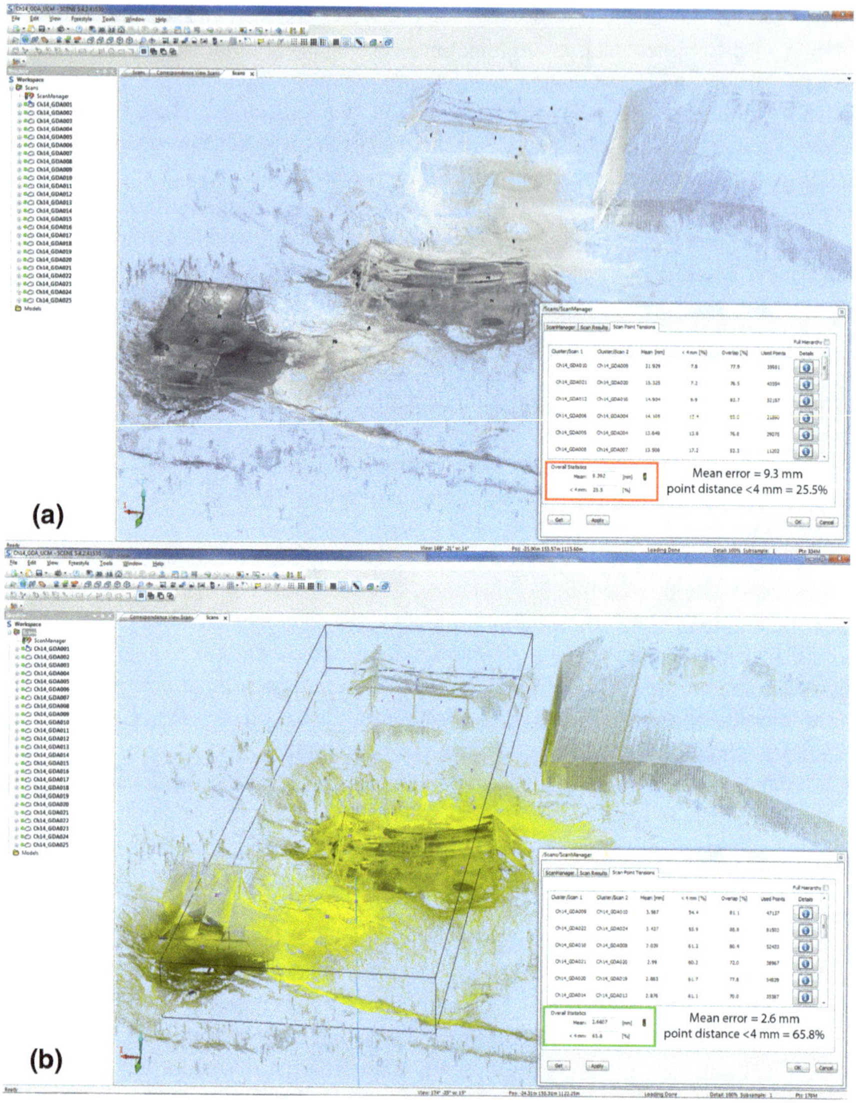

Fig. 10 Comparison of **a** top view and **b** cloud to cloud target-less automatic registration of point clouds of Çatalhöyük GDA and TPC areas in FARO SCENE 5.4 showing improvement in mean error and point distance <4 mm

References

3D Reshaper (2015). http://www.3dreshaper.com/en1/En_PointCloudProcess.htm. Accessed 3 May 2015.

3DRobotics (2015). http://3drobotics.com/. Accessed 1 Mar 2015.

3D Systems (2015). http://www.rapidform.com/products/xor/overview/. Accessed 25 Feb 2015.

Aitelkadi, K., Tahiri, D., Simonetto, E., Sebari, I., Boulaassal, H. (2014). Automatic Extraction of Facade Details of Heritage Building Using Terrestrial Laser Scanning Data. *Journal of Architectural Engineering Technology*, 3(133), 2.

Andrews, D., Bedford, J., Blake, B., Bryan, P., Cromwell, T., Lea, R. (2009). *Measured and Drwan. Techniques and practice for the metric survey of historic buildings* (2nd edition). Swindon: Historic England Publishing.

Autodesk Recap (2015). http://www.autodesk.com/products/recap/overview. Accessed 28 Mar 2015.

Behr, J., Eschler, P., Jung, Y., Zöllner, M. (2009). X3DOM: A DOM-Based HTML5/X3D Integration Model. In Proceedings of the 14th International Conference on 3D Web Technology, 127–135. ACM. http://dl.acm.org/citation.cfm?id=1559784.

Berggren, Å., Dell'Unto, N., Forte, M., Haddow, S., Hodder, I., Issavi, J., Lercari, N., Mazzucato, C., Mickel, A., Taylor, J. S. (2015). Revisiting reflexive archaeology at Çatalhöyük: integrating digital and 3D technologies at the trowel's edge. *Antiquity*, 89 (344), 433–448.

Bryan, P., Blake, B., Bedford, J. (2009). *Metric Survey Specifications for Cultural Heritage*. Swindon: Historic England Publishing.

Callieri, M., Dell'Unto, N., Dellepiane, M., Scopigno, R., Soderberg, B., Larsson, L. (2011). Documentation and Interpretation of an Archeological Excavation: an Experience with Dense Stereo reconstruction Tools. In *Proc. of VAST The 11th International Symposium on Virtual Reality Archaeology and Cultural Heritage* (pp. 33–40). Eurographics.

Callieri, M., Dellepiane, M., Scopigno, R. (2015). Remote visualization and navigation of 3D models of archeological sites. In *Proc. of 3D-ARCH Conference* (pp. 147–154), ISPRS Archives, Vol. XL-5/W4.

Campana S., Morelli G., Catanzariti G., Krishopher, S., Forte, M., Lercari, N. (2013). 4D Surveys at Çatalhöyük (Turkey): Magnetometry, sa-GPR & hr-GPR, Laserscanning. Paper presented at *Space2Place, Digital Heritage International Congress 2013*, Marseille, France, Oct. 28- Nov 01, 2013.

Casula, G, Fais, S., Ligas, P. (2009). Experimental Application of 3-D Terrestrial Laser Scanner and Acoustic Techniques in assessing the quality of stones used in monumental structures. *Int. Journ. of Microstructure and Materials Properties*, 4(1), 45–56.

Chien, C. H., & Aggarwal, J. K. (1986). Volume/surface octrees for the representation of three-dimensional objects. *Computer Vision, Graphics, and Image Processing*, 36(1), 100–113.

Christofori and Bierwagen (2013). Recording cultural heritage using terrestrial laserscanning – dealing with the system, the huge datasets they create and ways to extract the necessary deliverables you can work with. *Intl. Arch. Photogrammetry, Remote Sensing and Spatial Information Sciences*, Vol. XL, part 5/W2. XXIV. International CIPA Symposium, 2–6 Sept. 2013, Strasbourg, France.

Cignoni, P., Callieri, M., Corsini, M., Dellepiane, M., Ganovelli, F., Ranzuglia, G. (2008). MeshLab: an Open-Source Mesh Processing Tool. In *Proc. of Eurographics Italian Chapter Conference* (pp. 129–136), The Eurographics Association.

CloudCompare (2015). http://www.cloudcompare.org/. Accessed 30 Mar 2015.

CloudPro (2015). http://www.leica-geosystems.com/en/Leica-CloudPro_105688.htm. Accessed 28 Apr 2015.

Crutchley, S., & Crow, P. (2009). *The Light Fantastic – Using airborne LiDAR in archaeological survey*. Swindon: Historic England Publishing.

Cyclone (2015). http://hds.leica-geosystems.com/en/Leica-Cyclone_6515.htm. Accessed 27 Apr 2015.

Cyclone SERVER (2016). http://hds.leica-geosystems.com/downloads123/hds/hds/cyclone/brochures-datasheet/Cyclone_SERVER_DS_en.pdf. Accessed 25 Aug 2016.

Dellepiane, M., Dell'Unto, N., Callieri, M., Lindgren, S., Scopigno. R. (2012). Archeological excavation monitoring using dense stereo matching techniques. *J. of Cult. Herit.*, 14, 201–210.

De Reu, J., Plets, G., Verhoeven, G., De Smedt, P., Bats, M., Cherretté, B., De Maeyer, W., Deconynck, J., Herremans, D., Laloo, P., Van Meirvenne, M., De Clercq, W. (2013). Towards a three-dimensional cost-effective registration of the archaeological heritage. *Journ. of Arch. Science*, 40, 1008–1121.

De Reu, J., De Smedt, P., Herremans, D., Van Meirvenne, M., Laloo, P., De Clercq, W. (2014). On introducing an image-based 3D reconstruction method in archaeological excavation practice. *Journ. of Arch. Science*, 41, 251–262.

Doneus, M., & Neubauer, W. (2005a). 3D laser scanners on archaeological excavations. In *Proceedings of the XXth International Symposium CIPA*, Torino.

Doneus, M., & Neubauer, W. (2005b). Laser scanners for 3D documentation of stratigraphic excavations. In M. Baltsavias, A. Gruen, L. Van Gool, M. Peteraki (Eds.), *Recording, Modeling and Visualization of Cultural Heritage* (pp. 193–203), Taylor & Francis.

DroneDeploy (2015). https://www.dronedeploy.com/. Accessed 11 Nov 2015.

DJI (2015). http://www.dji.com/. Accessed 1 Mar 2015.

DPI8 (2015). http://dotproduct3d.com/DPI8.php. Accessed 27 April 2015.

Evans, A., Romeo, M., Bahrehmand, A., Agenjo, J., Blat, J. (2014). 3d graphics on the web: A survey. *Computers & Graphics*, 41(0), 43–61.

FARO Focus3D X330 (2015). http://www.faro.com/en-us/products/3d-surveying/faro-Focus3d/overview. Accessed 25 Feb 2015.

FARO Warranty department (2015). http://www.faro.com/en-us/support/faro-maintenance/warranty-maintenance-renewal. Accessed 28 Mar 2015.

FLIR&DJI (2015). http://dronexpert.nl/dronexpert-thermical-packages/. Accessed 25 Feb 2015.

Forte, M., Pescarin, S., Pietroni, E., Dell'Unto, N. (2005). The Appia antica project. In M. Forte (Ed.), *Archaeological Landscapes through Digital Technologies: Proceedings of the 2nd Italy-United States Workshop*. British Archaeological Report (BAR).

Forte, M., Dell'unto, N., Issavi, J., Onsurez, L., Lercari, N. (2012). 3D Archaeology at Çatalhöyük. *Int. Journ. of Heritage in the Digital Era*, 1, 351–378.

Forte, M., Dell'unto, N., Jonsson, K., Lercari, N. (2015). Interpretation Process at Çatalhöyük using 3D. In I. Hodder and A. Marciniak (Eds.), *Assembling Çatalhöyük*, Maney Publishing.

Forte, M., Danelon, N., Biancifiori, E., Dell'Unto, N., Lercari, N. (2016). Archive Report 2015. B89 and 3D Digging Project. *Çatalhöyük 2015 Archive Report*. Çatalhöyük Research Project – Stanford University.

Francovich, R. & Campana, S. (2005) Introduzione. Laser scanner e GPS: paesaggi archeologici e tecnologie digitali, 1: I workshop, Grosseto, 4 marzo 2005. *Quaderni del Dipartimento di archeologia e storia delle arti. Sezione archeologia/Università di Siena*, Edizioni All'insegna del giglio. p. 63. doi:10.1400/184085.

Geomagic Design, X. (2015). http://www.rapidform.com/products/xor/overview/. Accessed on 25 Aug 2016.

Girardeau-Montaut, D. (2011). CloudCompare-Open Source project. OpenSource Project.

Grinzato, E., Bison, P.G., Marinetti, S. (2002). Monitoring of ancient buildings by the thermal method. *Journ. of Cultural Heritage*, 3(1), 21–29.

Guidazzoli, A., Liguori, M. C., Felicori, M. (2012). Collecting, sharing, reusing geo and time-variant 3D models of the City of Bologna: An open project. In *Proc. of Virtual Systems and Multimedia (VSMM) 2012* (pp. 611–614). IEEE.

Guidi, G., Remondino, F., Morlando, G., Del Mastio, A., Uccheddu, F., Pelagotti, A. (2007). Performances evaluation of a low cost active sensor for cultural heritage documentation. In *Proc. of VIII Conference on Optical 3D Measurement Techniques*, ETH.

Guidi, G., F. Remondino, M. Russo, F. Menn, and A. Rizzi. 2008. 3D Modelling of Large and Complex Site Using Multi-Sensor Integration and Multi-Resolution. Paper presented at *The 9th International Symposium on Virtual Reality, Archaeology and Cultural Heritage (VAST) 2008* (pp. 85–92), 3-5 December 2008, Braga, Portugal.

Helion Eucaliptus (2015). https://www.eucalyptus.com/eucalyptus-cloud. Accessed Apr 29 2015.

Hess, M., Meyer, D., Hoff, A., Rissolo, D., Guillermo, L. L., Kuester, F. (2014). Informing Historical Preservation with the Use of Non-destructive Diagnostic Techniques: A Case Study at Ecab, Quintana Roo, Mexico. Paper presented at *International Conference EuroMed 2014* (pp. 659–668). Limassol: Springer International Publishing.

HxIP (2015). http://www.leica-geosystems.com/en/HxIP-Hexagon-Imagery-Programme_106454.htm. Accessed 29 Apr 2015.

Intel RealSense (2015). http://www.intel.com/content/www/us/en/architecture-and-technology/realsense-overview.html. Accessed 25 Feb 2015.

Kacyra, B., Dimsdale, J., Brunkhart, M. (1999). U.S. Patent No. 5,988,862. Washington, DC: U.S. Patent and Trademark Office.

Kersten, T. P., & Lindstaedt, M. (2012). Image-based low-cost systems for automatic 3D recording and modeling of archaeological finds and objects. In *Progress in cultural heritage preservation* (pp. 1–10). Springer Berlin Heidelberg.

Khoshelham, K. (2011). Accuracy analysis of kinect depth data. In *ISPRS workshop laser scanning*, 38(5), W12.

Khoshelham, K., & Elberink, S. O. (2012). Accuracy and resolution of kinect depth data for indoor mapping applications. *Sensors*, 12(2), 1437–1454.

Koller, D., Frischer, B., Humphreys, G. (2009). Research challenges for digital archives of 3D cultural heritage models. *Journ. on Computing and Cultural Heritage* (JOCCH), 2(3), 7.

Lambers, K., H. Eisenbeiss, M. Sauerbier, D. Kupferschmidt, T. Gaisecker, T. Hanusch. (2007). Combining Photogrammetry and Laser Scanning for the Recording and Modelling of the Late Intermediate Period Site of Pinchango Alto, Palpa, Peru. *Journ. of Arch. Science*, 34(10), 1702–1712.

Lavoie, C. and Lockett, D. (n.d.). Producing HABS/HAER/HALS Measured Drawings from Laser Scans: the Pros and Cons of Using Laser Scanning for Heritage Documentation. Heritage Documentation Programs, National Park Service. Retrieved from http://www.nps.gov/hdp/standards/laser.htm. Accessed 27 Mar 2015.

Leica ScanStation (2015). http://hds.leica-geosystems.com/en/Leica-ScanStation-P30-P40_106396.htm. Accessed 28 Mar 2015.

Leoni, C., Callieri, M., Dellepiane, M., O Donnell, D., Rosselli Del Turco, R., Scopigno, R. (2015). The Dream and the Cross: A 3D Scanning Project to Bring 3D Content in a Digital Edition. *Journ. on Computing and Cultural Heritage* (JOCCH), 8 (1).

Lercari, N. (2010). An Open Source Approach to Cultural Heritage: Nu.M.E. Project and the Virtual Reconstruction of Bologna". In M. Forte (Ed.), *Cyber-Archaeology* (pp. 125–133). Oxford, UK: Archaeopress, BAR International Series 2177.

Lercari, N., Toffalori, E., Spigarolo, M., Onsurez, L. (2011, October). Virtual heritage in the cloud: new perspectives for the virtual museum of bologna. In *Proceedings of the 12th International conference on Virtual Reality, Archaeology and Cultural Heritage* (pp. 153–160). Eurographics Association.

Lercari, N., Forte, M., Onsurez, L., Schultz, J. (2013). Multimodal Reconstruction of Landscape in Serious Games for Heritage. An insight on the creation of Fort Ross Virtual Warehouse serious game. In *Proceedings of Digital Heritage International Congress 2013*, Marseille, France, Oct. 28 - Nov 01, 2013.

Lercari, N., Mortara, M., Forte, M. (2014). Unveiling California history through serious games: Fort Ross Virtual Warehouse. In A. De Gloria (Ed.), *Lecture Notes in Computer Science: Games and Learning Alliance* (pp. 236–251). Berlin: Springer.

Lercari, N. & Lingle A. (2016). Çatalhöyük Digital Preservation Project – Field season 2015 Report. *Çatalhöyük 2015 Archive Report*. Çatalhöyük Research Project – Stanford University.

Luzi, G., Monserrat, O., Crosetto, M. (2012). Real Aperture Radar interferometry as a tool for buildings vibration monitoring: Limits and potentials from an experimental study. Paper presented at *10th International Conference on Vibration Measurements by Laser and Noncontact Techniques 2012* (pp. 309–317). Ancona, Italy, 27-29 June 2012.

Matterport (2015). http://matterport.com/technology/. Accessed 26 Feb 015.

MeshLab (2016). http://www.3d-coform.eu/index.php/tools/meshlab. Accessed 25 Aug 2016.

Mills, J., & Andrews, D. (2011). 3D Laser Scanning for Heritage (second edition – First Edition authors: Barber, D., Mills, J.): *Advice and guidance to users on laser scanning in archaeology and architecture*. Swindon: Historic England Publishing.

Montuori, A., Luzi, G., Stramondo, S., Casula, G., Bignami, C., Bonali, E., Bianchi, G.M., Crosetto, M. (2014). Combined use of ground-based systems for Cultural Heritage conservation monitoring. In *Geoscience and Remote Sensing Symposium (IGARSS), 2014* (pp. 4086–4089). IEEE International.

Neubauer, W., Doneus, M., Studnicka, N., and Riegl, J. (2005). Combined high resolution laser scanning and photogrammetrical documentation of the pyramids at Giza. In *Proceedings of CIPA XX International Symposium* (pp. 470–475).

Nurmi, D., Wolski, R., Grzegorczyk, C., Obertelli, G., Soman, S., Youseff, L., Zagorodnov, D. (2009, May). The eucalyptus open-source cloud-computing system. In *Cluster Computing and the Grid, 2009. CCGRID'09. 9th IEEE/ACM International Symposium* (pp. 124–131). IEEE.

Oculus Rift (2016). https://www.oculus.com/. Accessed 25 Aug 2016.

Pix4D (2015). https://pix4d.com/. Accessed 1 Mar 2015.

Pescarin, S. & Pietroni, E. (2007). La documentazione digitale: Le tecnologie integrate. In Forte, M. (Ed.), La Villa di Livia. Un Percorso di ricerca di archeologia digitale (pp. 101–110). Roma: L'Erma di Bretschneider.

Photoscan (2015). http://www.agisoft.com/. Accessed 8 Nov 2015.

Pollefeys, M., Van Gool, L., Vergauwen, M., Cornelis, K., Verbiest, F., Tops, J. (2001). Image-based 3D acquisition of archaeological heritage and applications. Paper presented at Conference on *Virtual Reality, Archeology, and Cultural Heritage 2001*. ACM.

Potenziani, M., Corsini, M., Callieri, M., Di Benedetto, M., Ponchio, F., Dellepiane, M., Scopigno, R. (2014). An Advanced Solution for Publishing 3D Content on the Web. In N. Proctor and R. Cherry (Eds.), *Proceedings of Museums and the Web 2014*, Silver Spring.

Ranzuglia, G., Callieri, M., Dellepiane, M., Cignoni, P., Scopigno, R. (2012). MeshLab as a complete tool for the integration of photos and color with high resolution 3D geometry data. Paper presented at *Computer Applications and Quantitative Methods in Archaeology*, Southampton, UK, 26–29 March 2012. Pallas Publications - Amsterdam University Press (AUP).

Rajendra, M. Y., Mehrotra, S. C., Kale, K. V., Manza, R. R., Dhumal, R. K., Nagne, A. D., Vibhute, A. D. (2014). Evaluation of partially overlapping 3d point cloud's registration by using ICP variant and CloudCompare. *ISPRS-International Archives of the Photogrammetry, Remote Sensing and Spatial Information Sciences*, 1, 891–897.

Raluca Popescu, C., & Lungu, A. (2014). Real-Time 3D Reconstruction Using a Kinect Sensor. Computer Science and Information Technology, 2(2), 95–99. doi:10.13189/csit.2014.020206.

REAL (2015). http://www.real2015.com/. Accessed 24 Feb 2015.

Remondino, F., & Menna, F. (2008). Image-based surface measurement for close-range heritage documentation. *The International Archives of the Photogrammetry, Remote Sensing and Spatial Information Sciences*, Vol. XXXVII. Part B5. Beijing 2008.

R&D, E. D. F. TP (2011). CloudCompare (version 2.3)(GPL software).

Samsung Gear VR 2016. http://www.samsung.com/global/galaxy/gear-vr/. Accessed 25 Aug 2016.

SCENE (2015). http://www.faro.com/en-us/products/faro-software/scene/overview#main. Accessed 25 Apr 2015.

SCENE WebShare 2Go (2016). http://www.faro.com/faro-3d-app-center/scene-plug-in-apps/scene-webshare-2-go. Accessed 25 Aug 2016.

SCENE WebShare Cloud (2016). http://www.faro.com/products/faro-software/scene-webshare-cloud/our-offer#main. Accessed 25 Aug 2016.
SCENE WebShare Server (2016). http://www.faro.com/faro-3d-app-center/stand-alone-apps/scene-webshare-server. Accessed 7 Nov 2015.
Scollar, I., & Giradeau-Montaut, D. (2012). Georeferenced orthophotos and DTMs from multiple oblique images. *AARGnews*, 44, 12–17.
Seek Thermal (2015). http://www.thermal.com. Accessed 3 Mar 2015.
Sequoia (2015). http://sequoia.thinkboxsoftware.com/. Accessed on 25 Feb 2015.
Silvestre, I., Rodrigues, J. I., Figueiredo, M. J. G, Veiga-Pires, C.. 2013. Cave Chamber Data Modeling and 3D Web Visualization. In *Proc. of the 2013 17th Int. Conf. on Inf. Vis. (IV '13)* (pp. 468–473). Washington, DC: IEEE Computer Society. doi:10.1109/IV.2013.103.
Siotto, E., Dellepiane, M., Callieri, M., Scopigno, R., Gratziu, C., Moscato, A., Burgio, L., Legnaioli, S., Lorenzetti, G., Palleschi, V. (2014). A multidisciplinary approach for the study and the virtual reconstruction of the ancient polychromy of Roman sarcophagi. *Journ. of Cultural Heritage*, 15.
Sixense Razer Hydra (2016). http://sixense.com/razerhydra. Accessed 25 Aug 2016.
Smith, N.G., Cutchin, S., Kooima, R., Ainsworth, R.A., Sandin, D.J., Schulze, J., Prudhomme, A., Kuester, F., Levy, T.E. (2013). Cultural heritage omni-stereo panoramas for immersive cultural analytics—From the Nile to the Hijaz. Paper presented at *8th International Symposium on Image and Signal Processing and Analysis (ISPA)*. IEEE.
Smith, N., Passone, L., al-Said, S., al-Farhan, M., Levy, T. E. (2014). Drones in Archaeology: Integrated Data Capture, Processing, and Dissemination in the al-Ula Valley, Saudi Arabia. *Near Eastern Archaeology*, 77(3), 176–181.
Structure Sensor (2015). http://www.structure.io/. Accessed 25 Feb 2015.
Thinkbox (2015). http://sequoia.thinkboxsoftware.com/. Accessed on 6 Nov 2015.
TruView (2015). http://hds.leica-geosystems.com/en/Leica-TruView_63960.htm. Accessed 28 Mar 2015.
Vergauwen, M., & Van Gool, L. (2006). Web-based 3d reconstruction service. *Machine vision and applications*, 17(6), 411–426.
Verhoeven, G. (2011). Taking computer vision aloft archaeological three- dimensional reconstruction from aerial photographs with Photoscan. *Archaeological Prospection*, 18, 67–73.
Verhoeven, G., Taelman, D., Vermeulen, F. (2012). Computer vision-based orthophoto mapping of complex archaeological sites: the ancient quarry of Pitaranha (Portugal-Spain). *Archaeometry*, 54, 1114–1129.
Visionary Cross Project (2015). http://www.visionarycross.org/. Accessed on 29 Apr 2015.
WebGL (2015). https://www.khronos.org/webgl/. Accessed 28 Mar 2015.
X3D(2015).http://www.web3d.org/sites/default/files/page/About%20Web3D%20Consortium/What_is_X3D_201206.pdf. Accessed 28 Mar 2015.
X3DOM (2015). http://www.x3dom.org/. Accessed 28 Mar 2015.
Zhang, Z. (2012). Microsoft Kinect sensor and its effect. *MultiMedia*, IEEE, 19(2), 4–10.

Airborne Laserscanning in Archaeology: Maturing Methods and Democratizing Applications

Rachel Opitz

Abstract Archaeologists have been using airborne laserscanning (ALS) for over a decade in projects ranging from heritage management schemes for post-industrial uplands in the UK or state-managed forests in Germany to research on cities now obscured by tropical jungle canopy in Central Mexico. The basic methods for the analysis and interpretation of this data have matured considerably and data is increasing available. Building on this increasing accessibility and an established basic methodology, archaeologists are addressing a growing variety of ground conditions and research and heritage management objectives through this technology. With this diversification comes the need to adapt the basic methods used to new landscapes and types of archaeological remains, and to integrate the practice of working with ALS with diverse fieldwork and research practices.

Introduction

This chapter aims reviews the current state of practice in the use of Airborne Laser Scanning (ALS), or LiDAR, in archaeology, highlighting areas of ongoing research and open questions where further work is needed. Happily, the use of Airborne Laser Scanning as a method has reached a state of relative maturity. It is broadly accepted as a basic source of extensive and detailed topographic data (Opitz and Cowley 2012) and is widely used to promote the integrated study of past and present land use and landscape modification in archaeology. The general workflow is one of data acquisition, classification, surface and bare earth terrain model creation, visualization, and interpretation. Different projects and applications may introduce variations at some or all of these steps to suit their needs, but the general workflow remains the same.

R. Opitz (✉)
Center for Virtualization and Applied Spatial Technologies (CVAST) and Department of Anthropology, University of South Florida, Tampa, FL 33620, USA
e-mail: ropitz@usf.edu

M. Forte and S. Campana (eds.), *Digital Methods and Remote Sensing in Archaeology*, Quantitative Methods in the Humanities and Social Sciences,
DOI 10.1007/978-3-319-40658-9_2

Looking at the specific steps in the processing chain, the current classification methods used throughout the ALS industry produce results that can be used in archaeological research. Several papers have emerged comparing different visualization methods and making recommendations for their use under various archaeological and ground cover circumstances (Štular et al. 2012; Bennett et al. 2012). Software toolkits collecting the most commonly used Digital Terrain Model (DTM) visualizations, and targeted for an archaeological audience, have been produced, and are available through the ArchaeoLandscapes Europe network (http://www.arcland.eu/outreach/software-tools) or directly from individual or institutional creators (RVT—http://iaps.zrc-sazu.si/en/rvt#v and LiVT—http://sourceforge.net/projects/livt/). In short, the strictly methodological work in this domain has moved from the general case to special circumstances in which advanced processing of the data or customized visualizations are needed.

Having said that the methodology is more or less settled, if we broaden out our definition to include the fieldwork routines, image interpretation, and map-reading, then we see a rapidly evolving methodology among a growing ALS-using archaeological community. This second level of methodological development is not about the technical developments that interest specialists, but rather about the use of the data provided through this technology by the broader archaeological community. In this sense, the practice of working with ALS in archaeology has moved decisively off the desk and out into the field, and its integration into fieldwork practice is an important area of innovation. The democratization of the use of ALS in archaeology, as it moves from a specialist tool to one integrated into many projects, as are aerial photographs and maps, is the driver behind most current methodological progress.

The impact of the availability of ALS data to a growing pool of researchers, and the subsequent re-focus on topography as a source of archaeological information, raises a number of interesting questions about specific classes of features and the impacts of more recent land use. These points have been raised elsewhere in this volume, notably by Limp in his discussion of the broader impacts of High Density Survey and Measurement, of which ALS is a subset (see also Opitz and Limp 2015). Along similar lines, the use of ALS is driving an interesting expansion in the practice of image interpretation, as the terrain model visualizations are just unfamiliar enough to force careful image reading, while being just familiar enough to encourage many archaeologists to undertake their own interpretive work.

Established Methods and Practices for Archaeological ALS

Data Acquisition

The acquisition of ALS data, while generally not carried out by archaeologists, does require that they involve themselves in the specification of the survey parameters, or—in the case of pre-existing datasets—that they are aware of the limitations imposed by the survey parameters chosen by someone else. The desired spatial resolution of the final data product, the number of returns per pulse, the tradeoff between aerial coverage and resolution (essentially a choice in terms of helicopter vs. fixed wing platform, with implications for cost and survey area size), the scan angle, and occasionally the instrument choice must be specified. For most archaeological applications, an x-y planimetric resolution of 8 pts/m2 is considered to be the minimum for archaeological work, with 16 pts/m2 or greater being a desirable range. Smaller scan angle ranges, minimizing returns lost to vegetation are likewise considered desirable, with the exception of mountainous landscapes or areas of steep-sided valleys or canyons, where a larger scan angle might help avoid occluded areas. The number of returns per pulse is perhaps less crucial as modern instrumentation generally support the minimum of four returns per pulse recommended for archaeological surveys. The general guidelines for data acquisition can be found in Opitz and Cowley (2012, Chap. 2), and the recommendations remain largely unchanged.

Classification Routines

The method for the classification of ALS point cloud data to remove vegetation and reveal the 'bare earth' and traces of archaeologically relevant topography is well established for most situations (Fig. 1), and the areas where it is likely to be problematic are known. Broadly, areas with low scrub vegetation, low scrub within a multistory canopy, and close coniferous plantation remain problematic, as do areas with large amounts of surface debris e.g. boulder fields. There are two approaches taken to improving classification in these cases. First is the collection of full waveform data. The number of projects using full waveform data remains comparatively small. Where it is employed, the echo width of each peak in the signal is used to help distinguish between low scrub, sloping terrain, and terrain returns (Chauve et al. 2007). The improved classification achieved using echo width is seen as the main advantage of full waveform over discrete return data. Second, are alternative classification routines that are generally more computationally expensive, but may be effective for treating small areas (e.g. Brodu and Lague 2012; Opitz and Nuninger 2014). These alternative approaches to classification are discussed below, in the *Special Cases in Classification* section.

Outside of these environments, the algorithms implemented in industry standard software packages including Terrascan (www.terrasolid.com) and LAStools (http://rapidlasso.com/lastools/), and in the open source MCC-Lidar (http://sourceforge.net/projects/mcclidar/) will produce terrain models of sufficient quality to support archaeological interpretation and to be the basis for a survey project. Projects receiving pre-classified data from larger industry providers will likely be working with the results of the modified Axelsson algorithm (Axelsson 1999) implemented in Terrascan, the dominant software package for companies in the business of ALS data acquisition. As LAStools increased in popularity in the archaeological community, with its lower cost compared to Terrascan and (in this author's opinion)

Fig. 1 Profile view of a classified ALS point cloud. Terrain returns are shown in *orange* and vegetation returns in *green*

more user-friendly implementation, its classification routine is increasingly employed by those producing their own classifications from a raw point cloud, or reclassifying subsets of their data in hopes of improved results.

Visualizations

The majority of ALS related methodological discussion in recent years has centered in the panoply of visualization methods available, and how to select the most effective ones for a given type of landscape or features. Archaeological ALS DTM visualizations are dominated by simple hillshades, local relief models (and variants thereon) (Hesse 2010), sky view factor (and variants thereon) (Kokalj et al. 2011), and elevation ramps. Broadly, hillshades are beneficial in that they highlight low reliefs by simulating raking light from a single direction across the terrain surface. Sky view factor is useful for identifying small pits and depressions, or narrow areas of relief, as it highlights the difference between tightly enclosed spaces (where one would have little ability to see the sky) and topographically prominent ones (with clear views all around). Elevation ramps, while only showing anomalies well in exceptionally flat terrain, or by stretching the available values to match the elevation range present in a small localized area, are useful in providing an absolute sense of the size of anomalies. These visualizations are commonly combined in layers, and different combinations are employed depending on the circumstances (Fig. 2).

At its core the problem with visualizations is that while producing 16 different renderings of each part of the landscape is relatively simple and may be automated, looking through sixteen different visualizations for each chunk of landscape is not feasible for most projects. Somehow, an individual interpreter or team must select a basic subset of all possible visualizations with which they will work for each project. In multi-interpreter teams, the likelihood of achieving some level of consistency between individuals will be increased if everyone works from the same set

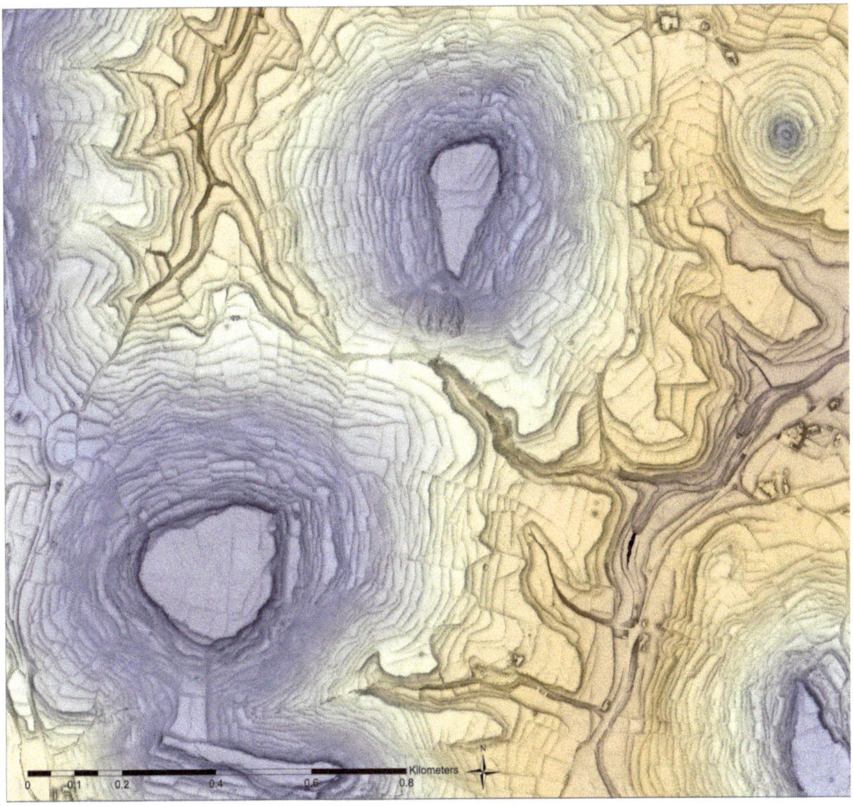

Fig. 2 Terraces and field systems in Gozo are visualized using a combination of Sky View Factor (16 look directions, 10 m radius) and an elevation ramp. These layers provide a sense of absolute elevation change while highlighting terrace edges and boundary walls. (Data provided with kind permission of the University of Malta FRAGSUS team The *Digital Terrain Model* used in the process of model building was derived from the LiDAR data made available through an agreement signed between the University of Malta and the Malta Environment and Planning Authority in 2013. We acknowledge the source: *ERDF LIDAR data, 2012, ERDF156 Developing National Environmental Monitoring Infrastructure and Capacity, Malta Environment and Planning Authority*)

of visualizations. Best practice has emerged as selecting around three visualizations to be used in combination. This represents a compromise between techniques that do well at low relief, techniques that distinguish open and closed, and techniques not dependent on lighting/illumination, without requiring interpreters to engage with an overwhelming number of visualization combinations. This question has been discussed by both Štular et al. (2012) and Bennett et al. (2012) in detail and suggestions for good combinations of visualizations for certain ground cover conditions and types of archaeological features are made in these articles.

Ongoing First Level Methodological Development

Special Cases in Classification: Low Scrub

Low, dense vegetation masking both the detail of surface topography and any low standing structures e.g. field boundary walls is a longstanding problem in ALS-driven archaeological landscape studies. Two problems converge in situations where the ground cover is dominated by low scrub. First, at the moment of data collection very few returns will reach the ground and return to the sensor, as energy is scattered by the complex canopy structure of this vegetation. Second, classification algorithms dependent on changes in slope or curvature between neighboring points struggle to differentiate between returns from low scrub and those from small changes in the topography. The result is an entire class of land use where archaeological visibility is remarkably poor, though in long-standing scrub areas preservation is likely to be better than average. While little experimental work has been carried out, lower flying elevation, smaller beam footprints, a different wavelength outside the NIR range reflected strongly by vegetation and a smaller scan angle range are all methodologically feasible possibilities for improving the number of returns from the ground in areas with this kind of ground cover.

Here we can touch briefly on the rapidly evolving suite of UAV ALS platforms. Notably, these platforms may transform the quality of data collection in scrub areas, as they facilitate lower flying elevations and smaller footprints, which should improve data quality (see Nex and Remondino 2014 for an overview of applications and future directions). UAV ALS platforms will likely produce data requiring processing strategies more like those used for mobile lidar or Terrestrial Laserscanning Survey (TLS). In particular, given the fine resolution of the data relative to the precision available from the combined DGPS/IMU platforms deployable on a UAV, tight alignment between flight strips is likely to remain a challenge (Watts et al. 2012). As with TLS survey, the quality of the alignments between individual scans are often more than sufficient for the reliable interpretation of archaeological features, but larger sets of scans/flightstrips may result in propagating errors to the point that visual mis-alignments interference with the identification of features.

Barring fundamental changes in data collection strategies, improved classification of the point cloud following the kinds of methods employed for the classification of TLS scenes (Brodu and Lague 2012) have been pursued by the author (Opitz and Nuninger 2014). Classification strategies that depend on variations in local density, roughness, and 3D rather than profile curvature, measured across multiple scales have proven, in limited tests, to produce better results in complex scenes with low scrub and standing remains or surface microtopography.

Special Cases in Classification: Stony Surfaces

Closely related to situations in which a landscape is dominated by low scrub that masks the archaeological topography and standing remains of interest, are 'cluttered' landscapes where surface debris including boulders and large stones are common. Here we face several complications. First, the stones themselves sometimes form part and parcel of the archaeology, as they are gathered into field boundaries, small cairns, and other forms. Second, there is zero penetration to the ground—any removal of returns from the stones will results in a visually jarring gap in the terrain model. Therefore the aim of this kind of classification of stony surfaces into 'surface objects' and 'background terrain' in not to remove the stones from the digital model, but to highlight or obscure them, aiding in the visual interpretation of topography and surface features. Further, one may attempt to separate individual or small clusters of surface debris from larger clusters or alignments, essentially moving to an object oriented classification approach. Software including eCognition (www.ecognition.com/) supports this kind of classification routine. Barring a formally object oriented approach, these regions might be flagged as concentrations of higher variability in local curvature or roughness—metrics paralleling those used to remove vegetation from low scrub scenes.

As with low scrub classification, the size at which the kernel is set is critical to the effective separation of different classes of features. The kernel size must be calibrated to a scale at which surface stones are showing a high degree of variability but the background topography is not. As with vegetation classification routines, the results of this kind of classification will not be 100 % accurate, and 80 % accuracy in separation of the surface scatters of stones and background terrain would be considered a good result.

Special Cases in Visualization: Direct Viewing of Point Cloud Data

Most archaeologists working with ALS data take the Digital Terrain Model and Digital Surface Model (DTM and DSM respectively) as their starting points, and build additional visualizations of the data on top of these models. The DTM and DSM, of course, are themselves abstractions of the point cloud data (itself an abstraction of the waveform data) that is the native format for ALS. In many cases the visualizations based on the DTM and DSM represent the best choice for interpretive work, as features appearing in them look somewhat familiar to those with experience reading aerial imagery or topographic maps. However, under some circumstances it is useful to engage directly with the point cloud data. For example, in situations where standing structures in overgrown areas, particularly where these structures have multiple parts e.g. a ruined castle or farmstead, need to be understood and situations where strongly vertical cuttings form part of the feature of interest studying the point cloud in addition to the surface model can be productive.

Closely related to the discussion below on *Communication Issues,* while the visualizations effective for reading DTMs and DSMs are largely agreed upon, as are the effect of various parameters e.g. light source direction, hillshade altitude, number of segments in a sky view factor calculation, the effects of light source, view angle, icon size, view clipping, or color schemes on point cloud visualizations remain poorly studied. The development of a methodology for the representation of standing archaeological remains and surface topography in point cloud data represents a necessary next step, as archaeologists use it to study increasingly complex remains and landscape areas.

Second Level Methodological Development

Integration with Fieldwork

Perhaps the most important development in the use of ALS in archaeology in recent years is the increasing use of—primarily—raster visualizations in fieldwork (Fig. 3). Both Heritage Management agencies, e.g. Historic England (formerly English Heritage) and research projects, e.g. The Survey and Landscape Archaeology on Montserrat (SLAM) Project, are using ALS visualizations to guide fieldwork. For Historic England, in internal projects like the study of the miner-farmer landscapes in the North Pennines (Ainsworth et al. 2013; Schindling and Gibbes 2014), archaeologists took ALS visualizations into the field on tablets or paper, using the data to guide the in-person interpretation of landscape features in an uplands former industrial region. For the SLAM Project (Cherry et al. 2012; Ryzewski and Cherry 2015; Opitz et al. 2015) the ALS visualizations primarily guided both the planning of the survey in terms of areas to visit on the ground. As in the Pennines, ALS visualizations were taken into the field, and aided in the identification of more subtle landscape features.

These two projects can be used as examples of quite different fieldwork engagements with ALS data. For the Pennines Project, the field archaeologists using ALS imagery were already familiar with the territory and had spent

Fig. 3 Fieldwork in Montserrat (*left*) and Chailluz (*right*) involves bringing visualizations of the ALS data into the field. This practice allows direct visual comparison between features seen on the ground and those recognized in the data. (Image Credits: Cherry and Ryzewski—SLAM Project and Daval—LIEPPEC)

substantial time on the ground. This, then, is a situation of 'new eyes' with ALS visualizations helping field archaeologists to see a relatively familiar landscape in new ways. The unfamiliar look of ALS becomes beneficial in many ways, forcing visual attention on areas of a known region that might otherwise be missed. Further, the 'overview' character of an ALS visualization can aid archaeologists in the field in recognizing extensive micro-relief features, particularly in open areas, and this is one of the main contributions of the technology for projects like the that in the Pennines (Ainsworth et al. 2012).

The archaeologists in the field with the SLAM Project face a very different situation. Because of the extremely difficult conditions on the ground, dominated by relatively young tropical forest, rugged and steeply sloping terrain, and boulder fields and unusual geological formations due to recent volcanic activity, ALS visualizations first and foremost were used to target on-the-ground survey. The archaeologists on this project do not have the benefit of long periods in the landscape because of its inaccessibility. Further, while there is a well-established visual and mental template of features likely to be seen in an English landscapes and what they look like in aerial imagery, no such template exists for the colonial landscape of Montserrat. As a relatively young project in a nation without a substantial history of landscape survey or aerial image interpretation, the starting point for the reading of ALS while in the landscape is quite different. Visual 'clutter' when working on the ground, including complex patterns of light and shadow, dense vegetation and surface scatters readily obscure more subtle micro-topographic features. Consequently, the kinds of features identified using the ALS visualizations were different (Opitz et al. 2015).

These projects are representative of the use of ALS under quite different circumstances. They illustrate the general trend toward an on-the-ground engagement with ALS visualizations, and a continued blurring between what once were the distinctly separate worlds of desk-based work and fieldwork, and of heads-up interpretation of aerial imagery on the one hand and field interpretation of topography on the other.

Communication Issues: Aerial Imagery Interpretation, Dynamic Visualizations and the Retreat from the Hachure Plan

Archaeology of the scientific bent, as practiced today, values the publication of data and a clear explanatory chain detailing how one arrived at an interpretation based on that data. The interpretation of ALS sits somewhat uncomfortably in this respect, as the precise and detailed metric models carry an air of the scientific, and the publication of the data itself is reasonably common practice, but the interpretation is purely visual and individual. Like the identification of a ceramic sherd or lithic tool by an expert, photos and drawings can be published, but the interpretation and identification of the type of object remains only loosely linked to these. But where the need for specialist expertise is broadly recognized for the identification of ceramics or lithics, the reading of aerial imagery is often treated as self-evident (thanks in some small part to the prevalence of google maps & co., bringing aerial imagery to the general public). The hachure plan through which topographic data was communicated prior to the advent of lidar, much like the drawing of a lithic fragment or ceramic sherd, had the interpretation pre-packaged, and this reified communication of form was not really questioned. After all, you would have to go to the place and take detailed measurements yourself, a laborious process, to produce a new plan with which to question the extant one. In most cases the shape of the thing was its shape. The multitude of techniques used to visualize ALS terrain models, and the open discussion within the practitioner community about their relative merits and the difficulties of features that 'appear' and 'disappear' in different visualizations, has created a situation in which the visual interpretation is always up for debate. The all too evident ease with which the visual appearance of different shapes is manipulated means an expert's interpretation is readily questioned. The shape of a terrain feature, as seen in the ALS terrain model, shaded in some way, is not necessarily its shape. Further, we cannot simply retreat to just showing the 'raw' terrain model, as it is clear that many real features are only recognizable when visualized in certain ways. So what to do when trapped between the good and bad of data projections and visualizations? They are sometimes the only thing that shows up the real archaeology, but they are also the primary source of visual artifacts and illusions. The answer brings us back to expertise, and the expert interpreter's ability to choose an appropriate visualization technique and differentiate between real features and illusory ones.

In practice, most expert interpreters use multiple terrain model visualizations side-by-side to make an interpretation. The appearance or absence of a feature in any given visualization provides additional information. It is this collection of visual impressions that results in a final identification and interpretation. In conclusion, in spite of the highly metric and precise nature of ALS data and models derived from it, it is unreasonable to ask for a scientific-style chain of evidence from data to interpretation, because dynamic viewing and visual impression sit squarely in the middle of the process. But having left the reified topographic language of the hachure plan behind in favor of a dynamic suite of terrain visualizations, the inherent problem of interpretation based on visual gestalt—an issue not limited to ALS—is pushed to the fore.

Beyond Methodology

Topography as Archaeological Information: Reading the Impact of Recent Land Use

Any archaeological feature recognized in ALS data is viewed in the context of the contemporary landscape, and the impacts of current or recent land use are often abundantly apparent. Barrows are seen in and out of ploughed fields and limekilns are seen in and out of the forest. The effect of mechanized agriculture, in particular deep ploughing as practiced in much of Europe, while bringing buried ceramics and building materials to the surface and creating a golden age for fieldwalking survey, caused evident deterioration to topographic features (Darvill 2014). This is most evident when single features are located on a border between land use zones, as a direct comparison can be made. Even where direct comparison is not available, the general appearance of a class of features of which there are many scattered across the landscape which are likely from roughly the same period allows for the impact of subsequent land use to be assessed. This situation opens the possibility for a close study of the impacts of management on earthwork and topographic features of various types.

The afforestation of areas under plough until the first half of the 20th century and then abandoned after a population decline or transfer have the converse effect, resulting in pockets of unusually well preserved remains of the 18th and 19th centuries. (Though some forest management practices are as destructive to archaeological topography as ploughing; see Crow and Moffat 2005) Separately, it provides the means to engage with a landscape archaeology of the recent past. The later of these has already begun, primarily by archaeologists working on WWI and WWII landscapes (e.g. Hesse 2014; Gheyle et al. 2014), whose entrenchments, battlegrounds, and abandoned towns can be found in well preserved states within today's forests. While this work has primarily been carried out in an European sphere, the potential for new work in the archaeology of the recent past, taking advantage of excellent preservation and visibility afforded by ALS data collection in woodlands, exists across almost all regions where data is now being acquired.

Topography as Archaeological Information: Interpreting Other 'scrapes' in the Landscape

In the European context, where topographic survey has the longest history, we may divide traditional interest in archaeological topography into three areas: monumental earthworks, e.g. mounds, barrows, ring-forts; extensive organized features i.e. field boundaries; and standing monuments e.g. ruined cities and estates (as an instructive example, see Collis 2013). In Northern Europe monumental earthworks have been particularly dominant, particularly in the discourse on prehistoric landscapes and settlement, while the Classical Mediterranean region developed a strong tradition of survey of standing monuments in urban contexts. The uptake in use of ALS has encouraged archaeologists to move outside these areas of study and pay attention to the topographic signatures of a much wider variety of features. These features generally are smaller in size, and related to the economy and use of the forests themselves, or other aspects of the rural economy. Thus we have a proliferation of studies incorporating or depending primarily on mapping and characterization of limekilns, charcoal burning platforms, surface quarries, animal trapping pits, field clearance cairns, and ditches, with which we can build a picture of the rural landscape (Fig. 4). Raab et al. (2015), in their work using ALS regional survey data processed by the state heritage agency, specifically point to the enormous quantity of small topographic features, charcoal burning platforms in southeast Germany being a case in point, as calling for another form of interpretation, as the distribution of these features and their emplacement in the landscape is of interest, rather than any single feature.

The shift in focus from the individual monument or organized system of features to the dense palimpsest of small features reflects both the character of the ALS data and the coincidence of its development in parallel with the rise of landscape archaeology as a theoretical approach. Along these lines, Mlekuž offers several papers

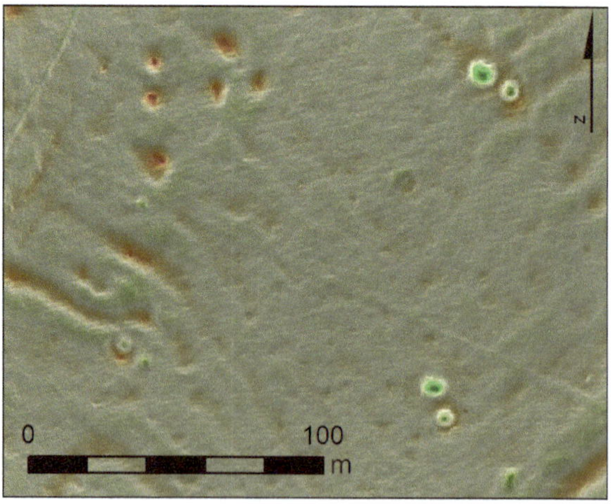

Fig. 4 Charcoal burning platforms and other evidence of rural economic activity, seen in the lidar data from the Besançon lidar survey. Data kindly provided by the LIEPPEC Project, Université de Franche-Comté

reflecting on topics from braided hollow ways representing fluctuating routeways, to collections of kilns, charcoal burning platforms, and natural karstic depressions forming complex landscape assemblages (2012, 2013). Mlekuž, among others, emphasizes the use of lidar to look not just at a single feature or collection of features, but to tie more closely together the features and the landscape in which they are embedded, referring to the remains of various activities as 'scrapes' in the landscape (Mlekuž 2012). The expansion to look at all the activity, all the 'scrapes', in the landscape provides a means of tracking the intensity of activity over time, in a general sense.

While we can speak broadly about trends in the intensity of activity based on a broad look at all the features appearing in the landscape associated with a given activity or activities, in engaging with the many 'scrapes' of a densely inscribed archaeological landscape, the perennial problem of establishing even a relative chronology remains. Mlekuž's work on hollow ways illustrates the constant cutting and re-cutting of these paths, seemingly seasonally or ad hoc, resulting in a buildup of archaeological features. This kind of sequence highlights the perhaps obvious fact that a relatively large number of archaeologically visible events does not necessarily equate to a long time period, and the difficulties that can arise in reading the succession of cuts and overlays from the topographic data.

Topography as Archaeological Information: Diffuse Urbanism in Neo-tropical Rainforests

The advent of ALS for archaeological projects studying the phenomenon of urbanism in now heavily forested regions of Central and South America and Southeast Asia has resulted in a substantially improved understanding of the structure of these cities. Notably, survey results show that 'diffuse' urban areas are more extensive than previously thought. In some cases, the suburban and ex-urban surroundings of the densely urbanized core stretch almost to the limits of the ex-urban zone around another city. The scale of urbanized areas was extremely difficult to grasp prior to the use of lidar, due to the poor visibility and difficulties of accessibility on the ground in dense jungle environments. In addition to the shift in scale, the details of the patterns of occupation outside the dense cores of these cities are emerging through systematic study of the topographic models, combined with targeted survey work on the ground. Notable projects working on this topic include the Cambodia Archaeological Lidar Initiative (http://angkorlidar.org/), whose study areas include Ankor Wat, Sambor Prei Kuk, Koh Ker, and Phnom Kulen (Evans et al. 2013), studies of Caracol in Belize, led by Chase and Chase (Chase et al. 2011), and work at Angamuco in Mexico by Fisher (Fisher and Leisz 2013). A discussion of the impacts of ALS on this area of study can be found in Chase et al. (2012) wherein they note that lidar data, "are changing commonly held interpretations of societal development profoundly." as, "[t]he more complete

LiDAR data demonstrate that some ancient Mesoamerican sites are far more extensive and complex than was thought possible following popular sociopolitical models". For questions of diffuse urbanism, the large scale at which ALS data allows archaeologists to observe patterns in detail is crucial. As surveys in these regions rapidly proliferate, the impact of the availability of ALS as a tool for rapid mapping of sprawling built up areas will continue.

Conclusions

Over the past decade ALS has become an important archaeological survey technique, used in a variety of landscape and regional scale research and heritage management projects. The methodology used to analyze and interpret ALS data has matured, and the general approach is now widely agreed upon. Archaeology is seeing a rapid democratization of the use of this data, and methodological development to treat difficult landscapes or classes of features continues as the use of ALS proliferates. Outside of the cases where basic methodological work is still needed to improve archaeological visibility, notably in areas dominated by low scrub or a profusion of objects on the surface, the main methodological developments are now focused on the combination of LiDAR and field survey to improve the interpretation of broad areas of landscape and accelerate landscape and regional survey.

The majority of archaeological research projects or Heritage Management schemes employing ALS continue to be found in Northern and Western Europe (Opitz and Cowley 2012), unsurprising as much early work with archaeological ALS was carried out in these regions. However, the success of ALS in elucidating urban plans in tropical forest regions such as at El Caracol in Belize and other Mesoamerican sites (Chase et al. 2011), and the increasing availability of and growing interest in the publicly-accessible USGS LiDAR data in the United States (Pluckhahn and Thompson 2012; Randall 2014) are encouraging the use of ALS in these regions. While the method has spread from its roots in Europe to use in the Americas and Southeast Asia, its use for archaeological survey in Africa, where forest dominates large swathes of landscape, and much of Asia, where land cover also obscures much of the archaeological landscape, has not yet begun. The adaptation of archaeological ALS methods to the ground conditions and research questions relevant in these places is an area with great future potential.

Reflecting on more than a decade of increasing research activity, the greatest impacts of ALS are that by allowing archaeologists to more efficiently survey expansive regions obscured by woodland or scrub vegetation, ALS has reintegrated these areas into research based on regional survey data, and by opening forests to systematic study it has reinvigorated lines of research focused explicitly on archaeology in woodlands. Even more exciting, the results of these surveys are beginning to support the study of broader regional patterns, allowing new approaches to fundamental questions on rural land use and economy, diffuse urbanism, and others yet to be explored.

References

Ainsworth, S., Oswald, A. and Went, D. (2013) Remotely acquired, not remotely sensed: using lidar as a field survey tool. In Opitz, R. S. and Cowley, D. C. (eds.) Interpreting Archaeological Topography. 3D Data, Visualisation and Observation. AARG Occasional Paper 5. Oxbow Books: Oxford, 206–22.

Axelsson, P., 1999. Processing of laser scanner data—algorithms and applications, ISPRS Journal of Photogrammetry and Remote Sensing, 54:138–147.

Bennett, R., Welham, K., & Ford, A. (2012) A comparison of visualization techniques for models created from airborne laser scanned data. *Archaeological Prospection, 19*(1), 41–48.

Brodu, N., & Lague, D. (2012) 3D terrestrial lidar data classification of complex natural scenes using a multi-scale dimensionality criterion: Applications in geomorphology. *ISPRS Journal of Photogrammetry and Remote Sensing, 68*, 121–134.

Chase, A. F., Chase, D. Z., Weishampel, J. F., Drake, J. B., Shrestha, R. L., Slatton, K. C., & Carter, W. E. (2011). Airborne LiDAR, archaeology, and the ancient Maya landscape at Caracol, Belize. Journal of Archaeological Science, 38(2), 387–398.

Chase, A. F., Chase, D. Z., Fisher, C. T., Leisz, S. J., & Weishampel, J. F. (2012) Geospatial revolution and remote sensing LiDAR in Mesoamerican archaeology. *Proceedings of the National Academy of Sciences, 109*(32), 12916–12921.

Chauve, A., Mallet, C., Bretar, F., Durrieu, S., Deseilligny, M. P., & Puech, W. (2007) Processing full-waveform lidar data: modelling raw signals. In *International archives of photogrammetry, remote sensing and spatial information sciences 2007*, 102–107.

Cherry, J. F., Ryzewski, K., & Leppard, T. P. (2012) Multi-period landscape survey and site risk assessment on Montserrat, West Indies. *The Journal of Island and Coastal Archaeology, 7*(2), 282–302.

Collis, J. (2013) The Development of Archaeological Thought as Evidenced in the*Yorkshire Archaeological Journal. Yorkshire Archaeological Journal* 2013; 85(1), 5–26.

Crow, P. and Moffat, A.J. (2005). The management of the archaeological resource in wooded landscapes: an environmental perspective. Conservation and management of archaeological sites. 7, 103–116.

Darvill, T. (2014) *Ancient monuments in the countryside: an archaeological management review.* English Heritage Publishing.

Evans, D. H., Fletcher, R. J., Pottier, C., Chevance, J. B., Soutif, D., Tan, B. S.,& Boornazian, G. (2013) Uncovering archaeological landscapes at Angkor using lidar. *Proceedings of the National Academy of Sciences, 110*(31), 12595–12600.

Fisher, C. T., & Leisz, S. J. (2013) New Perspectives on Purépecha Urbanism Through the Use of LiDAR at the Site of Angamuco, Mexico. In *Mapping Archaeological Landscapes from Space* (pp. 199–210) Springer New York

Hesse, R. (2010) LiDAR-derived Local Relief Models–a new tool for archaeological prospection. *Archaeological Prospection, 17*(2), 67–72.

Hesse, R. (2014) Geomorphological traces of conflict in high-resolution elevation models. *Applied Geography, 46*, 11–20.

Gheyle, W., Dossche, R., Bourgeois, J., Stichelbaut, B., & Van Eetvelde, V. (2014) Integrating Archaeology and Landscape Analysis for the Cultural Heritage Management of a World War I Militarised Landscape: The German Field Defences in Antwerp. *Landscape Research, 39*(5), 502–522.

Kokalj, Z., Zaksek, K., & Ostir, K. (2011) Application of sky-view factor for the visualisation of historic landscape features in lidar-derived relief models. *Antiquity*, *85*(327), 263–273.

Mlekuž, D. (2012) Messy landscapes: lidar and the practices of landscaping. In Opitz, R.S. and Cowley, D.C. (eds.) Interpreting archaeological topography: lasers, 3D data, observation, visualisation and applications, AARG Occassional Paper 5. Oxbow Books: Oxford, 90–101.

Mlekuž, D. (2013) Skin Deep: LiDAR and Good Practice. In C. Corsi, B. Slapsak, and F. Vermeulen. *Good Practice in Archaeological Diagnostics: Non-invasive Survey of Complex Archaeological Sites*, 113–129. New York, Dordrecht and London: Heidelberg.

Nex, F., & Remondino, F. (2014) UAV for 3D mapping applications: a review. *Applied Geomatics*, *6*(1), 1–15.

Opitz, R.S. and D.C. Cowley (eds.) 2012. Interpreting archaeological topography: Lasers, 3D data, observation, visualisation and applications. Oxford: Oxbow.

Opitz, R., & Limp, F. (2015) Recent Developments in HDSM for Archaeology: Implications for Practice and Theory. *Annual Review of Anthropology*, *44*(1).

Opitz, R., & Nuninger, L. (2014) Point Clouds Segmentation of Mixed Scenes with Archeological Standing Remains: A Multi-Criteria and Multi-Scale Iterative Approach. *International Journal of Heritage in the Digital Era*, *3*(2), 287–304.

Opitz, R., Ryzewski, K., Cherry, J. and Moloney, B. (2015) Using Airborne LiDAR Survey to explore Historic-era archaeological landscapes of Montserrat in the Eastern Caribbean. *Journal of Field Archaeology,* 41(1): 523–541.

Pluckhahn, T. J., and Thompson, V. (2012) Integrating LiDAR Data and Conventional Mapping of the Fort Center Site in South-Central Florida: A Comparative Approach, *Journal of Field Archaeology* 37(4): 289–301.

Raab, A., Takla, M., Raab, T., Nicolay, A., Schneider, A., Rösler, H., & Bönisch, E. (2015) Pre-industrial charcoal production in Lower Lusatia (Brandenburg, Germany): detection and evaluation of a large charcoal-burning field by combining archaeological studies, GIS-based analyses of shaded-relief maps and dendrochronological age determination. *Quaternary International*, *367*, 111–122.

Randall, A. R. (2014) LiDAR-aided Reconnaissance and Reconstruction of Lost Landscapes: An Example of Freshwater Shell Mounds (ca. 7500-500 CAL B.P.) in Northeastern Florida, *Journal of Field Archaeology* 39(2): 162–179.

Ryzewski, K., & Cherry, J. F. (2015) Struggles of a Sugar Society: Surveying Plantation-Era Montserrat, 1650–1850. *International Journal of Historical Archaeology*, *19*(2), 356–383.

Schindling, J., & Gibbes, C. (2014) LiDAR as a tool for archaeological research: a case study. *Archaeological and Anthropological Sciences*, *6*(4), 411–423.

Štular, B., Kokalj, Ž., Oštir, K., & Nuninger, L. (2012) Visualization of lidar-derived relief models for detection of archaeological features. *Journal of Archaeological Science*, *39*(11), 3354–3360.

Watts, A. C., Ambrosia, V. G., & Hinkley, E. A. (2012) Unmanned aircraft systems in remote sensing and scientific research: Classification and considerations of use. *Remote Sensing*, *4*(6), 1671–1692.

Part II
Image and Digital Processing

Terrestrial Lidar and GPR Investigations into the Third Line of Battle at Guilford Courthouse National Military Park, Guilford County, North Carolina

Stacy Curry, Roy Stine, Linda Stine, Jerry Nave, Richard Burt and Jacob Turner

Abstract A joint geophysical and archaeological field school was conducted near the third line action at the battle of Guilford Courthouse, located at the Guilford Courthouse National Military Park, Greensboro NC. The location of the third line is under debate by historians and archaeologists. A ground penetrating radar (GPR) survey revealed a linear feature approximately 50 cm in depth, varying in width and trending north south for approximately 68 m before entering a heavily wooded area. Excavation of a narrow trench towards the end of the field season revealed a colonial surface, possibly a road or gully, covered in fill dirt. Both a road and a gully have been discussed in the literature, and their discovery would yield important clues to the location of the third line. The surface of this buried feature was slightly concave. A team from Auburn University joined UNCG and NC A&T SU researchers with a terrestrial laser scanning (TLS) survey to see if a highly detailed elevation map could trace the surface manifestation of the feature into and through the wooded area. The results of the research demonstrate the successful exportation of GPR data into three dimensional point clouds. Subsequently, the converted GPR points in conjunction with the TLS were explored to aid in the identification of the colonial subsurface. The TLS dataset has the capacity to discern the concave surface found in the dense overgrown and obstructed wooded area which could be a continuation of the subsurface feature seen in the GPR data.

S. Curry (✉) · R. Stine · L. Stine · J. Turner
The University of North Carolina at Greensboro, Greensboro, NC, USA
e-mail: securry@uncg.edu

J. Nave
NC A&T University, Greensboro, NC, USA

R. Burt
Auburn University, Auburn, AL, USA

M. Forte and S. Campana (eds.), *Digital Methods and Remote Sensing in Archaeology*, Quantitative Methods in the Humanities and Social Sciences,
DOI 10.1007/978-3-319-40658-9_3

Introduction

Guilford Courthouse National Battlefield Park (GUCO) is the site of a pivotal 18th century Revolutionary War battle. In March of 1781 General Cornwallis leading the British army engaged American forces made up of militia units from North Carolina and Virginia as well as Continental line troops near Greensboro North Carolina (Fig. 1).

The courthouse was pivotal in the action over the contested land. There is agreement as to the general location of the first two lines of battle (Fig. 2).

The last action of the battle or third line was located near the courthouse. From this location General Greene directed the battle and finally had his army retreat along a north south trending road. While technically a victory for the British army the losses suffered in the battle caused its commander General Cornwallis to leave the Carolinas and move into Virginia were was later defeated at Yorktown. The landscape of the Park from colonial settlement and county courthouse to battlefield to farm to historic preserve surrounded by housing developments has seen modification and reuse. The exact locations of the courthouse and the "retreat" road are an ongoing debate by various scholars and would help enhance the interpretation of this site (Babits and Howard 2009; Baker 1995; Coe and Ward 1973; Durham 2004; Cornelison et al. 2007; Hatch 1970; Hiatt 1999; Ward and Coe 1976; Stine and Stine 2013; Stine et al. 2003). The discovery of the courthouse location, the retreat road or other subsurface features may lead to an accurate placement of the third line of battle.

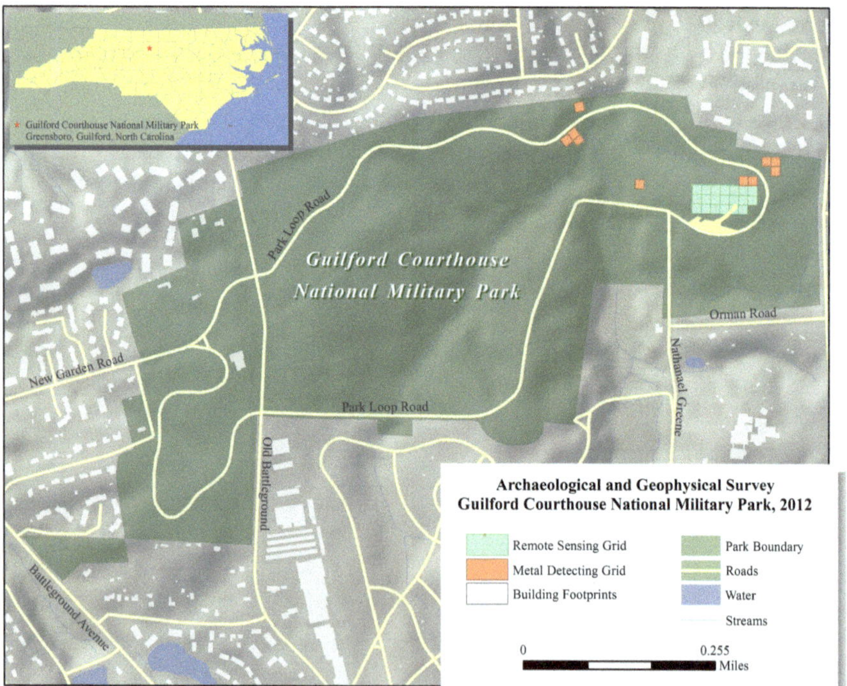

Fig. 1 Guilford Courthouse National Military Park (Dr. Elizabeth Nelson)

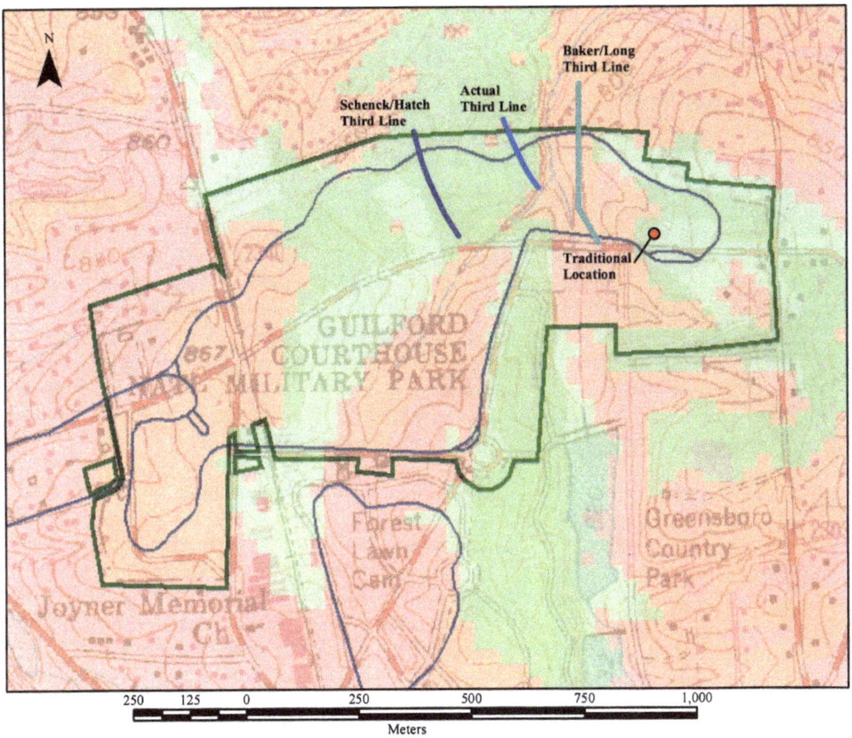

Fig. 2 Disputed location of the third line (Cornelison et al. 2007)

The environment surrounding one of the suspected locations of the courthouse is partially accessible, with mowed grass transitioning into secondary growth brush and trees with an undulating topography. Aerial and satellite imagery and traditional airborne Lidar have proved ineffective at determining the microtopography in this type of environment. Guilford Courthouse's unique blend of environmental conditions, both woody and grass provide a testing ground for utilizing other such methods of mapping similar forested sites. Applying the terrestrial laser scanning to certain subsets of the site can begin to answer questions about the landscape obscured by the woody environment (Fig. 3).

The overarching question of the larger research project involves the potential to combine multidimensional datasets from multiple sensors to produce an effectively fused above and below ground dataset. Drawing on historical archaeological data, GPR, TLS point cloud, and Total Station datasets, this paper focuses on the methods and results of the digital data fusion. In addition, the discovery and implementation of the most effective strategies to handle research sites with heavy vegetative cover and/or obstruction with regards to sensors selections and data fusion methodology are explored. Discovering the most beneficial way to visualize fusion datasets to aid in understanding historical landscapes is a major thrust of this study.

Fig. 3 Wooded study site

Literature Review

Geophysical surveys are one of the critical sources of subsurface data in this research. The roots of archaeological geophysics lies in its ability as a prospection tool to locate map and produce images of buried cultural materials (Conyers 2010). Non-invasive investigations of subsurface anomalies through geophysical surveys can provide archaeologists with valuable information prior to, or in-place of, the non-reversible processes of excavation (Yu-Min Lin et al. 2011). The continued application and development of geophysical coverage for archaeological assessment has begun to introduce an alternative perspective into regional, or landscape archaeology (Kvamme 2003). Such surveys provide information on the structure and organization of a site enabling the study of spatial patterns and relationships relevant to research questions. In addition to the large-scale perspective of the site, geophysical survey results also provide a high-resolution focus on individual site features (Watters 2012). Applying advanced acquisition and processing techniques can not only map the spatial extent of buried features precisely in three-dimensions, but potentially can determine specific material properties of subsurface features such as stone, earth or brick. When these types of analysis are incorporated within a historical framework, ideas about the past can be tested and studied in ways not possible before (Conyers 2010).

Ground Penetrating Radar (GPR) was chosen from the geophysical surveys employed in this research to be used as the subsurface dataset. GPR transmits an electromagnetic pulse and measures a reflected signal that is dependent upon the dielectric properties of subsurface material. With GPR, the potential for the reconstruction of high-resolution 3D data visualizations of the composition of the subsurface is possible (Yu-Min Lin et al. 2011). Identifying discontinuities in the subsurface, including stratigraphic contacts, walls, house or pit floors, rubble, or midden deposits, causes the radar energy to be reflected back to the surface (Kvamme 2007). The velocity of this energy varies greatly, depending on dielectric properties of the subsurface materials. If velocity can be estimated, then return times of echoes from pulses give information on depth, while amplitudes indicate something of the nature of subsurface changes (Kvamme 2007).

An additional source of data for the visualization of above and below ground surface features includes the exploitation of point cloud data. A point cloud is a collection of discrete three-dimensional locations (points) that can have additional metadata associated with each record. Point clouds appear realistic to even the most casual observer because of their three-dimensional nature. Active scanning technologies generate their own scanning energy and can record and even discover archaeological features at both site and landscape scales. These systems send out discrete pulses of light and record both how long it takes those pulses to return and how much of the original energy comes back. That information, when combined with data about where the sensor is positioned and how it is oriented with respect to the real world is used to construct the point cloud. Each point in the cloud represents a location where the light pulse reflected off of a surface (White 2013). The active system that is used in this research is the terrestrial laser scanner (TLS). The term "laser scanning" describes any technology which accurately and repeatedly measures distance using laser pulse, by precise measurement of time needed for the laser pulse to travel from the object and back and transforms these measurements into a series of points, or a point cloud, from which information on the morphology of the object being scanned may be derived. (Mlekuz 2013) Terrestrial laser scanning (also known as ground-based LiDAR) is increasingly used as a method of collecting spatial data, and when supported by digital photogrammetry, can render quantitatively accurate and visually impressive representations of land surfaces (Entwistle et al. 2009). Terrestrial laser scanning enables the researcher to quantify and integrate previously implicit knowledge-based field observations of topographic setting into a framework for interpreting an archaeological site and its characteristics (Entwistle et al. 2009).

Ultimately, given enough observations of a densely-covered landscape by an active scanning system, some inevitably come from the ground beneath or next to the cover and can be used in conjunction with an extrapolation process to reconstruct the ground surface. The more ground observations you have, the better the surface reconstruction (White 2013). The 3D laser scanning data and GPR survey information also share common characteristics in that both can be broken down into a series of spot readings or sample rates, in other words the data can be treated as points. This is most familiar as the basic form of laser scan data, the point cloud.

However for GPR the archaeological deliverables mostly come in the form of 2D images. By producing the results of the GPR as a list of X and Y coordinates based on the relative grid positions and sample spacing, and treating the calibrated depth as a Z the data could also be interpreted like a point cloud. In this case the signal response then becomes the Intensity value just like the reflection of the laser from the scanner (Watters 2012).

An essential part to this research is the data fusion and integration of all data collections. Construction of multi-scale models can be time-consuming, but this is offset by the following advantages: much improved regional context that is immediately accessible visually when analyzing and interpreting more localized field datasets (Jones et al. 2009). Employing a combination of methods over a survey area can help provide information as to the nature, or material, of an anomaly, thus providing insight for site interpretation. Mapping the distribution of disturbances over a site can assist in the recognition of such disturbances generated through cultural activities revealing the spatial distribution and association with site features (Kvamme 2003). These independent data sets are combined in 3D space through their geospatial orientation to facilitate the detection of physical anomalies from signatures observed across various forms of surface and subsurface surveys. The data types are variable in nature and scale, ranging from 2D imagery to massive scale point clouds (Yu-Min Lin et al. 2011). The data fusion process is able to establish interrelationships and patterns between multidimensional data sets, and therefore improve the identification and interpretation of surface and subsurface traces, that may otherwise go unnoticed (Ogden et al. 2009).

Geophysical surveys have been employed on a variety of locations at GUCO (Cornelison et al. 2007; Cornelison and Groh 2007; Stine and Stine 2013). A variety of subsurface anomalies and features have been located. Because of its protected status as a National Park few of these items have been excavated. Most recently Stine and Stine (2013) conducted a magnetometer survey which covered 4675 m^2 and the GPR survey that covered 2714 m^2 in an area thought to be the courthouse. Almost 160 anomalies were recorded and mapped. Stine and Stine were granted a permit to excavate in a specific location within the park. It is highly probable that 2–4 new structures (foundations) were located; one was excavated and showed to be a stone foundation. One of the most interesting features located by the GPR was a subsurface anomaly between 45 and 50 cm in depth and trended north/south for over 30 m before entering a heavy shrub and forest area with dense secondary growth. In the open area there was a slight depression on the ground surface. This area was near what Ward and Coe (1976) reported to be the Americans' retreat road. The small trench was excavated over the anomaly. There was a light scattering of recent material on the surface of the excavation then sterile clay fill for a depth of over 45 cm. The excavation revealed a tannish brown lens of sandy soil with Revolutionary War period ceramics such as pearlwares and creamwares as well as lead sprue, copper disks; and a piece of swan shot all falling within the colonial period (Stine and Stine 2013). It could not be determined if this was the historic retreat road based on the results of the 2011 field season.

At this same location the goal is to examine the microtopography to search for any deformation related to a possible retreat road and/or gully that were prominent features in the battle but have since disappeared from the landscape. A comprehensive geophysical survey using ground penetrating radar (GPR) combined with a terrestrial laser scan (TLS) helps identify key elements in modeling this historic landscape. This provides not only provide a more comprehensive view above and below the surface of this feature, but demonstrates a new method of fusing datasets from differing sensors.

The discovery of the third line would help place other military units and ultimately lead to the location of the courthouse, a major goal of the 2011 project. Using various remote sensing and geophysical surveying techniques the road or gully may have been identified in a comprehensive three dimensional visualization. The fusion of datasets from very different sensors provides a new way of examining the cultural and physical landscape thought to be the third line. As an emerging research topic this investigation demonstrates the capability to discover landscape features through nondestructive means. The implementation of methodology for the visualization of three dimensional data from different types of sensors; Ground Penetrating Radar (GPR) and Terrestrial LiDAR (TLS) begins to illustrate the usefulness of combining such data.

Methodology

The GPR survey was conducted using a GSSI ground penetrating model DC 3000 equipped with a 400 MHz antenna was used to conduct the site survey. The total area coverage for the entire study site was 2714 m^2 with standard transects. Transects were collected in 50 cm with a dielectric constant of 8 and in 16 bit format. All pre-fusion processing was completed in Radan 7 software. A linear feature approximately 50 cm in depth, varying in width and trending from north to south for approximately 68 m before entering a heavily wooded area was identified in profile. In Fig. 4 the red box indicates the area of interest. The higher the amplitude the return from the GPR signal the more intense the coloration shown. A linear feature that extends to the north begins to emerge with a high amplitude signature (Fig. 5).

For the TLS scanning help was provided by the team from Auburn University using a Leica C10 laser scanner. The scanner ran six 360 × 270° scans. Scan setups were spaced anywhere from 60 to 150 ft apart, depending on the density of the forest surrounding the scanner. The scans were registered together using seven targets, a number of which were entered into the scanner at each setup. In order to improve accuracy of the terrain measurements, the scanner was placed on a seven foot high tripod. The increased height reduced the angle of the return laser and lessened shadows from low-lying ground cover. The data were initially preprocessed in Leica Cyclone software. The point cloud that is created was interpreted into x, y, z coordinates (Fig. 6).

Fig. 4 Excavation (2011) of road/gully potential location

In addition to GPR and TLS data collections, standard total station mapping was also conducted for registration of the two datasets. The deployment of a traditional total station survey provides accurate positioning for both data collections and for successful data fusion. A survey grade Topcon GR-3 Global Positioning System (GPS). The GPS antenna is capable of Real-Time kinematic (RTK) survey. The RTK survey method utilizes two GPS antennae: a stationary base that is set up over a point with known coordinates, and a manned, moveable, rover that optimally

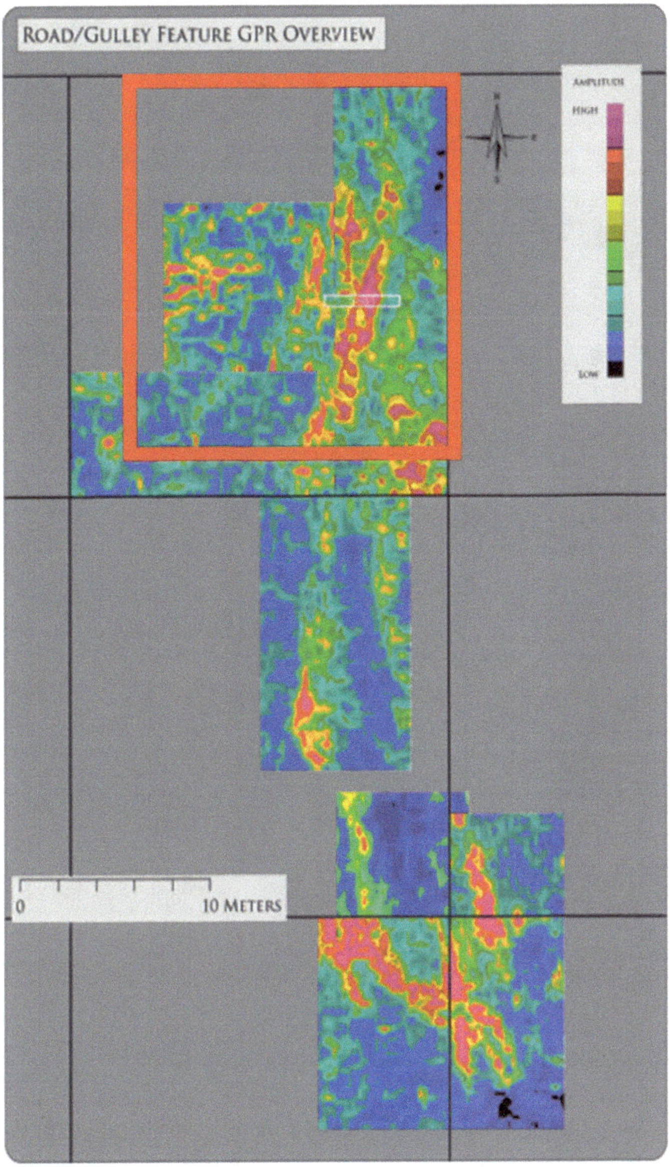

Fig. 5 GPR data collection

receives the same satellite signals as the base, but also receives instant correction via a radio link to the base antenna. This method enables a high level of positional accuracy that other GPS units cannot achieve. A traverse was begun by setting a GPS base station over Lincoln Monument—a brass disk established by the North Carolina Geodetic Survey—using the Lambert Conformal Conic State Plane (feet)

Fig. 6 Leica C10 TLS

coordinates referenced to NAD83/86. A new datum point was then established with the rover positioned over semi-permanent marker such as a nail. The National Park Service requires all coordinate information in be completed in the Universal Transverse Mercator (UTM) projection and shown in meters. The projection and coordinates (including X, Y and Z) were, therefore, shifted to the UTM Zone 17, NAD83/86 using ArcGIS 9.3. Once the datum was established all additional datums, grid layouts and location points were completed using a Topcon GTS 233W total station with a Recon data collector equipped with Survey Pro 4.1.5.

The ground penetrating radar data is processed using GSSI Radan 7 software to normalize surface, velocity, and other standard corrections. After examining in the profile, an area of interest emerge indicating the road/gully feature previously discussed. These areas are then isolated by depth and are exported in the three dimensional formatting of xyz. Where each depth slice of 10 cm to 1.50 m is exported with UTM NAD83 coordinates are represented as the x and the y with z being the elevation and a further attribute of amplitude return from the GPR antennae. In Fig. 7 the yellow box indicates the area of the potential gully/road. In Fig. 8, the area is isolated to show the point cloud derivative from the GPR used for exploration of fusion methods.

The terrestrial laser data was preprocessed at Auburn University in Leica Cyclone propriety software package. Once receiving the dataset from Auburn, the

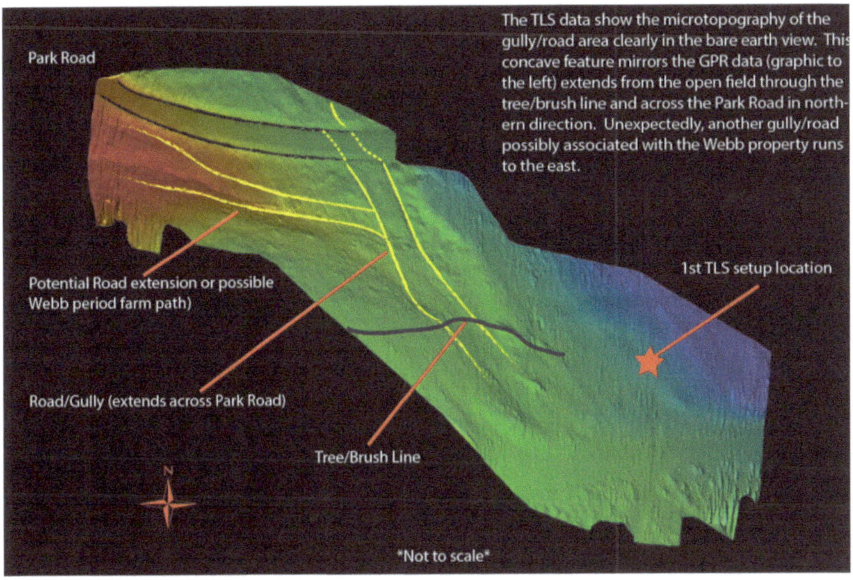

Fig. 7 TLS derived digital elevation model and interpretation

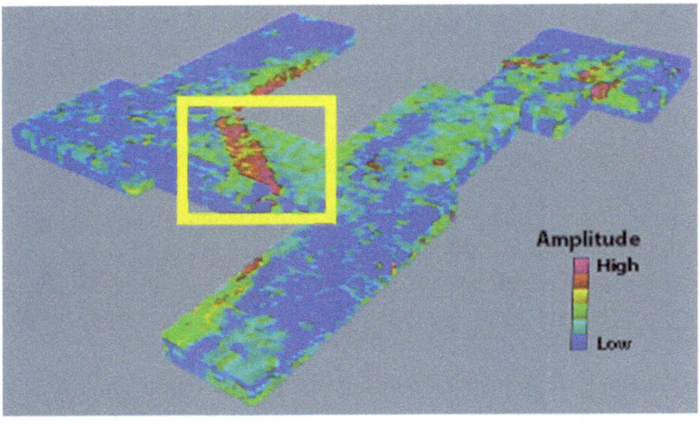

Fig. 8 GPR in point cloud

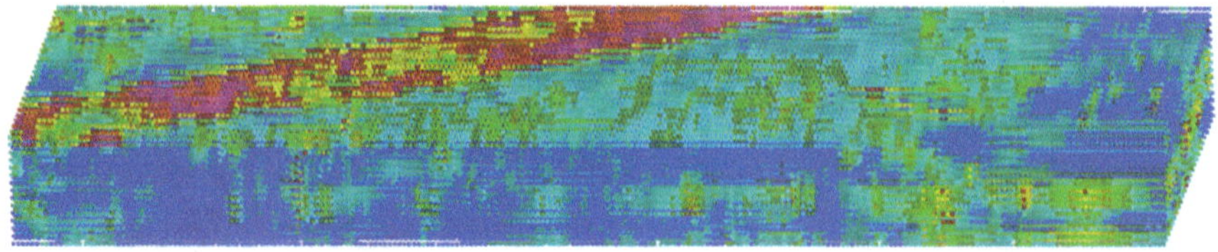

Fig. 9 GPR in point cloud of possible portion of road/gully

data were converted to xyz formatting using Bentley Pointools. Since the fusion is based on the geographic coordinates of both the TLS data needed both georeferencing and registration of coordinates in order to fuse with exported GPR datasets. The TLS data had to be further clipped, gridded, and divided into multiple smaller subsets in order to be able to work within the computing power restraints. Georeferencing results in ESRIs ArcMap and LAStools proved unattainable due to computing power and software capability to handle such point clouds. Further attempts were taken Civil 3D CAD software and proved difficult. However, using the opensource software Cloudcompare allowed for partial alignment of small sections using previously collected total station ground points. Figure 9 illustrates the potential road area in the TLS point cloud.

The research goal was to determine if we could visualize in the datasets the road and attempt to fuse the GPR and TLS data together. Using Golden Software's Voxeler software package, both datasets can be imported multiple individual files to create three dimensional point cloud. Taking a small area of the identified road feature and adding both sets of point clouds a preliminary proof of concept is achieved.

Results

Initial results for the development of methods to export and fuse GPR and TLS data and create three dimensional files for modeling using Voxeler software proved successful. After exportation and alignment procedures were completed, Voxeler provides quick and easy to use visualization tools. The subsurface colonial road/gully can be visualized along the more open area of the site with its surface manifestation (the slight depression) mapped. Using the TLS data to follow the concave surface into the wooded area also proved successful. Figure 10 depicts the preliminary results from the data fusion using the coordinates and elevation as the attributes to match each point. The yellow box indicates the road/gully area of interest that appears in both datasets (Fig. 11).

Working in wooded areas are challenging for these surveys. GPR data are attenuated by trees roots and moving the antenna through thick brush is not possibly. In some instances cutting brush is an option but not on a protected site. The wooded areas surveyed contain dense brush and leaf litter, the methods using laser scanning resulted in an highly effective strategy for tackling such obstructed sites. The TLS

Fig. 10 TLS point cloud highlighting the potential road/gully

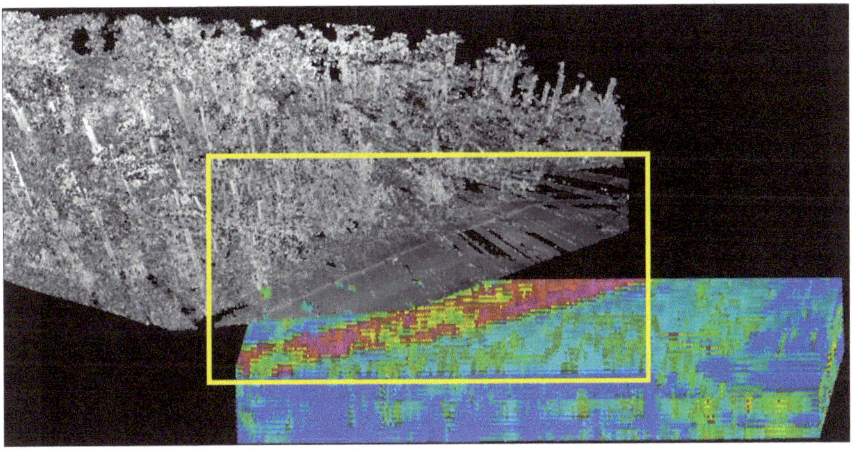

Fig. 11 Data fusion of GPR and TLS datasets

data and post-processing measures did show a north south trending concave surface within the wooded area. The authors cannot of course state that this is a surface manifestation of the subsurface feature without excavation. The data and methods do however point to specific locations to test in the future. Methods from this research highlight the ability to take two different sensors and use them to examine subsurface and above ground landscape simultaneously. A further benefit from the research is the ability to achieve results from enormous datasets while operating with low level computing power found in traditional computer labs. Also, the results show what can be gained while working with opensource and low cost computer packages.

Discussion

This research was to design to develop methods to fuse subsurface data collected by GPR with surface information provided by the TLS. Literature is lacking in methods to take these two widely used data sets and combine them to visualize the landscape above and below ground. Difficulties encountered by the authors included learning and integrating the variety of software used by the different researchers. The size and resolution of the datasets created, seamless transfer of the data created processing and storage issues on our computers. Each sensor required a variety of differing preprocessing software before the datasets can be exported for fusion to occur. The processes derived are considered as an initial step which we hope to develop in the future.

The research reveals that new and improved methods are needed to enhance future similar endeavors. Repeat collects and subsequent point cloud collects are needed to generate the needed coverage for the data fusion process. Alignment issues need to be further accounted for due to the lack of proper software and georeferencing. Difficulties arose due to the numerous software packages and multiple iterations were needed in order to export, fuse, and visualize all data. Topographic correction of the surface layer from the GPR data are needed to better represent the nature of the surface. Future efforts will involve building on the methods developed during this research and applying to other historic sites with spatial research questions. A critical component of future work would assess the accuracy of the point to point data fusion through the application of geostatistical methods. The value of future research would be to develop additional methods to address in the field registration, and enhanced processing of datasets through access to more powerful possibly supercomputing opportunities. Ultimately the authors would like to create an immersive dataset creating a virtual landscape of the historic site where the researcher and community can virtually navigate the site and examine all the features above and below ground.

The second research goal was to investigate the extent of the subsurface feature as in was seen to the open area of the site. Walking into the wooded area the concave feature quickly disappeared, thus ruling out the use of traditional total station survey, (it's hard to map what you cannot see!) The use of the TLS and the generated point cloud allowed the researches to identify areas that seemed to be a continuation of the subsurface road/gully. Hopefully future test excavations will be able to verify or reject this possibility.

Conclusion

This research investigated methods to fuse GPR and TLS data. The data are quite different one is generate from a radio wave the other from a light source. One arrives with discrete x, y, z coordinates the other must have the coordinates generated from a time slice. The initial work in this area has proved successful resulting in a fused dataset showing below ground, surface and above surface 3D points. The research also was successful in delineating a surface feature, easily seen in the open area but hidden by dense shrubs and leaf litter in the wooded part of the site. The TLS data collection and post-processing indicated the possible continuation of the feature and will hopefully be verified by future excavation.

This research investigated methods to fuse GPR and TLS data. The data are quite different; one is generated from a radio wave the other from a light source. One arrives with discrete x, y, z coordinates the other must have the coordinates generated from a time slice. The initial work in this area has proved successful resulting in a fused dataset showing below ground, surface and above surface 3D points. The research also was successful in delineating a surface feature, easily seen in the open area but hidden by dense shrubs and leaf litter in the wooded part of the site. The TLS data collection and post-processing indicated the possible continuation of the feature and will hopefully be verified by future excavations.

The data fusion of the sensors allowed for detailed three dimensional above and below ground surfaces. The techniques have shown the ability to document archaeological features from more than one perspective and where traditional techniques (shovel testing and pedestrian survey) have proven less successful. The identification of a buried surface feature (road/gully) combined with the vague surface elements of the feature continuing in the woods creates an historic landscape. The potential of this fusion means that future excavation of the area should reveal the exact nature and direction of the feature. Both the gully and the road are keys to unlocking the location of the elusive third line of battle at Guilford Courthouse; giving archaeologists, historians and geographers a more complete picture of the battlefield landscape.

The authors are continuing the application of multidimensional data fusion methodology from GPR and TLS to a variety of other archaeological and historical sites. The techniques are transferrable to any location that is looking to view above and below ground archaeological features and make them visible for interpretation in the context of the landscape. For example, current research is being conducted at the House in the Horseshoe (Alston House) State Historic Site located in Sanford, NC. The Alston house is an 18th century property with a complex history of land use. The property was the scene of skirmish between North Carolinians loyal to the British crown and those in favor of independence. Unlike Guilford Courthouse Battlefield, a still extant structure is present with the original bullet holes. Current work suggests that the visible topography has been altered. In the 19th century, the site was a robust plantation of a NC governor, including his household and the enslaved, encompassing much more acreage. The site provides a unique opportunity to study the landscape changes brought about over time by these varying scales of the property's uses.

Fusion techniques at the House in the Horseshoe include an extensive geophysical survey using GPR, gradiometer and resistivity/conductivity. This survey has already provided insight into the buried features located on the property and results were coordinated with archaeological testing. In addition to the geophysical survey methods, the House in the Horseshoe site presents an opportunity to examine the historic structure of the Alston House through passive scanning. The Alston House is used to test the hypothesis that using a SLR digital camera to capture multiple images of the Alston House can provide an accurate point cloud. The structure is imaged through acquisition of multiple photos taken of the house from multiple angles. Using software such as Structure for Motion (SFM), AGIsoft Photoscan, and Meshlab a three dimensional point cloud can be created to create a realistic model. The goal is to implement this technique and then test the accuracy of the point clouds to the real world points from a total station survey. Goals of the project would be to then compare the digital photography techniques to a traditional TLS collection, perform accuracy assessments, and ultimately conduct the data fusion process incorporating the geophysical survey data. The specific techniques defined in this research are being refined for different historical landscapes with different research questions. Archaeologists, geographers, and remote sensors interested in landscape analysis will find these techniques informative and relatively inexpensive. Fusing a wider selection of sensing data will hopefully allow for the discovery, identification and interpretation of below ground features and their surface interactions.

References

Babits, Lawrence and Joshua Howard (2009). *Long, Obstinate, and Bloody the Battle of Guilford Courthouse* The University of North Carolina Press, Chapel Hill, NC.

Baker, Thomas E. (1995). *Redeemed From Oblivion: An Administrative History of Guilford Courthouse National Military Park.* Manuscript on File, Guilford Courthouse Military Park, Greensboro.

Coe, Joffre L. and Trawick Ward (1973). *Preliminary Archaeological Tests Guilford Courthouse.* Research Laboratories of Anthropology, University of North Carolina, Chapel Hill.

Conyers, L. B. (2010). Ground-penetrating radar for anthropological research. *Antiquity*, 175-184.

Cornelison, Jr., John E. and Lou Groh with Greg Heide, Tammy D. Cooper, Guy Prentice and David Lowe contributors (2007). *Battle Lines and Courthouse: Archaeological Survey and Testing at Guilford Courthouse National Military Park Greensboro, North Carolina The 1995, 1997, and 1998 Field Projects.* National Park Service, Southeast Archaeological Center, Tallahassee, FL.

Durham, John Lloyd (2004). *Historian's Report on the Location of Guilford Courthouse.* MS on file, Guilford Courthouse National Military Park. August 2004.

Entwistle, J. A., McCaffrey, K., & Abrahams, P. W. (2009). Three-dimensional (3D) visualisation: the application of terrestrial laser scanning in the investigation of historical Scottish farming townships. *Journal of Archaeological Science*, 860-866.

Hatch, Charles E., Jr. (1970). *Guilford Courthouse and Its Environs.* United States Department of the Interior, National Park Service, Washington, D.C.

Hiatt, John (1999). *Guilford Courthouse National Military Park Cultural Landscape Report.* Manuscript on file, National Park Service, Guilford Courthouse National Military Park, Greensboro.

Jones, R. R., McCaffrey, K., Clegg, P., Wilson, R., Holliman, N., Holdsworth, R. E., . . . Waggott, S. (2009). Integration of regional to outcrop digital data: 3D visualisation of multi-scale geological models. *Computers & Geosciences*, 4-18.

Kvamme, K. L. (2003). Geophysical Surveys as Landscape Archaeology. *American Antiquity*, 435-457.

Kvamme, K. L. (2007). Integrating Multiple Geophsyical Datasets. In J. Wiseman, & F. El-Baz, *Remote Sensing in Archaeology* (pp. 345-374). New York: Springer.

Mlekuz, D. (2013). Skin Deep: LiDAR and Good Practice of Landscape Archaeology. In C. Corsi, *Good Practice in Archaeological Diagnostics* (pp. 113-129). Switzerland: Springer International.

Ogden, J., Keay, S., Earl, G., Strutt, K., & Kay, S. (2009). Geophysical Prosepection at Portus: An Evaluation of an Integrated Approach to the Interpretation of Subsurface Archaeological Features. *Computer Applications to Archaeology*, (pp. 1-17). Williamsburg, VA.

Stine, Linda F. and Roy S. Stine and Kristen S. Selikoff (2003). Multidisciplinary Landscape Research at Tannenbaum Historic Park, Guilford County, North Carolina. *North Carolina Archaeology* vol. 52:20-52.

Stine, L., & Stine, R. (2013). Archaeological and Geophysical Research at Guilford Courthouse National Military Park (GUCO) Greensboro, North Carolina.

Ward, Trawick and Joffre L. Coe (1976). *Archaeological Excavations at the Site of Guilford Courthouse*. Research Laboratories of Anthropology, University of North Carolina, Chapel Hill.

Watters, M. S. (2012). *Geophsyical and Laser Scan Surveys at the Longfellow House – Washington's Headquarters National Historic Site.* NPS.

White, D. A. (2013). LIDAR, Point Clouds, and Their Archaeological Applications. In M. J. Harrower, & D. C. Comer, *Mapping Archaeological Landscapes from Space* (pp. 175-197). New York: Springer.

Yu-Min Lin, A., Novo, A., Weber, P. P., Morelli, G., Goodman, D., & Schulze, J. P. (2011). A Virtual Excavation: Combining 3D Immersive Virtual Reality and Geophsyical Surveying. *Advances in Visual Computing: Lecture Notes in Computer Science (LNCS): 8th International Symposium* (pp. 229-238). Las Vegas: Springer.

Applying UAS Photogrammetry to Analyze Spatial Patterns of Indigenous Settlement Sites in the Northern Dominican Republic

Till F. Sonnemann, Eduardo Herrera Malatesta and Corinne L. Hofman

Abstract When in December 15, 1492 Christopher Columbus' ships passed the hilly north of an island he had named La Hispaniola, today divided into Haiti and the Dominican Republic, he noted in his captain's log smoke of many fires rising from the hill tops. While it remains questionable if these were meant as signals or simply represented daily domestic activities, his description provides evidence of indigenous occupation on the slopes of the island's northern mountain ranges. What evidence can be found through photogrammetrically surveying pre-colonial settlements in northern Hispaniola? Archaeological topographic studies show that the first inhabitants of this island significantly transformed the landscapes they lived in —these 'footprints' now serving as one of the criteria that indicate an archaeological habitation site. The indigenous settlements were located on hill tops, flanks, slopes or ridges, their settings provided a view over a valley or towards the sea, and good inter-visibility between various settlements. Depending on the location, the intra-site topography can either be characterized as a group of shallow mounds (i.e. montículos described by the Spanish) or flattened areas, each of which having been dug into the natural slope to level the base for wood-supported round houses—Excavations of post hole features in a circular array in the north-western Dominican Republic support this theory. Around these house structures ceramic, lithic and food waste was dumped creating shallow mounds, which were additionally used for other domestic and ritual purposes. The recent development of affordable, small, camera-mounted UAS, has made it possible to record these sites by photogrammetric means. The resulting orthophotos and the DEM analysis complement the archaeological finds of the site of El Manantial in the Montecristi province. They highlight the slight changes in soil patterns and topography, and reveal the existence of levelled mounds. The technique shows great potential for fast and precise recording of archaeological sites in difficult terrain. Digital reconstruction could provide answers how a village was spatially structured and organized at its time of occupation. In addition, the technique provides an opportunity to map and measure

T.F. Sonnemann (✉) · E.H. Malatesta · C.L. Hofman
Leiden University, Leiden, The Netherlands
e-mail: tillsonnemann@hotmail.com

M. Forte and S. Campana (eds.), *Digital Methods and Remote Sensing in Archaeology*, Quantitative Methods in the Humanities and Social Sciences,
DOI 10.1007/978-3-319-40658-9_4

more recent changes to the landscape, caused by excavations, illegal looting or ploughing.

Introduction

Archaeological research focuses on evidence of human activity, observed predominantly where people settled: where most actions occur, most time is spent, resulting in most likely residual of dwelling. For a number of reasons, in tropical environments settlements have left fewer traces than in moderate climates. Due to the climate, decay is accelerated and the material culture is often limited to non-perishables. There may also have been less need for the indigenous population to use masonry, as long as the house was structurally firm to provide shelter against rainfall and the sun. Where year-long harvesting of fruits and vegetables is possible, and access to water, fish and seafood in abundance, food storage has not the same importance. In that sense, the use of light material for habitation in the pre-Columbian Caribbean was similar to other contemporary tropical cultures, e.g. by the Khmer at Angkor (e.g. Bâty and Seng 2012; Sonnemann 2011: 87–92), or the Maya in Central America (e.g. Healy et al. 2007), which used stone for mostly ceremonial, water management and defensive functions. In addition, in the Caribbean is particularly prone to natural disasters such as earthquakes and hurricanes; houses made of wood are less affected or quickly rebuilt (Samson et al. 2015; Hofman and Hoogland 2015a). Nevertheless, orientation and arrangement of post holes show that pre-colonial houses may have had aesthetic attributes (Samson 2011). Stone was used in places of worship or gathering, such as ball courts, often framed by large boulders, sometimes as base for house platforms (Ulloa Hung 2014), but not used for wall construction (Veloz Maggiolo 2003; Ortega 2005; Hofman and Hoogland 2015a). The Spaniards experienced very early the ineffectiveness of European construction style, when their first town, La Isabela, was severely damaged by hurricanes in 1494 and 1495 (Ortega 1988; Deagan and Cruxent 2002), and the city of Concepción de la Vega was destroyed in the earthquake of 1562 (Ortega and Fondeur 1978); as a result both were later abandoned.

Non-destructive research in these environments has focused predominantly on the identification of stone remains (Sonnemann 2015); in the Caribbean on ball courts (Curet et al. 2005) or colonial archaeology (e.g. Finch 2013; Ulloa Hung and Sonnemann, forthcoming). While due to improved equipment or filtering and processing methods, there is a growing number of studies worldwide that map subsurface evidence of perishable material (Sonnemann 2011:87–90), such as postholes, successful results highlighted very large postholes (e.g. Nishimura 2008: 263) with distinguishable change in electromagnetic property to the surrounding soil (e.g. Kvamme 2003) where the wooden base remained, was burnt to the ground, or the postholes were cut into the bed rock (Pincus et al., forthcoming).

However, the identification of underground features is not the only challenge for Caribbean archaeology. Archaeological sites have been mainly identified and registered by traditional walk-over surveys. A number of non-destructive techniques have been tried to identify indigenous activity in the Caribbean; from remote sensing (Capobianco 2005; Lyew-Ayee and Conolley 2008; Samson et al. 2012; Sonnemann 2016), to subsurface geophysics (Welch 2010; Medica et al. 2010). The remote sensing surveys however were interpreted as strongly affected by atmospheric interference (Samson et al. 2012), geophysics by soil indifference in karstic environment (Medica et al. 2010), and neither postholes nor settlements could be firmly identified, displaying the need for other investigation methods for this particular region.

A solution to study the topography and potentially intra-site organization of single settlements could be UAS platforms with high-resolution cameras. The use of drones to record archaeological sites has skyrocketed over the last years (Sauerbier and Eisenbeiss 2010; Casana et al. 2014; Block-Berlitz et al. 2014) with flight stability improving and prices having dropped considerably (Nex and Remondino 2014). The recording opportunities of multi-copters, oriented through an internal GPS, the possibility to change the camera angle, and precise acquisition of pictures and videos due to gimbal mounted cameras, is rapidly replacing traditional and more static methods for archaeological recording, such as kites and balloons, or pole-photography. The possibility exists to program a flight plan prior to the survey by using GPS information and a virtual globe, map and geographical information program. In addition, development in photogrammetric software and ever faster processing speed, allow the use of non-calibrated cameras with a higher amount of pictures, meaning that little experience is necessary to set up a workflow. Photogrammetry has become an equal rival to laser scanning regarding the recording of structures (Remondino 2014) but also for landscape features (Ouédraogo et al. 2014), and is the method used here to virtually extract archaeological features in the northern Dominican Republic.

Indigenous Settlements on Hispaniola

Significant geographical varieties are observed over the 200 km long Cordillera Septentrional or Sierra de Montecristi. Due to differences in altitude, and variation in rain fall, micro-climate regions are created ranging from semi-arid, dominated by cacti and other succulents, to rain forest on some mountain tops and the northern slope of the higher mountains. From north to south, mangrove forests at the coast are replaced by sloping hills to a single mountain range of karst limestone that tops to 1.229 m height [Pico Diego de Ocampo], separating the coastal region from the Rio Yaque del Norte in the Cibao Valley, the breadbasket of the Dominican Republic (Moya Pon and Pérez Ceballo 2004).

The indigenous peoples who created the smoke rising from the hill tops that was seen by Columbus (Dunn and Kelly 1989), were living in nucleated villages, from small to very large communities (Rouse 1992; Wilson 2007: 2), in dirt-floored structures made of timber and thatch, which are archaeologically attested by circular alignments of postholes (Samson 2013: 12). Groups of villages formed districts (Wilson 2007: 125) of a combined population in the tens of thousands, that were presided over by one chief, or cacique (Wilson 2007: 110) who had varying degrees of centralized authority (Oliver 2009: 26). Early 16th century sources, such as the historian Bartolomé de las Casas, O.P. (c. 1484–1566) estimated a total of 1.1–3 million people on Hispaniola alone, while modern estimations speak of below half a million people as a realistic number (Vega 1980; Henige 1978; Denevan 1992). Upon the arrival of the Spaniards, the indigenous population is reported to have rapidly declined due to war, mistreatment and disease (de las Casas 1552; Cruz Mendez 1999; Cook 1993).

Archaeological Knowledge on Settlements

There is no above-surface remains of indigenous architecture or clear settlement pattern that underlies the distribution of archaeological sites in the landscape. In Haiti about one thousand pre-colonial sites, varying agglomerations of shells, lithic materials, and ceramics, have been recorded over the past 50 years, mostly by Clark Moore (Moore and Tremmel 2002). An overview of known indigenous sites in the Dominican Republic was published by Ortega (2005), its map showing that large portions of the country remained terra incognita regarding archaeological research. Dominican institutions, e.g. the Museo del Hombre Dominicano, concentrated over a long period on sites of possible ritual activities, such as stone circles and other non-perishable evidence. An investigation of El Atajadizo in the southeastern DR in the 1970s revealed habitation and funerary mounds that surrounded a rectangular plaza (Luna Calderon 1975:87). Large scale excavation of settlements have been undertaken in the south-eastern Dominican Republic (Hofman et al. 2006, 2008; Samson 2010), and most recently in the northern provinces of María Trinidad Sánchez (López Belando 2012) Puerto Plata, and Valverde (Hofman and Hoogland 2015a). The site of El Flaco at the foot of the Cordillera Septentrional in Valverde Province revealed that the natural slope was levelled to build round houses and cooking structures. These structures have diameters varying between 4 and 10 m (Hofman and Hoogland 2015b).

Since 2007 new surveys were conducted to provide more information on the location of sites as well as on their structure. To record registered sites and identify non-registered sites, field-walking surveys geo-referencing probable site extent from surface finds, were conducted in the provinces of Puerto Plata, Montecristi and Valverde (Fig. 1).

Generally following local sources, 96 of these have been documented in Puerto Plata since 2007 (Ulloa Hung 2014; Ulloa Hung et al. 2014; de Ruiter 2012) while since the beginning of 2013, 100 new sites were recorded in Montecristi (Herrera Malatesta et al. 2014; Ulloa Hung and Herrera Malatesta 2015). Most sites were found on hill tops or mountain flanks, with visibility of the sea or the valley and neighboring settlements.

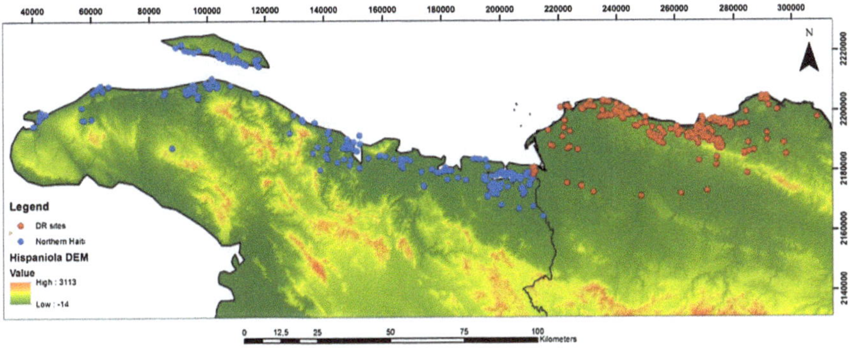

Fig. 1 Location of the sites in the northern Dominican Republic and northern Haiti (*blue* data from Moore and Tremmel (2002); *red* from Ulloa Hung and De Ruiter (2011), Ulloa Hung and Herrera Malatesta (2015), Leiden University team working on the area since 2012)

Indigenous settlement pattern apparently responds to an interconnection of sites from the coast into the mainland (Ulloa Hung and de Ruiter 2011: 62). Also, at the coastal area of Montecristi (Herrera Malatesta et al. 2014), the sites can be roughly divided into three types: sites close to the coast and the mangroves for marine resource exploitation; the hills seem to have been dominated by "small" sites (less than 1 ha), at varying distance to the coast, providing good visibility of the surroundings, while most "large" sites (more than 1 ha in size) are generally found in the intermountain plains. There seems to be a connection between site size, topography and visibility: while the inner mountain plains provided space suitable for larger settlement, they do not provide good visibility (de Ruiter 2012; Herrera Malatesta et al. 2014) and may have had different functions. The sites located on flat terrain possibly supported a larger population; their archaeological record, distinct ceramic assemblages (Ulloa Hung 2014) shows evidence of larger variety of activities, while hilltop may have been connecting the coast and the mainland. These assumptions are currently being tested in a fieldwork project that focuses on the coast of the Montecristi Province (Herrera Malatesta et al. 2014).

The Site of El Manantial in the Montecristi Province

The site of El Manantial in the province of Montecristi is one of the "large" sites located four kilometres from the coast, on a shallow hill on agrarian land, and about 100 km direct distance from La Isabela, the first European town in the Caribbean that was founded by Christopher Columbus in 1493 (Deagan and Cruxent 2002).

At an altitude of 170 m on its highest point, the site is delimited to the north by a steep decline from where the Atlantic Ocean is visible in between the hills of the Cordillera Septentrional. According to the local owners of the field, the site has been known for a long time, and, while visited extensively in the 1960s–80s by

Fig. 2 Overview image of the El Manantial landscape in Google Earth (Digital Globe, February 23, 2012) (looking North-East, passed the Cordillera Septentrional towards the Atlantic)

"private collectors", artefacts are still abundant on the site. Recorded officially during the field campaign of 2014, archaeological material was found distributed over about 8 ha, showing a potentially very large pre-contact village, while its location on a hill and little vegetation made it particularly suitable for the following case study (Fig. 2).

Ten year aerial coverage data of Google Earth from 2004 to 2014, shows that the site was for some time protected by bushes, until the field was cleared from most vegetation and ploughed to be used for growing maize (Fig. 3). If the lighter shades of the soil represented material distribution, about 40 possible larger agglomerations were identified (Fig. 4). Ploughing has had considerable impact on the site, levelling the areas and making the actual mounds difficult to identity through inspection from the ground or in aerial photos (Fig. 5).

In 2014 a sample of ceramic, lithic and shell materials was collected (see Fig. 6), providing an overview of the extensive material found at El Manantial. The majority of decorated ceramic material is culturally affiliated with the so-called Meillacoid series (Rouse 1939; Veloz Maggiolo et al. 1981), which started around AD 600 in Puerto Rico (Sinelli 2012: 221) and AD 800 in the Dominican Republic (Ulloa Hung 2014). Some fragments have been identified as the more decorative, but thicker Chicoid ceramics (Veloz Maggliolo 2003; Ulloa Hung 2014), found from around AD 1000 (Ulloa Hung 2014) until after the European encounter (Deagan and Cruxent 2002). In some sites the two ceramic styles possibly existed contemporaneously, and both ceramics styles have been found on sites with mounds (Ulloa Hung and de Ruiter 2011: 65). Current research has revealed both styles on settlements with mounds as well as flattened areas (Hofman and Hoogland 2015a), and in some cases there is a mixture of both styles combined in one ceramic sherd (Ulloa Hung 2014).

The lithic tools found at the site of El Manantial are mainly hammer stones, flakes, perforators, axes, net weights, and stone cores. They were mostly made from

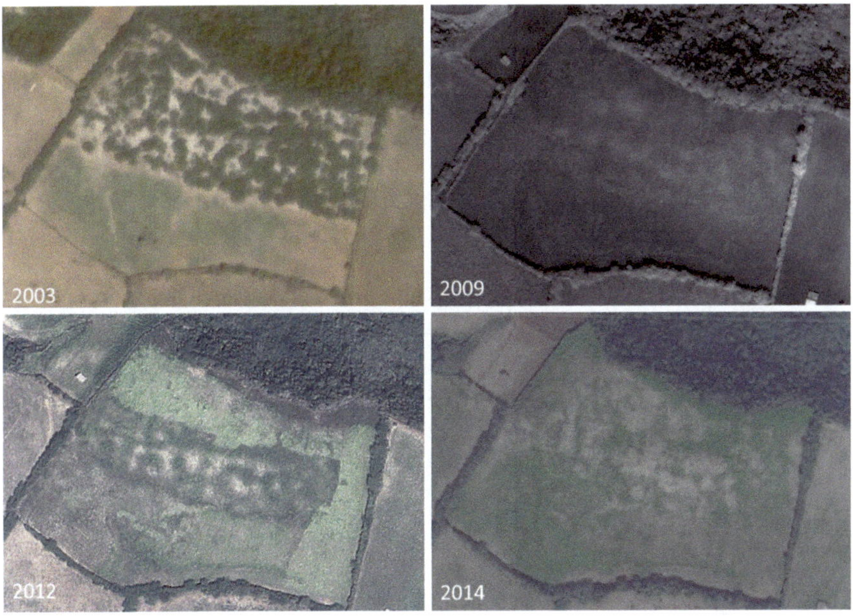

Fig. 3 Four images of Google Earth over a period of four years show the change of use and vegetation cover that provide possibilities to measure the spread of material over the site

Fig. 4 Potential distribution of finds using the time series of imagery from Google Earth plotted over the image of 2012

local materials found near the site with a few lithic materials of more distant origin (A. Knaf pers. com. February 2015). Objects from shell are predominantly gubias. The gubia is an object made from the core and side of the *Lobatus gigas*, and was possibly used to dig holes for agriculture (Vargas Arenas et al. 1997; Ortega 2001). Other objects are beads, preforms and hammers. While the final analysis of the finds is still in progress, the presence of non-local objects, the considerable number of shell preforms and decorated ceramics could indicate that this settlement played an important role in the interaction networks of this region. The presence of ceramic griddles, hammer stones and gubias suggests the production and consumption of maize, manioc and other root crops regarded as key food resources in pre-colonial times (Rodríguez Ramos and Pagán Jimenez 2008; Vargas Arenas et al. 1997).

Besides its abundance in material remains, the prominent location of El Manantial indicates a possible importance within a network of past activities between exploitation areas on the coast and settlements at various distances to the sea. However, there is poor inter-visibility from El Manantial to other so far identified archaeological sites in direction of the Atlantic. View-shed calculations and ground observations show that only one site is in direct view, with three more sites in the proximity of visible areas. All four sites are Meillacoid, and of the category 'small' (Fig. 7).

Photogrammetry Data Acquisition

Ten sites with pre-contact archaeological material and relatively sparse vegetation were flown by UAS, some recently been cleared for farming, or missing

Fig. 5 UAS fish eye camera still of El Manantial from Southwest in October 2014, showing a view direction similar to Fig. 2. The Atlantic Ocean is visible at the horizon on the left

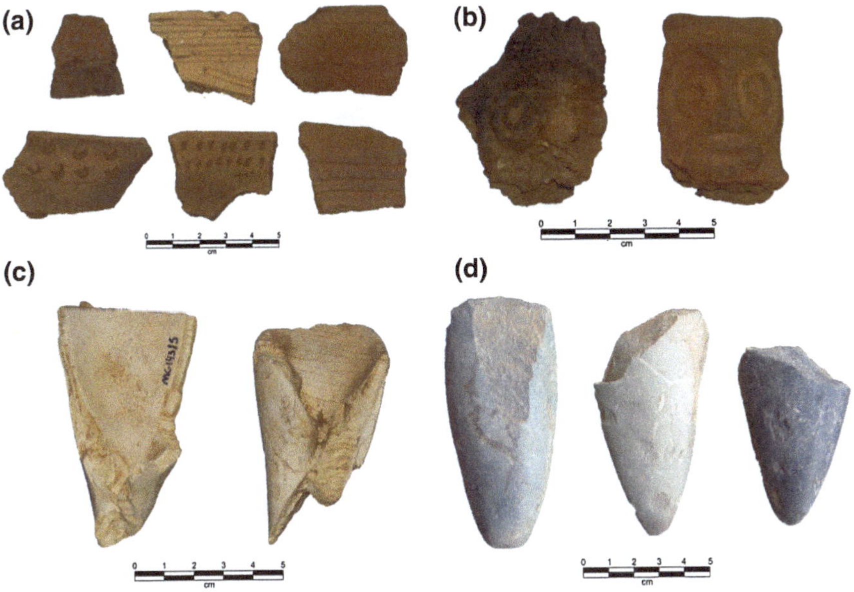

Fig. 6 **a** Meillacoid ceramics, **b** chicoid adornos, **c** gubias, **d** stone axes fragments

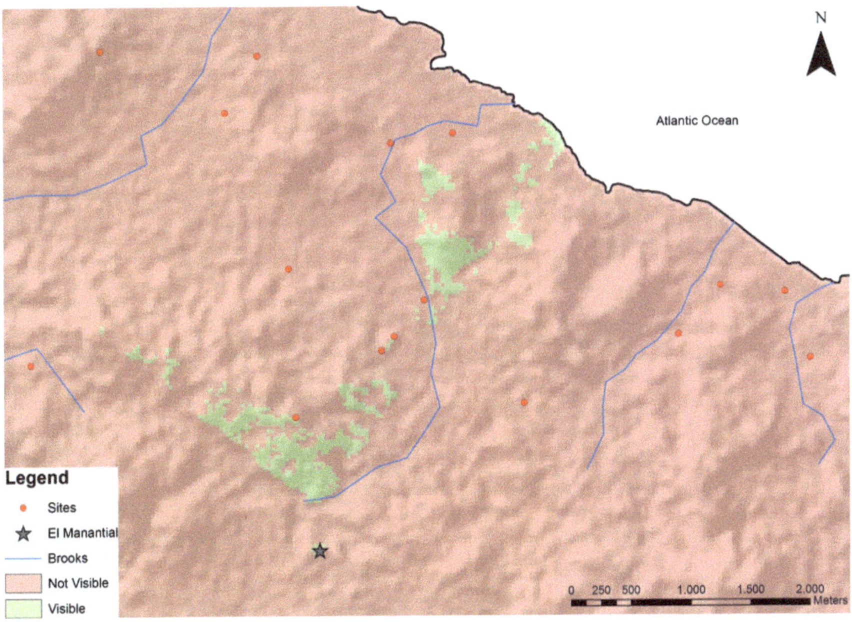

Fig. 7 View shed from El Manantial (*green star*) on basis of ASTER GDEM data to analyse intervisibility between sites (*red dots*)

underbrush, providing the possibility to map below trees. Particularly important was a smooth ground surface, provided by grassing areas. The areas were overflown at least twice with one flight to take photos and another to record a video, providing the opportunity of using additional imagery from the video, where photo coverage was missing or sparse. For the fly-overs, the camera was set in both horizontal position for traditional aerial view as well as in a ~45° angle to take oblique photographs, providing the opportunity to collect data sideways from beneath larger trees. A small quadrocopter, equipped with a 12MPixel camera, was set to take one image per second, in 4 min flight a total of 240 images excluding start and landing. The use of a fish eye camera lens was seen as a potential problem for photogrammetric recording of slight topographic differences without structural elements for co-recognition. However, processing of the imagery was mastered by the commercial SFM software without any problems and produced an even topographic surface for orthophoto and DEM without major errors. Ground control points were measured by total station (see Fig. 8), referenced in a local grid to provide correct slope, orientation and differential height information and were further mapped by handheld GPS for integration into a GIS environment.

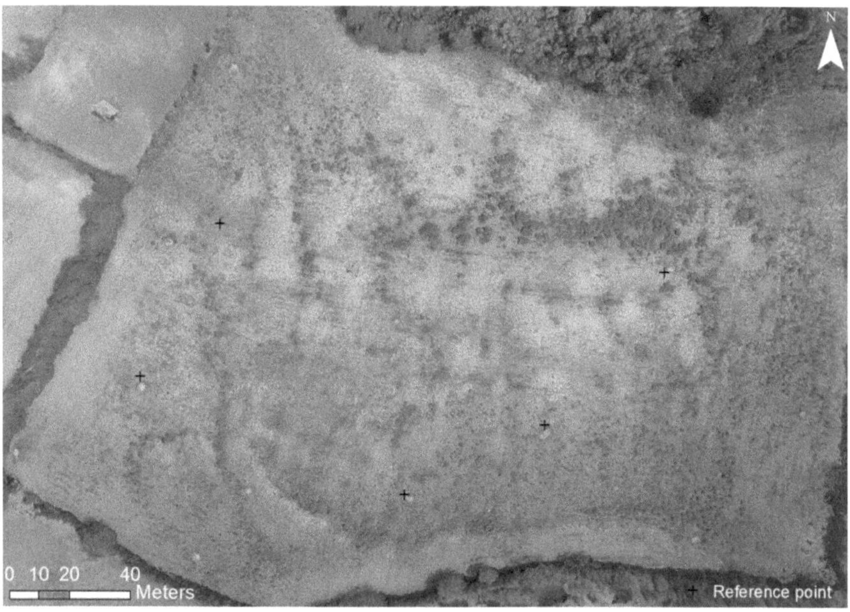

Fig. 8 Orthophoto of El Manantial using 240 images and ground control points

Interpretation of the Photogrammetric Data

In several field campaigns mound features have been identified in the Valverde and Puerto Plata regions (Veloz Maggiolo et al. 1981; Ulloa Hung and de Ruiter 2011; Hofman and Hoogland 2015a), while very few had been known in the eastern Montecristi region. Due to the dense concentration of material at El Manantial, it is impossible to define a clear distribution pattern on the ground. The material is spread on the surface like a carpet, appearing more or less clustered, making it difficult to clearly separate the agglomerations. Slightly visible elevation changes in relation to the clusters provide some evidence for mounds. Plough marks show the heavy impact of recent agricultural activity (see Fig. 9). Debris is easily identifiable in the south and west part of the field, but diagnostic archaeological material is predominantly located in the area surrounding the major clusters (purple line on Fig. 9). This could mean that although the site has suffered severe damage, the majority of materials is still concentrated in the core area, which may have been the actual habitation site.

The resulting DSM (see Fig. 10), particularly with a seven times exaggerated height factor and its visualization using hill shade (see Fig. 11), enhances the subtle height change in the relatively vegetation free site in high resolution to define and measure the clearly visible mounds. Low vegetation cover has decreased the quality of the results insignificantly; more affected is the quality of the point cloud and resulting DSM by missing photo coverage in the east.

Fig. 9 Interpretation of the El Manantial orthophoto

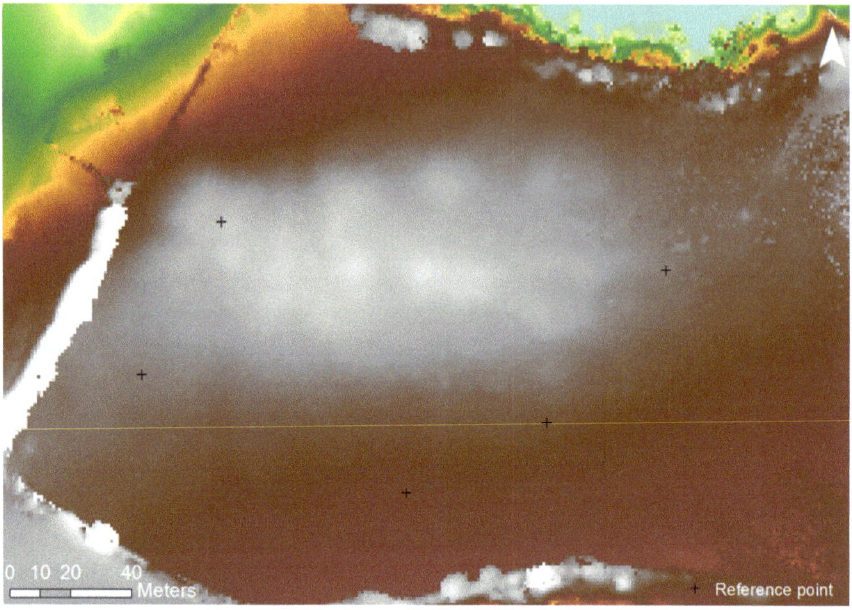

Fig. 10 DSM at 1 m resolution from photogrammetric UAS survey

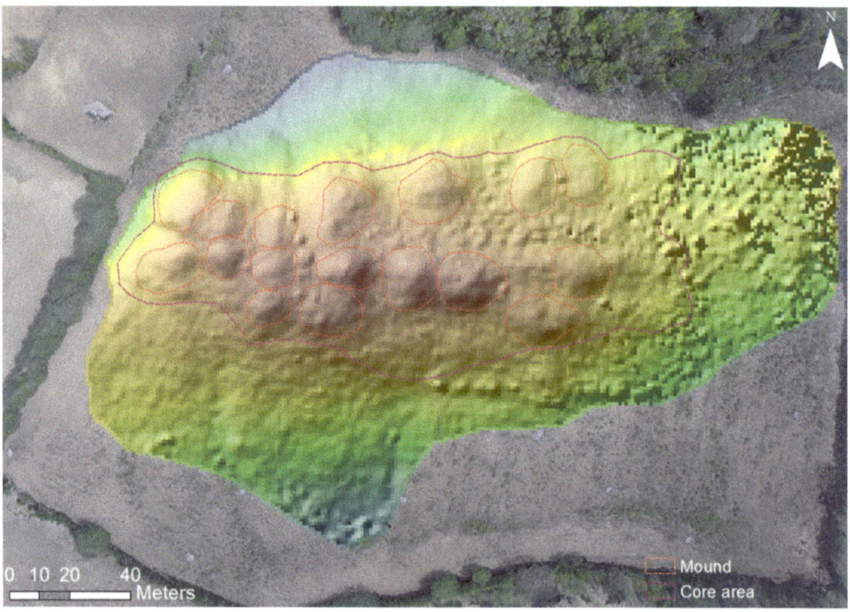

Fig. 11 Interpretation of DSM results over hill shade model (height factor = 7)

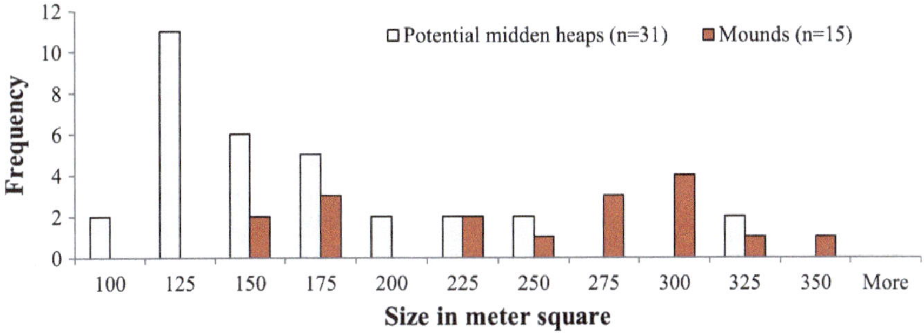

Fig. 12 Midden heaps as mapped from aerial photo in comparison to actual mounds from topographic information

The topographic vs orthophoto results show that the area of the monticulos is limited to the hill top only, covering an area of about one hectare, while total material distribution is about 2.5 ha. It clearly depicts the misinterpretation from solely using the aerial (40 potential mounds counted) or drone photos in number (31 counted vs. 15 extracted mounds), showing also that the mounds are about double the size, centering around 300 m^2, compared to the potential midden heaps identified in aerial pictures. Ploughing has contributed to a widening and lowering of the original shell mounds, making any height-width ratio calculation difficult. Just based on the topography and imagery information, there is no hierarchical structure definable in the mound distribution of El Manatial, be it from size or location. The mounds are loosely aligned in two rows on the upper most part of the hill (as suggested as a settlement pattern by de Ruiter (2012) for sites in the Puerto Plata region) (Fig. 12).

Conclusion

The example shows that not only does photogrammetry serve as a precise recording tool of a cultural landscape, an excavation site, or a building at a certain point in time, but can provide important supplementary information for the analysis of archaeological sites. UAS surveys provide a rapid and thorough collection of data limited by the restricted flight time due to battery power and payload. Affordable hard- and software has progressed significantly to make small-UAS recording a rapid and important asset in archaeological surveys, adding new information to field surveys. Immense advances have been made in drone technology over the last years, previously only possible from planes or satellites: the use of thermal cameras to identify non-structural features by soil moisture differences (Casana et al. 2014), or micro-size multispectral cameras that can provide imagery in the near-infrared spectrum.

Concentrating on the analysis of monticulos with new techniques contributes to the debate of distribution and configuration of settlements in the Greater Antilles. De Ruiter (2012) and Ulloa Hung (2014) focused on identifying pre-colonial mound features in field surveys, providing overviews of the distribution, while Hofman and Hoogland (2015b) have presented results on the interior structure. High resolution topographic information and the interpretation of the photogrammetric results in a GIS environment, allow a visual analysis to measure amount, size and distribution of archaeological material (e.g. shells, ceramics). More photogrammetric DTM models of different type of sites in various settings, may provide new information on intra-settlement configurations. Slight topographic anomalies, even if further changed by recent human activities, can still be recognized, providing new criteria to identify archaeological sites and distinguish them from non-site areas in the Caribbean, where artefacts may indicate human activity but not particularly a settlement. Ongoing research in the Montecristi province shows evidence that the presence of these mounds is a common element of sites located in the mountains. In the southwestern part of the province, a site with these features was registered close to the Chacuey Dam. This means that the photogrammetric DTM models could aid both, settlement pattern analysis and regional archaeological research. The example also shows the potential of new techniques to mark out hazards affecting on these types of archaeological sites, such as deep ploughing or looting, of which the impact can be monitored.

Acknowledgments The authors would like to thank Menno Hoogland, Jorge Ulloa Hung and Sony Jean for their contributions in the field work, Angus Mol and Julijan Vermeer for their contributions to the script. The research leading to these results has received funding from the Europe-an Research Council under the European Union's Seventh Framework Programme (FP7/2007-2013)/ERC grant agreement n° 319209 NEXUS1492-project.

References

Bâty, P. and C. Seng. 2012. *Homes, fields and temples - Excavations at the airport in Siem Reap.* Paper presented at EurASEAA 14th Int. Conf., Sept. 18-21, 2005, Dublin.

Block-Berlitz, M., Ducke, B., Kroefges, P., Martinez Mora, E. and R. Rojas. 2014. *Low-cost and efficient UAV-based 3D documentation in Tamtoc (Mexico)*, Poster at the CAA2014, Paris.

Capobianco, K. 2005. *Excavation Site Prediction Using High Resolution Satellite Imagery and GIS Data Development of Archaeological Deposits for En Bas Saline, Haiti.* Unpublished Master Dissertation, University of Florida, Gainesville.

Casana, J., Kantner, J., Wiewela, A. and J. Cothrenc. 2014. Archaeological aerial thermography: a case study at the Chaco-era Blue J community, New Mexico. *Journal of Archaeological Science*, 45, 207–219.

Casas, Bartolomé de las. 1552. *Brevísima relación de la destrucción de las In-dias.* Reprinted in Alicante: Biblioteca Virtual Miguel de Cervantes, 2006.

Cook, N. D. 1993. Disease and Depopulation of Hispaniola, 1492-1518. *Colonial Latin American Review,* 2, 213-46.

Cruz Mendéz, M. 1999. *Historia Social Dominicana.* Santo Domingo: Impresora Soto Castillo, S. A.

Curet, L. A. Welch, D., Hamilton, W.D. and L.A. Newsom. 2005. Search of the Lost Features: A Progress report of a Geophysical Study at the Civic-Ceremonial Center of Tibes, Puerto Rico. In: G. Tavares María * M. Garcia Arévalo (eds.) *Proc. of the 20th Int. Congr. for Caribbean Archaeology*, 65-74. Museo del Hombre Dominicano, Santo Domingo.

Deagan, K., and J. M. Cruxent. 2002. *Columbus's outpost among the Taínos Spain and America at La Isabela, 1493-1498.* New Haven: Yale University Press.

Denevan, W. M. 1992. *The native population of the Americas in 1492. 2nd rev.ed.* Madison, WI: The University of Wisconsin Press.

Dunn, O. and Kelley, J.E. (ed. and trans.) 1989. *The diario of Christopher Co-lumbus' First Voyage to America 1492-93, Abstracted by Fray Bartolomé de las Casas.* Norman, OK. University of Oklahoma Press.

Finch, J. C. 2013. Surveying Caribbean cultural landscapes: Mount Plantation, Barbados, and its global connections. *Internet Archaeology*, 35, 35.

Henige, D. 1978. *On the contact population of Hispaniola: History of Higher Mathematics.* Hispanic American Historical Review 58.

Healy, P. F., Helmke, C. G. B., Awe J. J. and K.S. Sunahara. 2007. Survey, Settlement, and Population History at the Ancient Maya Site of Pacbitun, Belize. *J. of Field Archaeology* 32, 1, 17-39.

Herrera Malatesta, E., T. Sonnemann, and C. L. Hofman. 2014. *Approaches to Amerindian Landscapes in the Northern Dominican Republic.* Poster at the 3rd Landscape Archaeology Conference (LAC), Rome.

Hofman, C.L. and M. L. Hoogland. 2015a. Beautiful tropical islands in the Caribbean Sea. In W. J. H. Willems & H. P. J. van Schaik (eds.) Water & Heritage – material, conceptual and spiritual connections. Sidestone Press, Leiden, 99–120.

Hofman, C.L. and M. L. Hoogland. 2015b. Investigaciones arqueológicas en los sitios El Flaco (Loma de Guayacanes) y La Luperona (Unijica). Informe pre-liminar. *Boletin del Museo del Hombre Dominicano.*

Hofman, C. L., Hoogland, M. L. P., Oliver, J. R., and Samson, A. V. M. 2006. Investigaciones Arqueológicas En El Cabo, Oriente de La República Domi-nicana: Resultados Preliminares de La Campaña de 2005. *El Caribe Arqueológico,* 9, 95–106.

Hofman, C. L., Hoogland, M. L. P., Samson, A. V. M., and Oliver, J. R. 2008. Investigaciones Arqueológicas En El Cabo, Oriente de La República Dominicana: Resultados Preliminares de Las Campañas 2005 y 2006. *Boletín Del Museo Del Hombre Dominicano,* 42, 307–16.

Kvamme, K. L. 2003. Geophysical Surveys as Landscape Archaeology. *American Antiquity* 68, 3, 435-457.

López Belando, A. 2012. *El sitio arqueológico Playa Grande. Río San Juan, María Trinidad Sánchez. Informe de las excavaciones arqueológicas campaña 2011-2012*, Unpublished Manuscript, Museo del Hombre Dominicano.

Luna-Calderon, F. 1975. *Preliminary Report on the Indian Cemetery "El Ata-jadizo", Dominican Republic.* Unpublished report. Muséo del Hombre Dominicano.

Lyew- Ayee, P. and I. Conolley. 2008. The Use of Imagery to Locate Taino Sites in Jamaica in a GIS. In: Reid, B. A. (ed.) *Archaeology and Geoinformatics: case studies from the Caribbean*, 137-154, Tuscaloosa: University of Alabama Press,.

Medica, D., Mušič, B., and A. Samson. 2010. Recognition and Interpretation of Shallow Microrelief Features in Limestone Bedrock using the Ground Penetrating Radar Method, the case of El Cabo, Dominican Republic (In Slovenian with English abstract). *Arheo,* 27, 15–43.

Moore, C. and N. Tremmel. 2002. Internet Source: http://www.nilstremmel.com/haiti/images/ Clark_Moore_Haiti_sites.kmz. Accessed August 15, 2014

Moya Pon, F., and M. Pérez Ceballo. 2004. *Atlas de los recursos naturales de la República Dominican.* Santo Domingo: Secretaría de Estado de Medio Ambiente y Recursos Naturales.

Nex, F. and F. Remondino. 2014. UAV: platforms, regulations, data acquisition and processing. In: Remondino, Fabio and Stefano Campana (eds.) *3D Re-cording and Modelling in Archaeology and Cultural Heritage.* BAR Int. Ser. 2598, 73-88.

Nishimura, Y. 2008. Comparative geophysical survey results in Japan: Focusing on kiln and burial remains. In: Campana, S., and S. Pirlo. *Seeing the Unseen. Geophysics in Landscape Archaeology*. Boca Raton: Taylor & Francis 2008.
Oliver, Jose R. 2009. *Caciques and Cemi Idols. The Web Spun by Taino Rulers between Hispaniola and Puerto Rico.* Tuscaloosa: University of Alabama Press.
Ortega, E. 1988. *La Isabela y la Arqueología en la ruta de Cristóbal Colón.* San Pedro de Macorís: Universidad Central del Este.
Ortega, E. 2001. *Los objetos de conchas de la prehistoria de Santo Domingo.* Santo Domingo: Fundación Ortega Alvarez, Inc.
Ortega, E. 2005. *Compendio General Arqueológico de Santo Domingo* Vol. 1. Santo Domingo: Academia de Ciencias de la República Dominicana.
Ortega, E., and C. Fondeur. 1978. *Estudio de la cerámica del período indo-hispano de la antigua Concepción de La Vega.* Vol. 1. Taller.
Ouédraogo, M., Degré, A., Debouche, C., and Lisein, J. 2014. The evaluation of unmanned aerial system-based photogrammetry and terrestrial laser scanning to generate DEMs of agricultural watersheds. *Geomorphology*, 214, 339-355.
Pincus, J., Carr, R. and D. Whitman. *Imaging Sub-surface Features of the Miami Circle with Ground Penetrating Radar*, University of Arizona's Archaeological Sciences of the Americas Inaugural Symposium, Tucson, Arizona. Forthcoming. Also found at: http://www.mnemotrix.com/geo/mcircle/m_circ.doc (accessed 02/05/2015)
Remondino, F. 2014. Photogrammetry – Basic theory. In: Remondino, F. and S. Campana (eds.) *3D Recording and Modelling in Archaeology and Cultural Heritage*. BAR Int. Ser. 2598, 63-72.
Rodriguez Suarez R., and J. R. Pagan-Jimenez. 2008. The Burén in Precolonial Cuban Archaeology: New Information Regarding the Use of Plants and Ceramic Griddles during the Late Ceramic Age of Eastern Cuba Gathered through Starch Analysis In: Hofman, C. L, Hoogland, M. L., and A. L. van Gijn (eds.): *Crossing the Borders*. Tuscaloosa: University of Alabama Press.
Rouse, I. 1939. Prehistory in Haiti: A study in method. Yale Publications in Anthropology, 21, Yale University Press, New Haven, USA.
Rouse, I. 1992. *The Tainos: Rise and Decline of the People Who Greeted Columbus*. New Haven, CT: Yale University Press.
de Ruiter, S. 2012. *Mapping History An analysis of site locations in the north-western Dominican Republic*. Unpublished Master Thesis. Leiden University.
Samson, A., Rogab, C. and T. Küster. 2012. *Informe preliminar sobre investigaciones de teledetección arqueológica realizadas en la costa norte de la República Dominicana, Octubre 9-16, 2012*. Unpublished manuscript. Museo del Hombre Dominicano, Santo Domingo.
Samson, A. 2010. Renewing the House. PhD Dissertation. Leiden: Sidestone Press.
Samson, A. 2011. The most beautiful house in the World. In: Hofman C.L. & A. van Duivenbode (eds.) Communities in Contact. Sidestone Press, Leiden: 421-43.
Samson, A. 2013. Household Archaeology in the Pre-Columbian Caribbean. In: Keegan, W. F., Hofman, C. L. and R. Rodriguez Ramos (eds.). *The Oxford Handbook of Caribbean Archaeology,* Oxford University Press.
Samson, A.V.M., Crawford, C.A., Hoogland, M.L.P., and C.L. Hofman. 2015. "Resilience in Pre-Columbian Caribbean House-Building: Dialogue Between Archaeology and Humanitarian Shelter." *Human Ecology* 43: 323–37.
Sonnemann, T. 2011. Angkor Underground – Applying GPR to Analyse the Diachronic Structure of a Great Urban Complex. Unpublished PhD Dissertation, The University of Sydney. Accessed online: http://ses.library.usyd.edu.au/handle/2123/8069.
Sonnemann T. 2015. Classification of geophysical data of Angkor, Cambodia, and its potential as an online source. In: A. Traviglia (ed.) Across Space and Time - Papers from the 41st CAA, Perth 25-28 March 2013. Amsterdam University Press: 64–69.
Sonnemann, T. 2016. Oportunidades por el uso de la teledetección en la arqueología del Caribe. *Boletin del museo del Hombre Dominicano*, 47, Santo Domingo.

Sauerbier, M. and H. Eisenbeiss. 2010. UAVs for the Documentation of Archaeological Excavations. In: *Int. Arch. of Photogram., Rem. Sens. and Spat. Inf. Sci.*, Vol. XXXVIII, 5 Commission V Symposium, Newcastle upon Tyne: 526–531.

Sinelli, P. 2012. Meillacoid and the Origins of Classic Taino Society. In: Keegan, W. Hofman, C. L, and R. Rodriguez Ramos: *The Oxford Handbook of Caribbean Archaeology*, Oxford University Press.

Ulloa Hung, J. 2014. *Arqueología en la Línea Noroeste de La Española. Paisajes, cerámicas e interacciones.* Instituto Tecnológico de Santo Domingo, Santo Domingo.

Ulloa Hung, J. and S. de Ruiter. 2011. Arqueología en la línea noroeste de la República Dominicana. Un esbozo del paisaje arqueológico y las interacciones. *El Caribe Arqueológico* 12: 60–75

Ulloa Hung, J. and E. Herrera Malatesta. 2015. Investigaciones arqueológicas en el norte de La Española, entre viejos esquemas y nuevos datos. *Boletín del Museo del Hombre Dominicano, 46,* Santo Domingo: 75–107.

Ulloa Hung, J. Herrera Malatesta, E. and T. Sonnemann. 2014. *Resumen de trabajos de campo - Segunda campaña de 17 enero al 17 febrero del 2014.* Unpublished Manuscript. Museo del Hombre Dominicano, Santo Domingo, and Leiden University (Nexus1492).

Ulloa Hung, J. and T. Sonnemann. Forthcoming. Exploraciones en la ruta de Colón de 1494.Una nueva aproximación arqueológica a la fortaleza de Santo Tomás de Jánico y su entorno. *Boletín del Museo del Hombre Dominicano*, Santo Domingo.

Vega, B. 1980. *Los Cacicazgos de la Hispaniola*. Santo Domingo: Ediciones Museo del Hombre Dominicano.

Veloz Maggiolo, M. 2003. *La isla de Santo Domingo antes de Colón.* Santo Domingo: Banco Central de la República Dominicana.

Veloz Maggiolo, M.; Ortega, E. y Ángel Caba. 1981. Lo*s modos de vida Meillacoides y sus posibles orígenes*. Santo Domingo: Editora Taller,.

Welch, D. 2010. Geophysical prospection at the ceremonial site of Tibes, 1998-2001. In: Luis Antonio Curet and Lisa M. Stringer (eds.) *Tibes, People, Power, and Ritual at the center of the cosmos*, 60-79. Tuscaloosa: University Alabama Press.

Vargas Arenas, I, Ismenia Toledo, M., Molina, L., Montcourt, C.E. 1997. Los Artifices de la Concha. Museo Arquéologico de Quíbor, Venezuela.

Wilson, S. 2007. *The Archaeology of the Caribbean.* Cambridge: Cambridge University Press.

Part III
Landscape Representation and Scales

Towards a Holistic Archaeological Survey Approach for Ancient Cityscapes

Frank Vermeulen

Abstract Recent technological advancement in ways of comprehensive data acquisition, processing, analysis and fusion, allow today to look at the archaeological record without extensive excavation. Non-invasive technologies of remote sensing, well integrated with a full comprehension of all legacy data and the application of integrated geo-archaeological approaches, scan large and complex sites in ways never attained before. Even in fields where the results of large-scale digs have since long dominated the academic debate, such as the study of ancient urbanism, intensive intra-site surveys and remote sensing operations allow now to fully reconsider old assumptions and to tackle exciting new questions. Based on the experience of directing intensive fieldwork in and around a number of abandoned Roman city sites in the western Mediterranean, a re-examination is proposed of approaches towards abandoned classical cities and of the leading role the archaeologist needs to keep on playing in order to understand, address and manage new technologies for the study and reconstruction of the Past. This paper proposes also some good practice in the total non-invasive investigation of such large sites by way of field survey, and suggests the creation of a 'total package' of methods that allow, in an integrated way, to study and visualise ancient cityscapes and landscapes in the near future.

Introduction

As a result of increasing sophistication in the technology of remote sensing, in its broadest meaning, and the responsibility to cope with the persistent human pressure on our landscape, archaeologists from all over the planet are more and more inclined to use non-invasive survey in their field strategies. Data capture in the field, from pushed and pulled geophysical instruments to laser scanning, and in the air, from low altitude drone photography to the spectacular resolutions obtained with satellite aperture radar, is moving the limits of our observation of the Past, in a spectacular way. Technological

F. Vermeulen (✉)
Department of Archaeology, Ghent University, Ghent, Belgium
e-mail: frank.vermeulen@ugent.be

M. Forte and S. Campana (eds.), *Digital Methods and Remote Sensing in Archaeology*, Quantitative Methods in the Humanities and Social Sciences,
DOI 10.1007/978-3-319-40658-9_5

advancement, in particular during the last decade, in ways of comprehensive data acquisition, but also in processing, analysis and fusion, allow today to locate and meaningfully study an important part of the archaeological record without extensive excavation. A new age of sensing and imaging has begun. Now we are able to advance knowledge and understanding in a spectacular way, without having to destroy what we want to observe.

Very recent developments of a geopolitical nature in eastern parts of the Mediterranean and the Near East, where the cradle of urban civilization is located and where ancient sites with ideal characteristics for non-invasive approaches abound, have added extreme urgency to the current debate about applying 'sensible' technologies in field archaeology. The ongoing destruction and large-scale looting of archaeological sites throughout countries like Iraq, Syria and Libya is an even greater threat to the world's cultural heritage than any natural erosion or bad quality excavation can ever produce. The homelands of the Sumerian, Assyrian, and Babylonian empires, and the regions where relics and sites of classical Greek and Roman civilisations are by far the best preserved, are today strewn with small and large sites pockmarked with looters' pits as the result of war and uncontrolled upheaval. Added to this, so-called 'cultural cleansing' by Islamic State militants that are actively campaigning to purge ancient relics they say violate their fundamentalist interpretation of Islamic law, using bulldozers and explosives in the process, have destroyed large parts of some of the most important ancient sites in the region. These actions are currently erasing the most valuable standing buildings and connected archaeological record of pilot sites like Nimrud, Khorsabad, Hatra, Apamea and Palmyra for ever, before full documentation with the best technologies at hand was ever achieved.

The great urgency forced upon scientists by these geopolitical events, as well as the extraordinary nature of results from always wider archaeological survey applications urges us, however, to design carefully our new methodological frameworks. The technology which we all wish to use and apply on a grand scale is not without financial cost. And let us admit, the risk of just obtaining awesome looking imagery without really answering meaningful questions of historical importance, is never far away. To meaningfully advance archaeological research and its applications in an efficient and healthy manner, we need to base our activity today and in the future on good practice. It is of the utmost importance to take into account not only successful experience, but also failures, as it is also from the "worst of times in survey" that we are able to learn (Corsi et al. 2013).

It cannot be emphasized enough that there is a need for very smart strategies for the processing, interpretation and archiving of the many types and quantities of field data. While designing new strategies we must also keep our eyes wide open for the many matters of scale and resolution that we have to deal with when surveying large and multi-period sites and landscapes (Johnson and Millett 2012). Real understanding of the complexity of our object of investigation, demands constant confrontation with the results, identification of visibility or interpretation problems and most of all a lively debate, based on bringing together experiences from different environments and landscapes. We must also be fully aware that we do not necessarily need the latest or most expensive technology, but that we just have to ask the right questions and use the best suited approach (Vermeulen et al. 2012a) (Figs. 1 and 2).

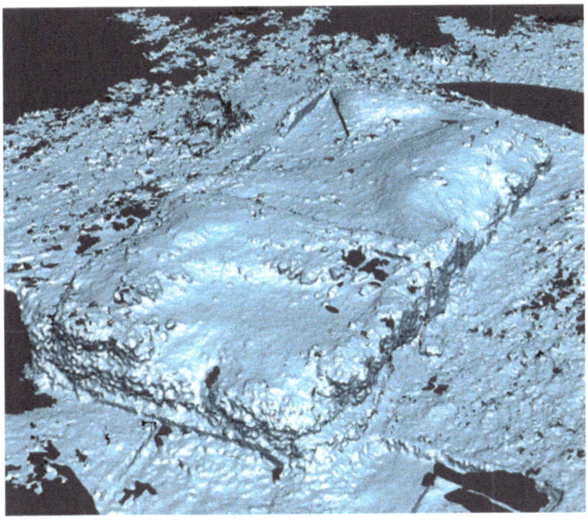

Fig. 1 Terrestrial Lase Scanning (TLS) point cloud image of the preserved base of the Roman forum temple at *Ammaia* in Portugal (image The Discovery Programme). TLS and also low cost photography-based solutions for 3D registration of still standing structures are today becoming part of the basic package of non-invasive survey of abandoned town sites

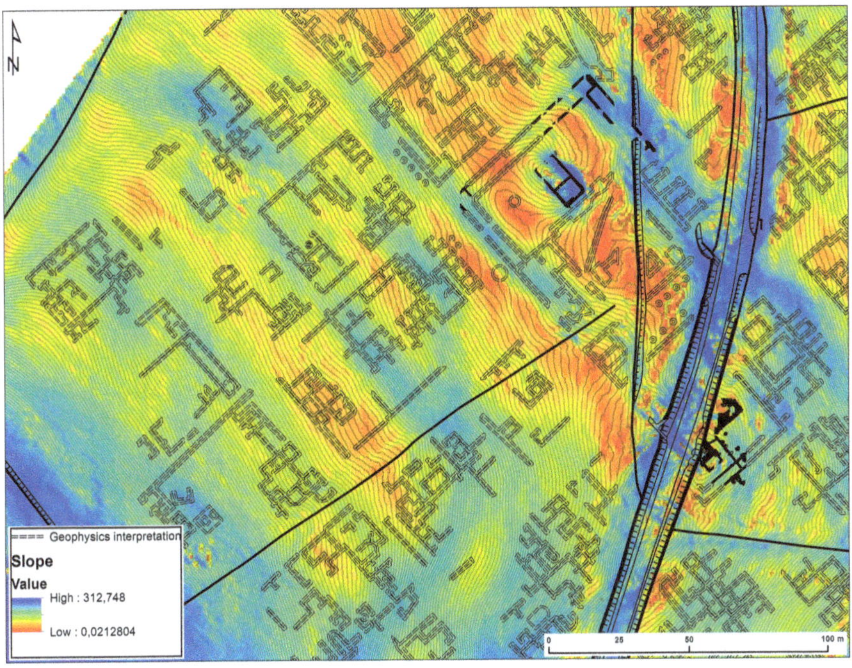

Fig. 2 DTM obtained with combined field measurements with a total station and DGPS instruments, confronted with data from geophysical prospections and punctual excavations in the central area of the buried urban site of *Ammaia* (elaboration E. Paliou)

Urban Survey Experiences in the Western Mediterranean

Today, it can be comfortably demonstrated that, even in fields where the results of large-scale digs have since long dominated the academic debate, such as the study of ancient urbanism, intensive and non-invasive intra-site surveys allow now to reconsider old assumptions and to answer many exciting new questions. At the same time, however, we need to stress that a certain success in this quite traditional field of studying large and complex proto-urban or urban sites, is the result of constant adaptations and flexibility in strategy. The progress made here during the past two decades is a tale of some good results by way extensive application of certain field techniques, but also of trials and patience. It is often about a process of learning from wrong choices made, and even now and then from mistakes or erroneous first interpretations. But it is clear that the growing realization from the late 1980s onwards that the same techniques of non-invasive fieldwork which had revolutionised rural landscape approaches also held out the promise of making a major contribution to our understanding of ancient urban sites (Millett 2012). Particularly the large towns, cities and ports of the Classical Mediterranean were first seen as a suitable category to tackle. These huge and complex, diachronic sites, which until then were almost solely approached with archaeological excavations and traditional topographic work, typically centred on the more monumental or visible structures, often also had a well preserved buried patrimony which could now be revealed by ever more sophisticated sensing equipment, especially based on geophysics. This new interest for the category of urban sites, which today are often partly or even fully abandoned, was spurred by seminal projects such as the study of Boeotian towns by Bintliff and Snodgrass (1985), as well as by the refinement of geophysical survey techniques and aerial photography that could be used for the fine-grained analysis required to bring out details of urban layout (e.g. Scollar et al. 1990; Doneus 2004; Barber 2011).

The consequence of these developments, including the widespread use of GIS in archaeology, has been an upsurge in the non-destructive survey of urban sites, in the Mediterranean and beyond. Archaeologists of the classical periods have been quick to realize the potential offered by this technique (Christie and Augenti 2012). Large and complex urban sites which had hitherto been studied in a piecemeal approach that was largely predicated upon the monument-based interests of earlier scholars are the past decade increasingly being "scanned" with survey techniques to rapidly generate plans of partial, or in some cases, complete townscapes. Parallel with some developments of large scale integrated town survey in more northern regions of the Roman World, such as at the pilot-sites of *Wroxeter* (White et al. 2013) and *Carnuntum* (Doneus et al. 2001), these Mediterranean urban surveys were more and more dominating the large scale survey efforts of whole teams of researchers, such as is well illustrated at the classical urban sites of *Falerii Novi* (Patterson 2004), *Portus* (Keay et al. 2005), and *Ephesos* (Groh 2012) to name but a few. This has also led to a revolution in how classical archaeologists approach urban sites, with survey techniques being used increasingly often to generate a plan of a town site prior to excavation as a way of ensuring that the excavation can be used to address site-specific questions in a way that had not been possible before. Cultural heritage management authorities have also benefited from this approach, with urban surveys providing them with a very effective tool for gauging the degree of archaeological survival on major urban sites in their care and choosing appropriate conservation strategies (Figs. 3 and 4).

During the last decade, such research has begun to reveal the advantages of intensively integrating a whole range of different non-destructive techniques on urban sites, instead of only applying one method of geophysical prospections. When integrating different non-invasive techniques and especially choosing those suites that are most appropriate for the nature of the town in question, great efficiency could be attained. When little more than ten years ago I started to direct research in and around a number of abandoned proto-urban and urban sites in the western Mediterranean, mostly dating to the Roman period, this integration of new and old sensing was put high on the research agenda (Vermeulen 2004, 2013). Within the framework of a series of landscape archaeological projects, developed since around 2000, six ancient town sites, situated in different areas and landscape contexts of the western Mediterranean were chosen for intensive new fieldwork operations. Apart from a whole series of historically driven research questions, often connected with 'longue durée' settlement dynamics in their respective

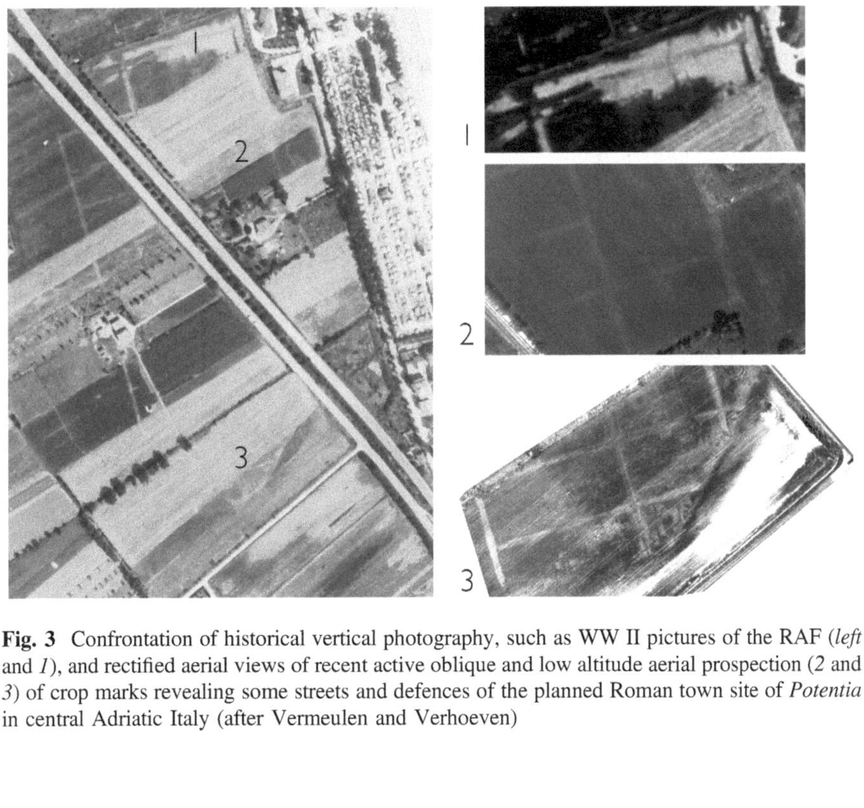

Fig. 3 Confrontation of historical vertical photography, such as WW II pictures of the RAF (*left* and *1*), and rectified aerial views of recent active oblique and low altitude aerial prospection (*2* and *3*) of crop marks revealing some streets and defences of the planned Roman town site of *Potentia* in central Adriatic Italy (after Vermeulen and Verhoeven)

Fig. 4 Near vertical image taken from a light airplane over distinct crop marks of a buried Roman amphitheatre discovered in 2011 through regular site monitoring from the air in the central area of the former Roman town of *Ricina* in central Italy (photo F. Vermeulen)

regions, a set of objectives were more of a methodological nature. Some of these were particularly connected with a more holistic approach to field and desktop based survey methods, partly filling the need to develop guidelines for good practice in this domain (Figs. 5 and 6).

The sites chosen for this wide area study have a main occupation history somewhere between the 3rd century BCE and the 5th century CE, but some structures belong to the immediately preceding Iron Age or protohistoric phases of the regions concerned, when certain towns were still in an embryonic stage. The ancient sites, located in Portugal (*Ammaia*), southern France (*Mariana* on Corsica) and central Adriatic Italy (the Roman towns of *Potentia*, *Ricina*, *Trea* and *Septempeda*, and the pre-Roman village of Montarice) belong to the category of quite large and complex, diachronic sites, which until recently were almost solely approached with very limited punctual archaeological excavations and traditional topographic work. This earlier attention was typically centred on a few monumental or visible intra-mural structures, or on presumed defensive elements. As these abandoned town sites are today mostly devoid of modern habitation and are essentially reduced to agricultural land, they are suited for new investigations of the non-invasive kind. They are ideal for being "scanned" with survey techniques to quite rapidly generate plans of partial, or in some cases, complete townscapes. Certain new technologies did not have to be applied here. As the sites are not

Fig. 5 NIR image taken with a helikite of crop marks with part of the circuit wall and a newly discovered gate at the Roman town site of *Septempeda* in Italy (*photo* G. Verhoeven). The NIR imagery was able to bring out much more archaeological detail than the traditional photography of the site

Fig. 6 Detail of a mosaic composed of rectified and enhanced oblique aerial photographs with remarkable crop marks of the Roman forum and town centre of the abandoned urban site of *Trea* in Adriatic Italy (elaboration Vermeulen and Verhoeven)

covered with impenetrable vegetation and show almost no archaeologically significant relief features or archaeological structures at the surface, archaeologically driven high resolution LiDAR, was not a necessary research option, even if the still somewhat costly use of this technique might have facilitated rapid mapping of the current topographic background of the sites (Figs. 7 and 8).

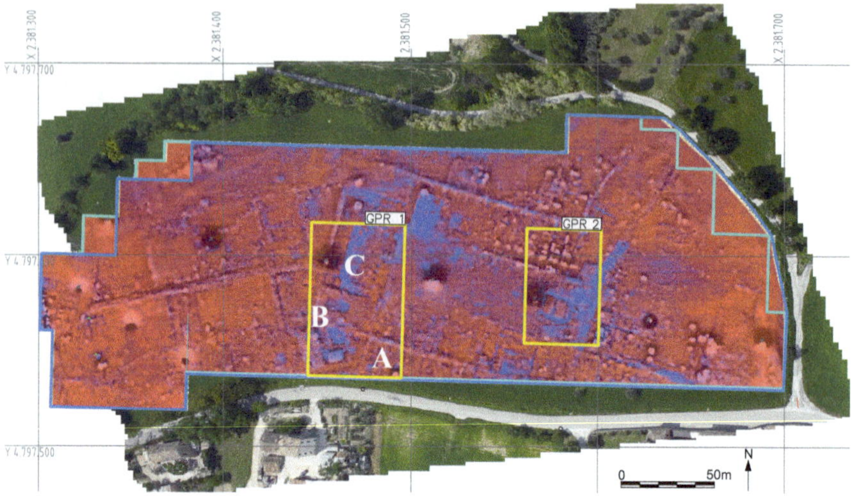

Fig. 7 Fusion of imagery obtained from geomagnetic and earth resistance surveys, on a background of rectified aerial photos in the same central area of Roman *Trea*. GPR surveys over parts of this area revealed almost no traces, while those picked up by the two other methods were very convincing in identifying the main public and private monuments and streets (elaboration B. Music)

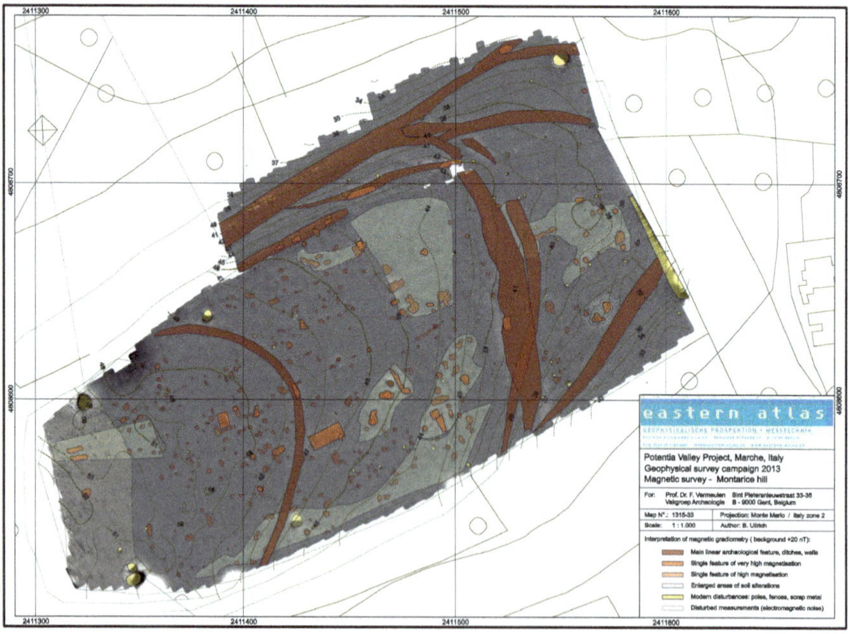

Fig. 8 Overlay of topographic map and the imagery and interpretation of the 2013 magnetic survey of the protohistoric village site of Montarice near Porto Recanati (elaboration Eastern Atlas). Clearly visible are remains of multi-phased systems of defences, and a whole series of pits, post-holes and floors relating to buried remains of pre-Roman buildings

But also when, as in many cases of large scale survey efforts today, LiDAR is not really needed or available, because of financial reasons or because one works in landscapes where this laser technique cannot be applied yet for reasons of 'national security', the production of a high-resolution Digital Terrain Model or DTM is a basic requisite. This DTM is needed not only to contextualize better the underground (and above ground) features detected with certain non-invasive prospections, but also to understand the "phenomenology" of complex sites and landscapes. The DTM supports the 2D and 3D mapping and volumetric reconstructions, but also allows spatial analysis with a full understanding of the site and its post-depositional processes. Whether such a terrain model is built with the help of total stations or a DGPS survey, or nowadays also based on aerial photography and the production of derived photogrammetric models, it is of crucial importance for the GIS-based operations of interpretative mapping that will be achieved thereafter. Where possible, such digital terrain modelling can be linked to the higher resolution and more and more rapidly achieved ground-based scanning of still standing or already excavated structures that today can be observed above the ground surface, as we could realise in the case of the site of *Ammaia* in central Portugal where certain remains of a gate building and a temple platform allowed such operations (Corsi and Vermeulen 2012) (Figs. 9 and 10).

The urban surveys on the chosen western Mediterranean sites, as on some comparable research elsewhere in Europe, have revealed the advantages of intensively integrating a range of different non-destructive techniques. Among the wide array of approaches at hand, active aerial photography of the sites and their surrounding landscape was often chosen as a first approach, as it remains a very potent technique of recovery of the buried evidence of classical town sites. As was well exemplified at the site of the Roman coastal colony of *Potentia,* founded in 184 BC in central Adriatic Italy near present-day Porto Recanati, the active aerial survey needs a symbiosis with the more "passive" use of existing vertical photography (Vermeulen and Verhoeven 2004; Vermeulen 2012). The importance of certain available, older vertical photography, even if taken for other purposes than archaeology, can surely not be underestimated in such research. This applies as well for historical photography, such as the excellent World War II imagery taken in large parts of Europe by RAF or Luftwaffe pilots, as for the now widely available and high resolution aerial views and satellite images (e.g. QuickBird, IKONOS, WorldView) from websites such as Google maps or Bing maps. The aerial photography and satellite imagery of always better quality and resolutions, available via these commercial websites, is particularly useful in regions where aerial archaeology flights have never or seldom been achieved (Figs. 11 and 12).

Specific to the study by way of active aerial photography of abandoned Roman urban sites is that parts of these ancient towns are often already discovered in the past. The contribution of a systematic reconnaissance of these large sites lies, therefore, not so much in their initial finding, but more in their full comprehension as an urban landscape, including a first appreciation of their total size, their planned layout (wall circuits, street network, …), their relation to the general landscape (roads, field systems, …), their suburban areas, etc. A recurrent phenomenon in the

Fig. 9 Magnetic survey results of the western intramural part and *suburbium* of the Roman colonial town *Potentia* on a background of vertical aerial imagery (elaboration B. Ullrich, Eastern Atlas and S. Hay, BSR). Outside the gate area (*A*) many remains of funerary monuments (*D*, *E*) placed near the outgoing Roman road (*C*) were discernible, but geo-archaeological coring operations at greater distances from the road revealed that part of the burials remained hidden from the geophysical prospections

aerial survey is, however, the first discovery of new features thanks to prolonged flying at different moments of the year, under ever changing conditions of ground visibility. The specificity of Roman townscapes with their good architectural visibility, and often the homogeneity of the application of certain architectural models and plans, is of course a great help in this. The example of the now only partly abandoned town of *Ricina* shows this well, as the regular flights in the most rewarding seasons (mostly Spring and Summer) of almost every year between 2004 and 2013 have revealed here a whole series of totally unknown urban features, such as the circuit wall, the main forum temple, the best town residences and a large amphitheatre, all spared from modern habitation and infrastructures that partly cover the still mainly rural valley landscape near Macerata (Vermeulen 2011) (Figs. 13 and 14).

My own experience from several of the urban surveys where active aerial photography has been applied from a "classic" manned aircraft learns that very useful imagery of complex urban sites can also best be acquired from low-altitude

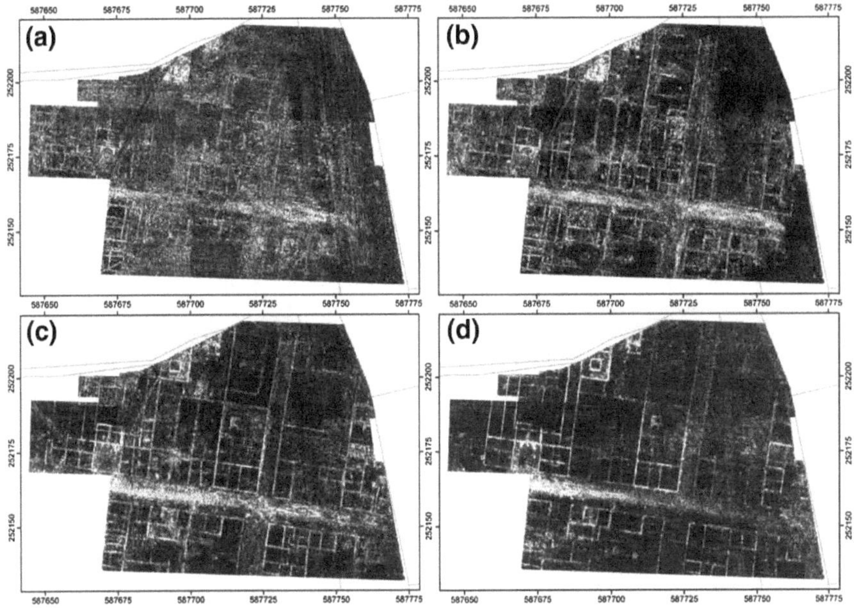

Fig. 10 'Time slices' with imagery at different depths obtained during the recent ground penetrating radar prospections on the central area of the abandoned Roman town of *Mariana* on Corsica. The crossroads of four town streets and remains of many building and housing structures were revealed with great detail (imagery L. Verdonck)

unmanned platforms. Today all kinds of devices (kites, balloons, drones, …) are being used to take still cameras aloft and remotely gather aerial imagery. Especially radio-controlled (multi-)copter platforms are at present popularizing aerial photography over large and complex sites. Their remarkable speed and image quality and their development into increasingly low-cost solutions make these platforms potent instruments for the aerial study and monitoring of ancient urban sites, wherever the national legal context allows archaeologists to use them. In addition, they stimulate the further exploitation of new imaging techniques, such as close-range near-infrared (NIR) photography and near-ultraviolet imaging (Verhoeven 2008, 2012). These techniques, involving a wider use of the spectrum, seem particularly useful when surveying Roman urban landscapes, as the omnipresence of durable building materials in the subsoil guarantees good results when the moment of photography is well chosen. This is particularly well illustrated by recent examples from our surveys in the central-Italian town of *Septempeda*, near modern San Severino Marche, where parts of the Roman infrastructure were much better revealed in the late Spring grain using NIR digital cameras from a drone (Verhoeven 2012) (Fig. 15).

With the very successful aerial survey of the town of *Trea*, still in central Adriatic Italy, again the importance of repetitive surveys should be stressed. Between 2003 and 2009 almost all significant urban features of the Roman town

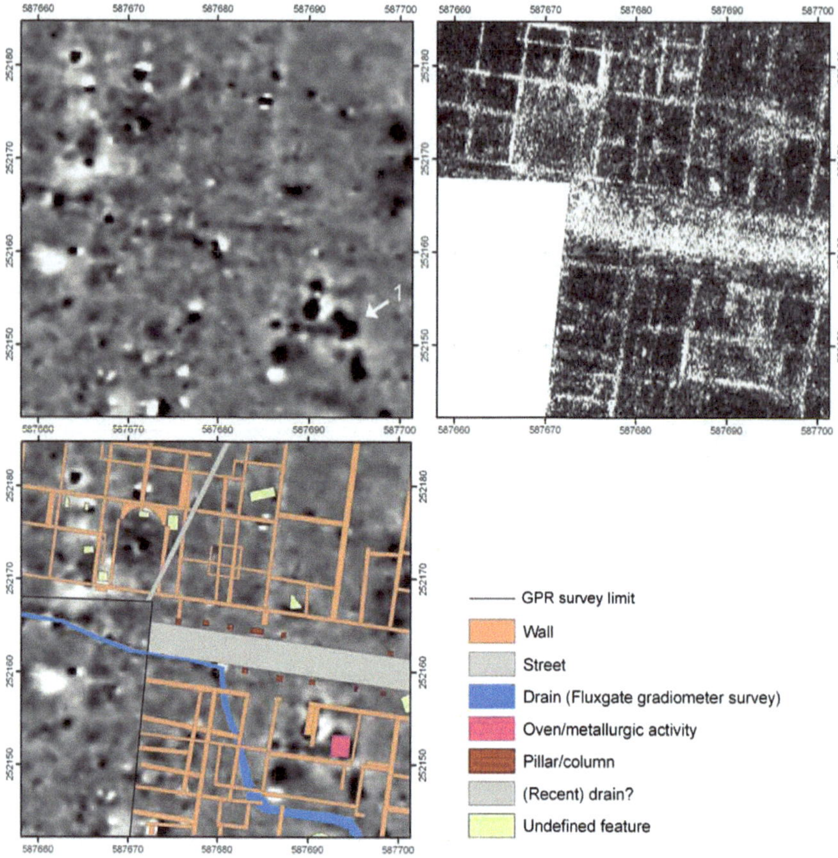

Fig. 11 Integration of GPR (*right*) and magnetic (*left*) prospection imagery allowed certain pushed interpretations of the fused imagery of the buried evidence, such as of a probable workshop for metallurgic activity along a porticoed street (elaboration L. Verdonck)

were revealed in the crops covering the area of the circa 12 hectare large intra-mural town: the forum with all essential public buildings (temples, *basilica, curia*, …), the circuit wall and a number of gates and towers, many houses, shops, a porticoed market place, workshops, etc. (Vermeulen et al. 2012b). Whichever the specific climatic conditions or season, and whichever the type of crops or vegetation covering the many different parts of the cityscape, regular and numerous flying were necessary to arrive at this result. We are convinced that this monitoring of the urban sites is an important element in a full comprehension of the many archaeological structures present in the soil, and the only possible approach to its full complexity. Indeed, one encounters many examples of truly remarkable 'evolutions' of urban archaeological sites due to totally different detection opportunities over different moments, seasons or years. These repetitive observations conducted over a period of several years draw new details again and again from the soil. All best results can

Fig. 12 General plan of the intramural part of Roman *Ammaia* with integration of the interpreted geophysical prospection data (*grey* and *black*), and the excavated structures (*red*). Central in the regular grid of housing blocks and streets lie the forum and an adjoining bath complex. The hilly part of the intramural town could not be surveyed due to dense vegetation overgrowth

then be joined, like pieces of a puzzle, into an extensive overall view of the urban landscape at one time in its development. With luck, or if the conditions of soil contrast and plant stress are very favourable, also diachronic evolutions of the town plan can sometimes be observed. The latter are particularly evident when features cross each other with different orientations or when the pattern of detected wall structures in an area is so dense that they can only be the result of regular rebuilding on the same spot. As with all types of aerial archaeology, reading aerial photographs from an ancient town site does not mean trying to identify only the elements that indicate past human activities related to the urban phase in Antiquity, but involves using all present day landscape features as elements of contrast that help to bring out the residual components of the ancient landscape. Furthermore, it

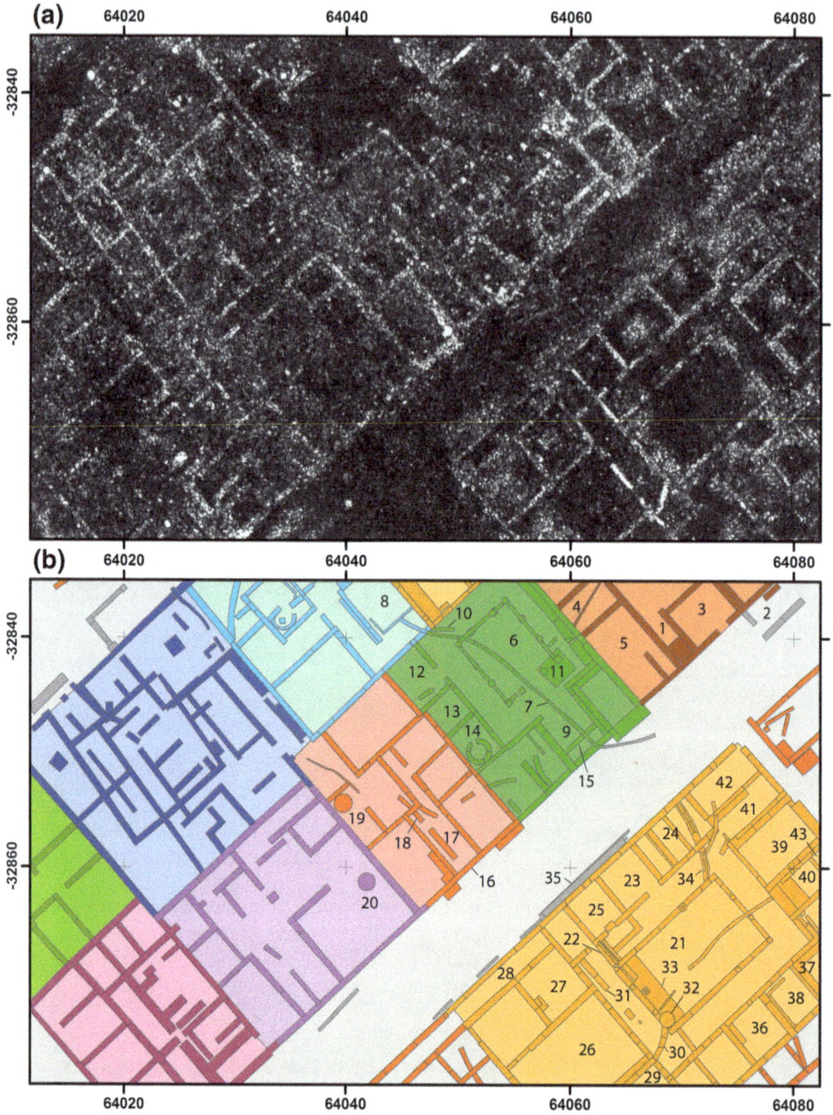

Fig. 13 Detail of high resolution GPR results and their interpretation in part of the Roman housing area of *Ammaia* (elaboration L. Verdonck). The sensing results allow unprecedented details for the study of simple Roman town houses in the Iberian peninsula

is of great importance that the sometimes easily recognizable and regular Roman features are not the only ones to be filtered out, but also old structures and marks that might belong to other phases of the Past. These might be of important interpretative value to the 'longue durée' understanding of the site or help to explain the deterioration of Roman elements in the soils and therefore, in surface visibility. To

Fig. 14 Transformation procedure of the integrated geophysical data into a 3D-modelled visualisation of the Roman town of *Ammaia* during its early Imperial heyday (elaboration M. Klein, 7Reasons)

Fig. 15 Visualisation of early Imperial *Ammaia* in its reconstructed landscape context (elaboration M. Klein, 7Reasons)

push this diachronic interpretation enough far, in order to make it meaningful, further field operations are necessary, such as corings of the features at specific intersections or geophysical mapping.

Despite the current large variety of means to process and ortho-rectify the oblique imagery (Verhoeven et al. 2012), their archaeological information will not be exploited efficiently as long as the image is not thoroughly interpreted, meaning interpretatively mapped and integrated with other data sources. A thorough understanding of ancient cityscapes is based on combining the interpreted evidence from various prospection methods and the approach to the interpretation of aerial images must be comprehensive. Here the application of a suite of geophysical prospections, today the main approach for such urban surveys, comes in the picture. Geophysical prospection has long been applied for (better) locating and evaluating parts of sites or buried structures rather than fully investigating them (Gaffney et al. 2002; Campana and Piro 2009). Most surveys of non-threatened urban Roman sites are however, carried out in the context of academic research so they should be expected to do precisely the latter. These research-driven surveys aim to clarify the town lay-out and increasingly map the ancient city as extensively as possible. This presents a logistical issue of balancing speed of survey with resolution of the data collected in order to permit the observation of buried features. Therefore, a correct balance has to be found between the selection of survey parameters to detect the expected archaeological structures, and to allow for the size of the survey area given the time and budget available. The resulting data plots are then conceived as proxy maps of the subsurface rather than merely visualisations of measurements with specific geophysical properties. Increased sampling rates, and a more precise location of the data, result not only in better visualisations, but also permit accurate data combination and modelling of the buried archaeological features through their geophysical signature.

As magnetic survey results generally detect most types of archaeological features within Roman urban sites, other geophysical survey techniques were only seldom applied on a large scale, mainly because they were for long more time consuming and hence expensive. However, feature detection through multiple parameters enhances interpretative validity. In addition, recent developments, such as DGPS equipped multi-instrument platforms and towed arrays, allow relatively rapid collection of more extensive but also more intensive geophysical datasets. It is clear that for the total survey of buried Roman towns sites we need to apply or use as much as possible all relevant geophysical approaches: so apart from geomagnetic survey, if available and possible also earth resistance prospections and georadar, to name but the most effective today. Therefore, on a number of sites, such as on the cited small Roman town of *Trea*, we applied these three geophysical prospection techniques evenly over most of the intra-mural area. Experience learned here that only two techniques really worked well, while the third produced no results at all. While a full coverage georadar survey of the site revealed no structures whatsoever, most likely because of the very high clay content of the soil, both magnetics and earth resistance prospections were highly successful. Both techniques not only complemented each other well, but also confirmed and sometimes refined the already spectacular aerial survey results. Ancient floors were now better mapped—including a more reliable appreciation of their state of preservation—areas of workshops clearer defined, functions better attributed to certain public buildings, etc.

This experience of fieldwork in the mid-2010s demonstrated well that if geophysical surveys will in the future be more and more multi-method, we must also put good effort in the integration of our data and in the quality of interpretation, achieving first of all real data fusion. The latter is more than mere integration, it is interpretation in the same workspace, allowing for more objective assessments and often for new and more transparent approaches to the buried evidence. In the case of *Trea* the field team fused the raw resistivity imagery with magnetic data from the forum area, which helped considerably in deciphering the structures of the monumental town centre, to levels unseen before in non-excavation field work on such 'well known' Roman urban models. Integration with the aerial data, and eventually also with legacy data from older ill located sources, such as some earlier not well located 19th century excavations, reveal now also an understanding of functions and even chronologies of the individual structures and certain components of the town centre.

One might observe with reason that the Roman normative system of town building helps a lot in this process and indeed when we studied certain pre-Roman sites discovered in the same area of the central Adriatic towns, much more interpretative problems arose. Buildings, partly or fully in perishable materials and of indistinct types are difficult to spot, streets are irregular and generally unpaved, and defensive systems have intricate features not easily to be categorized without excavation. The example of the pre-Roman proto-urban site of Montarice, predecessor of the colony of *Potentia* near the mouth of the River Potenza, demonstrates this complexity. Nevertheless, a good combination of systematic aerial monitoring of the site since its discovery in 2001, and focussed geophysical sensing by way of crossing total coverage magnetic survey with electrical resistance tomography as a complementary technique to obtain 3D profiles, with additional geo-archaeological field controls through coring operations and artefact surveys, has proven to be quintessential in mapping and at least partly understanding the complexity of this diachronic site. Especially the vertical profiling of buried structures, detected first via combined aerial surveys and magnetometer prospections, proved to be a way of understanding some of the diachronic phases of the defended settlement between its Bronze age origin and late Iron age heyday, roughly between 1200 and 300 BC. Further non-invasive surveys by way of more of these focussed geophysical operations might well add new crucial information about the character, size, and development of a type of protohistoric village, otherwise very difficult to grasp in its spatial organisation and multi-period phasing.

Such integrated geo-archaeological approaches are also much needed when we are confronted with so-called 'hidden landscapes', hard to approach because of the geological processes of deep burial of the archaeology. In the suburban area of the nearby Roman colony *Potentia*, this applies to the zones of the cemeteries (Vermeulen et al. 2013). As these burial places outside the town walls lay originally on lower ground, now more deeply covered by alluvial processes, detection and interpretation is extra complex. In such a situation again augerings, geologically relevant deep geophysics, landscape evaluation and erosion modelling of the palaeo-landscape background, can be of great use. Therefore, this needs to be well inserted in the whole program and strategy of survey and field operations, enabling a constant dialogue between the geo-scientists and archaeologists in the field. The need for geo-archaeological approaches is really relevant if we want to deal with the sometimes very high stratigraphic complexity of buried multi-period sites and the very dynamic landscapes they are part of. In-site geo-archaeological analysis first and foremost focuses on the genesis of the archaeological sites, meaning the formation processes on the scale of the sites themselves, and on the factors leading to the fossilisation, preservation or reworking of the archaeological remains (Rapp and Hill 1998). It is thus complementary with the purely archaeological, traditional stratigraphic approach. It allows to establish the origin of the archaeological sediment and their evolution by highlighting what is linked with the anthropogenic, cultural and bio-pedological processes as well as the geological depositional (sedimentary) and/or post-depositional factors, which are so important in abandoned Roman towns sites (Vermeulen and De Dapper 2000).

Sometimes, as we learned in such an integrated survey of part of the abandoned Roman town of *Mariana* in the northern part of the island of Corsica (France), the search for a good geophysical approach needs just enough patience (Verdonck et al. 2012). Patience is also essential for convincing oneself as researcher that geophysical prospection really works in certain types of terrain, against the outcome of early stage trials. On this site in the very dynamic river valley of the Golo, buried features of the Roman town were not detected during the first magnetic and earth resistance prospections of the 2005 season. The instruments of the former method were initially not well tuned, while the soil was too hard during the mid-summer surveys when the latter method was first tried. Thereafter, new geomagnetic prospections met with more success. Amongst the range of geophysical techniques employed at this site, GPR survey definitely contributed in a spectacular way to our understanding the town plan, as well as the three-dimensionality of the buried structures. Archaeology has made increasing use of this radar-based approach in recent years in order quickly to identify the dimensions of buried features. First time-slicing enabled the comparison of archaeological and geological features in plan at varying depths. Most recently the truly volumetric analysis of these data has been possible, including voxel and iso-surface representations of the data as a mechanism for discerning variation within subsurface datasets. In *Mariana*, it proved immensely useful to go to higher resolutions with such a type of survey, as well as to integrate again well the results with those obtained with other techniques. To appreciate the details of a Roman house and not just understand the size and position of a housing block, means that high resolution GPR and magnetic survey need to join efforts in a well-chosen, and because of time and funding constraints necessarily somewhat smaller survey area. Here in *Mariana* the results pertain now only to part of the town site (circa 25 %), but the higher resolution prospections and crossing of data, produced quality interpretations of a series of housing blocks and associated streets, with the power of extrapolation to other areas of the site where only lower resolution was possible. The mapping of town residences and shops/workshops, with their individual rooms, floor dispositions, hydraulic systems and internal activity zonings reached here such fine results that a major contribution to Roman urbanisation of the island of Corsica could be achieved, opening avenues of new research questions and a major understanding of the spatial expansion and organisation of the ancient city site.

Sometimes the geophysics work really well not only because of useful contrasts between the buried archaeological remains and the soil, but also because of lack of recuperation of ancient building stones in Medieval times, when the ancient towns were being abandoned. This was clearly the case at the site of the lost Roman town of *Ammaia*, in the Alentejo region of Portugal (Corsi and Vermeulen 2012). Spread over only three campaigns of summer fieldwork, between 2008 and 2011, a full town plan was obtained here with remarkable detail, combining a total magnetic survey with earth resistance and GPR surveys in more focused areas of the site. The resistance, magnetometer and GPR data were not only fused here among each other to arrive at very detailed interpretations, but were also integrated with recovered and new punctual excavation data, modern topography and a survey and scanning of standing building evidence. This integrated approach has allowed direct comparisons to be made between corresponding datasets, as well as identifying locations that can be used to verify results, while assisting in the augmented interpretation of sub-surface structures. Due to the quality and high resolution of data in most areas of the intra-urban space, and to a certain repetitiveness of the mainly residential components of the ancient urban settlement, a full coverage detailed town mapping was achieved.

In a next step three-dimensional graphics have been employed as a means of recording and exploring data and their interpretation. The obtained models allowed for major understanding of the archaeological evidence in the field, but are considered neither definitive statements of fact, nor wholly imagined products of Virtual Archaeology. Viewing the data from integrated surveys in this way gives the archaeologist the opportunity to engage with hypotheses in a virtual physical environment. This step is useful in view of further discussion among scholars, but most of all to better inform and involve the wider public. It represents a shift from past research with a significant impact on how the material culture relating to such cityscapes is documented and understood. No more extrapolations based on too little information about one or two plots excavated in a full city, but visualisations based on intensive, total site surveys, integrated with all possible evidence from excavations, geo-approaches, legacy finds from this site and comparative research on other sites. However, even if the obtained VR-reconstructions of the Roman town during its Early Imperial heyday are the result of much interactive work and discussion among team specialists of different fields, using a wide array of local, regional and supra-regional data, we must be aware that the reconstructions are tentative and experimental. Much of the data remains non-stratigraphic in nature and some of the interpretations cannot disentangle the complexity of many centuries of settlement development. The reconstruction of ancient cityscapes is a challenging research activity implying the management of a high level of uncertainty, especially regarding the ecological context and the factors 'time' and 'evolution'. Better than any other approach to mapping and visualisation it reveals that the integrated study and interpretation of mostly buried town structures needs to be undertaken at different levels, i.e. at different scales, ranges of precision and with variety in the depiction of interpreted detail.

Concluding Remarks

When using mostly non-invasive methods to reveal some of the complexity of ancient town plans and their intricate phasing, one must be fully aware of the potential of integrating old and new technologies. If we master well all issues of scale and resolution, understand the influence on the results of factors such as geology, depth of deposition, presence of surface rubble, vegetation, types of building materials, if we can understand something of the phasing and dating of structures through integration with stratigraphic evidence, then they allow to answer exciting questions regarding e.g. size, demography, town functions, and integration with the territory. At the same time there is a need for patience and flexibility, rather than uniformity, when the wider use of non-invasive survey technologies is being achieved. It all starts with a thorough understanding of the options at hand, and of the problems we can realistically explore by non-invasive archaeology. Furthermore, at a time when the need for incorporation of technologies increases, it might also be crucial that the archaeologist as binding factor and guarantee for the historical meaning of the basic research questionnaire, needs to keep on playing a leading role in the whole process of studying the materiality of past landscapes and cityscapes. At the same time the challenges to understand, address and manage these new technologies are undoubtedly the object of critical teamwork and dialogue with specialists from different disciplines, and a good portion of common sense. Improving further the instruments, but also the approaches that we have today, can thus result in a better archaeology tomorrow.

Bibliography

Barber, M. 2011. *A history of aerial photography and archaeology. Mata Hari's glass eye and other stories.* Swindon: English Heritage.

Bintliff, John, and Anthony Snodgrass. 1985. The Cambridge/Bradford Boeotian expedition: the first four years. *Journal of Field Archaeology* 12: 123-161.

Campana, S., and S. Piro eds. 2009. *Seeing the unseen – Geophysics and landscape archaeology.* London: CRC Press.

Christie, Neil, and Andrea Augenti eds. 2012. *Urbes Extinctae. Archaeologies of Abandoned Classical Towns.* Farnham: Ashgate.

Corsi, Cristina, and Frank Vermeulen eds. 2012. *Ammaia I: The survey. A Romano-Lusitanian Townscape Revealed.* Gent: Academia Press.

Corsi, Cristina, B. Slapsak, and Frank Vermeulen eds. 2013. *Good practice in archaeological diagnostics. Non-invasive survey of complex archaeological sites.* New York: Springer.

Doneus, Michael. 2004. Aerial Archaeology in the Landscape of Carnuntum. *Archaeologia Aerea. Studi di Aerotopografia Archaeologica* I: 215-234.

Doneus, Michael, A. Eder-Hinterleitner, and Wolfgang Neubauer. 2001. Roman Carnuntum: prospecting the largest archaeological landscape in Austria. In *Archaeological Prospection: Fourth International Conference on Archaeological Prospection. Austrian Academy of Sciences, Vienna, 19-23 September 2001*, eds. Doneus, Michael, A. Eder-Hinterleitner, and Wolfgang Neubauer, 47-59. Vienna.

Gaffney, C., J. Gater, and S. Van Ovenden. 2002. *The Use of Geophysical Techniques in Archaeological Evaluations*. Institute of Field Archaeologists Technical Paper 6. Reading: IFA.

Groh, Stefan. 2012. Strategies and results of the urban survey in the Upper City of Ephesus. In *Urban Landscape Survey in Italy and the Mediterranean*, eds. Frank Vermeulen, Gert-Jan Burgers, Simon Keay, and Cristina Corsi, 85-96. Oxford: Oxbow.

Johnson, P.S., and Martin Millett eds. 2012. *Archaeological Survey and the City*. University of Cambridge Museum of Classical Archaeology Monographs 2. Oxford: Oxbow.

Keay, Simon, Martin Millett, Lidia Paroli, and Kris Strutt eds. 2005. *Portus. An Archaeological Survey of the Port of Imperial Rome*. Archaeological Monograph 15. London: The British School at Rome and Soprintendenza Archeologica di Ostia.

Millett, Martin. 2012. Understanding Roman Towns in Italy: Reflections on the role of Geophysical Survey. In *Archaeological Survey and the City*. University of Cambridge Museum of Classical Archaeology Monographs 2, eds. P.S. Johnson, and M. Millett, 24-44. Oxford: Oxbow.

Patterson, Helen ed. 2004. *Bridging the Tiber. Approaches to Regional Archaeology in the Middle Tiber Valley*. Archaeological Monographs of the British School at Rome 13. London: British School at Rome.

Rapp, G., and C. Hill. 1998. *Geoarchaeology: the earth science approach to archaeological interpretation*. New Haven and London: Yale University Press.

Scollar, Ian, A. Tabbagh, A. Hesse, and I. Herzog. 1990. *Archaeological Prospecting and Remote Sensing*. Cambridge : Cambridge University Press.

Verdonck, Lieven, Frank Vermeulen, Cristina Corsi, and Roald Docter. 2012. Ground-penetrating radar survey at the Roman town of Mariana (Corsica), complemented with fluxgate gradiometer data and old and recent excavation results. *Near Surface Geophysics* 10, 1: 35 - 45.

Verhoeven, Geert. 2008. Imaging the invisible using modified digital still cameras for straightforward and low-cost archaeological near-infrared photography. *Journal of Archaeological Science* 35: 3087-3100.

Verhoeven, Geert. 2012. Near-Infrared aerial crop mark archaeology: from its historical use to current digital implementations. *Journal of Archaeological Method and Theory* 19: 132-160. doi:10.1007/s10816-011-9104-5.

Verhoeven, Geert, M. Doneus, Ch. Briese, and Frank Vermeulen. 2012. Mapping by matching - A computer vision-based approach to fast and accurate georeferencing of archaeological aerial photographs. *Journal of Archaeological Science* 39: 2060-2070.

Vermeulen, Frank. 2004. Fotografia aerea finalizzata nelle Marche Centrali: Un progetto integrato. *Archeologia Aerea. Studi di Aerotopografia Archeologica* 1: 91-118.

Vermeulen, Frank. 2011. Reviewing 10 years of aerial photography in the valley of the River Potenza (Le Marche). *Archeologia Aerea. Studi di Aerotopografia Archeologica* 4: 259-264.

Vermeulen, Frank. 2012, Potentia: a Lost New Town. In *Urbes Extinctae. Archaeologies of Abandoned Classical Towns*, eds. Neil Christie, and Andrea Augenti, 77-95. Farnham: Ashgate.

Vermeulen, Frank. 2013. Interdisciplinary non-invasive survey approaches to ancient towns: some applications and visualisations from the Roman West. In: *Exempli Gratia: Sagalassos, Marc Waelkens and Interdisciplinary Archaeology*, ed. Jeroen Poblome, 165-182. Leuven: University Press.

Vermeulen, Frank, Gert-Jan Burgers, Simon Keay, and Cristina Corsi eds. 2012a. *Urban Landscape Survey in Italy and the Mediterranean*. Oxford: Oxbow.

Vermeulen, Frank, and Morgan De Dapper eds. 2000. *Geo-archaeology of the Landscapes of Classical Antiquity, International Colloquium Ghent 23-24 October 1998*. Babesch Supplement 5. Leuven: Peeters.

Vermeulen, Frank, B. Slapšak, and D. Mlekuž. 2012b. Surveying the Townscape of Roman Trea (Picenum). In *Archaeological Survey and the City*. University of Cambridge Museum of Classical Archaeology Monographs 2, eds. P.S. Johnson, and M. Millett, 261-282. Oxford: Oxbow.

Vermeulen, Frank, Dimitri Van Limbergen, Jésus Garcia-Sanchez, Devi Taelman, Sophie Dralans, and Debby Van den Bergh. 2013. Nuove ricerche nel suburbium di Potentia. In *Epigrafia e Archeologia romana nel territorio marchigiano. In memoria di Lidio Gasperini. Atti del Convegno (Macerata, 22-23 aprile 2013)*. Ichnia 13, ed. *Gianfranco Paci,* 69-92. Tivoli (Roma): Edizioni Tored s.r.l..

Vermeulen, Frank, and Geert Verhoeven. 2004. The contribution of aerial photography and field survey to the study of urbanization in the Potenza valley (Picenum). *Journal of Roman Archaeology* 17: 57-82.

White, R. H., Chris Gaffney, Vince L. Gaffney, with A. Baker. 2013. Wroxeter, the Cornovii and the Urban Process. Volume 2: Characterizing the City. Final Report of the Wroxeter Hinterland Project, 1994-1997. Oxford: Archaeopress.

Sensing Ruralscapes. Third-Wave Archaeological Survey in the Mediterranean Area

Stefano Campana

Abstract The present contribution discusses the so-called 'third wave' of archaeological survey, drawing attention to the wide gap between the development and implementation or archaeological research within townscapes as compared with rural landscapes in the Mediterranean area. The first part of the discussion summarises the development of landscape studies and survey methods during the last century, critically highlighting the outcomes and limitations of past experience. The paper then presents the initial results of the Emptyscapes Project, an interdisciplinary program of survey and interpretation work designed to stimulate changes in the way in which archaeologists, in Italy but also more generally within the Mediterranean world, study the archaeology of the rural countryside, moving from an essentially site-based approach to a truly landscape-scale perspective. The first results of the project have made it possible to challenge past landscape paradigms and to move towards a more complex and comprehensive understanding of a stretch of lowland rural landscape in southern Tuscany. In doing so the project has emphasised the extent to which choices about the methodological and technological framework of the work may to a certain extent predetermine the archaeological results and influence the archaeological questions that can be asked or answered.

Introduction

Among the historical sciences archaeology is unusual in one particular respect. History and Archival studies in their various manifestations—History of Arts, History of Architecture and so on—each established their own research methods a considerable time ago and, notwithstanding some improvements in their operational framework and workflow, could be considered relatively stable. This is definitely not the case for Archaeology, the changing methodological framework and

S. Campana (✉)
Ancient Topography, Head of LAP&T lab, Department of History and Cultural Heritage, University of Siena, Via Roma 56, 53100 Siena, Italy
e-mail: campana@unisi.it

M. Forte and S. Campana (eds.), *Digital Methods and Remote Sensing in Archaeology*, Quantitative Methods in the Humanities and Social Sciences,
DOI 10.1007/978-3-319-40658-9_6

procedures of which have over the time been progressively modifying our capacity to explore and understand the past. The history of Archaeology leaves no doubt about this. The definition of the discipline, as well as its methodological framework, has changed vastly within the last century alone. Indeed, at the end of 19th century, despite some outstanding exceptions, the subject was still closely associated with the figure of the Antiquary. Around the end of that century and the beginning of the next the archaeological community clearly developed an awareness and a capacity to engage more actively in the investigation of the past and to take part in historical debate. In 1927 O.G.S. Crawford, in his editorial for the first number of the journal Antiquity, defined archaeology as a branch of science attempting to recreate the past by observing and recording facts by means of excavations, fieldwork and comparative studies.

The same generation of scholars pushed forward the idea of landscape studies, which were at that time beginning to draw on technological and methodological improvements such as the application of aerial photography within archaeological studies. Closer to the present day, archaeology has made significant advances during and since the second half of the 20th century through the widespread application of improved techniques of archaeological excavation, through technological innovations (the most notable undoubtedly being radiocarbon dating) and through pioneering work in the sphere of systematic survey procedures. The last innovation lies at the root, of course, of the present discussion. A key role in this respect was played by John Ward-Perkins and the British School at Rome (Terrenato 1996), carrying out over the course of two decades between the 1950s and the 1970s a systematic survey of around 2000 km^2 of South Etruria with the aim of exploring and understanding the past landscape of the area through the discovery, documentation and interpretation of surface artefact scatters. From that time onward field survey gradually became one of the most potent approaches for analysing the human past. At roughly the same time a highly influential role was played in Greece and in the Mediterranean more generally by the Messenia Expedition of the University of Minnesota. This project is particularly interesting because of the inception in the late 1930s of attempts to resolve problems in the Homeric geography of the south-western Peloponnese (McDonald 1942). Over time this developed into a more general search for Mycenaean sites, then for sites of all periods from the Neolithic to the Middle Ages, culminating in an appreciation that the locational characteristics of such sites could not be understood without proper attention to environmental and anthropological questions such as coastal change, natural resources, soil fertility, agricultural economy and the social anthropology of the local farming communities, for instance (Cherry 2003). By the time of the final monograph in 1972 it was clear that there had been a major transformation in the research framework from its early beginning as an archaeological survey project to a developed form as a large-scale, multidisciplinary, strongly scientific investigation based on systematic survey and research work covering some 1500 km^2 of the Mediterranean landscape (McDonald and Rapp 1972).

From the 1970s onwards a new generation of increasingly intensive survey projects has been producing striking achievements. These so-called 'new wave' surveys marked a break with the pioneering projects of earlier years by introducing new methodological approaches to enhance the scope and reliability of the resulting information (Bintliff 2000). In this new phase of archaeological survey work has played a leading role over virtually all of the Mediterranean area, animated by a keen attention to methodological approaches so as to improve the reliability of the collected data while inevitably reducing the size of the landscape blocks that could be incorporated within the survey work (Terrenato 2004). This last point has been partially mitigated by developing and implementing representative sampling strategies (Orton 2000).

After about two decades of intensive survey work of this kind, straddling the 20th and 21st centuries, a series of remaining problems and limitations have began to be noted and debated in the archaeological literature (Francovich and Patterson 2000; Bintliff et al. 2000; Papadopoulos and Leventhal 2003; Alcock and Cherry 2004). Based on critical thinking aimed at tackling various problems within the field of landscape history some research workers have in the last decade been progressively introducing new approaches, in some cases substantially altering the methodological framework. This is exactly the point of the present paper, which is intended to discuss the opportunities and limitations of a 'third wave' of archaeological survey, emphasising the extraordinary liveliness of the archaeological discipline but at the same time remarking critically on the way in which technological tools and methodological procedures may to a certain extent influence or sometimes even predetermine the archaeological results. This is a focal point: we should never forget that all technologies should be regarded as 'cultural instruments' that are not truly neutral, any more than are the underlying strategies and methodologies that we implement within our research projects (Ihde 1993; Gillings 2000).

Some Remarks on Second-Wave Survey

By the end of 1990s leading experts in the field of landscape archaeology started identifying and pointing out some of the major limitations of the so-called 'Second' or 'New Wave' of intensive archaeological survey, focusing their attention on basic improvements that it would be useful to implement in the following years. In summary, among the major limitations most of scholars shared the general idea that field-walking survey and surface data were affected by inherent problems, which could only be reduced or partially overcome by integrating them with other survey methods. Particular hopes in this respect fell on non-destructive interventions such as remote sensing, geophysical prospection and geochemical studies (Keay et al. 2005). Moreover, the ever increasing role of computer and spatial technology, in particular GIS and GPS, was considered crucial within the sphere of landscape research. The improvements that they could make to data collection, analysis, synthesis and presentation were considered almost from the beginning as much more than new tools but rather as an opportunity to bring the archaeology of surface collection and observation to a qualitatively higher level (Gillings 2000; Cherry 2003).

Another issue demanding resolution was in their eyes one of theory and interpretation. For a long time most scholars had worked on the assumption that archaeological remains were concentrated in a finite number of locations or 'sites'. Systematic survey work, however, has amply demonstrated that the surface archaeological record is in fact much more widespread than this, including to a large extent the landscape as a whole (Bintliff 2000). Another gap that is worth recalling here was identified in the need to collect and incorporate within the archaeological record a fuller range of information that would facilitate a better understanding of geomorphological and biological transformations in the landscape across time, including fauna and wild flora as well domesticated animals and the evidence of agriculture.

At about the same time I and my colleagues at the University of Siena started a critical review of the long-lasting activity undertaken by the staff and students of the university's Department of Archaeology. Over the previous two decades and more from the late 1970s the Department's specialists in Medieval Archaeology had under the leadership of late Prof Riccardo Francovich undertaken an ambitious program of research aimed at the systematic mapping, in the *longue durée*, of the archaeology of Tuscany. After around 25 years of intensive research a critical appraisal of the results and the underlying research framework was definitely needed. The starting point for this review lay in some basic quantitative data on the chronological and geographical distribution of archaeological evidence revealed up to that date. Analysis of the 20,000 or so archaeological 'sites' entered into the database up to that stage showed that within Tuscany between 90 and 95 % of the all the evidence related to the timespan between the sixth century BC and the sixth century AD. Prehistory, the Iron Age and the early Middle Ages were poorly represented within the archaeological record assembled up to that time. A broadening of the geographical perspective revealed much the same result for other intensively studied parts of Italy such as Puglia and Lazio. Similar observations have been made about the Middle Ages in Greece (Bintliff 2000).

Another interesting pattern emerged from the survey work in Tuscany and from related excavation projects was the seemingly systematic abandonment of lowland at the end of Late Antiquity and the formation during the Early Middle Age of hilltop villages. A long process then showed the almost systematic transformation of hilltop villages into the castles which have played such a central role within the medieval and even present-day settlement patterns of the region. Both trends could be seen as representative of past patterns of settlement while at the same time being potentially biased by some underlying some strategic issues such as the methodological framework of the research, the material culture under examination and the environmental context of the time under review, and so forth. Our review started with a consideration of the quantitative impact of different sources of information, which showed the following pattern: 75 % from field-walking survey, 10 % from vertical aerial photography, 11 % from existing archaeological knowledge, 4 %

from documentary sources. The impact of field-walking survey was clearly extremely high, not in any sense a surprise but carrying with it both local and intrinsic limitations that constituted potentially distorting factors in the collected data and in the interpretations derived from them. A potentially important source of distortion, in our view, was the relatively limited consideration given up to that stage within the scientific community to the question of 'archaeological visibility' in the detectable remains of material culture.

Material culture, of course, changes over time, as do the materials that were is use at various stages in the past. The physical manifestation of settlements, communication systems and agricultural patterns can sometimes be substantial and long-lasting in character, at other times modest and ephemeral. The trappings of everyday life in one period may be highly durable, with well-made and hard-fired pottery for instance, but such things may have been radically different in the preceding or following phases, with the use of poor-quality clay, inadequate firing or even the abandonment of durable materials such as stone or pottery in favour of wood or other perishable materials. Different cultures may therefore present differing levels of archaeological visibility when viewed through the medium of material brought to the surface by ploughing or other forms of disturbance. The less intensive incidence of one culture in a particular context, and the less durable character of the materials used, can also give rise to fundamental difficulties in the archaeologist's recognition of crucial pieces of evidence. We noticed in our review that there was a clear correlation between those periods known to be characterized in the main by a less durable material culture (some Prehistoric phases, Proto-history and Medieval times) and seemingly poor results from field-walking survey. Moreover, our reflection on these issues led us to the conclusion that field-walking survey and artifact collection could be generalized as being relatively efficient for the identification of so-called 'positive evidence' (such as stone or brick walls, concrete floors, tiled roofs or other structural elements made of durable materials) but by contrast as virtually worthless for detecting 'negative evidence' (such as ditches, pits, postholes or features made of perishable materials). It was equally easy to show that there was a clear correlation between periods characterized by cultures producing mostly 'negative evidence' and those producing only weak or absent evidence in field-walking survey (Campana 2009).

Another obvious problem attracted our attention, as well as that of a number of eminent authors: the unbalanced geographical distribution of the archaeological evidence derived from field-walking survey. Fernand Braudel was perhaps the first to complain about this, in his monumental work on the Mediterranean (Braudel 1949). When based on this method the geographical coverage of landscape investigations is fragmentary and tends to focus on particular aspects to the exclusion of others. Some rural areas are not investigated on any significant scale and the technique is not ideal for application in pastureland, or in mountainous country and woodland (which often overlap, of course). These areas, therefore, were generally not studied as a whole but only locally, and infrequently, by archaeological excavation. As a result, in archaeological maps and landscape studies, woodland as well mountain and pastureland are apparently 'empty' or at the very least to demonstrate human activity that appears severely limited in its scale. As a result lowland areas have tended to dominate most analyses of Mediterranean landscapes and history (Barker 1995). For the open countryside this imbalance can be traced back to the key data-recovery method upon which much of the discussion has been based—field-walking survey. Since this technique relies on the capacity to collect material remains from the ground surface it is strongly influenced, geographically, by present-day land-use. As a result the source-data for archaeological and historical interpretation has come largely from arable land, and hence predominantly from low-lying parts of the landscape. Higher land, around the Mediterranean as elsewhere in Europe, is mostly given over to pasture or woodland, both of which are less responsive to field-walking survey and therefore less studied by this (or any other) method. This despite the fact that about 50 % of the north-Mediterranean landmass falls into these categories of present-day land use (FAO 2006).

Starting a 'Third Wave' of Landscape Survey

Around the turn of the millennium impatience with this situation, combined with discussion within the academic community about possible new approaches, were matched by a general improvement within the hard sciences and what I would define a 'second loss of innocence' in archaeology. I am alluding here to the awareness that every time archaeologists have pioneered a new technique or taken advantage of new technology (in collaboration, of course, with other branches of science) there has been a real breakthrough. A good example can be drawn from the field of geophysical prospection. This technique improved dramatically in the 1990s but the authentic revolution only came about in came in more recent years with the application of very large-scale geophysical prospection in both landscape and urban contexts. It is interesting to note that the jump in scale was initiated by archaeologists. Before the introduction of the large-scale devices of the present day the pioneers had relied on simpler equipment developed is essence for prospection on relatively small sites or parts of sites, principally within urban areas. Simon Keay, Martin Millett, and Frank Vermeulen, among others, played a primary role within the Mediterranean area. Dominic Powlesland, Chris Gaffney and Wolfgang Neubauer pioneered the same process in the UK and in central Europe. It matters little who did what first, or better. The key point is that the questions and needs of archaeologists are different from those of hard scientists, though the support of the latter is—and always has been—essential in the development of these new techniques.

Going back to the beginnings of what we might call this 'third wave' of archaeological survey, a substantial role has been played by refinements in remote sensing (Johnson and Millett 2013) and in new approaches to the application of GIS (Wheatley and Gillings 2002). As a consequence of these advances there has been a significant upsurge in the use of non-destructive survey, often (it is interesting to note) in a slightly unusual part of the rural countryside—the formerly urban context of 'vanished' historical towns. From the very first results it became clear that the opportunities and potential gains offered by these new techniques were enormous. Large and complex once-urban sites, previously studied for their monumental importance and historical or artistic value through field-walking survey, surface collection and exploratory or targeted excavation, could now be studied in the first instance through geophysical prospection, sometimes revealing the entire plan of the town before any intrusive method of investigation was put in hand. This was a truly significant revolution, allowing archaeologists to address specific questions in a way that had not been possible before. Unsurprisingly, important improvement in the understanding of urbanism followed many of these surveys. In particular the urbanism of the Roman Empire benefitted hugely from the integration of remote sensing methodologies in partnership with GIS-based archaeological mapping and of course field-walking survey, artefact collection and excavation. An important contribution has also been made in a variety of cases by aerial photography, both from targeted exploratory flights and through the analysis of 'historical' photographs already available in regional and national archives (Musson et al. 2013). The combined application of these essentially non-destructive techniques has greatly enhanced our knowledge of the scale, structure and chronology of specific buildings and the overall urban infrastructure within formerly urban contexts, allowing us to look at the wider phenomenon of urbanism from a valid and comparative viewpoint (Vermeulen et al. 2012).

It would be worthwhile here to provide a general overview of the incidence of large-scale geophysical surveys on urban contexts, as implemented so far within the Mediterranean area and other parts of Europe. The summary presented in Fig. 1 has been compiled through a systematic scanning of relevant national and international journals and publications in the fields of archaeological prospection, urbanism and topographic studies.[1]

It does not, of course, claim to be a comprehensive representation of the full geographical spread of survey-based research in urban studies. No doubt there are omissions, in some cases perhaps important ones, but the picture is probably a fair representation of the current state of affairs.

Even a cursory glance at the map leaves one clear impression: the application of this approach above all around the Mediterranean Sea but less so within continental Europe and the UK. However, to extract a general meaning the present sample would first need to be made more comprehensive and then to be normalised. But that is not our goal here. It is more relevant at this stage to focus on a different point: the chronological range of the targets explored. A substantial proportion of the survey work (over 85 %) has been aimed at the exploration of Roman towns and cities, with less than 10 % devoted to deeply stratified contexts across a wider

[1]The search also benefitted from personal contact with colleagues in Europe and in the United States through comments on my initial draft. This generous collaboration allowed me to add information that substantially enriched the survey results.

Fig. 1 Distribution around the Mediterranean Sea, in Continental Europe and in the UK of large-scale geophysical prospection in formerly urban contexts. The size of the *circles* is proportional to the size of the survey in hectares

chronological range, from the Iron Age to Late Antiquity or even beyond. Several reasons might perhaps explain why Roman cities are so highly represented, not least the fact that Roman society was primarily an urban phenomenon: leaving aside the megalopolis of Rome, there existed in Italy as a whole about 430 other urban centres, in addition to thousands more around the Mediterranean Sea (Launaro 2014). An additional factor may be that the buildings and street systems of Roman cities are inherently likely to produce good results from geophysical survey.

Accepting that the implementation of integrated remote sensing surveys has for the most part benefitted our understanding of abandoned *urban* contexts, it might be worth asking what has happened in *rural* contexts. Has the same improvements in technique and strategy had an equal impact in the open countryside? Generally speaking the answer is no, or at least not yet. It is true that some of the shortcomings and biases identified by scholars in the intensive surveys of the 'second wave' have been resolved or at least partially overcome. For instance most of these projects have incorporated the systematic use of spatial technologies: it is very unusual to hear or read about landscape studies that do not use GIS platforms to manage spatial aspects of the collected information, along with handheld or machine-mounted GPS devices to navigate their way around the landscape and to accurately map the fieldwork evidence. Over the past two decades, for instance, the systematic application of spatial technologies has predictably brought about a substantial improvement in the speed and documentation of field-walking survey, leading the discipline to a qualitatively higher level of implementation and utility (Campana and Francovich 2007). The collection of environmental data is also more common, though it cannot yet be considered as standard practice. The importance and value of the introduction alongside traditional landscape survey of geo-archaeological and bio-archaeological studies have been fully demonstrated in a number of important publications (see, for instance, Arnoldus and Citter 2007; Goethals et al. 2009; Vaccaro et al. 2013).

However, despite some obvious progress along these lines we must still acknowledge that there is still a wide gap between the general approach and intensity of research work focused on rural contexts as compared with formerly urban environments. This gap has a direct influence on our understanding of the past and of transformations across time, artificially divorcing city and countryside from one another and obscuring or denying their inherently symbiotic relationship. I would emphasise here that this methodological gap has even widened in the last decade. A clear example of this statement could be seen by comparing the implementation of non-destructive techniques in urban contexts (Fig. 1) as against those undertaken in the open countryside (Fig. 2) in terms of the incidence of large-scale or long-lasting projects of geophyysical prospection.

Another example that demonstrates the imbalance in the implementation of 'third-wave' survey revolves around the geographical focus within different parts of the countryside. Thus, as noted above, earlier waves of rural survey have been completely powerless to produce data for woodland areas, leaving a massive gap in our capacity to achieve a balanced understanding of the past. Around the beginning of the present millennium, however, the advent of airborne laser scanning (LiDAR or ALS) introduced a completely new opportunity to detect and map archaeological features and landscapes previously hidden beneath the woodland canopy. In the last ten years a fair number of LiDAR surveys have been implemented for archaeological purposes and it is now widely accepted that this technique represents the most efficient system for the exploration of wooded areas, and in some instances of open pastureland too. Within the framework of my own research it seemed worthwhile to implement a fairly detailed survey of projects making active use of LiDAR survey around the Mediterranean Sea and across Europe more generally. The result shows some interesting trends and some equally obvious limitations in the application of LiDAR technology within archaeology. Figure 3 shows the geographical distribution of 37 published case studies which have made use of LiDAR data. The virtual absence of experience in the Mediterranean area is both obvious and remarkable. The main reason may lie in differing methodological approaches, differing schools of thought or in cultural or environmental biases of various kinds. Some limited attempts have been made at the use of LiDAR survey in Italy but they have for the most part been fairly unsuccessful, usually (it has been argued) because of the quality and density of the Mediterranean woodland canopy

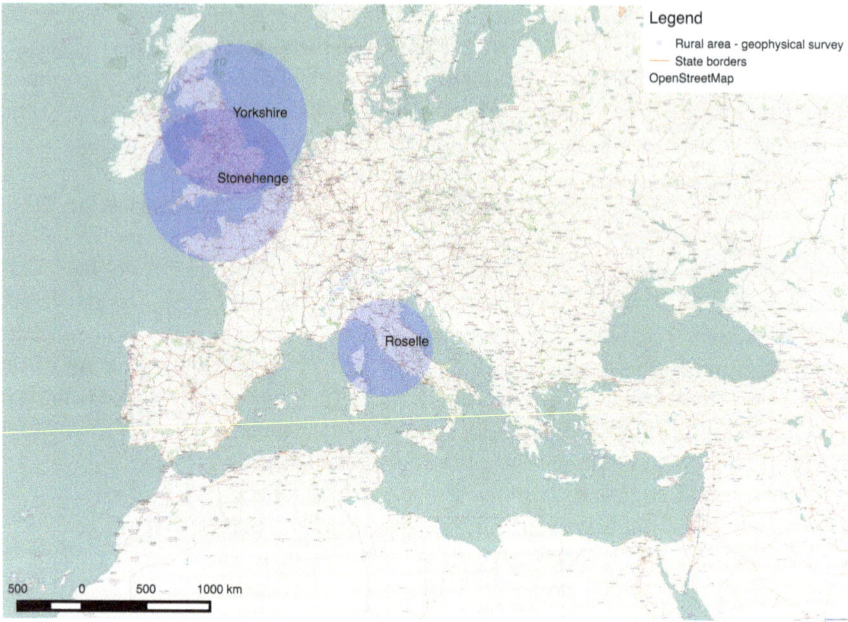

Fig. 2 The distribution around the Mediterranean Sea, in Continental Europe and in the UK of large-scale geophysical prospection in rural contexts, consisting of only two case studies in Britain and one in Italy. The size of the *circles* is proportional to the size of the survey areas in hectares

and scrub. Up until very recently, indeed, airborne LiDAR survey has proved substantially unproductive in Mediterranean contexts.[2]

The Emptyscapes Project: Intensifying 'Third-Wave' Survey in the Mediterranean Countryside

Emptyscapes is a two-year research project initiated by the present writer in July 2014 at the University of Cambridge and funded by the European Union under the Marie Curie scheme. However, the archaeological questions, methodological issues and research activities at the core of the project are deeply rooted in the work carried out over the previous two decades at the University of Siena, as well as in the context of wide-ranging international and academic partnerships over a substantial number of years (Campana and Forte 2001; Campana and Francovich 2003; Musson et al. 2013; Campana 2009). Moreover, in Italy as well in the

[2]New perspectives will probably be opened up in the near future through the development of lightweight LiDAR sensors that can be mounted on drones for very intensive survey work over woodland areas. See also Footnote 4 below.

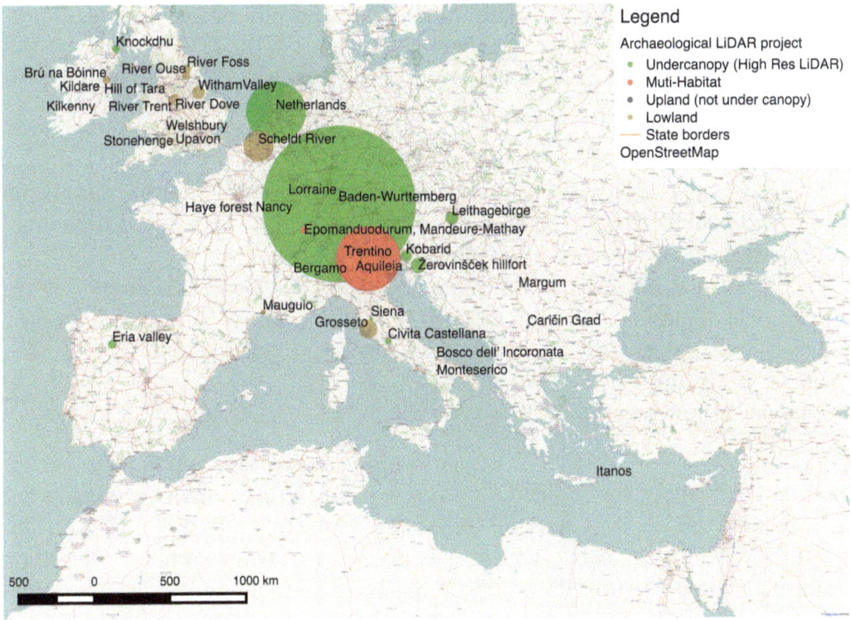

Fig. 3 The distribution around the Mediterranean Sea, in Continental Europe and in the UK of LiDAR prospection within archaeological projects. The size of the *circles* is proportional to the size of the survey in Km^2

Mediterranean more generally, this project forms one of the first attempts to apply a holistic approach encompassing many different methodologies to reconstruct the linked evolution of cultural and environmental landscapes within a Braudelian 'longue durée', spanning from late prehistory to the medieval period. This research project could perhaps be seen as an 'outrider' for what could be achieved in the Italian countryside by drawing upon approaches most extensively used up to now in the UK and in some parts of continental Europe, thereby encouraging more sophisticated approaches to landscape archaeology in the Mediterranean area.

Assuming success in the Emptyscapes project the resultant picture should be less about 'sites' than about a populated landscape in the social, economic and environmental context, including field systems, communication routes, trade networks and industrial and agricultural foci in addition to domestic settlements—all of these necessary to underpin the 'monumental' sites that have so far dominated the archaeological record. The project is aimed at finding a new balance—assuming there was ever a real balance in the past—between 'site' and 'off-site' archaeology, bridging the two categories or expanding the concept of 'site' to what might be called a 'archaeological *continuum*', a block of landscape perhaps varying in size from one area and time-period to another. Moreover, the extent and location of the landscape under investigation, and of any such blocks within it, will obviously be influenced by theoretical approaches, practical considerations and specific archaeological questions (on these see Powlesland 2009).

That said, within this context the Emptyscapes research strategy falls into four interlinked categories:

1. So called 'traditional' approaches, essentially based on the examination of archaeological literature, documentary sources, epigraphic sources, place-name evidence, iconography, technical, historical and thematic maps, geomorphology, field-walking survey and aerial photography.
2. Environmental studies based on geo-archaeology and bio-archaeology analyses.
3. New techniques in the form of high-precision, high-speed, large-scale geophysical survey and the collection and analysis of high-resolution LiDAR data aimed penetrating the woodland canopy.
4. Minimalist test-excavations.

Geographically, the project focuses on two sample areas in Central Italy (Fig. 3): the rural landscape between the hilltop Etruscan and Roman town of Rusellae and the Medieval town of Grosseto, in South Tuscany; and the now-rural but once-urban landscape of the ancient city of Veii in Central Latium, near Rome (Campana 2016). The Rusellae sample area can serve as a useful example to show the potential contribution of this 'third wave' approach to landscape studies.

The *Rusellae* Landscape

Rusellae was an important Etruscan and subsequently Roman town, which survived until the Middle Ages before finally being abandoned on 12th century. However, generic evidence of anthropic activity appears in the surrounding landscape around *Rusellae* from the upper Palaeolithic onwards. From the Chalcolithic there are the first signs of settlement in the area of the future city, probably attracted by the local mineral resources and the favourable position close to the docking and fishing opportunities of the *Prile* lake (Fig. 5). By the end of the Chalcolithic a hillfort was already in existence at Poggio di Moscona. The Bronze Age saw population growth and an increase in trade and socio-economic distinction, and from the start of the Iron Age there is increasingly clear evidence of villages; these were eventually abandoned, probably because of the expansion of *Vetulonium* on the opposite side of the *Prile* lake, a few kilometres west of *Rusellae* and a little later in its foundation (Fig. 4).

In the later phase of the Villanovian period the two hills of *Rusellae* seem to have been occupied by different groups, probably separated from one another by an area of pasture and open areas. Starting from the Orientalizing Period in around the mid 7th century there appear to have been city walls, a transformation in the topographic layout of the city and a generalised phase of public and private building activity. The process was continued and reinforced during the Archaic period, with

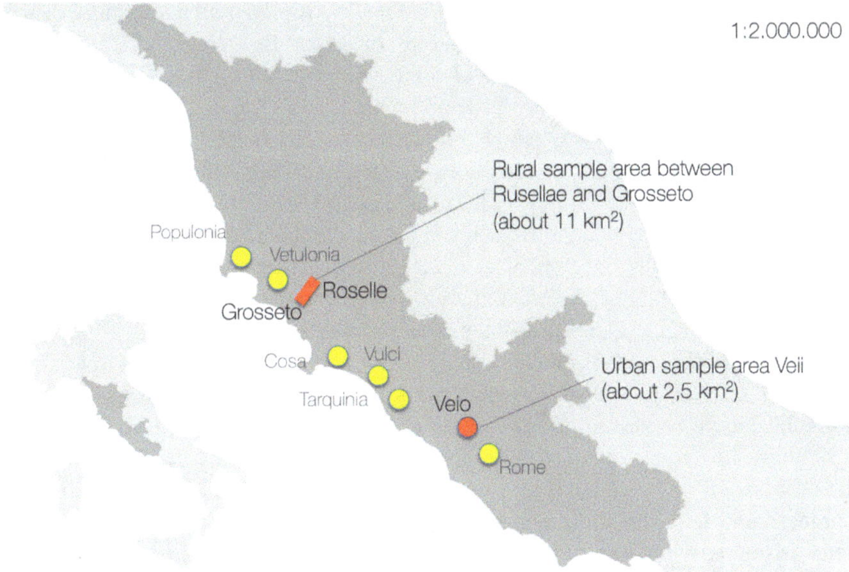

Fig. 4 Representation of the research areas, with *yellow circles* for other relevant sites

a general growth of the city and the progressive development in the surrounding landscape of a network of dispersed settlements which were probably related to agricultural production (from the outset the main vocation of the area, along with trade). There also grew up a network of roads and other communication, no doubt also exploiting the waterway of the Ombrone river to gain ready access to inland Etruria. In 294 BC the Etruscan city of *Rusellae* was conquered by the Romans and from the end of 3rd to the middle of the 1st century BC there ensued a long process of further building activity within the city. Roughly at the same period, *Vetulonium*, after an initial recovery from capture by the Romans, shrank to a secondary centre under the Empire, never to really recover. In the *Rusellae* area, as in the rest of Etruria, later centuries saw a major restructuring of the landscape, introducing Roman villa settlement and productive systems aimed at improving agricultural productivity. In the 1st century BC *Rusellae* was designated as a Roman Colony and from the 1st century AD there began a major phase of building activity, both within the city and in the surrounding landscape: forum, amphitheatre, temple, *domus*, further Roman villas in the countryside and maybe a pattern of centuriation (Nicosia and Poggesi 1998).

During the 4th century AD there is clear evidence of a further transformation including the conversion of public buildings into workshops and the abandonment and subsequent redevelopment of the public baths as a church. In the countryside, the pattern developed during the late Republican and the early Imperial phases fell into crisis during the Antoninian Age in the middle and later part of the 3rd century AD. From the 4th up to the mid 6th century AD a slight recovery is visible in the revival of a number of Roman villas, both close to *Rusellae* and in the more distant hinterland: at Aiali, Sterpeto and Casette di Mota for instance (Citter 2007; Campana 2009; Vaccaro 2012). It is also interesting to note that the vicinity of the future medieval city of Grosseto shows an increase in archaeological evidence from the 5th century AD onwards while the 'connective tissue' of fields and communication routes within the countryside was progressively transformed from the Roman structure to take on new patterns and scenarios (Sebastiani 2015).

However, *Rusellae* maintained its role as an administrative centre with a complex urban topography. Documentary evidence shows that from at least 499 AD the bishopric had its seat at *Rusellae*, remaining there until a move to Grosseto in 1138 AD (Celuzza 2011). The shift of the area's main urban centre and bishopric to Grosseto does not appear to have been a unitary and linear process and it would be extremely interesting to analyse whether any evidence of the process has been left in the landscape between these two central places: changes in the settlement and/or agricultural patterns, transformation and road systems, for instance. Nonetheless, the process will play a central role within the early Middle Age, involving first the Lombard and later the Carolingian lordships, the Papacy and a number of prominent aristocratic families, in particular the Aldobrandeschi. Evidence of urban activity in *Rusellae* is attested up to the 10–11th centuries, mainly in improvements to the fortifications. In the meantime in Grosseto, though still within a 'secondary' role, recent archaeological excavations have demonstrated a progressive development of the settlement area from the 9th to the 11th century, providing conditions for the transfer of the bishopric during the first part of the 12th century (Citter 2007). During this same general period, in this area as in the rest of Tuscany, there took place the process of 'incastellamento', the widespread development of hilltop villages. However, as Vaccaro recently emphasised: '*it should be stressed that in this area the population patterns of the 8th-early 10th centuries are extremely complex, diverse and rich in variants, to a much greater extent than hitherto thought*' (Vaccaro 2012).

Field Survey, Aerial Photography and the Detection of Archaeological Evidence

As noted in the introductory discussion, the University of Siena has from the late 1970s onwards fostered a systematic programme of landscape and archaeological investigation within the Maremma area.[3] As a result of this work the area now has a

---

[3]The first research project was established by Prof A. Carandini under the title 'Ager Cosanus-Valle dell'Albegna' (Carandini and Cambi 2002). The same period saw the start of work by the late Prof R. Francovich, surveying Grosseto, Scarlino and the area of the Colline Metallifere (Francovich 1985). In more recent decades archaeological mapping and field survey has been continued, particularly by the author (Campana et al. 2005; Campana and Piro 2009) as well as by Bianchi (Bianchi 2015), Arnoldus and Citter (2007) and Vaccaro (2012).

substantial database and GIS, developed mainly through the examination and analysis of the relevant archaeological and ancient literature, documentary and epigraphic sources, place-name evidence and systematic field-walking survey, along with a significant number of open-area excavations.

After about 35 years of rigorous research work it could be argued that this region is among the most intensively studied areas within the Mediterranean. However, despite the large amount of information assembled and examined over the years, it is undeniable that many important archaeological questions still remain unresolved. These may perhaps be categorised under two main keywords: ‘empty spaces’ and ‘empty phases’. For instance it quickly becomes obvious that the archaeological record assembled so far shows some very clear gaps both in space and time.

As in Tuscany as a whole, in this area too around 95 % of the evidence so far recovered relates to the time span from the 6th century BC to the 6th century AD, Prehistory, the Iron Age and the Middle Ages being poorly represented in the present archaeological record. Moreover, if we look again at the 1:100,000 distribution map in Fig. 4 and switch our attention from ‘sites’ represented as dots to the overall ‘background’ it is easy to appreciate how the ‘empty spaces’ predominate. This realisation raises a crucial question: what are we missing? The answer is before our very eyes: the landscape and its transformations across time. In an attempt to answer the conundrum of the ‘empty phases’ and to fill what at present appear to be ‘empty spaces’, the Emptyscapes project has been making an intensive study of a sample transect of the landscape between Grosseto and *Rusellae*, the size and location of which promises to fit these objectives. The area now under study is the mid-blue trapezoidal area in Figs. 5 and 6, placed so as to provide a diachronic perspective and to answer some major questions related to the three major cities of *Rusellae*, Grosseto and *Vetulonium*: for instance, the urbanisation of the Etruscan cities and in particular their relationship with the surrounding countryside, the Romanization of the cities and the landscape, the process of Christianisation and finally the impact on settlement, society and economy of the end of the Antiquity and the beginning of a new era, that of the Middle Ages. It should be clear, even from this short introduction that the area around *Rusellae* has the potential to illustrate the human and massive landscape transformations that characterise the many centuries of the 1st and early 2nd millennia AD.

In the attempt to improve our understanding of the sample area we have so far collected around 400 ha of geophysical survey (370 ha of magnetic and 30 ha of resistivity data); moreover, from as long ago as 2001 we have been undertaking exploratory aerial survey in the area and more recently we have commenced fieldwork with the purpose of collecting archaeological, geo-archaeological and bio-archaeological evidence within the chosen transect.[4]

---

[4]It should be noted here that in the spring of 2016 a sample area of the woodland was used as a test-bed for the effectiveness of a high-end Riegl lightweight LiDAR system mounted on an UAV. The collected datasets are currently under processing and interpretation. A detailed report of the outcomes will be published once the results have been properly assessed.

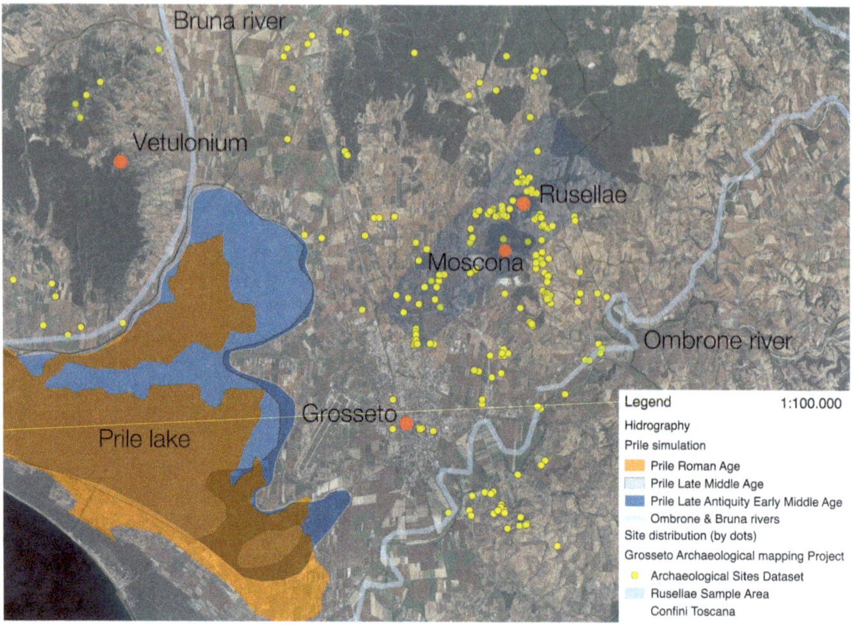

Fig. 5 Archaeological distribution map of the Rusellae area, showing the results of three decades of archaeological survey work and mapping. The gradually reducing extent of Lake Prile, which shrank in size from the first millennium BC up to the end of the Middle Ages, is shown in *solid colour*

At this point, if we wish to remain consistent with the critical approach set out earlier, especially as regards the limitations of the so-called 'second-wave' survey, we have to ask similar critical questions: after this amount of scientific effort have we in fact answered our research questions? Which new scenarios have been opened up? Which new questions are we now able to ask? Has our understanding of archaeological and landscape transformations within the sample area been substantially improved?

To answer these questions or at least provide a partial response, bearing in mind that the project is still in progress, the following paragraphs and illustrations will present some examples that will shed light on the present impact on our understanding of the area and the overall potential of the holistic approach developed and implemented within this study of a carefully chosen tract of ancient landscapes.

Let us start with a quantitative remark. The systematic examination of past archaeological research, documentary sources, epigraphic material, place-name evidence and historical maps, combined with a long-lasting programme of field-walking survey, has produced a substantial amount of information on the *Rusellae* area—including around 80 archaeological contexts of various kinds within the transect now undergoing more intensive study (Fig. 5). However, moving from

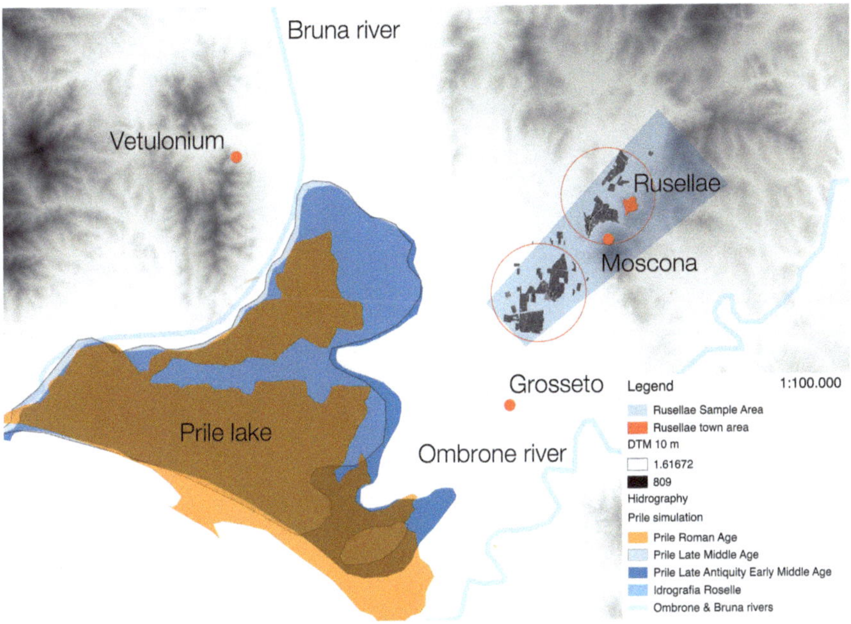

Fig. 6 Map of the Rusellae study area summarising (in *dark grey*) the present extent of large-scale contiguous geophysical survey (mainly magnetic but also some smaller areas of electrical resistance tomography, ERT) within the trapezoid sample transect. In the background map lowland is shown in *white* and upland in *mid-grey*

the 1:100,000 scale of Figs. 5 and 6 to a more detailed representation at 1:10,000 in Fig. 7, it is quite clear that even the most dense site concentration visible on 1:100,000 map displays large 'gaps' at the more detailed scale.

Caution is of course needed in making comparative quantifications of the results achieved by this 'traditional' research compared with the wider range of information that can now be collected through the use of sources such as remote sensing (mainly based on aerial photography) and magnetometer and electrical resistance survey. Nevertheless, the general increase in the 'visibility' of the archaeological evidence can be seen in the fact that these latter methods have so far produced 1886 previously undetected features within the sample transect. However, the aim here is not to make a simple comparison of numbers but rather to show that, taking all sources together, we have already collected a very substantial amount of information. With continuing survey and fieldwork we anticipate that this will soon reach what we might call a 'critical mass'. That said, we need to focus our attention not only on numbers but also on the *qualitative* contribution to the stated objective of holistic landscape interpretation. To illustrate the possibility of achieving that goal it might be useful to look at parts of the sample transect in closer detail, discussing some of the results achieved so far.

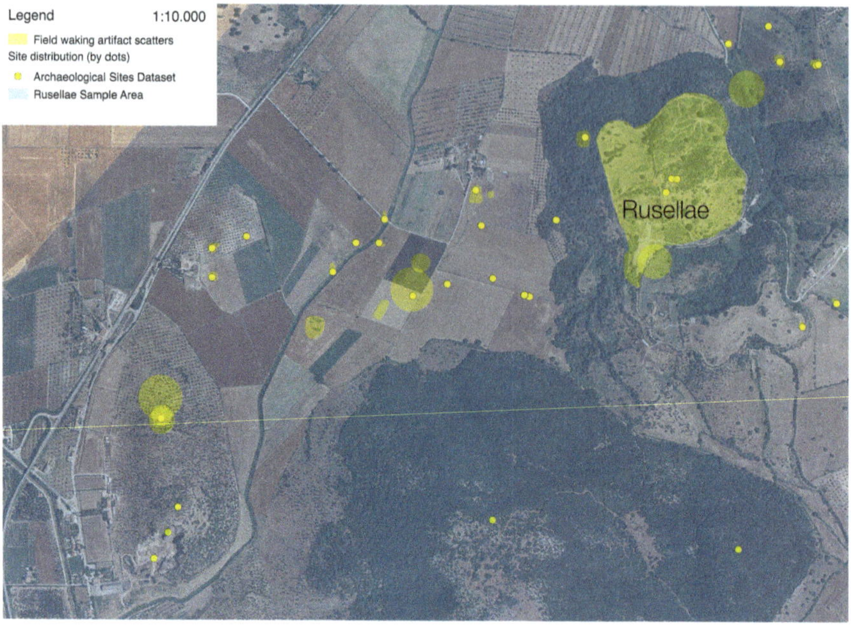

Fig. 7 Close up of the distribution map of sites detected by 'traditional' archaeological survey towards the north-eastern end of the sample transect, at a scale of 1:10,000

Archaeological Interpretation of the Medieval Landscape

The north-eastern part of the sample transect (Figs. 8, 9, 10, 11 and 12)—At present it is possible to recognize two main blocks of the sample transect within which we have so far been able to collect large-scale contiguous magnetic data, one in the south-west and the other to the north-east (Fig. 6). It will be useful to start the analysis in the north-eastern block. This is so close to the town of *Rusellae* that it could be viewed within the time range under consideration both as a suburban and as a rural area.

Aerial survey but especially magnetic prospection has thrown up entirely unexpected results. For instance, close below *Rusellae* itself, in an area of superficially undistinctive arable landscape, there is clearly visible in Figs. 8 and 9 a mass of magnetic features representing a major road connecting the countryside with the city; this is around 6 m wide at the bottom of the slope but known to expand to a width of 14 m wide as it approaches the city itself. Along both sides of this road the magnetic data shows a dense concentration of ring-ditches and rectangular anomalies that can without doubt be interpreted as burials, in effect the remains of a major cemetery probably dating to both the Etruscan and Roman periods. At present 34 ring-ditches and 37 rectangular anomalies have been recognized and mapped. The ring-ditches range from 13 to 43 m in diameter, with an

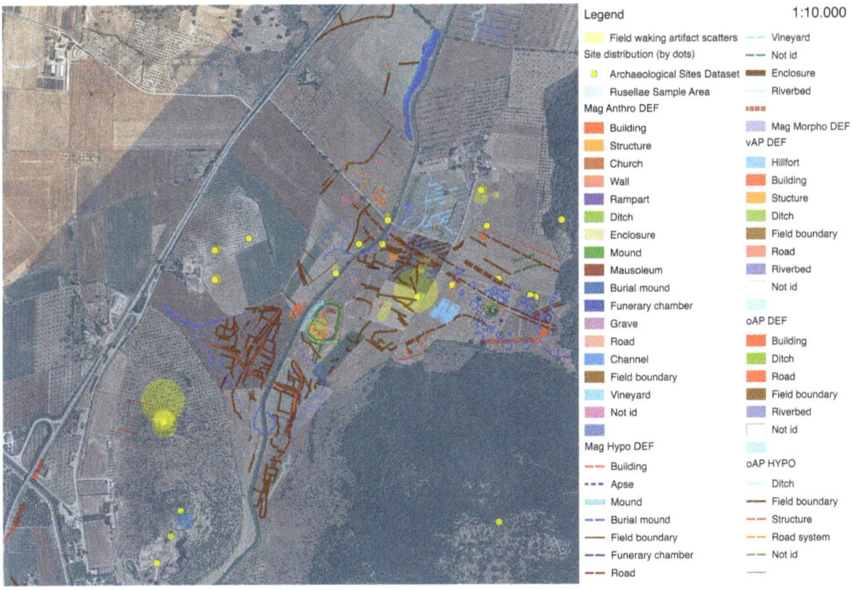

Fig. 8 Overlapping of the site distribution (from Fig. 6) and archaeological mapping of magnetic measurements (Mag) and oblique (oAP) and vertical (vAP) aerial photography towards the north-eastern end of the sample transect

average of 19 m, while the square features are more standardised at about 4 m by 6 m. On the basis of comparative studies on other Italian contexts such as Cerveteri (Tartara 2003: 157, 166) this is clearly a major and previously unsuspected funerary landscape placed along one of the main roads entering and leaving the city of *Rusellae*. Moreover on the southern (lower) edge of Fig. 9 it is possible to recognize another road and a peculiar structure showing as a round anomaly surrounded by a square of opposite magnetic polarity; the shape, articulation and size of this feature finds a convincing parallel in Roman mausolea (Johnson 2014).

If these interpretation were to be confirmed by excavation it would be possible to envisage an interesting scenario of a funerary landscape that with some slight changes might show continuity across a long time span from around the 7th to 6th centuries BC and on into the Roman period. This kind of continuity would be particularly interesting bearing in mind the various changes in the internal structure of the city between the Etruscan, Republican and Imperial phases of its development. Significantly, but surprisingly in the light of this very striking geophysical evidence, neither micro-morphological evidence nor field-walking survey in the past or in the summer of 2015 (apart from a very limited scatter of mainly off-site material) presented any interpretable evidence of this kind of road system or long-lasting funerary landscape.

Moving to a few hundred metres to the south-west within this part of the sample transect it is interesting to notice in Figs. 10, 11 and 12 how dataset continuity

Fig. 9 Geophysical data from the north-eastern part of the sample transect. From the *upper left corner*: general overview of the area; picture of the soil showing the extremely low concentration of artefact scatters; detail of the magnetic map showing the features interpreted as a possible mausoleum; general overview of the magnetic map displaying the large number of ring-ditches and square anomalies interpreted as a major cemetery lining either side of a road connecting the city of Rusellae to the surrounding countryside. The conjectured mausoleum appears at small scale in the *bottom left corner* of the *lower-right map*

corresponds to archaeological feature continuity in terms of the differing kinds of evidence relating to differing activities and diverse time frames. Indeed, the magnetic data shows a dense concentration of anomalies that can be readily interpreted as man-made functional elements and natural features within the local landscape: field systems, cultivation patterns, road systems, buildings, geomorphological features and so on. However, it may be profitable in the first instance to focus attention on a double-ditched enclosure alongside the river Salica, marked in green in Figs. 10, 11 and 12.

This feature was first identified in the magnetic data and then confirmed during field-walking survey. The magnetic data shows quite clearly a double-ditched enclosure with an internal area of about 0.8 ha and an overall area including the ditches of 1.63 ha. In the central enclosure it is possible to recognize several magnetic anomalies that coincide with artefact scatters, their size and shape suggesting interpretation as buildings. The larger size and east-west orientation of one of these, on the northern side of the enclosure, may well invite interpretation as a church. In this case field observation and artefact collection were very important in

Fig. 10 The mound double-ditched enclosure at La Canonica (in *green*). From the *upper left corner*: general overview of the area showing the magnetic measurements; ground-level view of the site morphology; mapping of the archaeological and geomorphological features including the double-ditched enclosure and its internal features along with geomorphology and the pattern of road and field systems

identifying a key feature of the site: a significant variation in elevation (of as much as 1.5 m) matching the features visible on the magnetic map. Putting the two types of evidence together the result is fairly obvious: an artificial mound, or alternatively a ditched enclosure occupying a natural area of slightly higher land in the local topography. Moreover, in the neighbourhood of the site, but mostly west of the present course of the river Salica, magnetic anomalies reveal a dense pattern of field boundaries, roads and palaeo-riverbeds.

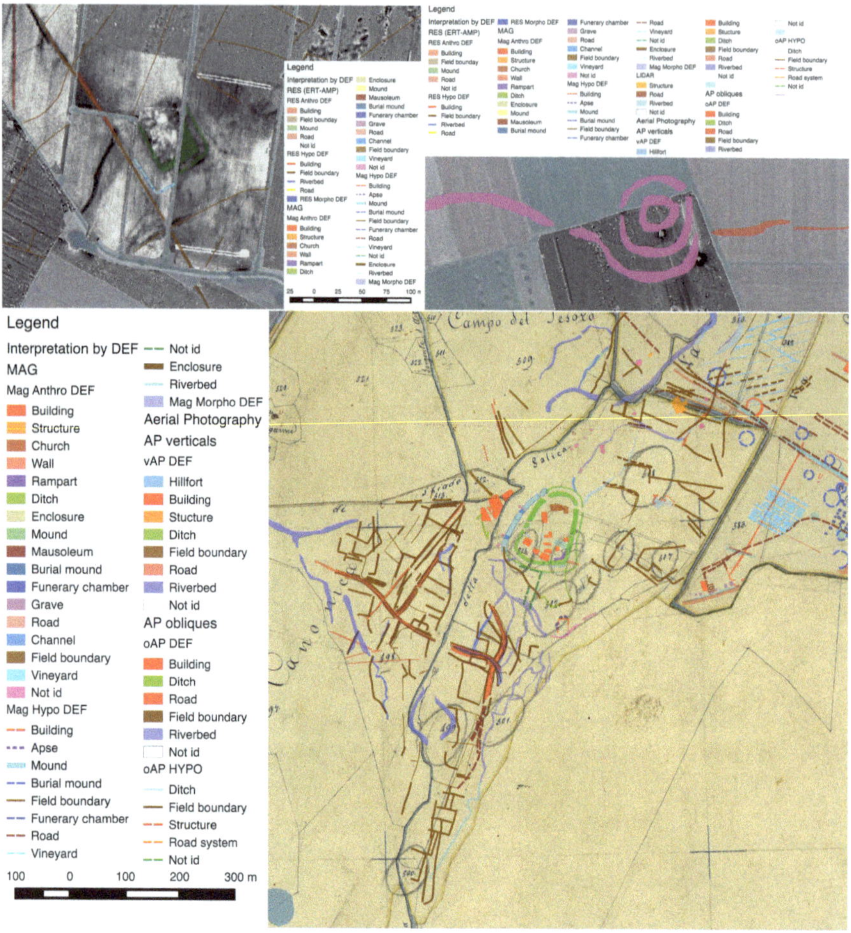

Fig. 11 Mounds and ditched settlements within the Rusellae sample transect. *Top left*, Brancalete, close to Aiali; *top right*, circular settlement; *bottom*, features mapped by archaeological interpretation of the aerial survey and magnetic data (ditched settlement, road and field systems) alongside the Salica river, superimposed on the historical cadastral map of 1817–1830

The outstanding character of the magnetic data and local topography prompted a decision to implement a borehole survey and an intensive programme of fieldwork based on artefact collection within a virtual grid of 10 m × 10 m ‘cells’.[5] This ‘targeted’ fieldwork was aimed at establishing the chronological range and function of the site and at providing a more detailed picture of the match between the magnetic measurements, micromorphology and artefact distribution. The survey

[5]For the method of artefact collection within a virtual grid see Campana (2005) and Campana and Francovich (2007).

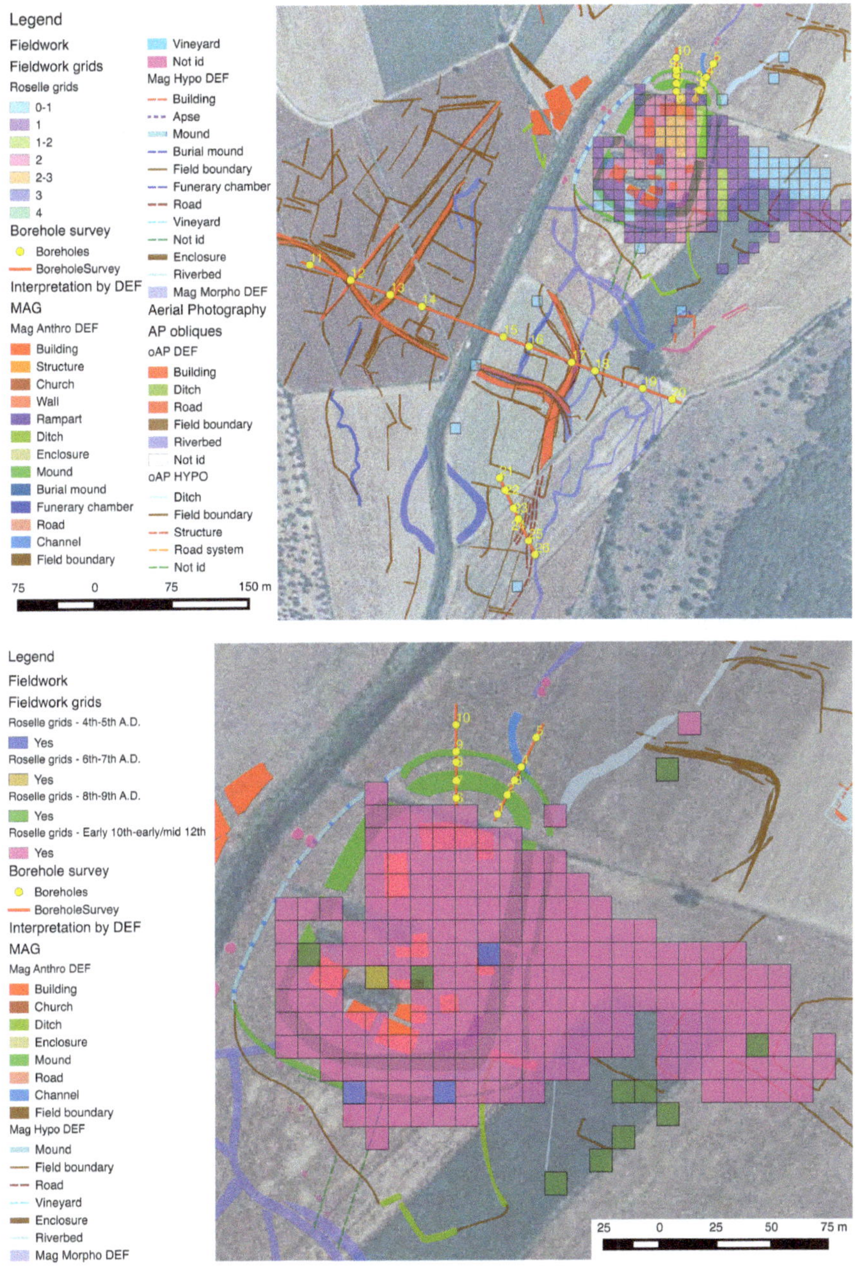

Fig. 12 The double ditched settlement at La Canonica. From the *top*: total density plot of surface artefacts in sherds per grid cell collected during field-walking and borehole survey; distribution of artefact scatters across time, from late antiquity to the late early 10th to early or mid 12th centuries AD

and the analysis of artefacts set out in Table 1 show a quite distinct pattern of intensive anthropic activity corresponding with the micro-morphological and magnetic evidence and deriving predominantly from the early 10th to early or mid 12th centuries AD. On the basis of comparative studies of shape, size, morphology, artefact assemblage and chronological range this site can confidently be interpreted as a medieval mound or ditched settlement that in significant respects resembles others identified in recent years within the landscape of the Grosseto lowland.[6]

Moreover, it is interesting to consider the possibility that the field and road systems in the immediate vicinity could be associated with the same cultural context and chronological range. It is noticeable that the parcels within the field system are characterized by a relatively consistent pattern of size, shape and boundary-type. The boundaries were clearly ditches which would have been used both to divide the land into functional sub-units and to provide indispensable drainage for arable land close against the river Salica.

The borehole survey undertaken in the summer of 2015 made it possible to distinguish stratigraphic layers, which could be cautiously associated with the filling of agricultural boundary ditches but unfortunately no dating evidence could be identified. Bureaucratic difficulties prevented implementation of the intended programme of test excavations but thanks to the support and collaboration of the Toscana Superintendency we expect to overcome these problems for the following years.[7] However, a comparative study based of size, shape and relationship of this group of features tends to support the interpretation as a medieval settlement and related field system. Among better-known case-studies in Italy, the Tavoliere in Puglia presents some very close parallels thanks to the effectiveness of aerial photography in that area: Casone San Severo, Motta della Regina, Masseria Petrullo and San Lorenzo in Carmignano all display a quite close resemblance in general appearance to our field system in terms of size, shape and overall pattern, and in some cases also to the shape of the settlements themselves (Guaitoli 2003: 106 119).

It is interesting to compare the field pattern under investigation with the cadastral map of the Grand Duchy of Tuscany surveyed in 1817–1830 (Fig. 11). Even at first glance it appears clear that the field patterns in the cadastral maps of this area (as well in the rest of the sample transect) are much closer to those of the present day, with almost no links with the patterns identified in the magnetic measurements.

As already mentioned above and illustrated in Fig. 11, two similar mounds or ditched enclosures have been located in recent years within the 25 km^2 area of the

[6]Within the *Rusellae* sample transect there are two other sites with the same basic characteristics, one close to Aiali and the other at Commendone. Along the coast in the area between Scarlino and Follonica two other medieval settlements of the same kind have been identified and are currently under investigation (Campana 2009; Campana et al. 2009; Marasco 2013).

[7]The present picture definitely needs to be improved by test excavation and further investigation, with particular regard to radiocarbon dating, pollen analysis and geo-archaeological survey for the medieval mound settlement and adjacent field system (is it an artificial mound or alternatively a ditched settlement taking advantage of an area of slightly higher land in the local topography?).

Table 1 Quantification and chronological distribution by grid cells of datable material recovered during artefact collection

Period	Number of cell grid with pottery
7th–6th B.C.	19
Mid 3rd–mid 1st B.C.	17
Late 1st B.C.–1st A.D.	10
2nd–3rd A.D.	1
4th–5th A.D.	3
6th–7th A.D.	3
8th–9th A.D.	12
Early 10th-early/mid 12th	218

sample transect, 10 km^2 of which consists largely of wooded upland and the rest of valley lowland, including 4 km^2 so far subjected to intensive survey. This represents a density of one such settlement for each 1.3 km^2 of the intensively surveyed area. It is accepted, of course, that a larger area of the landscape will need to be covered by this kind of survey before such a statistic begins to become truly meaningful. Nevertheless the identification of these lowland medieval settlements is significant in the sense that, as noted earlier in this discussion, over the past 40 years the archaeological development of Tuscany during this formative period of history has been intensively studied by excavation and traditional field-walking survey. A key concept resulting from this work has been the long-term development of hilltop villages (Francovich and Hodges 2003). Despite all of this work very few settlements of any kind had been identified in lowland Tuscany before 2005, and none of the type now coming to light in the *Rusellae* area (Campana et al. 2006; Vaccaro 2012).

The detailed reasons for this have already been discussed in the introduction to this article as well as elsewhere (Campana 2009). However, the discovery of this unexpected category of settlement is bound to stimulate discussion on how to integrate this new information into the historical concepts of 'incastellamento' in ways that will improve our understanding of landscape transformations in the centuries between Late Antiquity and the mature Middle Ages, not least in the interplay between the strength and strategies of the ruling classes and the continuing existence of functioning communities and settlement patterns within the Tuscan countryside (Bianchi 2015).[8]

Returning to the enclosure alongside the river Salica, the magnetic measurements reveal the general shape and size of the settlement, along with several conjectured buildings within it, while the artefacts from surface collection offer a chronological overview through the density and topographical distribution of the finds. On this basis we can envisage a new scenario for landscape transformation in this immediate area. The close match between the pottery distribution and the

[8]Recently, debate on this topic has been stimulated in Italy by the work of Settia et al. (2013) and Creighton (2012). Unfortunately much of the resulting discussion has been focused on the particularities of this kind of settlement rather than on the broader landscape as a cultural eco-system.

magnetic data for the ditches and interior of the enclosure prompts interpretation of the settlement as perhaps the result of at least two different processes. If we interpret the presence of few early medieval sherds as residual artefacts, the settlement could then be considered as a new foundation of the early 10th century rather than the outcome of a longer-term process of the kind envisaged within the 'incastellamento' model. In that case it would be possible to suppose that at this time the ruling classes invested resources in developing new settlements on the fertile lowland, perhaps moving the population from elsewhere. On the other hand, if—as the author would prefer—we interpret the presence of a small amount of early medieval pottery as deriving from a first phase of development initiated by some sort of community already living in or around this area, then the social and economic process could have been quite different. The answer to such speculation could only come from excavation, preferably on a fairly extensive scale.

However, the settlement and in particular the field system (assuming that their association can be confirmed by further investigation) illustrate an extraordinary vital stage of a society that had the capacity and/or need to reorganize settlement and landscape patterns, perhaps removing almost all vestiges of older patterns in the process. As a final remark in this context it is necessary to emphasize the complexity of the area under investigation. Past studies identified this as one of the most important areas within the Grosseto plain for agricultural production (Arnoldus and Citter 2007). However, the present phase of landscape study and the large-scale collection of remote-sensing data have produced clear evidence of a high level of hydrogeological instability in this part of the local landscape. Therefore, the creation of this new settlement and field system, whether financed initially by the ruling classes or undertaken of their own volition by an existing rural community, would have required advanced know-how of the local area, along with social resources in terms of labour and productive capacity, in order to fulfil the project in the first place and to keep it working as a viable social and productive concern over time.

The south-western part of the sample transect (Figs. 12 and 13)—Moving further to the south-west within the sample transect we encounter the second large block of intensive landscape survey (Figs. 6, 13 and 14). In this area, too, the quantitative results are remarkable. Previous surveys had identified 19 contexts (17 on-site and 2 off-site). Magnetic and electrical resistance prospection and aerial survey have now allowed us to detect a dense pattern of 883 features ranging chronologically from the Etruscan period to the Middle Ages and varying from settlements to field systems, enclosures, graves, road systems, geomorphology and so forth.

In this block a particularly important complex near Aiali was identified for the first time during the Aerial Archaeology Research School held at Siena in 2001 (Musson et al. 2013). Oblique aerial photography on that occasion revealed a large building complex that was subsequently further investigated through intensive artefact collection and various forms of geophysical survey including high resolution radar prospection to shed light on a massive rural settlement of about 4 ha in

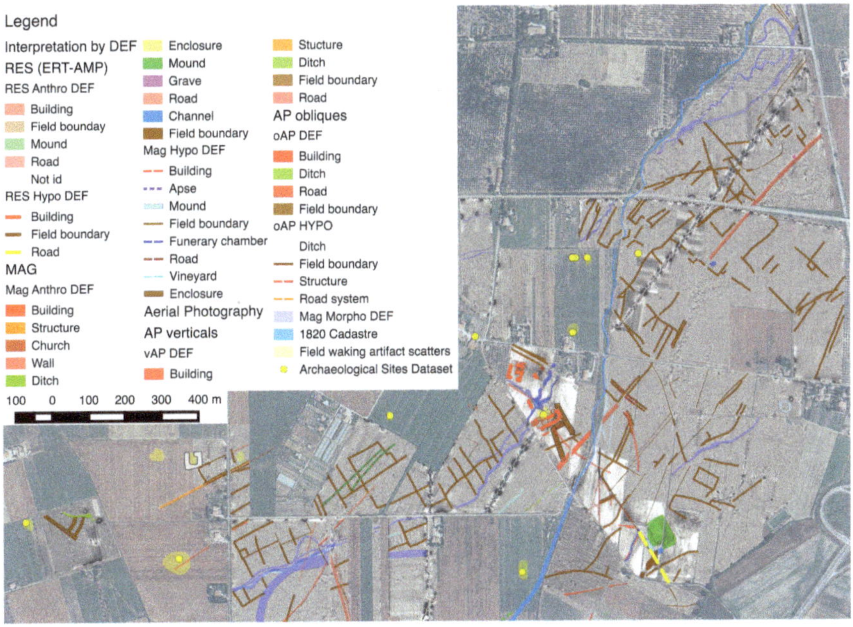

Fig. 13 The south-western part of the sample transect with *yellow dots* to show the results of 'traditional' archaeological investigation, superimposed on integrated remote sensing data and GIS-based data mapping. The *colours* show the various kinds of natural and archaeological features identified so far, with *red* for the site of the Roman villa complex and main road at Aiali. The medieval ditched settlement at Brancalete is shown in *green* alongside a branch road towards the *bottom right-hand corn*er of the figure. The quantitative and qualitative improvement in the landscape database is clearly visible, making it possible to ask, and eventually to answer, new archaeological questions about field systems, communication routes and settlement patterns as well as former water courses that may have conditioned the pattern of settlement and land-use within this local area

extent, including some open areas, covering a time range from the late Republic to the early Middle Age (Campana and Piro 2009).

The Roman complex, as illustrated in Fig. 13 and in closer detail in Fig. 14, lies either side of an apparent major road linking this area to *Rusellae* about 4 km away to the north-east. This is clearly an advantageous position within the landscape, along the Prile Lake, close to Rusellae and with the *mansio* at Hasta a further 18 km away to the south as the crow flies. Further significant features in this area include a clearly-organised and roughly grid-like field system oriented north-west to south-east. Artefact collection across the area has provided provisional dating for buildings and other features within the complex. The southern (and later) buildings in the complex (n.2 and 3 in Fig. 14), clearly parts of a villa complex, are oriented in keeping with the field system but the northern (and earlier) building (n.1) is not. There is also an early medieval building in much the same area (n.4) as well as an early medieval ditched enclosure, rectangular in this case, a few hundred metres to

Fig. 14 Extract from the south-western part of the sample transect. Close-up displaying the buildings of the Roman complex (*1*, *2* and *3*), the ancient field system (*6*, matching the orientation of Roman structures *2* and *3*), a medieval feature (*4*), and the Bracalete double-ditched medieval enclosure (*5*)

the south-east (green in Fig. 13; n.5 in Fig. 14). The ditched settlement is oriented in broad alignment with the field system, suggesting the possibility of some form of continuity in the organisation of the landscape from the Imperial Roman age to the early medieval period.

Within the field system the average size of the parcels is very close to the Roman *iugerum* or its multiples, but on the other hand the pattern is fairly uneven—as might perhaps have become the case if the system remained in use for many centuries after an original creation during the Roman period. Local historians have attempted in the past to identify a pattern of centuriation in this area but their hypothesis still awaits confirmation (Mazzolai 1960; Prisco 1998). At all events their supposed pattern does not match either the field system or other features described here. Admittedly, it is perhaps premature to claim, on the basis of orientation and parcel size, that the recently revealed fields represent a centuriation system set out in the mid Imperial age and remaining in use until early medieval times. But, when evidence of this kind has been revealed by 'third-wave' survey, historical question of this kind must be asked, and hopefully answered through further investigation. For the moment, however, the question of dating and derivation must remain unresolved, at least until test excavation or further remote sensing can provide a secure guide to the chronological range and possible extent of the field system so recently revealed.

Conclusions

These initial results of the Emptyscapes project, along with the earlier work by the University of Siena, clearly demonstrate that the kind of landscape investigation implemented within formerly urban contexts can be equally effective in the Mediterranean countryside, making the concept of exploring the archaeological *continuum* in such situations a tangible and achievable objective. The metaphor of 'waves' of archaeological survey seems to me quite effective in describing the history of studies in this field, in the sense that waves each have their own identity while at the same time being part of a broader pattern linked by reciprocal and sequential relationships. The final result, for archaeological survey, is not characterised only by the most recent (or even the next) wave but by a deeply integrated theoretical, methodological and archaeological sequence that encompasses all of the waves within the system.

The medieval landscape described in the body of this contribution provides an excellent example of the way in which, drawing together experience from different or successive 'waves', it is possible to enhance our understating by—in this case—recognising and dating settlements within a structured medieval landscape in lowland central Italy. The contribution cannot be looked upon as the mere addition of more dots on the map. These mounded or ditched settlements represent a quite unexpected form of lowland occupation within Tuscany, probably beginning to appear in the later part of the early Middle Age and continuing to proliferate during the central Middle Age (Settia et al. 2013). The excavation of the settlement at Scarlino in 2005 has tended to confirm this pattern (Campana et al. 2006; Marasco 2013).

As noted above, well-established archaeological concepts developed for this part of Tuscany in recent decades have pictured the lowland landscape as largely uninhabited (Francovich and Hodges 2003). We can now see that this was definitively not the case, and this in turn prompts new questions for future investigation. For instance, what kind of role was played by the ruling classes in the development and implementation of this kind of settlement structure and the related landscape? Was it perhaps, as suggested earlier in this discussion, mainly a product of initiative (or need) on the part of already existing rural communities? So it should be clear that, among major outcomes of this kind of survey work, new questions will arise and previous paradigms will need to be updated or even replaced. If we can attain the desirable critical mass of quantitative and qualitative data we will have created a new and effectiveness approach to the investigation and understanding of the landscape in southern Tuscany, and by extension perhaps in other parts of Italy and even beyond.

From the methodological point of view the sample area currently under investigation within the Emptyscapes project is quite small; additional research is clearly needed, for which new high-speed devices are now available (Campana and Dabas 2011; Neubauer et al. 2013). Nevertheless, the huge number of substantial pieces of evidence detected so far around *Rusellae* is without doubt highly impressive: a new form of aggregated medieval settlement, graves and burial mounds in a huge but previously unsuspected cemetery, field systems, an apparent mausoleum, a variety of different kind of buildings, a possible church, roads and other elements within a rural communication system—all of these have been revealed within quite a small area of superficially 'blank' landscape. Virtually none of these could have been securely identified through the traditional methods of survey and excavation that have dominated landscape archaeology in Italy until very recent times.

We now have the technology, the methodological framework and the enthusiasm to bring about a change in this situation, exactly matching that initiated by a select band of forward-looking archaeologists and landscape investigators in other parts of Europe where similar intensive approaches and methodologies have been applied. We must seize this opportunity, in Italy and the rest of the Mediterranean world, to adopt these methods and to challenge our present knowledge and understanding of rural landscape around the Mediterranean Sea.

Acknowledgments The research for this paper would not have been possible without the financial support of the Marie Curie action for the Emptyscapes project (FP7-PEOPLE-2013-IEF n. 628338) and the Culture 2007 ArchaeoLandscpes Europe project (Grant Agreement nr. 2010/1486/001-001). Stefano Campana carried on the research at the University of Cambridge as Fellow of the Faculty of Classics and fellow of the McDonald Institute for Archaeological Research. The author is particularly grateful to Prof Martin Millett, scientist-in-charge of the Marie Curie project, for his constant support, valuable advice and critical attitude. I would like to thank Ken Saito for his tenacity, rigour and outstanding work in the field and in the laboratory while gathering and processing hundreds of hectares of magnetic measurements in the *Rusellae* area and at *Veii*. The author is further indebted to Dr. Emanuele Vaccaro for artefact collection and pottery analysis. Grateful thanks also go to Prof Charles French for his support and guidance in geo-archaeological research, with particular regard to the organization of the borehole survey work in 2015.

Sincere thanks are also due to the Archaeological Superintendency of Tuscany, especially Dr. Andrea Pessina and Dr. Gabriella Poggesi, for their intellectual debate and problem-solving approach. The University of Siena spin-off company ATS srl played a crucial role, generously sharing use of the Foerster magnetometer system used in the survey work.

Special thanks are also due to two good friends who have followed and inspired so much of the writer's research work since early in his career, Chris Musson and Prof Dominic Powlesland. As ever, they have helped with constructive criticism and comments throughout all stages of the project. Grateful thanks also go to my colleagues of the University of Siena who contributed in a variety of ways in developing and fostering discussion, in particular Prof Giovanna Bianchi, Prof Emanuele Papi and Prof Gabriella Piccinni.

Finally, my thoughts turn inevitably to my mentor, the late Prof Riccardo Francovich, who gave me the cultural background and the intellectual vigour to face, time after time, new research challenges.

References

Alcock, S.E. & J.F. Cherry (ed.) 2004. *Side-by-Side Survey: Comparative Regional Studies in the Mediterranean*. Oxford: Oxbow Books.

Arnoldus A. & C. Citter 2007. Archeologia urbana a Grosseto. Origine e sviluppo di una città medievale nella "Toscana delle città deboli". Le ricerche 1997-2005, Vol.1, All'Insegna del Giglio: Firenze.

Barker, G., 1995. *A Mediterranean valley: landscape archaeology and annales history in the Biferno Valley*. Leicester: Leicester University Press.

Bianchi G., 2015. Analyzing Fragmentation in the Early Middle Ages: the Tuscan Model and the Countryside in Central-Northern Italy. In S. Gelichi, R. Hodges (eds.), New Directions in Early Medieval European Archaeology: Spain and Italy Compared. Turnhout: Brepols. 301-334.

Bintliff, J., 2000. Beyond dots on the map: future directions for surface artefact survey in Greece, in J. Bintliff, M. Kuna & N. Venclova (eds.) *The future of surface artefact survey in Europe*: 3-20. Sheffield: Sheffield Academic Press.

Bintliff, J., M. Kuna & N. Venclova (eds.) 2000. *The future of surface artefact survey in Europe*. Sheffield: Sheffield Academic Press.

Braudel, F., 1949. *La Méditerranée et le monde méditerranéen à l'époque de Philippe II*. Paris: Armand Colin.

Campana S., 2005. Mobile Computing in Archaeological Prospection, in proceedings of the 6th International Conference on Archaeological Prospection, (Roma 14-17 September 2005), Institute of Technologies Applied to Cultural Heritage, Roma, pp.427-431.

Campana S., 2009. Archaeological Site Detection and Mapping: some thoughts on differing scales of detail and archaeological 'non-visibility', in Campana S., S.Piro, *Seeing the unseen*, The Netherlands: Taylor & Francis. 5-26.

Campana, S. 2016. Filling gaps in space and time at Veii. 'Emptyscapes' project interim report, in O. Cerasuolo and J. Tabolli. *Veii – An Etruscan City*. Austin: University of Texas Press.

Campana, S. & M. Forte 2001. Remote Sensing in Archaeology, proceedings of the XI International School in Archaeology, (Certosa di Pontignano, 6-11 dicembre 1999). Florence: All'Insegna del Giglio.

Campana, S. & R. Francovich 2003. Landscape Archaeology in Tuscany: Cultural resource management, remotely sensed techniques, GIS based data integration and interpretation, in M. Forte (ed.), The Recontruction of Archaeological Landscapes through Digital Technologies. BAR International Series 1151. Oxford: Archaeopress. 15-28.

Campana S., Frezza B., Ghisleni M., Vaccaro E. 2005, Il popolamento tardoromano e altomedievale nella bassa valle dell'Ombrone. Progetto Carta Archeologica della Provincia di Grosseto, "Archeologia Medievale", All'Insegna del Giglio, Firenze, pp.461-480.

Campana S., R. Francovich & L. Marasco, 2006. Remote Sensing and Ground-truthing of a Medieval Mound (Tuscany - Italy), in S. Campana and M. Forte (eds.) *From Space to Place*, Proceeding of the 2nd International Conference Remote on Sensing Archaeology. Rome (4-7 December 2006). BAR International Series. Oxford: Archaeopress. 491-496.

Campana S. & R. Francovich 2007. Understanding archaeological landscapes: steps towards an improved integration of survey methods in the reconstruction of subsurface sites in South Tuscany, in J. Wiseman, F. El Baz (eds.), Remote Sensing in Archaeology. Boston MA: Springer. 239-261.

Campana S., Marasco L., Dabas M., Piro S., Zamuner D., 2009, Integration of remote sensing, geophysical surveys and archaeological excavation for the study of a medieval mound (Tuscany-Italy), Journal of Archaeological Prospection.

Campana S. & S. Piro, 2009. *Seeing the unseen. Geophysics and Landscape Archaeology*. The Netherlands. Taylor & Francis.

Campana S. & M. Dabas. 2011. *Archaeological Impact Assessment: The BREBEMI Project (Italy)*. Archaeological Prospection 18: 139-148.

Carandini A., Cambi F., 2002. Paesaggi d'Etruria. Valle dell'Albegna, Valle d'Oro, Valle del Chiarone, Valle del Tarone. Progetto di ricerca italo-britannico seguito allo scavo di Settefinistre, Roma.

Celuzza M.G., 2011. Roselle: la tarda antichità e il medioevo, in F. Nicosia, G. Poggesi (eds), *Roselle. Guida al parco archeologico*. Siena: Nuova Immagine. 43-48.

Cherry, J.F. 2003. Archaeology Beyond the Site: Regional Survey and Its Future, in J.K. Papadopoulos & R.M. Leventhal (eds.) *Theory and Practice in Mediterranean Archaeology: Old World and New World Perspectives*: 137-159. Los Angels: University of California.

Citter, C., 2007. *Archeologia urbana a Grosseto. Origine e sviluppo di una città medievale nella "Toscana delle città deboli"*. Le ricerche 1997-2005, Vol.2. Florence: All'Insegna del Giglio.

Crawford, O.G.S. 1927. Editorial notes. *Antiquity* 1: 1-4.

Creighton, O. 2012. *Early European Castles Aristocracy and Authority, AD 800-1200*. London: Bristol Classical Press.

FAO, 2006. *Global Forest Resources Assessment 2005*. FAO Forestry Paper 147. Rome: FAO Rome.

Francovich R. 1985. Scarlino I. Storia e Territorio, All'Insegna del Giglio, Firenze.

Francovich, R. & R. Hodges 2003. *Villa to Village*. London: Duckworth.

Francovich, R. & H. Patterson (eds.) 2000. *Extracting meaning from ploughsoil assemblages*. Oxford: Oxbow.

Gillings, M. 2000. The Utility of the GIS Approach in the collection, management, storage and analysis of Surface Survey Data, in J. Bintliff, M. Kuna & N. Venclova (eds.). *The future of surface artefact survey in Europe*: 105-120. Sheffield: Sheffield Academic Press.

Goethals T., M. De Dapper & F. Vermeulen 2009. Geo-archaeological implications of river and coastal dynamics at the Potenza river mouth (The Marches, Italy), in M. De Dapper, F. Vermeulen, S. Deprez & D. Taelman (eds.) *Ol' Man River: geoarchaeological aspects of rivers and river plains*. Ghent: Academia Press. 407-438.

Guaitoli M., 2003. *Lo sguardo di Icaro*. Roma: Campisano.

Ihde, C. 1993. *Postphenomenology: Essays in the Postmodern Context*. Illinois: Northwestern University Press.

Johnson, M.J., 2014. *The Roman Imperial Mausoleum in Late Antiquity*. Cambridge: Cambridge University Press.

Johnson, P. & M. Millett (eds.) 2013. *Archaeological Survey and the City*. Oxford: Oxbow Books.

Keay, S., M. Millett, L. Paroli & K. Strutt (eds.) 2005. *Portus. An Archaeological Survey of the Port of Imperial Rome*. Archaeological Monogrph of the british school at Rome: 15. Oxford: Alden Group.

Launaro, A. 2014. Survey archaeology in the Roman world (history, approach, methodology), in *Encyclopaedia of Global Archaeology*: 7146-7155. Berlin: Springer.

Marasco, L. 2013. La Castellina di Scarlino e le fortificazioni di terra nelle pianure costiere della Maremma settentionale, in A. Settia, L. Marasco & F. Saggioro (eds.), Fortificazioni di terra in Italia. Motte, Tumuli, Tumbe, Recinti. *Archeologia Medievale XL:* 57-68.

Mazzolai, A., 1960. *Roselle e il suo territorio*. Grosseto.

McDonald W.A. 1942. Where did Nestor live? American Journal of Archaeology 46: 538-545.

McDonald W.A. & G.R. Rapp (eds.) 1972. The Minnesota Messenia expedition: Recontructing a Bronze Age regional environment. Minneapolis: University of Minnesota press.

Musson, C., R. Palmer & S. Campana. 2013. Flights into the Past. Aerial photography, photo interpretation and mapping for archaeology. Available at: http://archiv.ub.uni-heidelberg.de/propylaeumdok/2009/1/flights_into_the_Past_2013.pdf (accessed April 5, 2016).

Neubauer, W., M. Doneus, I. Trinks, G. Verhoeven, A. Hinterleitner, S. Seren & K. Löcker 2013. Long-term Integrated Archaeological Prospetion at the Roman Town of Carnuntum/Austria, in P. Johnson & M. Millett, *Archaeological Survey and the City*. Oxford: Oxbow Books. 202-221.

Nicosia, F. & G. Poggesi, 2011. *Roselle. Guida al parco archeologico*. Siena: Nuova Immagine Editrice.

Orton, C. 2000. *Sampling in Archaeology*. Cambridge: Cambridge University Press.

Papadopoulos, J.K. & R.M. Leventhal (eds.) 2003. *Theory and Practice in Mediterranean Archaeology: Old World and New World Perspectives*. Los Angels: University of California.

Powlesland, D. 2009. Why bother? Large scale geomagnetic survey and the quest for "Real Archaeology", in S. Campana, S. Piro (eds.) *Seeing the Unseen. Geophysics and Landscape Archaeology*. The Netherlands. Taylor & Francis.167-182.

Prisco, G., 1998. *Castelli e potere nella Maremma grossetana nell'alto medioevo*. Grosseto.

Sebastiani, A. 2015. L'ager Rusellanus e la città di Rusellae nel periodo tardoantico (200-549 d. C.), in A. Sebastiani, E. Chirico, M. Colombini and M. Cygielman, *Diana Umbronensis a Scoglietto Santuario, Territorio e Cultura Materiale (200 a.C.-550 d.C.)*. Oxford: Archeopress. 60-67.

Settia, A., L. Marasco & F. Saggioro 2013. Fortificazioni di terra in Italia. Motte, Tumuli, Tumbe, Recinti, Scarlino. *Archeologia Medievale XL*: 9-187.

Terrenato, N. 1996. Field Survey Methods in Central Italy (Etruria and Umbria). Between local knowledge and regional traditions. *Archaeological Dialogues* 3-2: 216-230.

Tartara, P. 2003. Ortofotopiano storico IGM 1930 del territorio tra Cerveterei e la costa, in M. Guaitoli (ed.), Lo sguardo di Icaro. Roma: Campisano. 157-166.

Terrenato, N. 2004. Sample size matters! The paradox of global trends and local surveys, in S.E. Alcock & J.F. Cherry (ed.) *Side-by-Side Survey: Comparative Regional Studies in the Mediterranean*: 36-48. Oxford: Oxbow Books.

Vaccaro, E. 2012. *Sites and Pots: Settlement and Economic Patterns in Southern Tuscany (AD 300-900)*, BAR Int. Series 2191. Oxford: Archaeopress.

Vaccaro. E., M.. Ghisleni, A.C., Arnoldus-Huyzendveld, Grey, K., Bowes, M., Mackinnon, A.M., Mercuri, A., Pecci, M.A., Cau ontiveros, E., Rattigheri & R., Rinaldi 2013. Excavating the Roman peasant II: Excavations at Case Nuove, Cinigiano (GR). *Papers of the British School at Rome* 81: 129-179.

Vermeulen, F., G.J., Burgers, S., Keay & C., Corsi (eds.) 2012. *Urban Landscape Survey in Italy and the Mediterrenean*. Oxford: Oxbow books.

Wheatley, D.W. & M. Gillings. 2002. *Spatial Technology and Archaeology: the archaeological applications of GIS*. Taylor & Francis, London.

What Do the Patterns Mean? Archaeological Distributions and Bias in Survey Data

David C. Cowley

Abstract This paper explores aspects of the many factors that produce patterning in archaeological survey data, focussing on sources of bias that include topography, land use and environment; survey strategies and the roles of individuals; the problems of recognition, and those inherent in classifications and terminology; and settlement and land use histories. Examples of how such factors can be identified are given, illustrating the importance of a 'source-critical' approach to assessing the representativity of survey data.

Introduction

The archaeological data available for interpreting and understanding the past are heavily influenced by many factors, least of which may be the activities of people in the past. Evidence may be selectively destroyed or preserved by farming practice and settlement patterns, while the methods and sensors that archaeologists use to collect data, and their intellectual frameworks, influence what they see and record. The multivariate and highly complex patterning in known distributions of archaeological data caused by these factors not only distorts our perception of the survival and visibility of the archaeological remains of people's past activities, but by extension our interpretations. Understanding the nature of the patterning, and whether it is a representative reflection of past activity or the product of modern land use, for example, is central to moving from data collection to archaeological interpretations. This paper draws on survey data in Scotland to discuss some of the sources of inherent bias, and presents approaches to dealing with these biases.

D.C. Cowley (✉)
Aerial Survey & Remote Sensing, Historic Environment Scotland,
16 Bernard Terrace, Edinburgh EH8 9NX, UK
e-mail: dave.cowley@hes.scot

D.C. Cowley
Department of Archaeology, Ghent University, St. Pietersnieuwstraat
35—Ufo, 9000 Ghent, Belgium

M. Forte and S. Campana (eds.), *Digital Methods and Remote Sensing in Archaeology*, Quantitative Methods in the Humanities and Social Sciences,
DOI 10.1007/978-3-319-40658-9_7

Central to this discussion is the importance of trying to establish the reliability of archaeological distributions by seeking to understand the processes by which that data has been formed, its strengths and weaknesses and its representativity (or otherwise) of past activity. And while multi-sensor, multi-source integrated datasets can offset the impact of bias in any one data source, and is of course desirable to generate as full a view of the past as possible, rigorous analysis of the origins of patterning in survey data remains a vital building block for reliable archaeological interpretations.

Causes of Patterning in the Archaeological Record

The starting point for this paper is the observation that the enthusiasm of some archaeologists leads them to jump directly from an observed distribution of archaeological data to an interpretation, without due consideration of the influence of data collection strategies and of transformations in the landscape caused by ancient and modern land use. Often these factors are acknowledged in principle, but in practice they are rarely thoroughly considered. Indeed, relatively few archaeological distributions may stand up to rigorous analysis: things are seldom as they seem. This can be illustrated by the very uneven distribution of the over 1800 presently known examples of so-called ‘burnt mounds’ in Scotland (Fig. 1). These mainly Bronze Age deposits of fire-cracked stone are remarkably common in some upland areas where they are evident during field survey as C- and U-shaped earthworks. This has led some commentators to focus on their potential role in transhumant or hunting activities (e.g. Laurie 2004). That this aspect of the distribution maybe called into question is demonstrated by discoveries in lowland areas of burnt mounds with no surface expression during archaeological evaluation and monitoring of pipeline projects (Maynard 1993). Beyond such projects, which involve extensive topsoil stripping and archaeological monitoring, there are few opportunities for discoveries of burnt mounds in the lowlands. There is no general tradition of arable field-walking in Scotland that might identify characteristic spreads of heat-shattered stone and charcoal, and such ploughed-down features are unlikely to create recognisable features in the cropmark record, which is the main approach to survey in the lowlands. Thus, the known distribution of burnt mounds is a product of historic land use patterns creating a general zone of visibility and one of very limited visibility, and the different approaches to archaeological survey in these two zones. Moreover, the marked clustering observable in the known distribution can be traced back to the activities of specific field surveyors and the extents of upland survey projects (Halliday 1990, 2013, 73–74; Cowley 2011, 45–47). Burnt mounds in the uplands often occupy specific locations in the landscape beside water courses and marshy areas, but experience shows that they must be deliberately looked for by surveyors who have some experience or knowledge of these features; otherwise they are routinely missed or not recognised for what they are. Thus, while a superficial reading of the known distribution of burnt mounds (Fig. 1) might suggest marked local and regional differences in densities, and

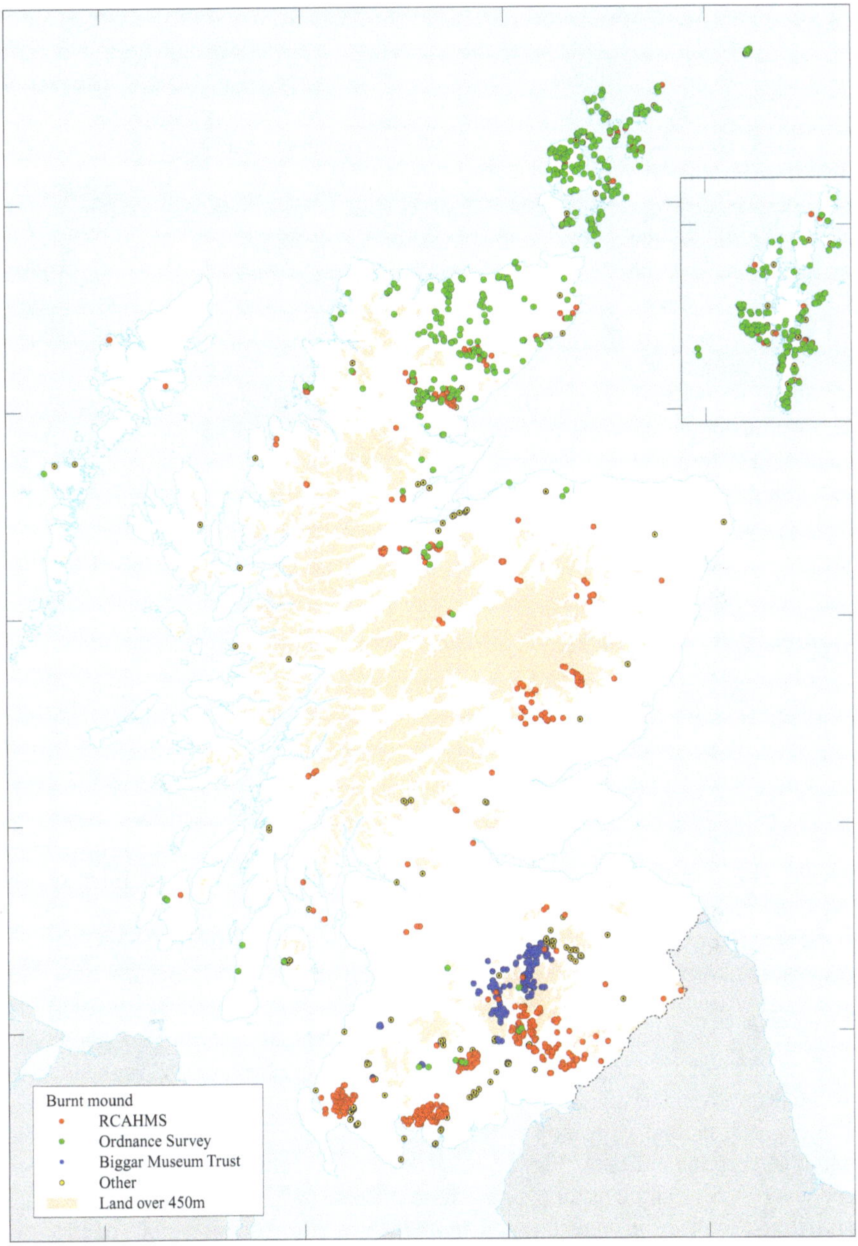

Fig. 1 The known distribution of burnt mounds in Scotland is not a true reflection of their occurrence in the past. Rather, it is a product of the survey work, experience and knowledge of particular individuals and agencies undertaking field work in various parts of the country. *GV004749 © Crown Copyright Historic Environment Scotland*

therefore their roles in settlement and land use systems, much of this patterning is due to historic land use and biased recognition. At present it is difficult to put aside these extreme biases to discuss what the apparent regional variations in distribution might have meant in the past.

While it might be tempting to dismiss the burnt mounds as an extreme example, most distributions of archaeological remains and sites suffer from the same general problems. Indeed, critical analysis of the origins of most site and monument distributions, and of their presentation in distribution maps, will quickly highlight inadequacies (Halliday 2011) and show how routinely misleading they are. Rather than the distributions of past activity that many archaeological maps claim to represent, most more effectively illustrate survey bias. These sources of bias and patterning are manifold and will vary from area to area, but generally include the following: topographical, land use and environmental factors; survey strategies and the roles of individuals; the problems of recognition, and those inherent in classifications and terminology; and settlement and land use histories. Each in its own way can be fundamental and illustrate in different ways the importance of a 'source-critical' approach to their interpretation (Cowley 2015).

Topography, Land Use, Soils and Weather

The general characteristics of a landscape are a principal influence on the survival and visibility of archaeological remains. As discussed above with reference to the burnt mounds, at a general level Scotland can be divided into upland and lowland zones, each with its own implications for what may survive and how it may be observed and recorded. In upland areas post-medieval and earlier remains routinely survive on the surface, lying beyond the highly destructive reach of 19th century and modern agriculture, while in lowland areas, generally below about 150 m above sea level, agriculture has severely eroded or entirely levelled most ancient monuments. Moving beyond such crude generalisation, many lowland areas are now widely used for arable crops and, where there are well-drained soils with the potential for high soil moisture deficits during the summer months, these can produce differential cropmarking revealing buried archaeological remains. The potential for cropmarking varies with the characteristics of the soils, with regional weather patterns and with crop types. Thus, at a national level the known distribution of cropmarked sites (Fig. 2) concentrates on sands and gravels in the lowland river valleys and coastal plains. These areas usually have lower than average rainfall and higher soil moisture deficits, which increase the potential for stress in growing crops. Some lowland areas, for example in northeast and west-central Scotland, are characterised by heavy, poorly-drained soils, and are dominated by grassland; they are much less likely to produce cropmarking, especially in the west which has generally higher average rainfall.

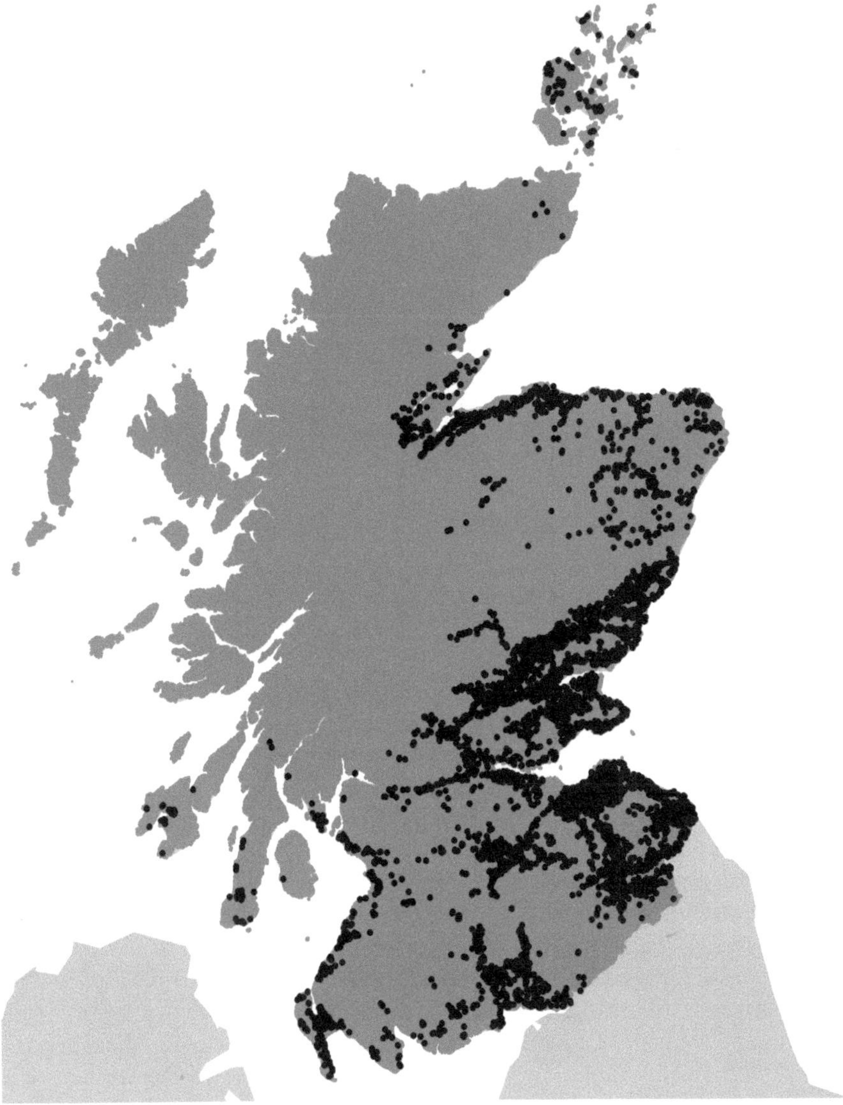

Fig. 2 The distribution of plough-levelled sites recorded as cropmarking in Scotland to 2014 reflects the locations of well-drained soils, arable crops and generally drier weather, as well as the way in which historical survey strategies have tended to focus on particular parts of the landscape. *GV004748 © Crown Copyright Historic Environment Scotland*

At a regional scale, even within areas with generally high potential for cropmarking, local variations in soils type are significant. Thus, an area of heavy clay soils in east-central Scotland (Fig. 3) has been unresponsive to aerial reconnaissance for many years (Cowley and Dickson 2007) and it is difficult to establish with

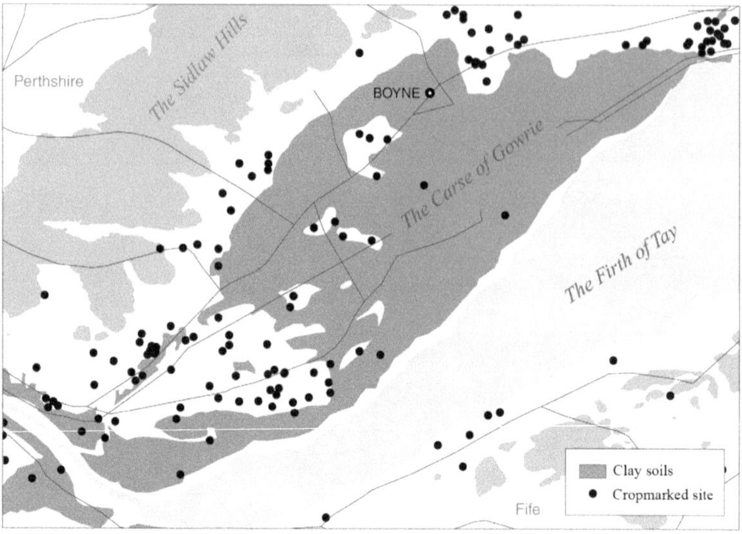

Fig. 3 Do the very few known sites on the clay soils of the Carse of Gowrie reflect past settlement distribution or not? Since these heavy soils do not produce cropmarking we cannot be certain whether the gaps in the distribution that we see here and on other unresponsive soils in Scotland are a real reflection of past settlement location or a result of bias of one kind or another in the survey strategy. *GV005205 © Crown Copyright Historic Environment Scotland*

any certainty whether the known sites reflect the extent and complexity of past activity. Further complication can be caused by variations in weather patterns, especially the amount of rainfall, from year to year. This can have a profound impact on the likelihood that arable crops will produce cropmarking, and therefore the potential productivity of aerial reconnaissance. The graph of average July rainfall on the eastern seaboard of Scotland (Fig. 4) illustrates some general trends, showing clearly some extremely dry years such as 1976 and 1977 when crop-marking was widespread. However, as a broad brush pattern it also hides other factors that may have a more local influence (data from Met Office 2015). For example, in 2013 parts of eastern Scotland were highly productive, while others had very limited returns, reflecting the interplay of regional weather patterns and potentially highly localised rainfall patterns. These factors influence the potential for productive aerial survey in any given year, and that variability can be seen from the distribution of sites recorded for the first time in 1976, 1977, 1978 and 1988 (Fig. 5). Comparing 1976 and 1977 there is some filling of gaps in coverage, while by 1978 the vast majority of work is repeat recording of already known sites. The three relatively dry years in the 1970s show the variability of results from generally 'good' years, while the very poor return of newly discovered sites from 1988 shows how profound the influence of a wet year can be. Returning to the pattern of rainfall and variable year to year weather (e.g. Fig.4), it is also worth noting that over the period since the 1970s there have been some major changes in farming practice that

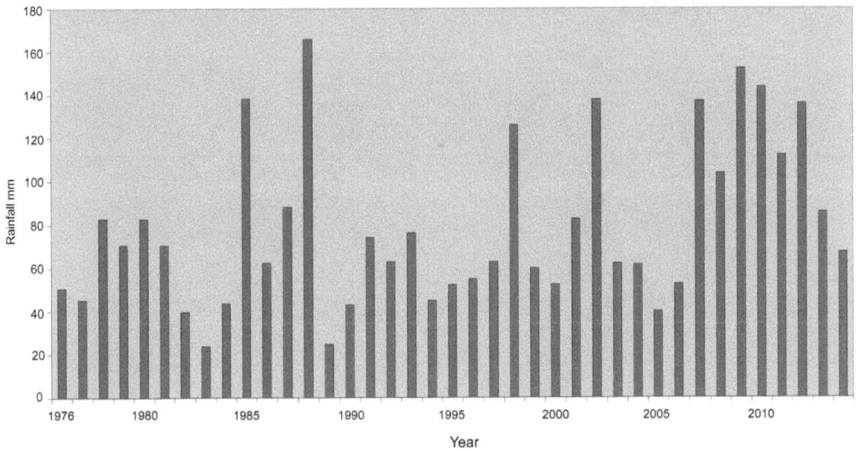

Fig. 4 Marked variation in rainfall across eastern Scotland at certain times of year can have a fundamental impact on the potential for cropmarking to develop and therefore on the returns from aerial reconnaissance. This is clear from the graph of averaged July rainfall, the variability in which relates in a general way to the effectiveness of survey in any one year. *Data from* Met Office 2015

add further complexity to the picture. Crop types have changed over the last 40 years and certain areas have seen the rapid development of highly mechanised farming, to name only two of a number of factors that influence the potential for cropmark formation (see De Guio 2015 for a wide ranging discussion of such issues).

These factors operate at many scales, from national annual rainfall patterns to local decisions about where to plant specific crops (Fig. 6); while in some places cereals are grown regularly in short rotations, in others they are rare and the emphasis of farming practice is on grass and fodder crops. These are all important factors which can profoundly influence the densities and extents of distributions, especially when survey is undertaken over limited periods of time (i.e. for a project) and assessment of annual variation in responses, for example, may be difficult. Here too, uncritical interpretation of site distributions will produce poor understanding and knowledge, which is also the case for survey strategies, the specification for which can also produce its own marked bias in the data collected.

Survey Strategies and People

The burnt mounds discussed above illustrate the potential influence on a national dataset of individual knowledge, experience and expectation. Other branches of survey, such as aerial reconnaissance, are no different (e.g. Cowley 2016). The first archaeological aerial reconnaissance in Scotland, undertaken in 1930 and 1939 by

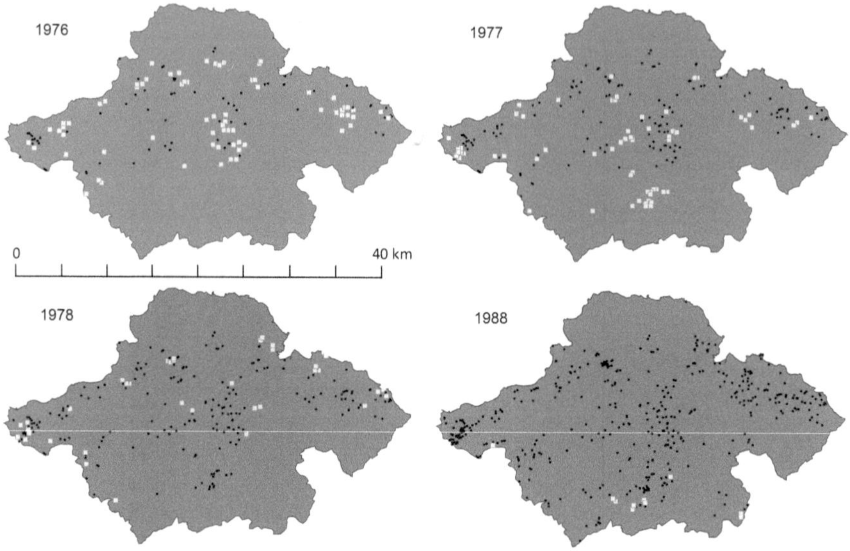

Fig. 5 The distributions of sites that have been recorded during different years in East Lothian, eastern Scotland. The white squares represent the sites recorded for the first time in a given year, shown against the cumulative record. The evident variability in 'new' sites in part represents the cumulative nature of the record, but also the variable weather conditions from year to year. *Site distribution from Historic Environment Scotland database*

Fig. 6 The decision of where to place the boundary between a crop that is likely to produce archaeological cropmarking and one that is not illustrates the potential impact of decisions by a modern farmer on the effectiveness of archaeological survey. *B73922 © Crown Copyright Historic Environment Scotland*

O.G.S. Crawford, aimed to explore the network of Roman roads and military sites (Crawford 1930, 1939). Crawford was explicit about this focus, stating that 'we followed Roman roads whenever possible' and that if there was 'more time to spare we could certainly have found many more sites' (Crawford 1930, 272, 1939, 285. See also Cowley and Gilmour 2005, 51–52; Jones 2005, 86–89). This direct statement of purpose is valuable because it informs users of the resulting data about a bias in how the information was collected, and therefore of some of its potential strengths and weaknesses.

Such statements, however, are relatively rare, and questioning survey strategy maybe received with hostility and suspicion. The reasons for this probably include how we tend to perceive the status of our roles and expertise in various forms of archaeological practice, and the implied criticism of such cross-examination. Nevertheless, we can contrast Crawford's approach to David Wilson of the Cambridge University Committee for Aerial Photography (CUCAP), who took a rather different position and suggested that those engaged in aerial reconnaissance 'should be sufficiently self-aware to see for themselves the biases in their own work and to declare them in contexts where this is relevant' (Wilson 2005, 72). While Crawford's declarations provide some helpful clues to inherent bias in the collection of his data, Wilson's self evaluation is less useful. Its emphasis on 'self-realisation' is not a good start as it does not speak of a reflexive approach to practice through which a level of self-critical reflection may be achieved. As importantly, it is also a view that seeks to discourage users from exploring the origins of data—this is highly undesirable as some weaknesses or inherent biases in survey data cannot be assumed to be self-evident and post hoc analysis may well reveal unanticipated insights into the structure of the data (Cowley and Gilmour 2005; Cowley 2015, 61–2).

In Scotland (as in other European countries), such analysis involves looking at both individuals, such as Crawford and J.K.S. St. Joseph (below), and at institutions undertaking national or regional survey. The work of St Joseph for CUCAP is a good case in point. From 1945 to the early 1980s he undertook aerial reconnaissance across much of the UK, like Crawford pursuing a particular interest in Roman military archaeology (Jones 2005, 2011, 19–24; St. Joseph 1976, 1977). The influence of known lines of Roman roads and campaign routes is evident in the distribution of the photographs taken between 1945 and 1977 (Fig. 7; Whimster 1983, 97–9). St. Joseph's Roman interests influenced *where* he flew, producing a marked geographical bias, the impact of which is still felt today in site distributions from aerial reconnaissance. He was fortunate to be flying during the exceptional dry conditions of 1949 when the large numbers of sites he discovered 'literally changed the map' (Evans and Jones 1977; St Joseph 1976, 7). Inevitably, the impact was greater for Roman sites than those of other periods because of his research focus and his expectations of where sites of interest might be located. While the distribution established by St Joseph has influenced the direction of later reconnaissance (below), the impact of his flying strategy can be identified in more subtle ways that are not necessarily self-evident. For example, the coincidence in some areas between the distribution of large Neolithic cursus monuments (ditched and banked

Fig. 7 The lines of Roman advance into Scotland are clear in this generalised map of sites photographed from the air by CUCAP, illustrating the impact St Joseph's particular academic interest had on the content of this collection. *After* Whimster 1983, Figs. 67–9 and Cowley 2013

ritual avenues) and Roman installations may owe something to the focus and quality of the images he took of these military monuments (Brophy and Cowley 2005, 16).

In addition to this, it can be shown that distributions of cropmark sites may often be reinforced by successive programmes of reconnaissance. For St Joseph, his spectacular Roman discoveries drew him back to the same locations again and again to improve and extend his record. This impact of known distributions on continuing reconnaissance has been discussed in a number of papers (see for instance Cowley 2002; Hanson 2005; Oakey 2005). Areas with a high certainty of cropmarking become attractive targets that aerial surveyors return to repeatedly ('honey-pots'), often recording the same sites over and over again regardless of whether or not additional information is visible. Well-known sites that come to be regarded as

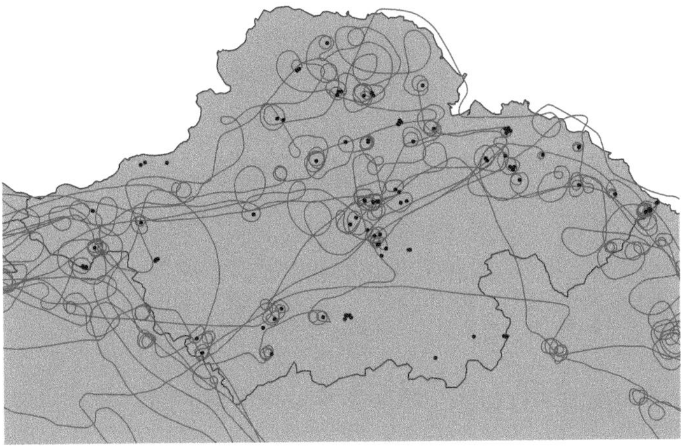

Fig. 8 Information about patterns of flying, such as GPS flight paths, give insights into areas that may have been examined during survey. In this example, the flight paths from 1993 are shown against the sites recorded during that year (note that several flight paths were not recorded). *Site distribution and flight paths from Historic Environment Scotland*

indicative of the general conditions in an area are also used by surveyors in a predictive way, but like the honey-pots these too can be self-reinforcing. Analysis of such issues benefits from as much information as possible about why and how survey was conducted. For Scotland, before 1993 this is limited to brief notes in flight notebooks and the distributions of sites recorded. GPS recorded flight paths for the majority of flights from 1993 can add important information about areas flown (e.g. Fig. 8), showing ground that may have been examined without seeing sites considered worthy of record. This is useful, but also limited as it does not provide any direct insight into other issues that might profoundly influence the effectiveness of a sortie, including the weather conditions at the time (i.e. turbulence), or the alertness of the airborne observer, or their interests (e.g. St Joseph above).

Looking Without Seeing

The discussion above has illustrated the roles of geographical factors and survey practice in creating patterning in the archaeological record. To these issues can be added the cognitive aspects of how archaeologists look (but may not see). Returning to the example of the burnt mounds, many field surveyors in Scotland (the author included) have walked past these monuments, failing to see them because they were not part of the observer's mental repertoire of known sites. In my own case, being 'introduced' to them by a more experienced colleague was a revelation, and I was able to 'train my eyes and mind' and see them and to recognise them for what they were. I took this knowledge into the field, but it also informed my understanding of mounds that I had seen previously but failed to interpret correctly (Cowley 1991). The point behind this example is that observation is conditioned by experience, and that we can subconsciously ignore features that do not fit expectations; that we can look without seeing (Halliday 2013). In this way archaeologists can create their own gaps in the data, ignoring some evidence as their knowledge and experience—and probably cognitive (eye/brain) framework—dictate.

The interpretation of the aerial photographic record for Newstead Roman fort in southern Scotland provides a good example of this issue. This site has been a focus for archaeological interest, including extensive excavations, for over 100 years (Hunter and Keppie 2012), and has been extensively documented on aerial photographs since 1945. Thus, it was with some surprise that a previously unrecognised temporary camp was discovered in 2012 on a photograph of the fort taken in 2006 (Fig. 9; Cowley 2012). Moreover, on examining the full archive of aerial images it became clear that the ditch of the 'new' temporary camp appeared on many photographs taken since 1976. So, why was it not seen before? In part at least, it was not recognised because of some underlying assumptions. Firstly, since the fort has been extensively excavated, those interested specifically in the fort have tended to rely on the excavation plans and not to use the aerial photographs as a primary source. Secondly, researchers interested in temporary camps have generally looked at images away from the fort. A third factor is that the large number of aerial photographs of this site, and the complexity of the visible features, requires time to work through systematically. The importance of reflexive practice has been mentioned above, and in illustrating the interplay of experience, knowledge, observation

Fig. 9 The cropmarked ditch and rounded corner of the Roman temporary camp that appears to project from under the later fort at Newstead was not identified as such until 2012, despite being clearly visible on many photographs taken in the years since 1945. *DP011649 © Historic Environment Scotland. Licensor canmore.org.uk*

and methodology in creating survey data, this example demonstrates their potential influence on knowledge. The contingent nature of observation and interpretation is one good reason to explore the potential of computer vision and artificial intelligence for feature detection and extraction—that such approaches should oblige researchers to better define what they are looking for and may highlight inconsistencies of approach (see Bennett et al. 2014 for a general discussion of this issue).

Classification and Terminology

The discussion above has illustrated how experience, observation and knowledge can create gaps in evidence and unreliable distributions in the archaeological record. How classification and terminology are applied can add further to these problems, as they underpin indexing and retrieval from databases and inventories. To a large extent this is inevitable when a database, such as one that might be held by a national heritage agency, has been created over a long period of time, by different individuals and institutions, with different briefs and objectives, and within an evolving framework of knowledge and understanding. The archaeological remains of buildings of potentially early medieval date provide an excellent example, where classifications used in the national database (https://canmore.org.uk/) reflect different sources of information (i.e. aerial reconnaissance and field survey), separating potentially similar buildings into a number of categories. In the cropmarked record the wall trenches and slightly hollowed interiors (often byres) of sub-rectangular timber buildings have long been recognised, but have been classified variously as 'timber halls' (implying high social status), 'timber buildings' or 'sunken-floored buildings', the use of each classification carrying with it some of the interpreter's assumptions about the status, date or cultural association of the structures. Excavation of one 'hall' at Balbridie in Aberdeenshire demonstrated that this particular example was Neolithic in date (Fairweather and Ralston 1993, 313–23), and has led some researchers to transpose the 'hall' terminology into Neolithic classifications, while further confusion is created by the similarity in basic morphology between the cropmarks of some of these buildings and those of Neolithic mortuary enclosures (Fig. 10).

For the broad topographic and land use reasons already discussed, the cropmark manifestations of such buildings are found only in lowland areas, but upland field survey identified the remains of another set of buildings which are now known to be of early medieval date. Sub-rectangular on plan and often tapering into a partly scooped area of the floor at one end, these came to be known as 'Pitcarmick' buildings after an exemplar site (RCAHMS 1990, 12–13), under which classification they appear in the national database. The sunken portion of the floor is remarkably similar to some of the cropmarked sites, suggesting that they be broadly similar buildings, while others have been uncovered by commercial excavations (Cook and Dunbar 2008, 149–65, 356–7; Driscoll 1997). Furthermore, the recognition of sunken-floored structures within the collection of oblique photographs is

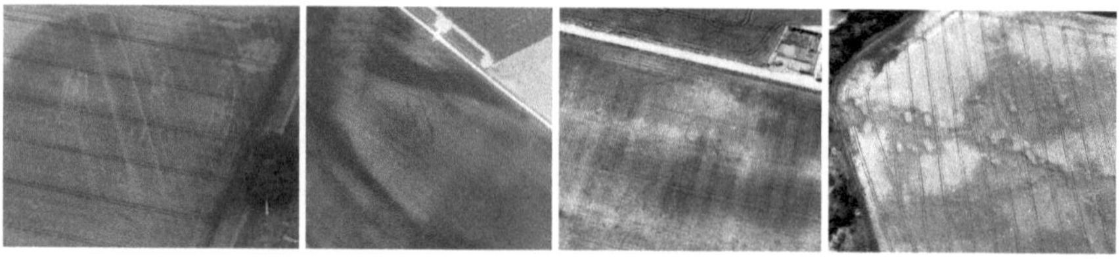

Fig. 10 The classifications and terminology used to describe cropmarked sites in the national database have been applied piecemeal over many years, and have never been systematically reviewed. This makes the identification of potentially similar sites very difficult without in-depth knowledge of the various terms used in describing and cataloguing them over a long period of survey history. Thus, while the sites recorded above may be elements of early medieval and medieval settlement it is not easy to retrieve them from the database. *DP050706 © Historic Environment Scotland. Licensor canmore.org.uk; AN3521, B22642, B79431 © Crown Copyright Historic Environment Scotland*

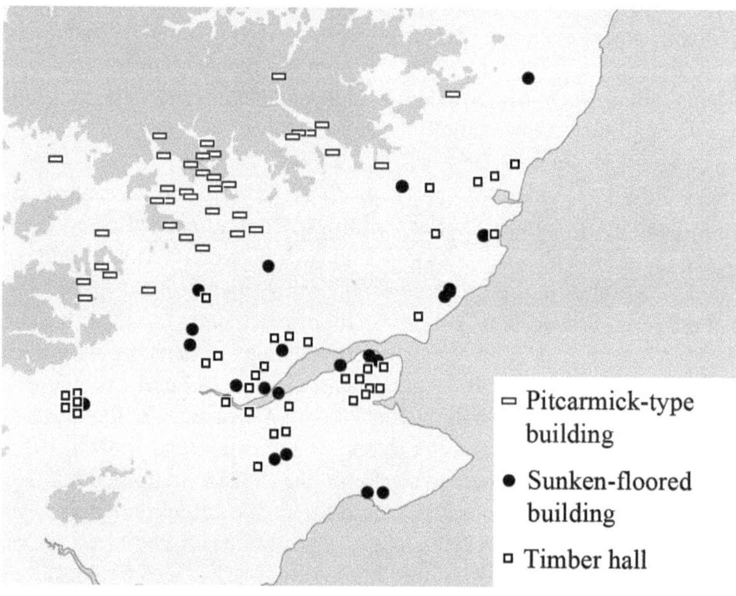

Fig. 11 This map of the distributions of sites classified as 'Pitcarmick-type building', 'Sunken-floored building' and 'Timber hall' in the national database shows the influence of terminology derived from different survey techniques in producing apparently different distributions in types of site that may in fact be similar. *GV 005208 © Crown Copyright Historic Environment Scotland*

very uneven. For all these reasons, retrieving the range of potentially early medieval settlement sites, let alone deriving some meaningful distribution, is almost impossible, even with a detailed knowledge of the history of recording and discovery and the range of classifications that might have been applied (Fig. 11).

This is a recurrent problem of many databases, which is to some extent inevitable. As knowledge grows, so new classifications may be added to deal with new discoveries, but there is seldom time to rework the whole database to apply them uniformly. This is the case for the Scottish database, which is the product of a long history initially as a card index system to document the depiction of antiquities on Ordnance Survey maps. It includes over 190,000 archaeological records (at January 2016) drawn from many sources of information and employs classifications developed in different intellectual contexts over at least the last 100 years. Like other sources of patterning, this type of 'administrative' bias needs to be understood and compensated for if it is not to create its own unrepresentative data.

Settlement and Land Use Histories

The discussion of sources of bias and patterning in archaeological data above has concentrated largely on the influence of topography, modern land use and archaeological survey strategies and classifications. To these can be added factors that more directly relate to past settlement and land use histories. In an excavation context, the survival of material for excavation depends on a wide range of processes that, broadly speaking, either lead to preservation of deposits or their destruction. The recent publication of an extensive excavation of an Iron Age hillfort at Broxmouth in south-east Scotland identifies a series of factors that condition the presence or absence of deposits (Armit and McKenzie 2013, 492–4). These include the observation that houses and other areas of the site were kept clean during occupation, preventing the build up of occupation deposits and perhaps indicating that refuse was stored in middens, presumably for subsequent removal to fields. Major landscaping of parts of the site during its occupation can also be identified, for example in the creation of large scoops on which stone-walled round houses were built, causing severe truncation of earlier features. More speculatively, the potential role of penned animals in the fort interior is also identified, as this would cause churning of the ground surface making it prone to erosion and lead to removal of earlier deposits along with dung. At Broxmouth, as may be the case in the majority of settlements in south-east Scotland, it appears that for a combination of reasons the norm is that earlier settlement traces are disturbed or removed by subsequent activity on site. This creates a highly fragmented record of the excavated site that is conditioned by factors such as past behaviours in keeping the settlement clean, and the pattern of site occupation, reworking and abandonment.

The same is true in the wider landscape around Broxmouth, where the amount of evidence for settlement varies considerably over time for a variety of reasons, the full complexity of which will not be discussed here. However, by way of example, earlier Iron Age settlement in the area is rare, known only from sparse evidence on sites that predominately date to other periods (Haselgrove 2009, 228; Lelong and Macgregor 2007, 125). This is partly due to the probability that earlier Iron Age settlement does not appear to be dominated by enclosures and is therefore difficult to recognise in the survey record, and possibly because of the characteristics of settlement at this date. There are hints that the settlement structure in some periods, including the earlier Iron Age, may be characterised by relative mobility. This, in combination with a general pattern of intermittent reoccupation of former settlement foci through much of the Iron Age, may ensure selective heavier damage to the evidence of some periods over others, compounding differing inherent visibility in the archaeological record (e.g. enclosure dominated, or non-enclosure settlement). A combination of such factors may also lie behind the limited evidence of settlement in this area during the 1st millennium AD, which is due to the lack of clear forms in the survey record, vulnerability to heavy plough attrition because such remains may lie at the top of settlement sequences, and perhaps due to the developing medieval settlement pattern by which a high proportion of such sites may be hidden. These factors create considerable complexity in interpreting site distributions, but for south-eastern Scotland it is clear that there are periods of time when enclosures are more prevalent than others. Because these are easily recognised in the survey record, they are disproportionately well represented in the archaeological record. This, and the periods of poor visibility outlined above, have the effect of creating chronological gaps in the evidence. Some of these are due to bias in the record, but, other aspects of which, at least, hint at past settlement processes.

Dealing with Bias

It would be easy to read the discussion above and despair that our archaeological data is hopelessly biased and unrepresentative. However, the real problem is not the truth of this position but what steps are taken to recognise the issues and deal with them. This paper began with an observation that often interpretation of data presented on distribution maps is developed with little understanding of how that data was formed. In these instances it is difficult to move beyond bias and 'false' patterning in archaeological datasets. Rather it requires a source-critical approach underpinned by rigorous analysis, the need for which remains paramount regardless of developments in survey sensors and technology.

Rigorous Analysis of Survey Data and Critical Reflection on Survey Practice

Central to a source-critical approach are the entangled factors of the origins, structure and strengths/weaknesses of survey data and the critical review of ongoing data collection (i.e. survey). This is a vast topic and only a limited commentary is offered here. A general issue is that the application of computer processing, whether for image (i.e. Lasaponara and Masini 2012) or data analysis, is uneven (i.e. Huggett 2013). Indeed, the statistical testing of spatial distributions is rare, and is sometimes regarded as the preserve of GIS specialists rather than a routine part of data collection and interpretation.

Critical reflection on survey practice and methodology is also uneven, and though there is now a good body of published work on this topic (for example papers in Brophy and Cowley 2005; Featherstone and Bewley 2000; papers in Mills and Palmer 2007; Whimster 1983), this too is often regarded as a specialist interest. This is, however, crucial to understanding data and plays a significant role in assessing and maintaining the effectiveness of survey. Analysis of the results of long-term aerial survey in Scotland has demonstrated the undesirable influence of indicator sites ('touch-stones') and previously established expectation on reconnaissance methodology (Fig. 12 left; Cowley 2002). The effectiveness of airborne observation during reconnaissance can be assessed through comparisons between what was recognised in the air and what appears on the photographs (Fig. 12 right). Weaknesses in survey that are identified can be offset, for example, by creating total coverage of busy areas by flying higher and photographing entire blocks of landscape and flying with two observers whenever possible. These analyses are driven by the recognition of the ever-present danger that survey strategy can quickly become self-reinforcing and that without regular critical analysis can become entrenched.

Presenting Data Effectively

Distribution maps are a ubiquitous means of presenting spatial data, and while they serve many purposes they are often labelled as a 'distribution of X type of monument' with little or no commentary. The distribution of burnt mounds discussed above (Fig. 1) is a case in point, where the map demands extensive commentary on the agencies and individuals responsible for it if it is to be used intelligently, and the same is true of any survey data. The map of rectilinear Roman Iron Age settlement enclosures in eastern Scotland is a good example of this point with reference to aerial photographic data (Fig. 13). Mainly known as cropmarks, in itself the distribution of these rectilinear settlements is difficult to analyse, but presented against an indicative extent of arable land and the overall distribution of sites known from cropmarking we can begin to judge the relationship between the particular monument type and the general distribution of cropmarked sites, as well as the potential influence of land use. This is vital to evaluating whether aspects of the distribution reflect past settlement patterns or are a product of survey methodology and modern land use. Thus, an often cited concentration of rectilinear enclosures in the area to the south of the great Roman Iron Age site on Traprain Law appears simply to reflect no more than a general concentration of cropmarked sites on soils that are prone to cropmarking. By the same token, the regular spacing of the enclosures in some areas, and the gaps apparent in the distribution in well-cropmarked areas elsewhere, are more likely to represent real aspects of the disposition of Iron Age

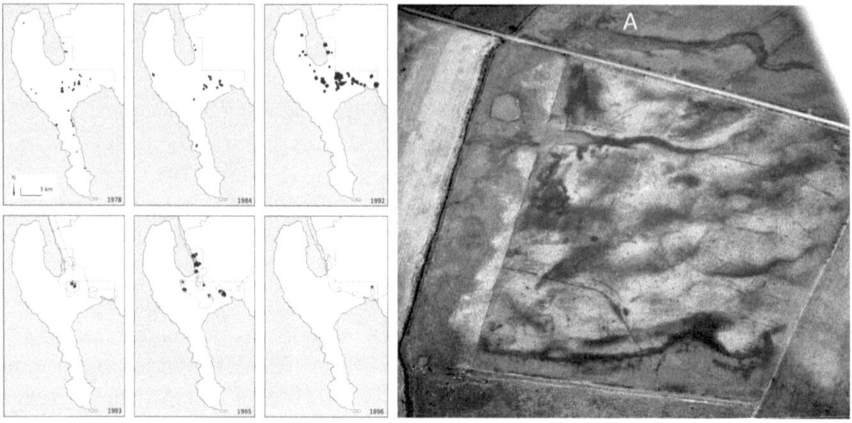

Fig. 12 *Left* The distribution of sites photographed in 1978 and 1984 shows an early stage of exploration in this part of south-west Scotland. However, the remarkable results in the very dry year of 1992 were a powerful influence in limiting the areas surveyed in future years, as can be seen in the GPS flight traces and sites photographed for 1993, 1995 and 1996. These show the use of indicator sites to decide whether or not to undertake further reconnaissance in an area. *Right* Very busy fields attract the attention of the aerial surveyor, which can be to the detriment of the record of adjacent sites such as the enclosure in the top of the photograph (*A*), for which this photograph is the only record. Post hoc analysis of survey returns is valuable in keeping reconnaissance effective and avoiding continually reinforcing known distributions. *GV005209, B72896 © Crown Copyright Historic Environment Scotland*

settlements (Cowley 2009, 216–217). As a result of this analysis, it is possible to challenge a long held belief that the rectilinear enclosures around Traprain Law represented a real concentration reflecting the regional importance of this area in the Roman Iron Age. This suggestion was initially based on limited survey data (Maxwell 1970) and has been uncritically reproduced so many times in the literature that it has it has become established as a 'fact' (Armit 1997; Armit and Ralston 1997, 179; Macinnes 1984, 183–186).

Beyond Bias?

The many sources of bias in archaeological data create complex patterning, within which it is often difficult to tell 'real' patterns of past activity from those produced by modern land use, survey methodology, perception and so on. Identifying these sources of bias is challenging, not least because it demands reflection on our own expectations and perceptions as these are some of the most significant sources of bias in our understandings of past landscapes. Indeed, contemporary attitudes to the contrasting landscapes of Scotland, for example, can have a major influence on whether archaeologists favour 'busy' past landscapes, or 'empty' ones (Fig. 14).

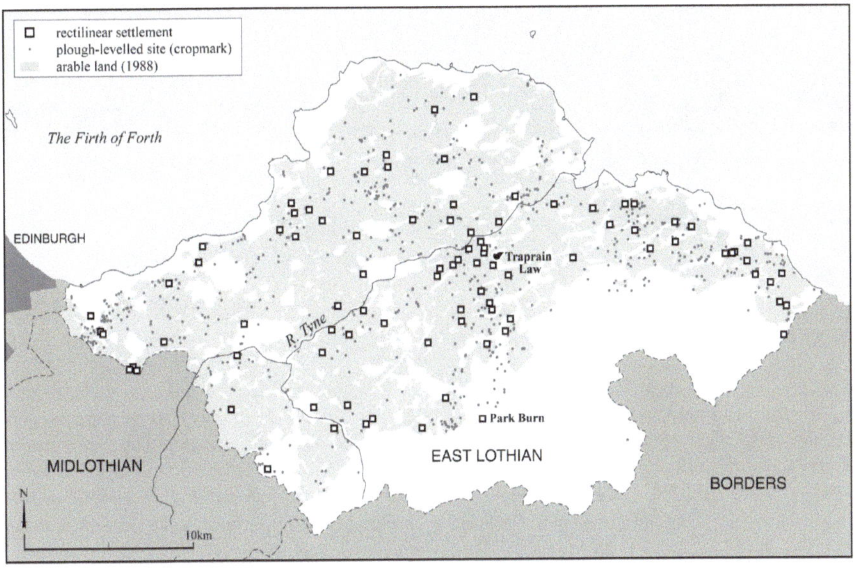

Fig. 13 It is only by presenting site distributions against relevant information that maps such as this one can be understood. In this example of the known Iron Age rectilinear settlements in East Lothian, knowledge of the distribution of arable ground and the other sites recorded as cropmarking is necessary in order to differentiate between survey bias and real evidence of the distribution of past settlement. *GV004486 © Crown Copyright Historic Environment Scotland*

Fig. 14 Archaeologists familiar with an intensively used landscape such as the one in East Lothian (*Left*) will often impose their contemporary expectations on to the past in these areas. Thus, wild land such as the north west of Scotland will also be viewed from a contemporary perspective, imposing concepts of 'emptiness' on the past where they may not be appropriate. *DP026198, DP156406 © Historic Environment Scotland. Licensor canmore.org.uk*

The lowland landscape of eastern Scotland (Fig. 14 left) is one of the best studied in Scotland (Haselgrove 2009) and it provides two examples where the available survey data can be used to identify markedly different intensities of past human activity. This area has not only been intensively surveyed from the air,

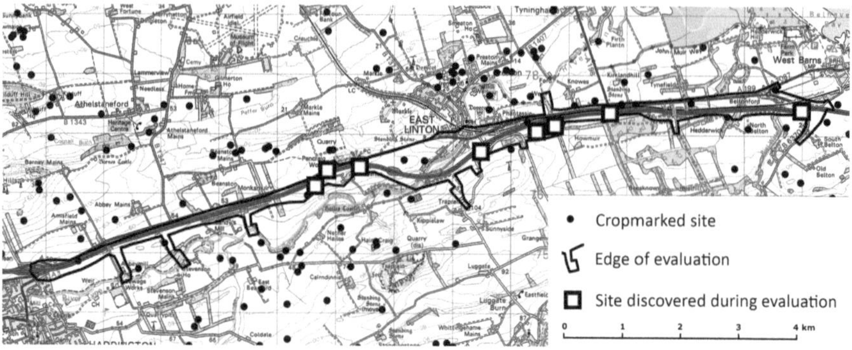

Fig. 15 The two very different distributions of sites discovered in this pre-development investigation can help to identify areas in past landscapes where there really were no remains of activities that might leave archaeological traces, such as the digging of ditches. The identification of varying intensities of past activity is an important foundation to reconstructing past settlement systems. *GV 005210 © Crown Copyright Historic Environment Scotland. Background map © Crown copyright and database right 2016. All rights reserved. Ordnance Survey licence number 100057073*

producing a large body of information, but it is also crossed by a linear road development within which there have been extensive sample excavations and invasive evaluations. This provides limited complementary data sources that can be used to assess the reliability of the aerial survey record (Lelong and MacGregor 2007). The A1 road line was designed to avoid known sites and was evaluated through trial trenching and topsoil stripping under archaeological supervision. These works produced two very different patterns along the road corridor (Fig. 15). An area 5.5 km in length in the west was entirely empty despite extensive monitoring, while to the east sites ranging from Neolithic to early first millennium AD in date were identified along the corridor. The soil maps show that the 'empty' western area lies on imperfectly drained soils that are prone to water logging (Cowley and Dickson 2007, 48–50). Here then, a gap in the cropmarks that might otherwise have been explained by environmental factors, perhaps reflects an underlying gap in earlier settlement patterns. From an early date, communities may have found that these imperfectly drained soils were unsuitable for intensive exploitation—they were perhaps always on the periphery of the main concentrations of settlement. Whether this analysis is correct in its bald detail should not detain us. The key point is that the use of complementary data sources shows the potential of integrating a spectrum of information to detect varying intensities of past activity in the landscape. In this way survey can inform discussion of settlement and land use systems at different periods, to create a textured understanding of the landscape in which mosaics of intensive land use, settlement foci and long term grassland can be further tested through the analysis of palaeo-environmental data.

Concluding Comments

Archaeological remote sensing is a rapidly developing field, with new data sources coming on stream, faster data collection, multi-sensor arrays and greater possibilities for analysis, manipulation and visualisation. However, my assumption when looking at distributions of archaeological information is that what we observe is more likely to represent data collection strategies or transformations by factors such as land use than it is to directly reflect what people did in the past, and this remains true, to varying degrees, no matter how technologically advanced our survey techniques. Thus, if our survey data is not accompanied by specific statements on the way in which the data was collected, either at the time of data collection or through subsequent analyses (and preferably both—to balance the different perspectives offered by self-reflection and external critical review), we routinely risk interpretations of the past based on constructs that have little to do with what happened in the past.

To move beyond the many sources of bias in survey data that could produce an *observed* distribution in the search for *real* distributions in past activity it is necessary to understand the nature of patterning. Addressing the representativity of landscape-scaled data is undoubtedly helped by analysis of complementary data sources (e.g. Powlesland 2011). And, while multi-sensor survey, for example, is desirable (e.g. Bennett et al. 2013; Verhoeven 2009; papers in Neubauer et al. 2013), it is expensive and because of cost is not ultimately an answer to addressing uneven or unrepresentative data at regional, let alone national, scales. One solution to this is the adoption of multi-scaled survey strategies incorporating a range of intensities of survey and varying details of information. The sampling strategy designed for the BREBEMI infrastructure project in northern Italy (Campana 2011; Campana and Dabas 2011) is an excellent example. To a multi-scaled approach to landscape survey can be added the overview that landscape characterisation provides. Such characterisations aim to provide a general mapping of the development of the landscape, covering both the contemporary and historic landscape (Crow and Turner 2009; Fairclough and Macinnes 2003; Fitch et al. 2011; Turner and Crow 2010). This is not an approach that has been widely adopted, but beyond characterising the landscape it also provides a framework for considering archaeological potential and then matching survey methodology to particular contexts (Cowley 2011, 43–44). Fundamentally, however, the key lies in deeper understanding of how our data has been collected and the factors that work together to structure it, of which what happened in the past is only one—and that remains true no matter how much our technology and data sources develop.

Acknowledgments Aspects of this paper were presented at '*MIND THE GAP—Absence, invisibility and emptiness in the interpretation of archaeological and landscape evidence*', an international seminar of the *ArchaeoLandscapes Europe* project,[1] organised by Stefano Campana,

[1]http://www.archaeolandscapes.eu

Chris Musson and Dominic Powesland at the Certosa di Pontignano, Siena, 22–24 April 2013. See also Cowley (2013, 2015) for further discussion of these topics.

Bibliography

Armit, I. (1997). *Celtic Scotland.* Edinburgh: Batsford.

Armit, I. and McKenzie, J. (2013). An Inherited Place - Broxmouth Hillfort and the South-East Scottish Iron Age. Edinburgh: Sosiety of Antiquaries of Scotland.

Armit, I. and Ralston, I. B. M. (1997). The Iron Age. In K. J. Edwards and I. B. M. Ralston (Eds). *Scotland: Environment and Archaeology, 8000 BC – Ad 1000.* Chichester: Wiley.

Bennett, R., Cowley, D. and De Laet, V. (2014). The data explosion: tackling the taboo of automatic feature recognition in airborne survey data. *Antiquity* 88, pp. 896-905. doi:10.1017/S0003598X00050766.

Bennett, R., Welham, K., Hill, R. and Ford, A. (2013). Using lidar as part of a multi-sensor approach to archaeological survey and interpretation. In Opitz, R. S. and Cowley, D. C. (Eds). *Interpreting archaeological topography: airborne laser scanning, 3D data and interpretation.* Oxford: Oxbow. pp. 197-205.

Brophy, K. and Cowley, D. (Eds). (2005). *From the Air – understanding aerial archaeology.* Stroud: Tempus.

Campana, S. (2011). 'Total Archaeology' to reduce the need for Rescue Archaeology: The BREBEMI Project (Italy). In Cowley, D.C. (Ed.). *Remote sensing for Archaeological Heritage Management.* Brussels: EAC/Archaeolingua. pp. 33-41.

Campana, S. and Dabas, M. (2011). Archaeological Impact Assessment: The BREBEMI Project (Italy). *Archaeological Prospection* 18: pp. 139-148. doi: 10.1002/arp.407.

Cook, M. and Dunbar, L. (2008). *Ritual, Roundhouses and Roman – Excavations at Kintore, Aberdeenshire 2000-2006.* Edinburgh: STAR.

Cowley, D.C. (1991). Some Burnt Mounds in Mid-Northumberland. *Archaeologia Aeliana* 5th Series, Volume XIX, pp. 119-121.

Cowley, D.C. (2002). A case study in the analysis of patterns of aerial reconnaissance in a lowland area of southwest Scotland. *Archaeological Prospection* 9(4), pp. 255-65.

Cowley, D.C. (2009). The Traprain environs in a regional perspective. In Haselgrove, C. (Ed.). *The Traprain Law Environs Project - Fieldwork and Excavations 2000-2004.* Edinburgh: Society of Antiquaries of Scotland. pp. 205-23.

Cowley, D.C. (2011). Remote sensing for archaeology and heritage management – site discovery, interpretation and registration. In Cowley, D.C. (Ed.). *Remote sensing for Archaeological Heritage Management.* Brussels: EAC/Archaeolingua. pp. 43-55.

Cowley, D.C. (2012). Newstead and environs from the air. In Hunter, F. and Keppie, L. (Eds). *A Roman Frontier Post and its People, Newstead 1911-2011.* Edinburgh: National Museums of Scotland. pp. 41-49.

Cowley, D. C. (2013). What kind of gaps? Some approaches to understanding bias in remote sensing data. *Archeologia Aerea* 7(13), pp. 76-88.

Cowley, D. C. (2015). Aerial Photographs and Aerial Reconnaissance for Landscape Studies. In Chavarria Arnau, A. and Reynolds, A. (Eds). *Detecting and Understanding Historic Landscapes.* Mantova: Post-Classical Archaeologies. pp. 37-66.

Cowley, D. C. (2016). Creating the cropmark archaeological record in East Lothian, southeast Scotland. In Crellin, R., Fowler, C. and Tipping, R. (Eds). *Prehistory without Borders: Prehistoric Archaeology of the Tyne-Forth Region.* Oxford: Oxbow.

Cowley, D. C. and Dickson, A. L. (2007). Clays and 'difficult' soils in eastern and southern Scotland: dealing with the gaps. In Mills, J. and Palmer, R. (Eds). *Populating Clay Landscape.* Stroud: Tempus. pp. 43-54.

Cowley, D.C. and Gilmour, S. M. (2005). Some observations on the nature of aerial survey. In Brophy, K. and Cowley, D. (Eds). *From the Air – understanding aerial archaeology*. Stroud: Tempus. pp. 50-63.

Crawford, O. G. S. (1930). Editorial Notes. *Antiquity* IV, pp. 273-8.

Crawford, O. G. S. (1939). Air Reconnaissance of Roman Scotland. *Antiquity* XIII, pp. 280-92.

Crow, J. and Turner, S. (2009). Silvri and the Thracian hinterland of Istanbul: an historic landscape. *Anatolian Studies* 59, pp. 167-81.

De Guio, A. (2015). Cropping for a better future: Vegetation Indices in Archaeology. In Chavarria Arnau, A. and Reynolds, A. (Eds). *Detecting and Understanding Historic Landscapes*. Mantova: Post-Classical Archaeologies. pp. 109-52.

Driscoll, S. (1997). A Pictish settlement in north-east Fife: the Scottish Field School of Archaeology excavations at Easter Kinnear. *Tayside and Fife Archaeological Journal* 3, pp. 74-118.

Evans, R. and Jones, R. J. A. (1977). Crop marks and soils at two archaeological sites in Britain. *Journal of Archaeological Science* 4, pp. 63-76.

Fairclough, G. and Macinnes, L. (2003). *Understanding Historic Landscape Character, Topic Paper 5, Landscape Character Assessment Guidance for England and Scotland*. Edinburgh: English Heritage/Historic Scotland.

Fairweather, A. D. and Ralston, I. B. M. (1993). The Neolithic timber hall at Balbridie, Grampian Region, Scotland. *Antiquity* 67, pp. 313–23.

Featherstone, R. and Bewley, R. (2000). Recent aerial reconnaissance in North Oxfordshire. *Oxoniensia* 65, pp. 13-26.

Fitch, S., Gaffney, V., Gearey, B., Ramsey, E. (2011). Between the Lines – enhancing methodologies for the exploration of extensive, inundated palaeolandscapes. In Cowley, D.C. (Ed.). *Remote sensing for Archaeological Heritage Management*. Brussels: EAC/Archaeolingua. pp. 173-204.

Halliday, S. (1990). Patterns of fieldwork and the distribution of burnt mounds in Scotland. In Buckley, V. (compiler). *Burnt Offerings – International Contributions to Burnt Mound Archaeology*. Dublin: Wordwell. pp. 60-1

Halliday, S. (2011). Pattern, perception and preconception: recognizing regionality in Scottish prehistory. In Jones, A. M. and Kirkham, G. (Eds). *Beyond the Core – Reflections on regionality in prehistory*. Oxford: Oxbow. pp. 19-32.

Halliday, S. (2013). I Walked, I Saw, I surveyed, but what did I see?… and what did I survey. In Opitz, R.S. and Cowley, D.C. (Eds). *Interpreting archaeological topography: airborne laser scanning, 3D data and interpretation*. Oxford: Oxbow. pp. 63- 75.

Hanson, W. S. (2005). Sun, sand and see: creating bias in the archaeological record. In Brophy, K. and Cowley, D. (Eds). *From the Air – understanding aerial archaeology*. Stroud: Tempus. pp. 73-93.

Haselgrove, C. (Ed.) (2009). *The Traprain Law Environs Project - Fieldwork and Excavations 2000-2004*. Edinburgh: Society of Antiquaries of Scotland.

Huggett, J. (2013). Disciplinary issues: challenging the research and practice of computer applications in archaeology. In Papadopoulos, K., Chrysanthi, A., Murietta-Flores, P., Romanowska, I., Earl, G., Sly, T. and Wheatley, D. (Eds). *Proceedings of the 40th International Computer Applications and Quantitative Methods in Archaeology Conference*, Amsterdam: Archaeopress. pp. 13-24.

Hunter, F. and Keppie, L. (Eds) (2012). *A Roman Frontier Post and its People, Newstead 1911-2011*. Edinburgh: National Museums of Scotland.

Jones, R. H. (2005). The advantages of bias in Roman Studies. In Brophy, K. and Cowley, D. (Eds). *From the Air – understanding aerial archaeology*. Stroud: Tempus. pp. 86-93.

Jones, R. H. (2011). *Roman Camps in Scotland*. Edinburgh: Society of Antiquaries of Scotland.

Lasaponara, R. and Masini, N. (2012). Image enhancement, feature extraction and geospatial analysis in an archaeological perspective. In Lasaponara, R. and Masini, N. (Eds). *Satellite Remote Sensing – A new Tool for Archaeology*. New York: Springer. pp. 17-63.

Laurie, T. (2004). Springs, Woods and Transhumance: Reconstructing a Pennine Landscape during Later Prehistory. *Landscapes* 5(1), pp 73-103.

Lelong, O. and Macgregor, G. (Eds) (2007). *The Lands of Ancient Lothian: Interpreting the Archaeology of the A1*. Edinburgh: Society of Antiquaries of Scotland.

Macinnes, L. (1984). Settlement and economy: East Lothian and the Tyne-Forth province. In Miket, R. and Burgess, C. (Eds). *Between and beyond the Walls: Essays on the Prehistory of Northern Britain in Honour of George Jobey*. Edinburgh: John Donald. pp. 176-98.

Maynard, D. (1993). Burnt mounds around a pipeline in Dumfries and Galloway. *Transactions of the Dumfriesshire and Galloway Natural History and Antiquarian Society*, Series 3, LXVII, pp. 33-52.

Maxwell, G. S. (1970). Early rectilinear enclosures in the Lothians. *Scottish Archaeological Forum* 1970, 86-90.

Met Office (2015) = http://www.metoffice.gov.uk/climate/uk/datasets/, Accessed May 2015

Mills, J. and Palmer, R. (Eds) (2007). *Populating Clay Landscapes*. Stroud: Tempus.

Neubauer, Trinks, Salisbury, Einwögerer: Neubauer, W., Trinks, I., Salisbury, R. and Einwögerer, C. (Eds) (2013). *Archaeological Prospection, Proceedings of the 10th International Conference – Vienna, May 29th – June 2nd 2013*. Vienna: Austrian Academy of Sciences Press.

Oakey, M. 2005. Patterns of aerial photography in the Central Midlands of England. In Brophy, K. and Cowley, D. (Eds). *From the Air – understanding aerial archaeology*. Stroud: Tempus. pp. 141-50.

Powlesland, D. (2011). Identifying the unimaginable – Managing the unmanageable. In Cowley, D.C. (Ed.). *Remote sensing for Archaeological Heritage Management*. Brussels: EAC/Archaeolingua. pp. 17-32.

RCAHMS (1990). *North-east Perth: and archaeological landscape*. Edinburgh: HMSO.

St Joseph, J. K. S. (1976). Air reconnaissance of Roman Scotland 1939-75. *Glasgow Archaeological Journal* 4, pp. 1-28.

St Joseph, J. K. S. (1977). Air reconnaissance in Roman Britain 1973-76. *Journal of Roman Studies* LXVII, pp. 125-61.

Turner, S. and Crow, J. (2010). Unlocking historic landscapes in the Eastern Mediterranean: two pilot studies using Historic Landscape Characterisation. *Antiquity* 84, pp. 216-29.

Verhoeven, G. (2009). *Beyond Conventional Boundaries. New Technologies, Methodologies, and Procedures for the Benefit of Aerial Archaeological Data Acquisition and Analysis*. PhD thesis Ghent University.

Whimster, R. (1983). Aerial reconnaissance from Cambridge: a retrospective view 1945-80. In Maxwell, G. S. (Ed.). *The impact of aerial reconnaissance on Archaeology*. London: Council for British Archaeology. pp. 92-105.

Wilson, D. R. (2005). Bias in aerial reconnaissance. In Brophy, K. and Cowley, D. (Eds). *From the Air – understanding aerial archaeology*. Stroud: Tempus. pp. 64-72.

3D Tool Evaluation and Workflow for an Ecological Approach to Visualizing Ancient Socio-environmental Landscapes

A Case Study from Copan, Honduras

Heather Richards-Rissetto, Shona Sanford-Long and Jack Kirby-Miller

Abstract Architectural reconstructions are the centerpieces of ancient landscape visualization. When present, vegetation is relegated to the background, resulting in an underutilization of plant data—an integral data source for archaeological interpretation—thus limiting the capacity to take advantage of 3D visualization for studying ancient socio-environmental dynamics. Our long-term objective is to develop methods of 3D landscape visualization that have value for examining changes in land use and settlement patterns. To begin to work toward this objective, we have (1) identified 3D tools and techniques for vegetation modeling and landscape visualization, (2) evaluated the pros and cons of these tools, (3) investigated biological and ecological approaches to simulate plant habitats, the data requirements of these approaches, and the pros and cons of these approaches for reconstructing archaeological landscapes, and (4) then built on these findings to propose a workflow to integrate archaeological, paleo-environmental, and ethnobotanical into Geographic Information Systems (GIS) for export into a virtual landscape for investigations of ancient socio-environmental interaction. To identify possible 3D digital tools and workflows to visualize plant distribution models alongside archaeological settlement, we keep in mind several key issues: capacity to handle georeferenced data, levels of detail, multi-scalar analysis, and availability of quantitative and qualitative data.

H. Richards-Rissetto (✉)
Department of Anthropology and Center for Digital Research in the Humanities, University of Nebraska-Lincoln, 840 Oldfather Hall, Lincoln, NE 68588, USA
e-mail: richards-rissetto@unl.edu

S. Sanford-Long · J. Kirby-Miller
Department of Geography, Middlebury College, Middlebury, VT 05753, USA

M. Forte and S. Campana (eds.), *Digital Methods and Remote Sensing in Archaeology*, Quantitative Methods in the Humanities and Social Sciences,
DOI 10.1007/978-3-319-40658-9_8

Introduction

Vegetation is a key source of data for investigating archaeological landscapes; however, in 3D visualization they often serve as a backdrop to architectural reconstructions rather than as an integral component of ancient landscapes. Our project's long-term vision is to develop methods of 3D landscape visualization that have value for examining changes in land use and settlement patterns in archaeological landscapes. We employ an ecological perspective that enables scholars to integrate cultural and environmental data in landscape reconstructions in order to facilitate a more holistic and active engagement with and exploration of ancient socio-environmental dynamics.

To begin to work towards this long-term goal, we have (1) identified 3D tools and techniques for vegetation modeling and landscape visualization, (2) evaluated the pros and cons of these tools, (3) investigated ecological approaches to simulate plant habitats, the data requirements of these approaches, and the pros and cons of these approaches for reconstructing archaeological landscapes, and (4) then built on these findings to propose a workflow to integrate archaeological, paleo-environmental, and ethnobotanical data into Geographic Information Systems (GIS) for export into 3D virtual environments. The case study is the UNESCO World Heritage Site of Copan, Honduras and 24 km^2 surrounding the site's main civic-ceremonial core. The workflow brings together 2D and 3D data with archaeological, environmental, ethnographic, and ethnobotanical data sets into a georeferenced 3D virtual environment for visual, spatial, and temporal analysis of ancient socio-environmental dynamics.

Case Study: Copan, Honduras

Located on the southeast periphery of the ancient Maya world, Copan's multi-ethnic history offers a unique perspective on social and environmental interaction (Fig. 1). For much of the Preclassic period (1300BCE-CE100/250), Copan's population was non-Maya. Ceramics indicate that populations from El Salvador, the Pacific Coast, and eastern Honduras occupied the valley (Canuto 2002; McNeil 2009, 2010, 2012; Viel 1993). Sometime between CE100–250, Maya populations immigrated to Copan (McNeil 2009, 2012). While ceramic evidence suggests that proto-Chorti (ancestors of present-day Chorti Maya) settled at Copan ca. CE100 (Sharer 2009), recent pollen data support also a later immigration ca. CE250 (McNeil 2009).

In CE 426 (the Early Classic period), Copan's 1st dynastic ruler, a foreign-born Maya noble, was seated (Bell et al. 2004; Stuart 2007). Until relatively recently, it was assumed that with the establishment of Copan's dynasty (which was to last nearly 400 years) Copan "becomes" Maya; however, new evidence continues to come to light to question this assumption. For example, El Salvadoran migrants,

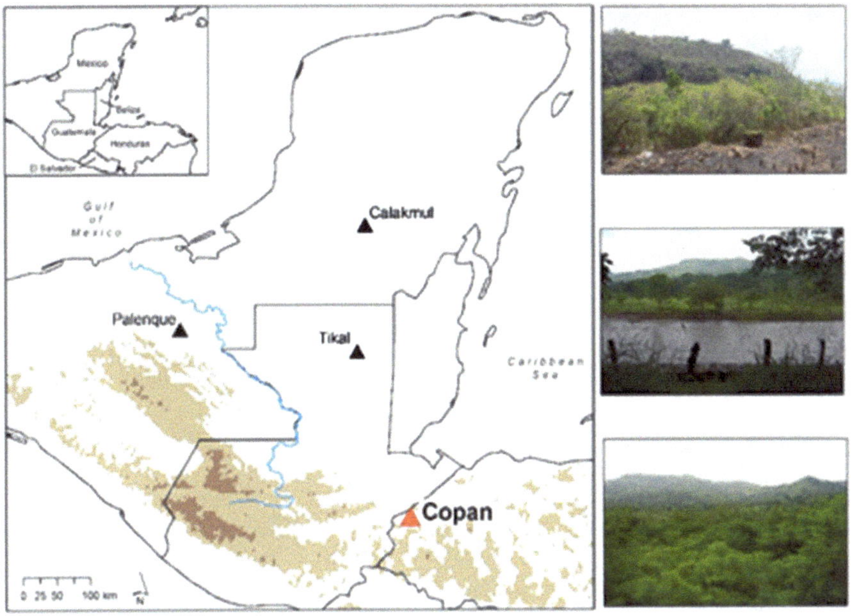

Fig. 1 Map showing location of Copan on southeast periphery of Maya area and photos illustrating Copan's diverse landscape

Table 1 Chronology of Copan, Honduras noting key events influencing ethnic composition

Time period	Key events	Cultural groups
Preclassic (1300BCE–CE100)	– 1300BC – 850BC	– Non-Maya (El Salvador, Pacific Coast) – E. Honduras, El Salvador
Protoclassic (AD100–400)	– AD100/250	– Immigration of proto-Chorti Maya
Early classic (CE400–600)	– AD 426: *Yax Kuk Mo* arrives – ca. AD430: Ilopango Volcano erupts	– Peten Maya (Caracol/Tikal) – Non-Maya Immigration, El Salvador
Late classic (CE600–822)	– AD 822: death of Ruler 16; dynasty ends	– Maya and Non-Maya?
Post-classic (CE822–950/1000)	– ca. CE900: burning at Principal Group	– Maya and/or Non-Maya?

who were displaced due to the eruption of Ilopango Volcano, settle in the valley ca. CE430 (McNeil 2009). This line of evidence along with ceramic, bioarchaeological, and settlement pattern data suggest that Copan contained a multi-ethnic population throughout the Late Classic period (CE250–850) (e.g., Canuto and Bell 2008, 2013; Gerstle 1987; Maca and Miller 2009; Price et al. 2008; Richards-Rissetto 2010) (Table 1).

While evidence of Copan's multi-ethnic polity is increasing, the scope of ethnic heterogeneity and its impact on cultural practices and interaction at particular time periods remains unclear. In other words, archaeologists know little about how the polity's multi-ethnic history may have impacted interaction with(in) the physical and social landscape. To address this issue from a unique perspective, we propose to explore the pros and cons of 3D visualizations (derived from paleoenvironmental, ethnographic, archaeological, and other data sources) and how these visualizations can be used to investigate changes through time in land use, particularly in relation to changing ethnic populations.

Advantages and Challenges to 3D Visualization for Landscape Archaeology

There are both advantages and challenges to using 3D visualization for archaeological research. While 3D data acquisition and 3D visualization varies in technology (e.g., terrestrial laser scanning, airborne LiDAR, photogrammetry, Computer Aided Design) and scale (ranging from artifacts to individual buildings to entire cities) there is a common thread—most 3D visualization focuses on artifacts and architecture. While the advantages and challenges of 3D modeling and visualization of artifacts and built environments are critical to archaeological practice (e.g., Barcelo et al. 2000; Dell'Unto et al. 2015; Frischer et al. 2008; Forte et al. 2003; Forte 2005; Richards-Rissetto 2013); our focus is on 3D visualization of vegetation in association with archaeological settlement data in *georeferenced* virtual environments because vegetation data (e.g., palynological, ethnobotanical) are an integral data source for archaeological interpretation and critical for understanding past landuse, environmental impact and cultural adaptation (e.g., Fedick 1996; McNeil 2012).

Some advantages of 3DGIS for landscape archaeology include: 3D models that provide a sense of ancient places to foster public appreciation and facilitate scholarly interpretation, clicking on plants to access information about plant communities (e.g., potential use, time range, habitat characteristics) in context of archaeological settlement patterns, ability to perform 3D analysis in a landscape context, for example, to calculate visibility or travel costs based on both ecological and settlement data in a 3D environment, and interaction with alternative 3D reconstructions of how individual buildings, architectural complexes, and entire cities may have a looked in their environmental contexts.

Some disadvantages of 3DGIS include: incomplete data sets requires "filling-in the blanks" to create realistic models that are not necessarily "accurate", difficult to convey uncertainty in 3D reconstructions, realistic models are often time-consuming and require expertise to build, current navigation tools are very basic, 3D plant modeling is typically limited to point data and models tend to be crude and unrealistic, GIS environments tend to exclude important qualitative data essential for understanding cultural complexities, and importantly modeling capabilities in GIS are limited potentially leading to reductive reconstructions (e.g. Forte et al. 2005a, b; Forte 2014; Pescarin et al. 2005).

Generally speaking 3DGIS has the potential to bring together experiential, quantitative and qualitative approaches into a single tool for comprehensive, holistic, and unique analysis; however, current 3DGIS systems still lack key functionality particularly in regard to capturing aesthetics and a realistic sense of movement within ancient landscapes. To address these concerns, several archaeological projects are working to develop 3DGIS tools such as the MayaArch3D Project (www.mayaarch3d.org/, von Schwerin et al. 2013) and Gabii Goes Digital (http://gabiiserver.adsroot.itcs.umich.edu/gabiigoesdigital/index.html); other projects use combinations of existing software for 3DGIS visualizations (e.g., Dell'Unto et al. 2015; Dylla et al. 2009; Opitz and Nowlin 2012; Saldana and Johanson 2013); however, currently these projects are not developing tools to incorporate ecological data.

To begin to move toward the development of a 3DGIS tool that integrates and analyzes georeferenced ecological and archaeological data, we devised a preliminary workflow (Fig. 2). The workflow allows us to simultaneously work linearly and iteratively. We adopt a linear—step-by-step—approach to identify the requirements for each component. However, we iteratively work back and forth among components to modify requirements and update steps as we learn and discover new issues, alternative tools, etc. in other components. Our goal is to use

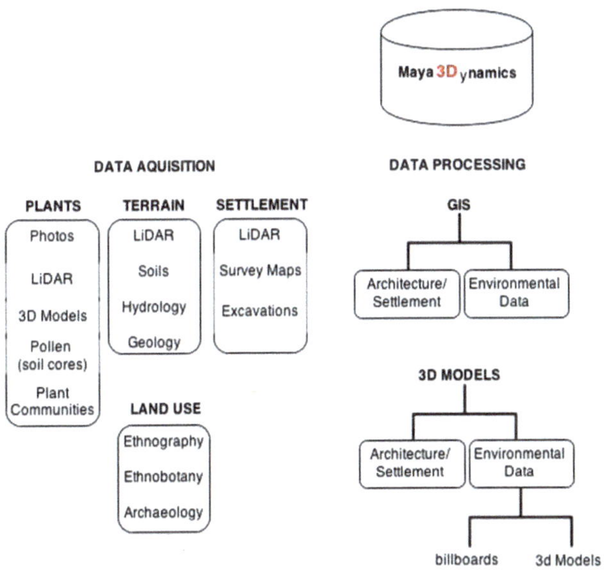

Fig. 2 Workflow for ecological approach to 3DGIS visualization

the workflow to identify required tasks, and then to evaluate existing 3D tools (particularly free and/or open source) that can fulfill these tasks.

Workflow: An Ecological Approach

The workflow has three main parts: (I) Data Acquisition, (II) Data Processing, and (III) 3DGIS Data Visualization and Analysis.

Data Acquisition

Data acquisition comprises four components: (1) Plants, (2) Terrain, (3) Settlement, and (4) Land Use.

Plants

Plant data serve two broad uses: identification and visualization. Identification: Palynological data (e.g., soil cores) and modern plant communities are two data sets that can be used to model plant species within a specified spatial extent (i.e., landscape). Visualization: Photos, airborne LiDAR (Light Detection and Ranging), and 3D model libraries can be used to generate 3D individual plant models and populate 3D landscape visualizations from those data sets.

Terrain

Terrain comprises vector (shapefiles) and raster data. GIS vector data for soils and geology provides attributes to describe terrain characteristics for discrete areas that influence the growth of plant communities. Elevation data from, for example, airborne LiDAR or rasterized contours (as continuous data), provides a Digital Terrain Model (DTM) that combined with hydrology and geomorphology data reconstructs the bare-earth surface.

Settlement/Architecture

Archaeological settlement data come from excavations, pedestrian surveys, aerial and satellite imagery, and airborne LiDAR. Excavations provide temporal data for diachronic visualizations.

Land Use

Ethnographic, ethnobotanical, and archaeological studies on land use provide context for interpreting and integrating plant, terrain, and settlement data for analysis and 3D visualization.

Data Processing

Geographic Information Systems (GIS)

GIS is a mainstay of archaeology for data management, creation, analysis, and visualization. Most importantly it affords archaeologists the tools to reveal and analyze complex patterns and trends in spatial data through an interactive mapping interface (Chapman 2006; Conolly and Lake 2006; Wheatley and Gillings 2002). In the workflow, GIS serves two main purposes: (1) integrate the settlement (building footprints) and environmental data including 2D (e.g., soils shapefile) and 2.5D (e.g., DTM) data sets into a common coordinate system (e.g., Universal Transverse Mercator) and (2) analyze spatial relationships among settlement/archaeological and environment variables to output new "analytical/interpretative" GIS data sets about ancient land use and settlement patterns. These GIS data also serve as inputs for 3D visualizations.

3D Modeling

3D modeling is rapidly becoming more commonplace in archaeology. 3D reconstructions are multi-purposed offering data preservation, public outreach, scholarly research and much more (Barcelo et al. 2000; Fisher 2012; Forte and Kay 2001; Forte et al. 2005a, b; Frischer and Dakouri-Hild 2008; Richards-Rissetto et al. 2013). The workflow converts photos, airborne LiDAR, and shapefiles into 3D models of plants, terrain, and hydrology. Billboards (to decrease data size and modeling time/cost) for plants are generated for Level of Detail (LOD) management and display to facilitate real-time visualization for faraway objects in a 3D scene (Kim et al. 2011). Excavation, settlement, and ethnographic data are integrated to generate 3D architectural models of buildings. GIS footprints (shapefiles) provide building locations for 3D architectural models.

A major challenge to 3D landscape visualization is the tools that best let researchers visualize architecture in 3D are not well-suited for visualizing vegetation in 3D, and even when they do a decent job of visualizing vegetation they are typically not good for landscape analysis. This challenge arises in great part because scholars have diverse research interests. As previously stated, we contend that to ensure appropriate 3DGIS data visualization and analysis, researchers must first define their goals and objectives. The second step is to devise specific methods to achieve those goals and objectives based on available data.

3DGIS Data Visualization and Analysis

Ecological Framework

Populating ancient landscapes with plant communities is not straightforward. The vast majority of sites are unexcavated, potentially leading to "biased" samples (from excavated sites), the preservation (quantity and quality) of paleoenvironmental data vary regionally and by type (species), and diversity in species pollen dispersal (e.g. range and amount) makes it challenging to "translate" paleoenvironmental data into discrete areas that can be mapped on the landscape, particularly at intermediate-level scales ranging (e.g., 20–50 km^2) (e.g., Franklin 2010; Franklin et al. 2015). The problems arises that if we cannot map ancient plant communities, then we cannot "accurately" visualize them in three-dimensions. To overcome some of these limitations, we propose an ecological model that uses proxies derived from modern and paleoenvironmental data to estimate the spatial extent of ancient plant communities. To properly derive proxies, it is necessary to step back and ask an initial set of questions to frame the ecological approach.

Initial Questions

We pose the following four questions:

1. What approaches do biologists and ecologists use to simulate plant habitats in landscapes?
2. What types of data do they use? How do they use these data?
3. Which of these approaches can we use given our data set for Copan, e.g., palynological, hydrological, soil, geological, and ethnobotanical?
4. What other data might we need? Can we get these data?

Based on our findings (and available data sets), we identified two, but not mutually-exclusive, approaches to simulate plant communities at ancient Copan. These two approaches are combined to create a model that employs data from the past and present to provide more holistic reconstructions of Copan's ancient landscape(s).

Approach #1

This approach employs ordination analysis of modern plant communities in order to determine plant communities that are most likely to occur in specific areas. Ordination analysis is a way to determine whether particular plant communities tend to be found in a consistent habitat type, and which communities are most similar to each other. This approach records attributes such as plant species, soil type, elevation, and water availability along transects in a study area and subsequently uses ordination analysis to develop a list of plant communities that tend to occur and the characteristics (e.g., soil type, elevation) of where they tend to occur. We can then assign plant communities to particular habitats if such patterns become apparent. While this method allows us to develop broader models of plant distribution throughout the site and to understand potential interactions between different plant species, the downside is that it assumes that modern communities have not changed from the communities that existed during the period of study.

Approach #2

This approach uses pollen data from soil cores to estimate the proportion of different plant species and functional groups within the landscape. The process involves five broad steps: (1) identifying ideal locations for sediment cores, (2) grouping the pollen data from cores into species as aquatic, terrestrial, or arboreal, to provide an initial sense of the proportions of each species and general potential for dispersal, (3) correct for amount of pollen produced by species, i.e., gymnosperms versus angiosperms, (4) establish ratios of each plant as part of the whole community, based on pollen core, and (5) model habitat preference categories, i.e., most likely locations of those plants based on habitat characteristics such as soil and elevation (factor in pollen dispersal distance, here or during analysis of ratios) (Fisher-Meerow and Judd 1989; Judd 1987; Higuera-Gundy et al. 1999; Holdridge 1945).

While this approach provides hard evidence of the existence of species dated to the time being studied, not all pollen has equal dispersal or durability and this method may lead to a model biased towards particular species. Additionally, potential locations for collection of pollen cores may be limited, increasing the bias towards plants near those locations.

To help counter some of the shortcomings of these two approaches, our methodology combines these two approaches into a single ecological framework to integrate data from the past and present in order to fill gaps in paleoenvironmental data and contextualize modern data (to the fullest extent possible given a particular data set) in prehistory. Archaeological, ethnographic, and ethnobotanical data (when available) are also incorporated into the model. Data integration occurs in a GIS to model potential plant distributions for specific time periods and then these georeferenced data are used to populate 3D visualizations with plant models in accordance with settlement data. We turn now to the case study at Copan, Honduras.

Case Study: Copan Honduras

The long-term goal of our project is to develop an innovative approach to investigate socio-environmental dynamics in ancient landscapes. The case study focuses on investigating how Copan's multi-ethnic history (900BCE-CE1100/1250) may have impacted interaction with(in) the physical and social landscape. Our main objective is to employ 3D visualizations (derived from paleoenvironmental, ethnographic, archaeological, and other data sources) to investigate this line of inquiry. To work towards achieving this objective, we have begun to test the workflow for Copan including evaluating potential 3D tools for plant modeling and landscape visualization.

Data Acquisition

The data for Copan are being acquired from a variety of sources and used for GIS and 3D modeling.

Plants

Photos

- Digital photos acquired from online databases (very few host data for Mesoamerica, particularly with high-resolution images necessary to generate 3D plant models or billboards)
- Digital photos acquired from field work. [When possible it is best to acquire photos of individual plants without background vegetation to facilitate subsequent 3D modeling (Fig. 3).]

Fig. 3 Digital photos of plants from field work at Copan, Honduras

Airborne LiDAR

- LiDAR data, acquired in May 2013 by the MayaArch3D Project, provides 3D point clouds of vegetation
- Forestry algorithms can be employed to generate 3D models of plants (Fig. 4) (Hu et al. 2014; Li et al. 2012).

3D Models

- Few 3D plant models for the semi-tropical environment of Central America are available from plant libraries or 3D warehouses.

Pollen (Soil Cores)

- Pollen and spore raw data from Petapilla pond sediment core, Copan Valley provide (when possible) species classifications from 46 levels (McNeil 2012)
- Percentages of pollen and spores classified as arboreal, herb, aquatic, or unknown from 46 levels (derived from raw data) (McNeil 2012).

Fig. 4 LiDAR data acquired for Copan, Honduras (courtesy MayaArch3D Project)

Plant Communities

- Maya Ethnobotanical Report—a quantitative ecological study for Copan Archaeological Park describes main ecosystems with data on species density, frequency, and dominance (House 2007).

Terrain

LiDAR

- Digital Terrain Model (DTM), 0.5 m resolution, generated from LiDAR.

Shapefiles (Vector Data)

- Soil
- Hydrology (Copan River, *quebradas*, ancient reservoirs)
- Geology

Settlement

- LiDAR provides georeferenced mound and building locations
- Pedestrian survey maps provide building footprints and type (Fash and Long 1983)
- Excavations provide attribute information for 3D reconstruction of buildings and additional social, political, ideological, and economic data.

Land Use

- Ethnographic data (e.g. Wisdom 1940)
- Ethnobotanical data (House 2007)
- Archaeological data (e.g., Baudez 1983; Webster 2005).

Data Processing

Data Processing is divided into two components, GIS and 3D Modelling. These two components overlap—GIS data are used to generate some 3D models and to spatially reference non-GIS derived 3D models such as individual 3D plant models—and the process is iterative, working back and forth between GIS and 3D modeling software. This is a critical phase in the workflow because it is in this phase that the ecological approach is applied to determine potential plant communities and their spatial locations within Copan's landscape.

GIS

Determining Potential Plant Community Locations

Step 1. Plant Classification: It is necessary to classify the plant data into broad types and then determine percentages of these types for specific time periods for analysis and visualization purposes. To begin this process, we are using McNeil's data (2012) pollen data derived from soil cores.

- Step 1a. We identified time periods (n = 7) that combine key cultural events with ecological trends at Copan in order to better understand the general trends and changes in specific plant species (Table 2).
- Step 1b. We compiled tables (n = 7) with percentage per species of total for the seven time periods listed in Table 2. Table 3 illustrates an example from the Early Classic Period (CE400–600).
- Step 1c. Using McNeil's (2012) data, we identified traits and trends in plant data for each time period by comparing to previous time period (e.g., comparing Early Classic Period data to ProtoClassic Period data).

Table 2 Time periods for the 3D visualization and analysis case study of Copan's socio-environmental dynamics

Period	Date range
Preclassic	900BCE–CE100
Protoclassic	CE100–400
Early classic	CE400–600
Late classic	CE600–780
Terminal classic	CE780–980
Proto-postclassic	CE980–1100
PostClassic	CE1100–1220/1300

Table 3 Illustrates species percentages classified by time period (e.g., Early Classic) [based on McNeil 2012]

Genus/Family	CE400–600	Total%
Acalypha (a) (Genus)		0.68
Acrocomia (a)-coyol palm	0.40	0.40
Alchornea (a)	0.48	0.48
Alnus (a)	0.15	0.15
Arecaceae (a)	0.35	0.35
Chamaedorea-type (a)-palms	7.08	7.08
Hedyosmum mexicanum (a)	0.35	0.35
Ilex (a)	0	0
Mimosa (type A) (a)	0	0
Mimosa pigra (type B) (a)	0	0
Pinus (a)	25.88	25.88
Quercus (a)	8.22	8.22
Urticales (a)	3.48	3.48
Piperaceae (a)	5.12	5.12
Liquidambar (a)	0.68	0.68
Burseraceae (a)	0	0
Myrtaceae (a)	0.15	0.15
Rhamnaceae (a)	0.13	0.13
Sapindaceae (a)	0	0
Osmunda-type (aq)	12.10	12.10
Typha (aq)	0.75	0.75
Cyperaceae (aq)	21.98	21.98
Pterid., monolete and psilate (aq)	11.15	11.15
Croton (aq)		
Begoniaceae (aq)		
Zea mays (h)	0.27	0.27
Asteraceae (h)	16.90	16.90
Chenopodiaceae/Amaranthaceae (h)	10.08	10.08
Poaceae (h)	18.97	18.97

Trees and Shrubs

- Diversity of tree species continues to increase, e.g., pine trees increase ~8 %; Oak stable; Urticales and Piperaceae stable

Aquatic Plants

- Typha stable, Osmunda increase from ~3.5 to 12 %, Cyperaceae decrease from 26 to 22 %, Pterid…increase from 6 to 11 %

Upland Herbs

- Zea mays slightly increases; daisies decrease from 37 to 17 %, Chenopodiaceae/Amaranthaceae increase from 6 to 10 %; grasses stable

  Step 1d. We subdivided each time period into two categories "trees and shrubs" and "upland herbs" based on McNeil (2012) (Table 4 illustrates the Early Classic Period).

Step 2. Compare all identified plant species (derived from sediment cores) to ecosystem shapefiles (generated from House 2007) to assign preliminary spatial locations to specific plant species

Step 3. Use GIS to generate "ecosystem" shapefiles for each time period (n = 7) based on total percentage of trees and shrubs and upland herbs

Step 4. Use GIS to calculate "potential" total area for each species within an ecosystem based on environmental variables such as soil and hydrology combined with percentages of individual plant species categorized as trees and shrubs, upland herbs, or aquatic; calculations stored as attribute of ecosystem (spatial location within each ecosystem assigned stochastically based on estimated area/plant species (stored as attribute) in 3D visualization).

Determining Archaeological Settlement Locations

LiDAR data combined with pedestrian survey maps provide settlement data (i.e. mound locations and building footprints) spanning the Late Classic and Terminal Classic Periods (ca. CE700–820). However, mapping settlement data for earlier time periods is not straightforward because the majority of Copan's nearly 3600 buildings remain unexcavated (Fig. 5 compares Preclassic to Classic Period settlement patterns).

To begin to generate building footprints for earlier time periods we turn to three data sources: (1) extensive excavations, (2) test excavations, and (3) settlement patterns derived from excavation and survey data at Copan and nearby valleys (e.g., Canuto and Bell 2008; Manahan and Canuto 2010).

Table 4 Table comparing percentage of trees and shrubs to upland herbs: example from Early Classic Period (derived from McNeil 2012 pollen data)

Early classic (CE400–600)	Total%
Trees and shrubs	53.5
Upland herbs	46.5

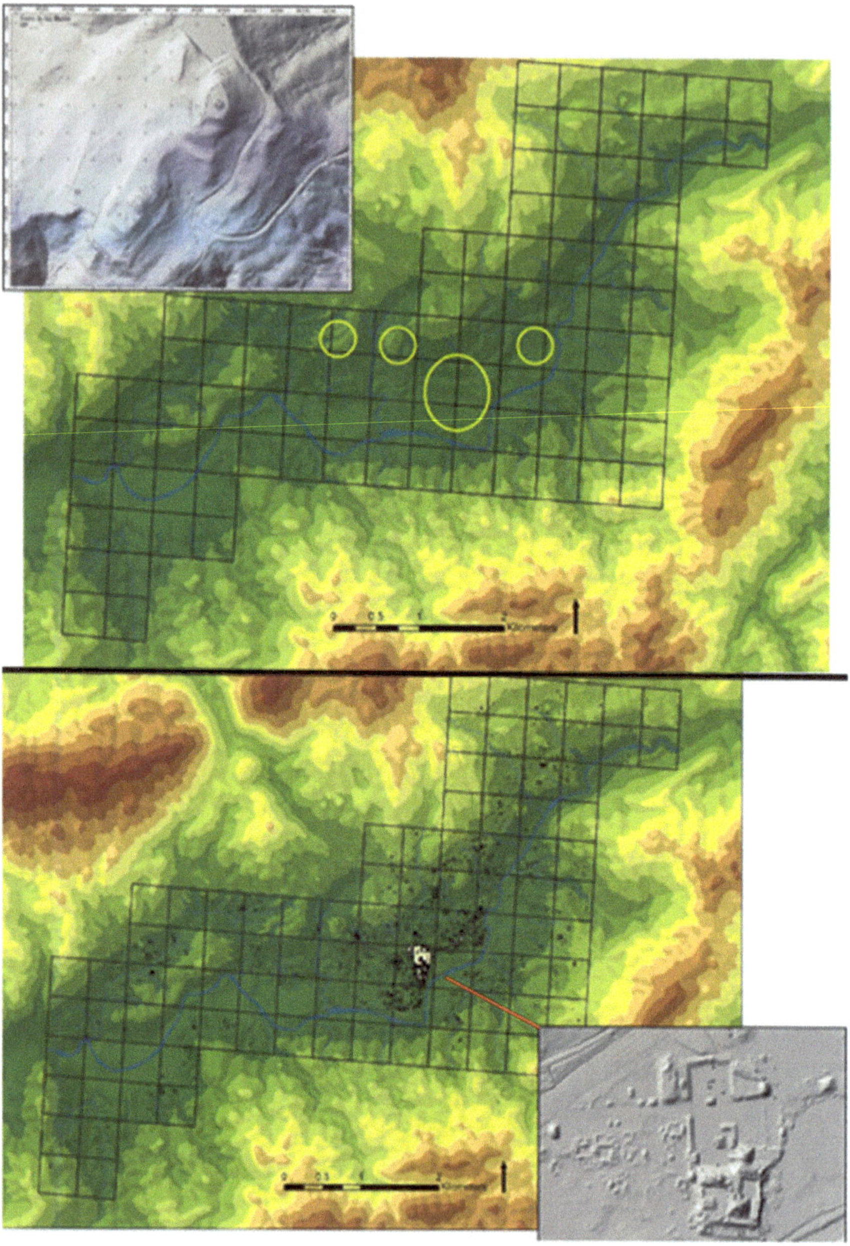

Fig. 5 Preclassic settlement pattern, Copan Honduras illustrating dispersed, low density settlement on hilltops prior to valley bottom settlement (*top*); Late Classic settlement pattern, Copan, Honduras illustrating high density settlement in urban core and into foothills (*bottom*)

Archaeologists have also identified four key indicators that differentiate Non-Maya and Maya settlements: (1) settlement pattern, (2) spatial organization, (3) architecture, (4) and use of space (e.g., Canuto 2002; Canuto and Bell 2008).

Non-Maya Patterns

Settlement Pattern:

- Hilltops overlooking floodplain
- Sites located approximately 30 m above agricultural fields

Spatial Organization:

- Standalone range structures
- Positioned along edge of hilltop
- Large, open, and accessible interior plazas

Architecture:

- Earthen platforms
- Pole-and-thatch superstructures
- Lower Buildings (<2 m)
- Residential platforms wider/longer than Maya
- Little intra-community variation

Use of Space:

- Higher P-accumulations in plazas
- Small refuse and garden areas
- Intensive in-field agricultural system
- Minimal household horticulture

Maya Patterns

Settlement Pattern:

- Settlement on/adjacent floodplains
- Not on hilltops

Spatial Organization:

- Clear boundaries defined by architecture
- Smaller and more tightly-packed buildings
- Residences w/smaller plazas
- Restricted access ceremonial structures

Architecture:

- Cobblestone platforms
- Taller buildings
- Large intra-community variation
- Superstructures (wattle-and-daub, cut-masonry)

Use of Space:

- Lower P-accumulations in plazas
- Large refuse and garden areas
- Fields distant from households
- Exclusive land-tenure pattern
- Intensive horticulture

We propose an approach that integrates these generalized non-Maya and Maya settlement patterns and site organization plans, arguably representative of distinct cultural groups, with paleoenvironmental data that McNeil (2009, 2012) has correlated to specific cultural groups in order to reconstruct land use, plant communities, and settlement patterns for each of the key time periods at Copan (n = 7). This strategy aligns with the ecological approach described above.

Admittedly, any spatial data sets derived for early settlement at Copan are approximations; however, we see this as a first but necessary step to begin to generate GIS data that can be ingested in a 3DGIS. Subsequently, these data are explored from a three-dimensional perspective that links underlying archaeological and ecological data to the 3D models to generate preliminary settlement and land use models for earlier time periods and importantly to initiate scholarly discourse within a collaborative, dynamic, and iterative framework about changing ancient socio-environmental at Copan that can ultimately be investigate in regard to ethnicity.

3D Models

The 3D models component of the workflow is divided into two parts: (1) identify options for modeling individual plant species and (2) evaluate existing software to visualize and interact with 3D landscapes that can integrate georeferenced architecture and vegetation data.

Generating Individual 3D Plant Models

Given that our case study is in Central America, existing 3D plant models, particularly free models, are limited to say the least. This circumstance requires that we devise a workflow to efficiently generate plant models. To this end, we have investigated and evaluated several data types and software to generate 3D plant models. Table 5 lists the evaluated software, primary function, pros, cons, and

Table 5 Summary table of 3D tools and data for generating individual 3D plant models

Software	Function	Pros	Cons	Comments
XFrog	Create 3D plant models	Full-featured plan modeling @$200–400	Steep learning curve; non-generative	Good option for creating models for GIS-based visualization
3D ArcStudio 3D TreeMaker	3D plant modeling	Easy to learn; inexpensive; customizable; parameters	Proprietary; tailored to work with SketchUp	Quick, easy, change tree templates
TreeGen	Procedural plant modeling	Open source, free; adaptable, e.g., to Blender	Python Script (requires coding to adapt; creates random trees)	Possible modify to generate non-random trees and generate other plants; designed for 3D StudioMax
Abaro	3D tree modeling	Implements tree generating algorithm; import into Blender; open source (Java)	Few trees in library	Open source; complicated; steep learning curve; time-consuming
SpeedTree	Middleware solution for modeling and real time rendering of plants	Procedural approach to modeling	Proprietary; expensive @ $895.00 ($19 monthly option)	Procedural approach with data interoperability compensates for its existing libraries having few Mesoamerican plants
SketchUp	3D modeling	Easy to learn; inexpensive; exports multiple formats	Proprietary; can create plugins but difficult to customize; manual process	Time-consuming to create models (existing libraries have few Mesoamerican plants)
LiDAR	Use 3D points to generate 3D plant models	Useful for present-day plants	Data acquisition and post-processing expensive; requires expertise, proper algorithms; data for species not in coverage not collected	Apply forestry algorithms to model tree species

comments. From our investigations, we have identified three general approaches to visualize plants in a 3D landscape context.

- "True" 3D models with x, y, and z coordinates—these have the advantage of being useful for analytical purposes but are rather large to store and visualize
- Billboards—2D object always oriented to face object—decreases data size and modeling time/cost and is particularly valuable for online visualization; however, not useful for 3D analysis
- Procedural Models—ultimately we prefer a procedural approach where the plants can be generated from a set of rules.

While all three options require researching plant species and collecting digital photos and/or drawings prior to 3D visualization, from our software evaluations we prefer software that employs procedural modeling such as TreeGen or SpeedTree because it permits flexibility in modeling that facilitates the generation of multiple 3D reconstructions, and hence an interactive, dynamic, and iterative exploration of socio-environmental simulations.

3D Settlement Data

While the focus of this chapter is not 3D modeling of architecture, it is important to provide an overview of the subject because the ultimate goal of the proposed workflow is to integrate 3D settlement and environmental data. Three methods of 3D modeling of architecture are briefly described.

- Extruding GIS footprints—streamlined, quick process using height attribute but results in blocky unrealistic representations that would hinder analysis, e.g., visibility analysis—an example is flat versus pitched roofs
- Manual Modeling—using software such as 3D Studio Max or SketchUp to model buildings—this process is useful for modeling individual buildings or small spatial extents; however, it is time-consuming to model vast landscapes
- Procedural Modeling—use a set of rules to generate and texture 3D buildings based on GIS attributes and automatically situate buildings to their spatially-referenced locations in a landscape.

Software for 3DGIS Landscape Visualization

Increasingly, more software for 3D landscape visualization are available. These software programs range in functionality (visualization vs. analysis), platform (desktop, web-based, mobile application), data interoperability, rendering capabilities, and ability to integrate and link to georeferenced data. Table 6 lists nine software programs along with their main function as well as pros and cons and comments relevant to an ecological approach for 3D visualization of plants and architecture in virtual landscapes.

Table 6 3DGIS Visualization Software evaluated for utility in ecological workflow

Software	Function	Pros	Cons	Comments
ESRI ArcScene	Visualizing 2.5/3D georeferenced data	Easy to use interface, file Compatibility (DWG, DXF, DGN), Interfaces with ArcGIS Desktop	Limited in use of full geometry for 3D Analysis and visualization; proprietary	Limited 3D rendering and analysis capabilities for web-based visualization
ERDAS Imagine Virtual GIS	Geological remote sensing, Image processing	Uses image data to create 3D representations; focus on visualization, rapid rendering and fly-throughs	Proprietary, expensive, and limited ability to include modeled 3D imagery. No ability for 3D analysis	Appropriate for quick visualization of existing landscapes, not ideally suited to integrating 3D plant models
GeoMedia Terrain	3D representation of 2D GIS	Visualizes 2D maps and runs fly-throughs of GIS data in GeoMedia platforms	Expensive, limited analysis and visualization capabilities	Intended for visualizing terrain, and expensive
GRASS GIS	3D representation of 2D GIS	Open Source, free, works with Quantum GIS; ALIDAR tools, Adaptable.	Limited in use of full geometry for 3D Analysis and visualization	Provides basic open source and free platform; however, for intensive visualization and integration of 3D vegetation models, additional plugins and adaptations required
Demeter Terrain Engine	Renders 3D Terrain using OpenGL	Open Source, free; uses OpenGL; adaptable; fast performance; integrate with Open Scene Graph to populate with objects, such as trees; adaptive mesh	Limited 3D analysis, needs to integrate Demeter into application to add 3D models (i.e., using Open Scene Graph)	Option as a basic platform, however, it is essentially only a visualization program, and would not be able to perform analysis

(continued)

Table 6 (continued)

Software	Function	Pros	Cons	Comments
Virtual Terrain Project	Project linking a suite of software to create 3D simulations of real world	Interfaces with Xfrog (Plant Modeler), Open source, free, has library of 94 species and 409 instances (as billboards)	Limited capacity for analysis, not well linked with common GIS platforms	Good potential, existing (limited) plant library; infrastructure for adding plants based on ecological classification
ESRI CityEngine	Procedural modeling of GIS data for architecture/plants	Rule-based (generative); exports to multiple formats; exports to WebGL; plant libraries; plants visualized based on probabilities	Proprietary, expensive; requires learning CGA Shape Grammar—steep learning curve	Procedural approach allows for flexibility and testing of visualizations; GIS platform enables vegetation placement based on ruler linked to attributes in geodatabase; permits subsequent GIS analysis
BioSphere 3D	Modeling and 3D visualization of real world on virtual globe	High visual quality, streamlined rendering, uses XFrog; used in visualization of landscape change; open source	Little analytical capability	Good potential for development; supports shapefile, satellite/aerial imagery, open source allows for modification to add analytical functionality
Visual Nature Studio 3	3D Visualization of architecture and plants using GIS data	Imports GIS (vector/raster) data; models ecotypes and foliage; dynamic modeling	Mid-range price (dongle); steep learning curve; exports do not store full scenes; no analysis	"Quick" option to model 3D landscape from GIS data but not a long-term solution for development

Initial tests were carried out using VNS3 (Visual Nature Studio 3). VNS3 performs 3D rendering of georeferenced data sets (raster and vector) taking advantage of GIS' ability to deal with locational data and overlay settlement and vegetation data. Moreover, because VNS3 was originally designed for forestry applications it enables users to set parameters for rendering vegetation data, for example, to set minimum and maximum size and density for ecosystems and foliage types in relation to underlying GIS data. This ability to integrate georeferenced settlement data as 2D vector and/or 3D architecture models (e.g., as OBJ) and render in conjunction with 3D plant models based on ecosystems defined by underlying GIS data enables users to test alternative landscape simulations (Figs. 6, 7, and 8).

Following these initial test simulations, we have also begun a procedural modeling approach using CityEngine in order to overcome the limitations of VNS3 including lack of flexibility in modeling to rapidly test alternative simulations, limited data interoperability for export, lack of interactive interface for rendered scenes, and steep learning curve (Richards-Rissetto and Plessing 2015).

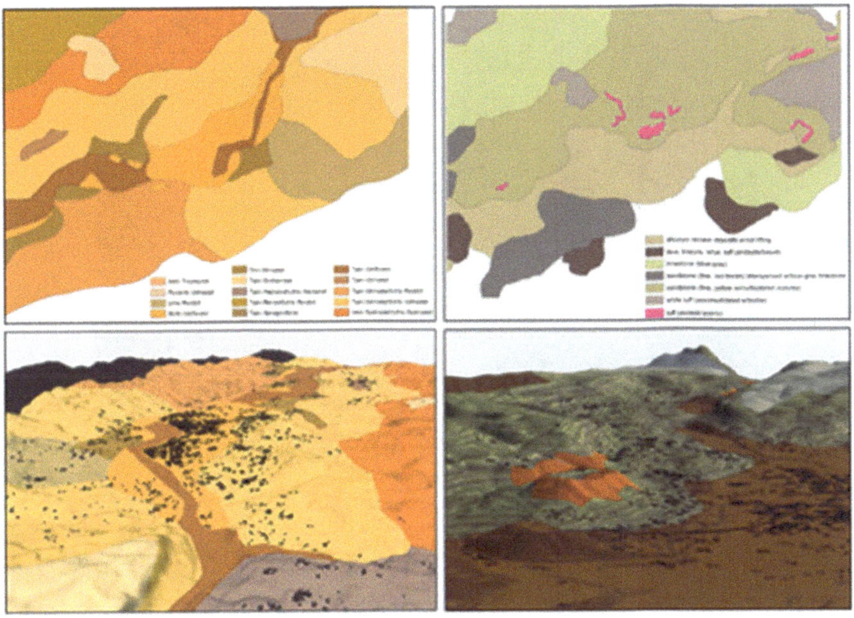

Fig. 6 GIS versus 3D Views of soil data (*left*) and geological data (*right*)—overlaid with Late Classic settlement data from Copan, Honduras in Visual Nature Studio 3

Fig. 7 A 3D simulation of Cerro Chino—a non-Maya Preclassic Period Site at Copan, Honduras illustrating hilltop settlement with downslope maize production using VNS3 in landscape

Fig. 8 A 3D simulation of Late Classic Period Sites at Copan, Honduras illustrating shift from hilltops to valley bottom with garden orchards using VNS3 in landscape

Future Direction: 3DGIS Data Visualization and Analysis

Implementation of the complete ecological workflow at Copan requires further collection of data on current plant communities, and possibly further assessment of paleological soils and microclimates. However, from these preliminary visualizations, we highlight some general advantages of visualizing archaeological and environmental data together in a three-dimensional environment including:

- Depth perception
- Explore multiple perspectives
- Visualize multiple variables
- Ability to portray movement
- Explore inter-relationships between archaeological and environmental features

Future research applications at Copan include:

- Visualize in 3D paleoenvironmental changes through time to understand processes of change in land use as different cultural groups immigrate to and emigrate from Copan
- Visualize in 3D ratio of trees to herbs through time to investigate (a) de(-forestation) rates and their impact on human behavior and (b) contrasting hypotheses about the nature of collapse at Copan

We close with the question: Is there a future for 3D landscape visualization in archaeological research? While this question is open-ended, we conclude from our initial investigations that 3D landscape visualization is a good way to bring different lines of evidence together—archaeological, geomorphological, paleoenvironmental, etc.—in order to investigate processes of past human behavior from non-traditional and alternative perspectives. We propose an iterative workflow that works back and forth between GIS and 3D visualization tools to come to see new and different things in a 3D environment that were not apparent in the GIS and vice versa—ultimately making past landscapes tangible objects of study.

Acknowledgments We thank the Undergraduate Collaborative Research Fund, Middlebury College, MayaArch3D Project, National Science Foundation (# 1064648), Honduran Institute for Anthropology and History (IHAH), and Dr. Cameron McNeil, CUNY. The MayaArch3D Project has generously provided airborne LiDAR of Copan.

References

Barcelo, J., M. Forte, and D. Sanders. 2000. Virtual reality in archaeology. Oxford: BAR Archaeopress.

Baudez, Claude-Francois. 1983. Introducción a la Arqueología de Copán, Honduras. Proyecto Arqueológico Copán, Secretaria de Estado en el Despacho de Cultura y Turismo, Tegucigalpa, D.C.

Bell, E, M. Canuto, and R Sharer. 2004. Understanding Early Classic Copan. University of Pennsylvania Museum of Archaeology and Anthropology, Philadelphia.

Canuto, M. 2002. A Tale of Two Cities: The Role of the Rural Community in the Socio-Political Integration of the Copan Drainage in the Late Preclassic and Classic Periods. Ph.D. Thesis, Department of Anthropology, University of Pennsylvania.

Canuto, M., E. Bell 2008. The Ties that Bind: Administrative Strategies in the El Paraíso Valley, Department of Copan, Honduras. Mexicon. Vol. XXX: 10-20.

Canuto, M., E. Bell and 2013. Archaeological investigations in the el Paraíso Valley: the role of secondary centers in the multiethnic landscape of classic period Copan. Ancient Mesoamerica 24:1-24. Cambridge University Press.

Chapman, H. 2006. Landscape Archaeology and GIS. Great Britain: Tempus Publishing

Conolly, J. and Lake, M. 2006. Geographical Information Systems in Archaeology. Cambridge: Cambridge University Press.

Dell'Unto N., G. Landeschi, A. Leander Touati, M. Dellepiane, M. Callieri, D. Ferdani. 2015. Experiencing Ancient Buildings from a 3D GIS Perspective: a Case Drawn from the Swedish Pompeii Project. Journal of Archaeological Method and Theory, 1-22.

Dylla, K, B Frischer, P Mueller, A. Ulmer, and S Haegler. 2009. Rome Reborn 2.0: A Case Study of Virtual City Reconstruction Using Procedural Modeling Techniques. Making History Interactive. 37th Proceedings of the CAA Conference 2009, Williamsburg, Virginia (Archaeopress: Oxford, 2010) 62-66.

Fash, W., Long K., 1983. Mapa Arqueológico del Valle de Copán. In Introducción a la Arqueologíade Copán. C.F. Baudez, ed, Vol. Tomo III. Tegucigalpa, Honduras: Instituto Hondureño de Antropología 475 eHistoria.

Fedick, S. 1996. The Managed Mosaic: Ancient Maya Agriculture and Resource Use. University of Utah Press: Salt Lake City.

Fischer, Lisa E. 2012. "↑irtually ↑isiting Williamsburg: The ← se of 3D Modelling to Enrich the Museum Guest Experience." Paper presented at the Computer Applications andQuantitative Methods in Archaeology Conference, Southampton, United Kingdom, March 26-29.

Fisher-Meerow, L. and W. Jud. 1989. A floristic study of five sites along an elevational transect in the Sierra de Baoruco, Prov. Pedernales, Dominican Republic. Moscosoa 5: 159–185.

Forte, M, and S. Kay 2001. Remote Sensing, GIS and Virtual Reconstruction of Archaeological Landscapes, in "The reconstruction of Archaeological Landscapes through Digital Technologies", Proceedings of the 1st Italy-United States Workshop, Boston, Massachussets, USA, November 1-3, 2001, BAR International Series 1151, Oxford, 2002, 109-116.

Forte, M., A. Guidazzoli, S. Pescarin, and M. Zane, 2003. The Aksum Project: A VR GIS for a 3D inclusive interaction with an archaeological landscape, "Enter the Past. The E-way into the Four Dimensions of Cultural Heritage", in CAA 2003, Computer Applications and Quantitative Methods in Archaeology. Proceedings of the 31st Conference, Vienna, Austria, April, 2003, BAR International series 1227 537-540.

Forte, M., L. Calori, C. Camporesi, A. Guidazzoli, and S. Pescarin, 2005. OpenHeritage: Integrated approach to web 3d publication of virtual landscapes, in Proceedings of ISPRS Workshop 3D-ARCH'2005: "3D Virtual Reconstruction and Visualization of Complex Architectures", Aug.05, Mestre-Venice (IT)

Forte, M., L. Calori, and S. Pescarin, 2005. Real-time interactive reconstruction of archaeological landscapes: an OpenSource approach: from GIS to Virtual Reality, in the series of Volumes of The International Archives of the Photogrammetry, Remote Sensing and Spatial Information Sciences from ISPRS Congress "3D Digital Imaging and Modeling: Applications of Heritage, Industry, Medicine and Land" (Padova 17-18 May 2005)

Forte, M. (ed), 2005. The reconstruction of Archaeological Landscapes through Digital Technologies, Proceedings of the 2nd Italy-United States Workshop, Rome, Italy, November 3-5, 2003. Berkeley, USA, May, 2005, Edited by Maurizio Forte, BAR International series, 1379

Forte, M. 2014. 3D Archaeology. New Perspectives and Challenges. The example of Çatalhöyük. *Journal of Eastern Mediterranean Archaeology*, Vol.2, n.1.

Franklin, J. 2010. Mapping Species Distributions: Spatial Inference and PredictionCambridge University Press, Cambridge, UK

Franklin, J., A. Potts, E. Fisher, R. Cowling, and C. Marean. 2015. Paleodistribution modeling in archaeology and paleoanthropology. Quaternary Science Reviews 110(15): 1-14.

Frischer, Bernard and Anastasia Dakouri-Hild. 2008. Beyond Illustration: 2D and 3D Digital Technologies as tools for Discovery in Archaeology. Oxford: Archaeopress.

Gabii Goes Digital; (last accessed March 24, 2015), http://gabiiserver.adsroot.itcs.umich.edu/gabiigoesdigital/index.html

Gerstle, A. 1987. Ethnic Diversity and Interaction at Copan, Honduras. In Interaction on the Southeast Mesoamerican Frontier, edited by E. J. Robinson, pp. 462. vol. ii. BAR International Series 327.

Higuera-Gundy, A., Brenner, M., Hodell, D. A., Curtis, J. H., Leyden, B. W., and Binford, M. W., 1999, A 10,300 14C yr record of climate and vegetation change from Haiti: Quaternary Research, v. 52, p. 159-170.

Holdridge, L. 1945. A brief sketch of the Puerto Rican flora. Plants and plant science in Latin America, ed. F. Verdoorn, Chronica Botanica, Waltham, Mass.

House, P. 2007. Etnobotánica Maya, Parque Arqueol'ogico Ruinas de Copán. Instituto Regional de Biodiversidad, Instituto Hondureno de Antropología e Historia IHAH.

Hu, B., J. Li, L. Jing, and A. Juda. 2014. International Journal of Applied Earth Observation and Geinformation 26: 145-155.

Judd, W. S. 1987. Floristic study of Morne La Visite and Pic Macaya National Parks, Haiti. Bull. of the Florida State Museum 32: 1–136

Kim, J., D. Shin, and E. Ahn. 2011. Image-based Modeling for Virtual Museum. T.h Kim et al. (Eds), Multimedia, Computer Graphics, and Broadcasting: Part I: International Conference, MulGraB 2011,pp. 108-119.Springer-Verlag Berlin Heidelberg.

Li, W., Q. Guo, M. Jakubowski and M. Kelly. 2012. A New Method for Segmenting Individual Trees from the Lidar Point Cloud. American Society for Photogrammetry and Remote Sensing 78(1): 75-84.

Maca, A., and K. Miller. 2009. Copan's Q'eqch'i Barrio? Paper presented at 74th Annual Society for American Archaeology Meeting, April 22-26, 2009, Atlanta, Georgia.

Manahan, K. and M. Canuto. 2010. Bracketing the Copan Dynasty: Late Preclassic and Early PostClassic Settlements at Copan, Honduras. Latin American Antiquity 20(4):553-580.

MayaArch3D Project; (last accessed March 24, 2015), http://www.mayaarch3d.org/

McNeil, C. 2009. The Environmental Record of Human Populations and Migrations in the Copan Valley, Honduras. B. Metz, C. McNeil, and K. Hull (eds) in Ch'ori Maya Area. University of Florida Press.

McNeil, C., D. Burney, and L. Pigott Burney. 2010. Evidence disputing deforestation as the cause for the collapse of the ancient Maya polity of Copan, Honduras. Proceedings of the National Academy of Sciences. Jan 14; 107(3):1017-22.

McNeil, C. 2012. Deforestation, agroforestry, and sustainable land management practices among the Classic period Maya. Quaternary International, vol. 249, pp. 19-30

Opitz, R. and J. Nowlin, 2012. "Photogrammetric Modeling + GIS: Better methods for working with mesh data". ArcUser Spring 2012:46-49.

Pescarin, S., L. Calori, and M. Forte, 200. Real-Time Interactive Reconstruction Of Archaeological Landscapes: An Opensource Approach From Gis To Virtual Reality, in Proceedings of Italy-Canada 2005 workshop, "3D Digital Imaging and Modeling Applications of: Heritage, Industry, Medicine & Land, CD ROM, Padova, 2005.

Price, D., J. Burton, P. Fullagar, L. Wright, J. Buikstra, and V. Tiesler. 2008. 87Sr/86Sr Ratios and the Study of Human Mobility in Ancient Mesoamerica. Latin American Antiquity 19 (2):167-180.

Richards-Rissetto, H. 2010 *Exploring Social Interaction at the Ancient Maya City of Copan, Honduras: A Multi-Scalar Geographic Information Systems (GIS) Analysis of Access and Visibility*. Ph.D. DissertationUniversity of New Mexico, Albuquerque.

Richards-Rissetto, H. 2013 From Mounds to Maps to Models: Visualizing Ancient Architecture across Landscapes. *Proceedings of Digital Heritage International Congress*, Marseille, France.

Richards-Rissetto, H., J. Robertsson, F. Remondino, G. Agugiaro, J. von Schwerin, and G. Galibri. 2013 Geospatial Virtual Heritage: An Interactive, Gesture-based 3D GIS to Engage the Public with Ancient Maya Archaeology. Proceedings of Computer Applications and Quantitative Methods in Archaeology, Southampton, United Kingdom, March 26-29.

Richards-Rissetto, H. and R. Plessing. 2015. Procedural modeling for ancient Maya cityscapes: Initial methodological challenges and solutions.Proceedings for Digital Heritage International Congress 2015, Granada, Spain, Sept. 28-Oct. 2, 2015.

Saldana, M and C Johanson. 2013. Procedural modeling for rapid-prototyping of multiple building phases. International Archives of the Photogrammetry, Remote Sensing and Spatial Information Sciences, Vol. XL-5/W1. 3DARCH 2013-3D Virtual Reconstruction and Visualization of Complex Architectures.

Sharer, R.J., 2009. The Ch'orti' past: an archaeological perspective. In: InMetz, B.E., McNeil, C. L., Hull, K.M. (Eds.), The Ch'orti' Maya Area. Past and Present. University Press of Florida, Gainesville, pp. 124-133.

Stuart, D. 2007. Maya Decipherment Blog: "Origin of Copan's Founder" https://decipherment.wordpress.com/2007/06/25/the-origin-of-copans-founder/

von Schwerin, J. , H. Richards-Rissetto, F. Remondino, and G. Agugiaro. 2013. The MayaArch3D Project: A 3D WebGIS for Analyzing Ancient Maya Architecture and Landscapes at Copan, Honduras.Literary and Linguistic Computing 2013, doi:10.1093/llc/fqt059. Oxford University Press.

Saldana, M. 2014. An Integrated Approach to the Procedural Modeling of Ancient Cities and Buildings. http://www.mariesaldana.com/dh2014/; Last accessed Dec. 3, 2014

Viel, R. 1993. Evolucion de la Cermica de Copán, Honduras. Instituto Hondureno de Antropología e Historia and Centro de Estudios Mexicanos y Centroamericanos, Tegucigalpa, D.C., Honduras and Mexico D.F., Mexico.

Webster, D. 2005. Political Ecology, Political Economy, and the Culture History of Resource Management at Copan, A. Wyllys and W. Fash. (Eds), In Copán: The History of an Ancient Maya Kingdom, pp. 33-72.School of American Research Advanced Seminar Series: Santa Fe, NM.

Wheatley, D.W. and Gillings, M. (2002). Spatial technology and archaeology: a guide to the archaeological applications of GIS. London: Taylor & Francis.

Wisdom, Charles 1940. The Chorti Indians of Guatemala. University of Chicago Press, Chicago.

Visualizing Medieval Iberia's Contested Space Through Multiple Scales of Visibility Analysis

Edward Triplett

Abstract Between the 12th and 14th centuries, Medieval Iberia's military-religious frontier was a volatile, fluctuating band of contested space that tenuously bisected the peninsula into Muslim and Christian controlled territory. On both sides, the structures that best exemplified the material efforts to retain newly gained territory were the hilltop and spur castles occupied by frontier settlers from the interior of each side of the conflict. These castles—which are nearly as ubiquitous as windmills in images of central Spain—are often photographed as crumbling bits of crenellation set sharply against the sky. Even images captured within these fortresses tend to use the extant masonry as a frame for sublime views of the landscape below. Rather than attempt to separate the indelible connection between landscape views, topographic siting and architectural fabric at these sites, this paper will alternate between looking away from, and within these fortresses through two methods of remote sensing. The first is a series of viewshed analyses set from the observation points of over seven hundred fortresses, monasteries and towns during a two hundred year period. This GIS process employs a digital elevation model of the surrounding landscape to ask not only which frontier institutions occupied which fortresses during any month between 1150 and 1350 CE, but also what portions of the surrounding landscape the occupiers could see at that time. The second remote-sensing method combines an array of digital light meters to calculate volumetric visibility for spaces within a 3D model of the 14th century fortress-monastery headquarters for the military order of Montesa. The combination of the volumetric and GIS viewshed analysis methods will reveal how frontier institutions valued landscape visibility as a measurement of security and surveillance, while also acknowledging how vision affected architecture-scale decisions at a military-monastic complex on the frontier.

E. Triplett (✉)
CLIR Postdoctoral Fellow, Duke University, Durham, USA
e-mail: edward.triplett@duke.edu

M. Forte and S. Campana (eds.), *Digital Methods and Remote Sensing in Archaeology*, Quantitative Methods in the Humanities and Social Sciences,
DOI 10.1007/978-3-319-40658-9_9

2D and 3D Viewshed Analysis

The debate over the chief influence on medieval Iberia's historical character still begins and ends with the constant presence of a shifting military-religious frontier.[1] In addition to the fluctuating legal, economic, and cultural rules of contact between religious groups, historians have turned to the built environment as evidence of cultural permeability and polemics on the frontier. Much of the scholarship on these structures has focused on urban, ecclesiastical and monastic architecture, or examples of regional frontier style, but less attention has been paid to the effects of fortifications on the greater surface of the frontier landscape. The hilltop fortresses —which are nearly as ubiquitous as windmills in the photographic record of central Spain—are typically captured as crumbling bits of crenellation cast against the sky. Even images captured within these fortresses tend to use the extant masonry as a frame for sublime views of the landscape below. The tendency to combine all of these fortresses into a vaguely crenelated environment results in a misrepresentation of each site's tactical influence on the frontier. This lack of precision also fails to capture how castles and watchtowers were interconnected in a broad, frontier network. Rather than attempt to separate the indelible connection between landscape views, topographic siting and architectural fabric at these sites, this paper will alternate between looking away from, and within these fortresses via two methods of remote sensing.

The first project—referred to as the ADIMO project hereafter—visualizes two hundred years of landscape visibility from the perspectives of virtual viewers standing at the highest point of over 800 frontier fortifications and fortified towns. Through GIS viewshed analysis, any month between 1150 and 1350 can be queried to reveal colored representations of not only which groups occupied which structures, but also what they could see from those locations. An additional aspect of this project reveals faction-specific lines of intervisibility between these sites that change colors based on who occupied each site. This illustration of shifting visibility networks reveals how remote sensing techniques can alter our historical perception of which sites had higher degrees of centrality on the frontier—regardless of their degree of survival.

The second project—referred to as the Montesa Modeling Project below—identifies a method for recording quantifiable levels of visibility within a 3D space using an array of digital light meters. The virtual space used to test this method is a 3D reconstruction of the 14th century fortress-monastery headquarters for the military order of Montesa. This model—which was built on top of a dense mesh of extant masonry captured using structure from motion—demonstrates how a combination of remote sensing technologies can enable scholars to test and visualize

[1]For an overview of works concerning the medieval Iberian frontier see: (Moreno et al. 1994; Castro 1971; Glick 1995; Feliciano and Rouhi 2006; Burns 1990; Forey 1984; Gerrard 2000; Pick 2004; Safran 2013; Bartlett et al. 1989; For medieval Iberian frontier studies in the fields of Art and Architectural history see: Robinson (2011), Dodds et al. (2008), Watt (2011).

levels of visibility in three-dimensional space. Finally, by merging landscape visibility with the architectural scale model in a 3D GIS environment, this project acts as an immersive translation for the original, more abstract 2D representations of viewshed.

ADIMO

The ADIMO project began as a response to a relatively benign statement made by Alan Forey—a historian working on Iberia's military orders—in a 1984 article titled "The Military Orders and the Spanish Reconquest in the Twelfth and Thirteenth Centuries." Forey's account of the spatial distribution of military order fortresses reads as follows:

> It is *impossible* to undertake a comprehensive survey of all the castles in frontier districts which passed into the hands of the military orders, partly because of inadequacies of evidence and difficulties of placename identification, and partly because of the incompleteness of research; but it is clear that in some frontier regions particular orders tended to predominate. (Forey 1984)

The word 'impossible' has been emphasized above because it inspired the decision to precisely locate these military order fortresses both spatially and temporally with the assistance of GIS and web technology that was not available in 1984. With assistance from the University of Virginia's Scholars' Lab, and the staff at the Institute for Advanced Technology in the Humanities (IATH), I built a custom relational database in Ruby on Rails that eventually contained over 1100 changing occupation events at more than 800 architectural sites. These sites included major cities and battle locations, but the bulk of the database was populated by hilltop fortresses with widespread views of the landscape. This ratio is largely due to the great abundance and reliability of evidence for individual fortress donations—as opposed to houses located in the many encomiendas of the military orders. When possible, I strove to identify these low-lying villages as well. This database, called ADIMO—Architectural Database of Iberian Military Orders—relied on a wide range of sources to precisely locate and date the structures. These sources included royal archives, bullariums of the different orders, a remarkably thorough online encyclopedia called castillos.net that is based on a 1997 three-volume print encyclopedia, and perhaps most usefully, the online publications by local and regional ministries of culture in Spain and Portugal.[2]

Even with faster, more comprehensive access to information about military order sites than Prof. Forey had in 1984, the database faced a number of topic-specific challenges. One obvious challenge was the wide range of architectural survival for fortresses held during the Christian reconquest. Many were built in topographic

[2](Calatrava et al. 1761; Santiago 1989; O'Callaghan 2002), (www.castilosnet.org), (Bernad et al. 1997; Rades y Andrada 1980).

locations that were as strategically attractive to Roman settlers as they were for soldiers during the Spanish civil war. Sometimes the only relevant medieval data at these sites is a place-name and the general location for a long since replaced structure. Most often, the fortresses entered in ADIMO were ones that had been left to ruin when the location of the Muslim-Christian frontier shifted further south. In some cases, the only remaining material evidence that a fortress existed is a semicircle of stone or an excavated cistern. Without the unexpected assistance of geo-tagged photographs taken by hikers and tourists that were uploaded to Google Earth via Panoramio and other photo hosting sites, many of these humble survivals would have been impossible to locate. On the opposite side of the spectrum, there are national Parador hotels that were built within the fabric of some of the most iconic castles associated with the military orders. The fortress-monastery of Alcañiz in Aragon is one such example of a Parador that retains a well-preserved church, tower, cloister and dormitory—all of which were constructed by the order of Calatrava in the early 13th century.

The second challenge for an architectural history of the Iberian military orders is the simple fact that a morphological analysis of their associated structures often reveals very little about the orders themselves. While the military orders did occasionally contribute new architecture to the frontier, their modus operandi was to occupy fortresses built by Almohads and other Muslim groups. As occupiers, they were either forced to adapt their newly gained structures or abandon them when—due to the nature of the switch from Muslim to Christian occupation—the directions of siege relief or invasion were inverted. With these hurdles in mind, the overarching strategy for ADIMO was to identify a strict set of criteria that could be enterable for every site despite its degree of survival, or whether it was substantially built or merely occupied by one of the military orders. There were initial hopes of including additional feature-specific metadata for each structure such as whether they contained monastic buildings such as full-scale churches, cloisters, refectories and dormitories, but it quickly became apparent that without a thorough archaeological survey, the number of structures with identifiable features would be so small as to be negligible as a source of comparison. As a result, ADIMO became chiefly concerned with occupation events. Precise latitude and longitude at the highest point within the footprint of a site, combined with affiliation (which military order was occupying the site) architecture type, and the dates of occupation, became the most significant fields in the database. It was also important to enter dates with a range of uncertainty for the beginning and ending of an occupation event. In many cases, the finest temporal grain for the initiation of a new occupation event was a year of donation or an offhand mention that a castle changed hands after a critical battle. Therefore, when a specific date such as January 1st, 1194 is queried in ADIMO, the GIS visualization reveals slightly transparent features if that date is between either the earliest and latest begin dates, or the earliest and latest end dates for that occupation event (Fig. 1).

ADIMO was also inspired by the repeated implication by historians that the military orders represented a predecessor to the modern standing army on the front lines of the Christian Reconquest. The evidence for this theory is ubiquitous in

Fig. 1 Viewsheds from fortresses and towns in southern La Mancha immediately *before* the Almohad Muslim victory against Castilian forces at the battle of Alarcos in 1195

Fig. 2 Viewsheds from fortresses and towns in southern La Mancha immediately *after* the Almohad Muslim victory against Castilian forces at the battle of Alarcos in 1195

documents describing the divvying out of land and privileges following a successful Christian campaign into Muslim territory (Santiago 1989). Very soon after a territory switched from Muslim to Christian control, Christian monarchs donated many of the most advanced fortified positions to the military orders. In exchange, the orders would garrison the border, protect new Christian settlers, subjugate the Muslim rural population and protect the interior of the Christian kingdoms. With this in mind, a spatial database striving to discover which military orders were in which castles at any particular time would also map the militarized edge between Christian and Muslim territory during the 12th through 14th centuries. This interactive map of the Reconquista is superior to the kinds of maps that historians

Fig. 3 Viewsheds in southern La Mancha after the order of Calatrava captured the fortress of Salvatierra and made it the new headquarters for the order after the disaster of Alarcos in 1195. The boldness of this decision is revealed through the orange viewsheds designating Almohad visibility that surrounds the landscape around Salvatierra in all directions

Fig. 4 A reversal of fortune for the Christian military orders and their Almohad Muslim rivals immediately after the Christian victory at Las Navas de Tolosa as revealed by changes in viewshed on the frontier. The fortress of Salvatierra had been a lone Christian outpost during the "years of crisis and survival" between the battle of Alarcos in 1195 CE and the battle of Las Navas de Tolosa in 1212 CE, but after Las Navas de Tolosa, this same fortress became a solitary Almohad possession residing in a sea of Christian viewsheds

working exclusively in print have used to describe the period. These maps, while illustrative of very broad strokes of territorial change, give a false sense of teleological inevitability due to their limited number. With ADIMO, the familiar, but inaccurate wavy lines bisecting religions on the peninsula over (at most) four monumental stages of the reconquest are replaced by a wide, shifting, insecure and volatile band of frontier space in which the military orders lay in greatest contact with their religious enemies. This kind of volatile dynamic change is only truly visible in an animated map in which many equally-timed frames diminish the tendency to privilege only the moments of greatest change. With this in mind I will encourage the reader to view the animated map linked in References (Triplett 2016).

Viewshed Analysis—Methodology

The point-data alone, which marks the locations of over 1100 time-stamped occupation changes, only begins the process of re-mapping the Christian reconquest. Once this data was entered it became possible to begin modeling the relationship between fortresses and their surrounding landscape, as well as the network of spatial and visual relationships between sites. The first example of remote sensing applied to this data was a thorough, iterated viewshed analysis for every site, during every occupation event. GIS viewshed analysis is defined as a visualization of all units of land surface that are visible for an avatar standing in a single point location, at a predetermined height above the map's surface. The relevant variables for this analysis can include architectural obstructions, landscape elevation, layers of forest cover, atmospheric visibility distances, or even weather. At its very basic level however, GIS viewshed analysis requires an observer point with a set distance above the surface, and a digital elevation model (DEM) whose individual pixels of raster data will be asked a binary question: Can this square of physical space be seen by the observer? The DEM used for the ADIMO project was downloaded from ASTER (Advanced Spaceborn Thermal Emission and Reflection Radiometer),—a partnership between NASA and Japan's Ministry of Economy, Trade and Industry. (Anon 2015) This open data is likely very familiar to GIS specialists and scholars interested in remote sensing. According to the ASTER website, the 2011 version of the DEM has a 30 m^2 posting/cell size, with a standard deviation of height between 7 and 14 m. ("ASTER Global Digital Elevation Map") While there will always be a desire to increase spatial accuracy and resolution of the DEM variable in spatial analysis projects—this resolution was more than sufficient for the scale of this project. This is partially the case because all sites were identified using satellite data rather than on-site GPS, but also because the analysis was more concerned with the impression of iterated, and later aggregated viewsheds from many sites, rather than testing specific, local instances of inter-site visibility.

In the ADIMO project, these viewshed analyses "painted" the landscape of Iberia with faction-specific colors based on the affiliation of the observer during a particular occupation event. For example, when the fortress of Salvatierra switched from Almohad control to the order of Calatrava in 1197, returned to the Almohads after a siege in 1211, and was finally reconquered and given back to the order of Calatrava in 1226, the color of the visible landscape for an observer standing at the highest point of this fortress changed from orange (Almohad) to green (Calatrava) to orange, and finally back to green. (Figs. 1, 2, 3, 4) Individually, these viewsheds fall unremarkably in line with previous uses of this technology. What eventually sets the ADIMO project apart is the ability to dynamically represent changes in landscape visibility as a result of iterating these viewshed analyses across the entire dataset of occupation events. This was accomplished with a relatively simple macro built with ArcGIS Model Builder.

Figure 5 illustrates how the model breaks the process down into 7 stages:

1. Iterate the point data for each occupation event
2. Execute the viewshed analysis tool using a buffer to clip the processing extents to 100 km and an offset height of 15 m
3. Extract the positive values of the binary raster
4. Convert this positive "visible" area into a scattering of polygons
5. Re-attach the time-enabled data and affiliation to the new viewshed polygon with a spatial join
6. Dissolve all the polygons into one feature
7. Merge all the polygon viewsheds into one table that matches the original point table.

Viewshed Analysis—Historical Context of the Reconquest

The specific historic episode that Figs. 1 through 4 represent has been described as the "years of crisis and survival" for the order of Calatrava by Prof. Joseph O'Callaghan. (1986) Figure 1 shows the fortresses and towns occupied in June, 1194 (as well as their viewsheds) just before the disastrous defeat of the Castilian King Alfonso VIII by Muslim Almohad forces at the battle of Alarcos in 1195. Figure 2 visualizes the changes in visibility immediately after Alarcos. The battle itself took place on a field just in front of a hastily constructed fortress that had been intended as a garrisoned possession of the order of Calatrava. While southern Castile, and especially the nearly "landscape-blind" frontier capital of Toledo was greatly affected by this loss, no frontier organization suffered greater losses than the order of Calatrava. Most members of the order died in the battle, and nearly all of their fortresses that had acted as a southern buffer for Toledo were taken back and garrisoned by Almohad soldiers. Figures 1 and 2 visualize how the green "Campo de Calatrava" reverted to an orange basin of Almohad visibility and surveillance. Figure 3 reveals how remarkable it was for the remnants of the order of Calatrava to choose the fortress of Salvatierra (which was well within the visual and spatial influence of Almohad forces) for their new headquarters and namesake.

The briefly named Order of Salvatierra epitomized the degree to which the military orders could define themselves through their proximity to their religious enemies. Salvatierra was a small and short-lived military order, yet it was clearly effective. As revealed in a statement by the Almohad Caliph Miramamolín after his armies finally dislodged this order from the fortress of Salvatierra, even a minor

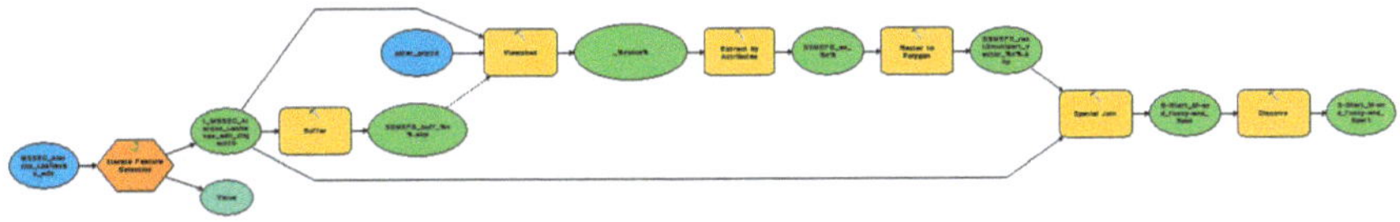

Fig. 5 An illustration of the 7 stages of the vectorized, time-enabled viewshed model

structure could have a profound impact on the frontier if it was advantageously sited:

> Salvatierra had fallen into the traps of the adorers of the cross and the presence of a bell on its church-tower was an insult to the Muslims, who, to the four points of the compass around this place, heard the muezzins glorify God and call them to prayer; it was a watch tower rising up against the sky on the bare plain, … an observatory which spied on us. This castle gave the Muslims no peace because the Christians made it the base for all their raids and organized it so that it was a sort of key guaranteeing the security of their strongholds and towns. — Caliph Miramamolín, 1211
>
> (Translated by D. Lomax 1978)

Above all else, this passage reveals that the ability to view the surrounding landscape from a topographically advantageous position was critical to territorial and religious influence on the frontier. It is in this capacity that viewshed analysis can offer new insights into the intentions of the military orders on the frontier.

One of the more omnipresent debates for historians of medieval castles is the degree to which a castle's "practical" or "defensive" features influence that castle's ability to express power.[3] In the case of the Iberian military orders, these debates are complicated by additional questions such as: What messages do castles communicate when ownership alternates between rival groups? Does a castle occupant communicate power differently than the original builder? If the "audience" for a castle remains the same, but he castle's allegiance fluctuates, how do we interpret that castle's meaning at different times?

The ADIMO project identifies vision as an agent of historical change in order to answer these questions, but it also must account for additional factors that determined a castle's siting. The first, most predictable reason for locating a castle on top of a hilltop is because these sites offer tactical advantages against attack. What is usually offered as the immediate secondary benefit of an elevated site is increased visibility of the surrounding landscape. Recently, historian Enrique Rodriguez-Picavea described this second benefit as "… the semantics of subordination: to behold the surrounding territory from the castle's heights produces a sudden feeling of dominion, whereas to behold the fortress from a lower site makes it grow to monumental proportions" (Rodriguez-Picavea 2012).Translating this dynamic into two dimensions via viewshed analysis enables a scholar viewing

[3]Castle-studies of medieval England have been particularly active in this debate (Johnson 2002; Platt 2007; Creighton and Liddiard 2008) but work on Iberian fortresses has been more concerned with their role in frontier settlement (Glick 1995; González Jimenez 1989).

ADIMO to associate the viewsheds with ideas such as security, domination, protection, authority, or rebellion through a temporal query of the database. It must be said that the attachment of these meanings to viewshed images is always at the discretion of the beholder, yet when associated with broader narratives of the reconquest, they can also evoke new questions about spaces and times that are only faintly described in the historical record. The long periods of visible interaction between Muslim and Christian forces allow historians to ask questions about the potential blurring of religious identities in defined areas and times of the reconquest, just as episodes of complete visual domination of a valley by one military order offer an opportunity to study how Christian minority groups attempted to secure wide swaths of Muslim-majority rural space. Put simply, dynamic viewshed analyses cast from frontier fortresses help to take the story of the Iberian reconquest beyond lists of conquered cities, and they distribute it liberally over the geography of Iberia.

Intervisibility Networks

viewshed analysis of the surrounding landscape is a useful measure of the tactical viability of a particular fortress site, but it partially obscures the importance of inter-site visibility networks on the frontier. Figure 6 reveals lines of intervisibility between Christian and Muslim fortresses and towns on both sides of the Sierra Morena mountain chain following the Christian victory at the battle of Las Navas de Tolosa in 1212 CE. The alternating dashed lines are used to signify sites that can see each other, but at the specified time, they are occupied by groups with different religious or military order affiliations. For example, the alternating orange and black lines in Fig. 6 show that there was a high degree of intervisibility between Christian nobles (Black) and Almohad Muslims (Orange) in frontier fortresses in September of 1213 CE. The technique used to build this network is essentially a spatial query of the previously iterated viewshed analyses. Figure 7 illustrates a five step modeling process:

1. Iterate a viewshed to select the polygon data of a single occupation event at a fortress
2. Use this polygon to select any point-occupations that fall within the viewshed —(these points are determined to be places/events that can be seen from a particular site)
3. Merge the time, latitude and longitude, and affiliation of the original observation point with the same information for each point that this observer can see into a single Table
4. Use the "XY to Line" function to draw a line between each set of latitudes and longitudes
5. Re-join fields from the original tables that get stripped by the XY to line tool

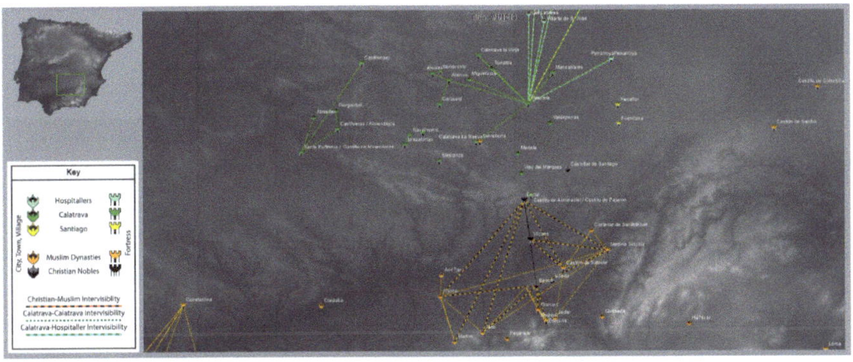

Fig. 6 Lines of intervisibility between Christian and Muslim-held fortresses and towns on both sides of the Sierra Morena mountain chain following the Christian victory at the battle of Las Navas de Tolosa in 1212 CE

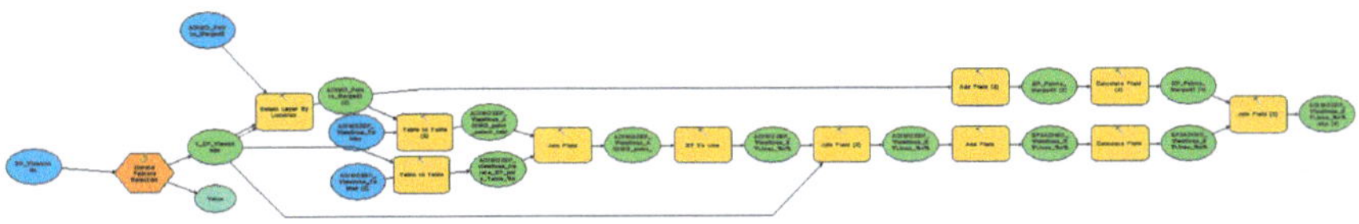

Fig. 7 Screenshot of an ESRI ArcMap model that begins with point data of fortress occupations, iterates a rasterized viewshed for each one, converts the positive values into polygon vectors, and re-attaches data such as architecture type, dates of occupation, and affiliation for each viewshed

One pattern that the lines of intervisibility can reveal is a surprising level of centrality for sites whose remote locations or advanced state of ruin cause them to be overlooked. The castles of Ferral and Mesnera shown in Fig. 6 are two such examples. Another somewhat surprising condition that emerges from the line-data is a general lack of intervisibility between sites belonging to the order of Calatrava in the valley region north of the Sierra Morena. With the exception of Mesnera, which lies on top of a cone-like volcanic outcropping in the center of the "Campo de Calatrava," very few of the Calatravan sites (in green) were visible to each other. Curiously, these same fortresses cast viewsheds that thoroughly covered the valley below. There are a number of ways to interpret this combination of a highly visible landscape with relatively few examples of intervisibility between fortresses—none of which are completely convincing without further research or archaeological survey. For instance, the Almohads who originally chose the sites for these fortresses may not have been as interested in creating a true fortress signaling-network as they were concerned with observing roads and mountain passes. It is disappointing that almost no evidence of the castle of Mesnera exists at the site, because it is tempting to hypothesize that this structure could have acted as a sort of "signaling station" due to its ability to see so many Calatravan and Hospitaller sites that were blind to each other. The number of shifting patterns seem endless as the lines of intervisibility change over time, and it is this ability to inspire further avenues of exploration that best represents the utility of the ADIMO project.

Calatrava La Nueva

Viewshed and intervisibility analysis represent two valuable methods for deriving meaning from the spatial relationships between frontier fortresses. As should be clear, vision was a contested commodity during the reconquest. Maximizing visibility in a contested landscape has strategic benefits, especially views of mountain passes that funneled possible invading armies. The fortress-monastery headquarters of Calatrava la Nueva offers an excellent example of a site chosen and modified by a military order to communicate visual dominance over specific enemies, while maximizing the strategic benefits of a southward viewshed. Without delving too deeply into the specifics circumstances of Calatrava la Nueva's foundation, it is important to note that the site and scale of this military-monastic complex were irrevocably tied to the circumstances of the order's near destruction at the battle of Alarcos in 1195, its brief but dramatic reconstitution at the fortress of Salvatierra, (1198–1211) and the complete rebirth of the order after the Christian victory at the battle of Las Navas de Tolosa in 1212 (Triplett 2009).

Calatrava la Nueva was built within and around the Almohad keep of Dueñas, which lay on a hill opposite of the castle of Salvatierra (Fig. 8). In 1211, the Almohad Caliph Miramamolín used Dueñas and the hills around it as the staging ground for an extended siege of Salvatierra and the military order who garrisoned it. After the battle of las Navas de Tolosa, Alfonso VIII of Castile donated Dueñas to the order of Salvatierra/Calatrava as the site of a new headquarters which was far closer to the edge of the frontier than their original fortress (later called Calatrava la Vieja) in the north. As can be observed in Figs. 9 and 10, this site held a very clear view of a mountain pass connecting southern La Mancha to Andalucía. This pass, known later as the "Calatravan Pass" was one of two parallel passes Almohad Armies had used in previous invasions of Castile. The second pass, further to the east, was called the Despeñaperros pass, and led directly to the battlefield of Las Navas de Tolosa in 1212. When photographic evidence is combined with a close observation of the viewsheds for Calatrava la Nueva and the fortress of Salvatierra (which remained in Almohad control even after the battle of Las Navas de Tolosa) the choice of site for the new headquarters seems to make strategic and symbolic sense. In comparison to the view from Salvatierra, the view from Calatrava la Nueva is distinctly southward in orientation (Figs. 10 and 11). It is interesting however, that due to an alteration to a hill just north of the fortress, Calatrava la Nueva's view to the north might have been more open than it appears in the viewshed. This hill (Fig. 12), appears to have been a quarry for the 13th century fortress. Flattening this hill allowed an observer at the top of the keep at Calatrava la Nueva to see a slightly larger portion of the landscape to the north, but it is more likely that this location could have contained a simple watchtower that would have allowed the order to view the north valley nearly as well as the south.

On a symbolic level, Calatrava la Nueva was also conspicuously visible for the early 13th century Almohad inhabitants of Salvatierra—and vice versa. Calatrava was larger, higher, and had a more widely dispersed viewshed than Salvatierra. In addition, each evening in the summer and spring, Calatrava la Nueva casts a shadow in the direction of Salvatierra until it covers the neighboring fortress (Fig. 13). As historian Christopher Gerrard wrote with regard to the military orders in Aragon, Christian shrines and monasteries such as the one at Calatrava la Nueva were designed so that the Christian "… religion could be signposted visually by its monuments. The pace of the day to day routine of rural life was regulated by the tolling of bells, an aural reminder of the presence of the church." (Gerrard 2000) When one considers how much the bells of Salvatierra had irritated the Almohad Caliph Miramamolín when that fortress had been occupied by the order of Calatrava, one can imagine that after the battle of Las Navas de Tolosa inverted the religious orientation of the region, there would have been motivation to ring the bells of Calatrava la Nueva for the nearby Almohad garrison at Salvatierra.. When combined with Gerrard's idea that the frontier landscape was a "…canvas on which to proclaim… (Christian) affiliation." Viewshed analysis can be interpreted as a mapping of religious proclamation (Gerrard 2000).

Volumetric Visibility at the Fortress-Monastery of Montesa

The second project (referred to hereafter as the Montesa modeling project) involves a novel approach to remote sensing that focuses on intra-site visibility rather than inter-site or landscape visibility. The project was completed by joining together open-source and industry standard software in order to capture extant masonry, reconstruct ruined walls to a hypothetical height, and visualize the degree to which volumetric areas were visible for a person walking through a specific path in the complex (Fig. 14). The castle-convent of Montesa was chosen for this experimental process for a variety of reasons. First, Montesa is a rare example of a true fortress-monastery that was built by a new military order to suit the needs of a mixed community of combative and non-combative monks, priests and lay laborers. Second, it was a highly symbolic military-monastic complex that was built as the frontier headquarters for the newly constituted military order of Montesa. Thirdly, Montesa was an excellent candidate for digital reconstruction due to the suddenness of its destruction during an 18th century earthquake, and a partial reconstruction in 2010. Montesa represents an opportunity to ask more specific questions about how a 14th century military order represented itself architecturally, and how an analysis of spatial and visual partitioning can lead to new conclusions about the meaning of "hybrid" architecture.

Fig. 8 South view of the fortresses of Salvatierrra (left) and Calatrava la Nueva (right) from the valley below

Fig. 9 South view of the "Calatravan Pass" from the top of the keep at Calatrava la Nueva

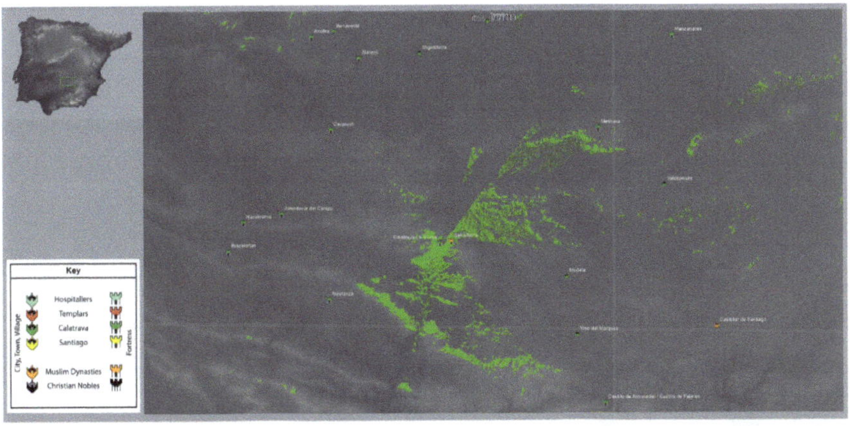

Fig. 10 Isolated viewshed of Calatrava la Nueva. Note the panoramic view to the south that includes the Calatravan pass

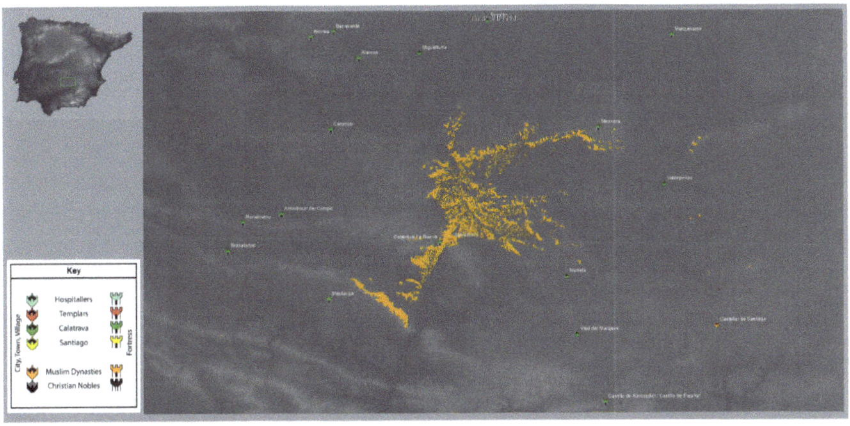

Fig. 11 Isolated viewshed of the fortress of Salvatierra immediately following Las Navas de Tolosa. Note that this castle is oriented more to the north than its neighbor, Calatrava la Nueva

Step 1: Photogrammetry

The decision to pursue a passive rather than active scanning technique at the data-capture stage was partially predicated by curiosity. After experimenting with DIY aerial photography and structure-from-motion in the fall of 2012, I decided to test the viability of an entirely photogrammetric scan of a large architectural site. The initial plan was to capture both Calatrava la Nueva as well as Montesa in order to compare hypothetical reconstructions of the two complexes. Over twenty thousand photographs were shot with hand-held DSLR cameras at Calatrava la Nueva

Fig. 12 North view from the top of the keep at Calatrava la Nueva. Note the flattened hill at the upper right that may have included a watchtower. Such a structure would have greatly added to the northern viewshed of Calatrava la Nueva

and processed using the open-source software VisualSFM, yet there was a critical element missing at this site. Unlike at Montesa, I was not able to secure permission to take aerial images at Calatrava la Nueva. As a result, the point cloud for this site contained minimal data of the ground after processing. Fortunately, at Montesa I was able to make extensive use of two important DIY aerial photography techniques. In addition to kite-aerial photography, I also attached a lightweight Canon T1i DSLR camera to a 16ft tall, telescopic painter's pole via a simple adapter. The camera was attached to my Android phone via a USB cord and controlled using an app called "DSLR Controller." With the phone attached to the pole at eye-height, and the end of the pole slipped into a belt-pocket, I was able to see through the camera lens, frame shots, adjust the exposure, release the shutter and move the camera while it was raised. This technique greatly improved the efficiency of the data-capture stage.

While it is generally preferable to select a single exposure for all images intended for photogrammetric matching, this is easier said than done. The cloud conditions at Calatrava la Nueva and at Montesa during three weeks of photography in March 2013 varied greatly. The first hurdle was recognizing the changes in light as clouds covered and uncovered the sun, and adjusting the exposure between shots. Fortunately, the increased control offered by the DSLR Controller app was also a great help in this regard. The second hurdle was a surprising lack of wind. By nature of their hilltop locations, many of Spain's fortresses lie in consistently windy locations, and having been to both sites previously, I believed there would not be

Fig. 13 East view of the ruined castle of Salvatierra from the top of the keep at Calatrava la Nueva. Note that the shadow of Calatrava la Nueva leans toward the neighboring castle in the summer months and eventually covers it

Fig. 14 A rendering of the Montesa modeling project. The colored voxels represent the number of times a hypothetical viewer walking from the gate (lower right) through the cloister (center) could see each unit of space. The experiment allows the viewer to look in all directions every two feet along their path

any issue getting the 10–18 Mph winds necessary to lift a kite and lightweight camera. Unfortunately, the first 12 days at Montesa were nearly free of wind. It was not until the final day of shooting that I was able to get the kite aloft. The entire setup required an 8 ft wide “sled” kite for maximum stability and verticality of the kite string, a “picavet” device which reacts to changes in the angle of the kite string to keep the camera in a horizontal position, and a lightweight camera with an intervalometer that shot pictures automatically while it was in the air. It cannot be overemphasized how important these photos were to the success of the data capture because even with the pole setup, the curtain walls made it impossible to bridge the exterior walls with the interior of the military-monastic complex. The tops of the walls would have been blank without the kite images, and the interior and exterior point-clouds would have had to be independently scaled and registered to one another in a post-process. Because kite-images such as Fig. 15 were taken on this final day, it was possible to drop all ten thousand photographs into a single matching process in VisualSFM.

A number of important lessons came out of this process. First, it is clear that aerial images are not optional when trying to capture a large site with photogrammetry. Second, the processing time that is theoretically saved by separating and processing images that are likely to match did not translate into saved real-world time. Attempts to register photogrammetry scans in post, or manipulate the match files for images that should “stitch” larger collections of similarly-spaced images were less successful than simply patiently waiting for the matching process to complete for all of the images. VisualSFM enables you to pause the process at any time, and restart without issue, yet it can be difficult to relegate a desktop computer for up to ten days depending on the number and size of images added for matching. In retrospect, SFM projects at this scale should be processed in a GPU cluster if possible.

Steps 2–5: Point-Cloud Cleaning, Meshing, Retopology and Modeling

One of the difficulties with the Montesa Modeling Project occurred because it was mostly constructed in 2013 and early 2014, rather than in 2015. The initial goal of capturing dense data at Montesa was to end processing at the point-cloud stage and complete the reconstruction by snapping to this cloud in 3D Studio Max. Staying in 3DS Max for the 3D viewshed analysis—rather than ArcScene—allowed for a more precise, granular analysis than ESRI’s 3D GIS toolset. This part of the project-plan gambled on Autodesk making their recently-revealed point-cloud indexing technology available in their 3D Studio Max software. Unfortunately, point-clouds that were cleaned and indexed in their ReCap software could be exported to AutoCAD and Revit by 2013, but this feature was not available in the educational license for 3D Studio Max Design until version 2015. Because the

Fig. 15 Kite-aerial photo of the fortress-monastery of Montesa in Valencia, Spain

end-goal was to use the light-meter planes available in 3D Studio Max to build the volumetric visibility analysis tool, the project was essentially wedded to this software. Consequently, rather than importing the photogrammetric point-cloud into 3DS Max and snapping to this cloud in the modeling stage, the point-cloud needed to be meshed, textured, and retopologized before the extant foundations could be brought into 3DS Max. After experimenting with a number of software solutions, this was eventually accomplished using a combination of MeshLab to mesh and texture the cloud, 3D Coat to retopologize the mesh into a more a manageable, low-polygon model, and the open-source X-Normal software to transfer textures, normals and bumps from the dense mesh to the retopologized model. While time-consuming, this process made it possible to port the photogrammetric mesh into the Unity3D game engine for later dissemination of the model.

Figure 14 illustrates the principles that governed the reconstruction phase of the Montesa modeling project. The photogrammetry-derived, retopologized mesh lacked the kind of metric accuracy that one finds with laser-scanning, or more controlled passive scanning processes, yet in the end, photo-realism was never a goal of the reconstruction. The polygonal walls that were extruded from the digitized foundations were left transparent in order to visually communicate the difference between the supposition and data-capture stages of the project. In cases where a ceiling height was suggested by horizontal gashes in the masonry walls, or evidence of vaulting was indicated by extant corbels, ribs and columns, those features were modeled, cloned and tested against the extant evidence. Other areas (such as the church) were left open if their extant masonry failed to suggest their height before the 18th century earthquake. This level of cautious supposition sets transparency as a core principle of the project, but more importantly, it eliminates any unnecessary modeling work that would not affect the 3D viewshed experiment. Renders of the model reveal transparent walls, but they were not transparent for the array of light meters inside the model. In the end, the main purpose of these reconstructed walls was not to visualize an earlier form of the site, but to act as obstructions for a viewer standing at a particular location.

The Light-Vision Metaphor

At the conclusion of the Montesa Modeling Project, no 3D visualization package contained a tool to measure the degree to which a unit of space was visible to one or multiple avatars. The volumetric viewshed measuring process proposed in this project is essentially an aggregate repurposing of a tool that was first added to 3DS Max Design in 2009. The "Light Meter Helper Object" is a plane with an adjustable number of vertical and horizontal subdivisions. The crossing of each of these subdivisions results in a light-sensitive node. Once the lights are laid out in the scene, this plane calculates light-values for each node/crossing and produces a heat-map on the surface of the plane. More importantly, each of these light-meter planes can export a .csv file with XYZ location information for each node, and how much direct and indirect light has reached that node.

The volumetric viewshed analysis proposed here is based on metaphorically reversing the active-passive relationship between light and viewer. If we imagine that a viewer could cast out their vision from a single point in all directions, and that only areas which are "struck" by this emanation of vision are "visible," then we can use a point-light as a stand-in for a viewer, and the light-meters as the measure of that visibility. This repurposing of a light-meter for recording of visibility was inspired by the work of Eleftheria Paliou, who has published extensively on the subject of 3D visibility analysis (Paliou 2011). The volumetric viewshed experiment takes Paliou's method of detecting the visibility of specific wall features and multiplies it into a grid. Just as Paliou experimented with multiple viewers of a single target-wall, this project added multiple lights, but instead of a grid of lights/avatar-viewers, the lights were aligned in a string as though they were a single person walking through the space. Of course, unlike light, our vision can only bounce off of non-reflective surfaces in a limited way, so other than the XYZ locations, the only relevant field exported from the Light Meter Helper is the "direct light" field. This is the column of the spreadsheet that identifies if a light/avatar-viewer could see the unit of space.

Predictably, because this method applied a digital tool for an unintended purpose, some cleaning of the data was necessary. On the positive side, it was possible to create a simple array of light meters so that the XYZ locations rested exactly on whole feet locations—creating a cubic lattice of light sensitive nodes with a $2 \times 2 \times 2$ ft resolution. This measurement was chosen as both a maximum granularity allowable in the scene before a memory failure, as well as a rough approximation of a person's shoulder-width. Consequently, a virtual person standing within the reconstructed space of Montesa would find it difficult to "hide" from the crossing nodes.

On the negative side, the light-meter helper is only light sensitive on one side. This means that to capture a cumulative measure of visibility projected from many view-locations, all of the lights/viewers in a scene either have to be placed at the same height for meter-planes stacked vertically, or on the same X or Y planes for meters stacked horizontally. Figure 16 shows that the irregular shape of Montesa's curtain walls required that the light-meters be stacked horizontally. The final experiment was to detect the degrees of volumetric visibility in Montesa for a person hypothetically walking from the lower ward through the cloister and into the church. This seemed to be a high-traffic pattern, as well as a route that passed through highly partitioned spaces occupied by different members of the community. Capturing the visibility of this path required that the light meters be processed for one viewer at a time. As the viewer moved beyond one of the meters, that plane had to be reversed so the light sensitive side was facing the viewer/light. After the meters were calculated and the .csv files were exported for each light meter, they were combined programmatically, and all values of direct light were switched to a binary of 1 for visible and 0 for invisible. The final stage of data cleaning was to count how many viewers could see each unit of space/spreadsheet row. A sample of the resulting spreadsheet is shown in Fig. 17.

3D Data Visualization

Once the visibility data was recorded, there were a number of software options to visualize the data. One of the most flexible options was Paraview. This open-source software will import any .xyz, .txt, or .csv file, plot it spatially, and apply filters to create colorized volumes or 3D glyphs from the values (Fig. 18). The Paraview visualizations were evocative, but the most useful images have come from the long-awaited ability to import point-clouds directly into 3D Studio Max Design 2015. Thanks to this new feature, the visibility values could be indexed in Autodesk ReCap and re-imported back into 3D Studio Max. Most importantly, 3DS max contains a very broad range of shaders that can be applied to the visibility point-cloud. This point cloud is now renderable, can receive and produce shadows, and has adjustable levels of specularity, opacity, bump, size and self-illumination. Figure 19 shows a render of the cumulative visibility for a 5′6″ tall person walking through the cloister toward the church. The floating colored voxels represent how many times each light-sensitive node was seen by the string of 25 point-light locations representing the simulated walker. As with most heat-maps, white and

Fig. 16 The stacking of light-meter planes (Blue) along the X-axis in 3D Studio Max Design 2014. These planes formed an arraqy of equidistant light-sensitive nodes. Each light meter can export a .csv file with the locations of each node and the amount of direct or indirect light that hit them

blue represent the least visible end of the scale, with red and purple representing the most visible spaces.

Broken into its essence—as a cubic mesh of exportable, light/vision-sensitive nodes—the visualization described above has great potential for projects that are less interested in the visibility of specific wall-features and more interested in the surveillance of spaces in which people can move. At a site such as Montesa, where wall paintings and other specific targets for sight are long since destroyed, there is less need for an experiment which selects a specific area of masonry for visibility analysis. On the other hand, Montesa is a site that housed a mixed community with a complex orientation with regard to exterior and interior visibility. There was a division at Montesa between monks who engaged in warfare—identified as brother knights and brother sergeants by Enrique Rodriguez-Picavea and others—and clerics whose purpose was to provide the sacraments to the former group (Rodriguez-Picavea 2012). This division was common to nearly all of the orders, yet it was only in true fortress-monastery headquarters such as Montesa and Calatrava la Nueva where these groups were expected to co-exist and represent the duality of their military-religious mission.

The term "hybrid" is often used to describe the military orders, as well as their architecture. This is understandable given that sites like Montesa and Calatrava la Nueva merged prototypical monastic spaces, including a chapter house, cloister, full-scale church, and dormitories with an exterior that clearly communicated in the language of frontier fortifications. Taken further, there would appear to be a crisis of orientation in such a structure, where practical surveillance of the exterior landscape would have to compete with a monastic plan that was designed to promote internal contemplation. It is with these concerns in mind that the Montesa modeling project attempted to test how well the architectural spaces, and potentially the communities

	A	B	C	D	E	F	G	H
1	Total Dire	Pos X	Pos Y	Pos Z	Direct_05	Direct_15	Direct_25	Direct_35
2	0	-2544	600	-120	0	0	0	0
3	0	-2544	600	-95	0	0	0	0
4	0	-2544	600	-70	0	0	0	0
5	0	-2544	600	-45	0	0	0	0
6	0	-2544	600	-20	0	0	0	0
7	0	-2544	600	5	0	0	0	0
8	0	-2544	600	30	0	0	0	0
9	0	-2544	600	55	0	0	0	0
10	1	-2544	600	80	0	0	0	0
11	1	-2544	600	105	0	0	0	0
12	1	-2544	600	130	0	0	0	0
13	1	-2544	600	155	0	0	0	0
14	1	-2544	600	180	0	0	0	0
15	0	-2544	624	-120	0	0	0	0
16	0	-2544	624	-95	0	0	0	0
17	0	-2544	624	-70	0	0	0	0
18	1	-2544	624	-45	0	0	0	0
19	1	-2544	624	-20	0	0	0	0
20	1	-2544	624	5	0	0	0	0
21	1	-2544	624	30	0	0	0	0
22	1	-2544	624	55	0	0	0	0
23	6	-2544	624	80	0	0	0	0
24	9	-2544	624	105	0	0	0	0
25	10	-2544	624	130	0	0	0	0
26	10	-2544	624	155	0	0	0	0
27	10	-2544	624	180	0	0	0	0
28	0	-2544	648	-120	0	0	0	0
29	0	-2544	648	-95	0	0	0	0
30	0	-2544	648	-70	0	0	0	0
31	1	-2544	648	-45	0	0	0	0
32	1	-2544	648	-20	0	0	0	0
33	1	-2544	648	5	0	0	0	0

Fig. 17 A screenshot of the spreadsheet that combined all of the data from each light-meter plane. Each plane is represented by one column of the spreadsheet

of knights and clerics were partitioned from one another. In some ways, this kind of consideration can be accomplished with a simple 2D plan. Counting the portals between transitional spaces and gauging their size, or forming a rudimentary network analysis of connected spaces can be accomplished in short order without the help of technology. Conversely, the impacts of terrain changes in the lower ward or the size of the ruined cloister doorway on a laborer's ability to see into the cloister

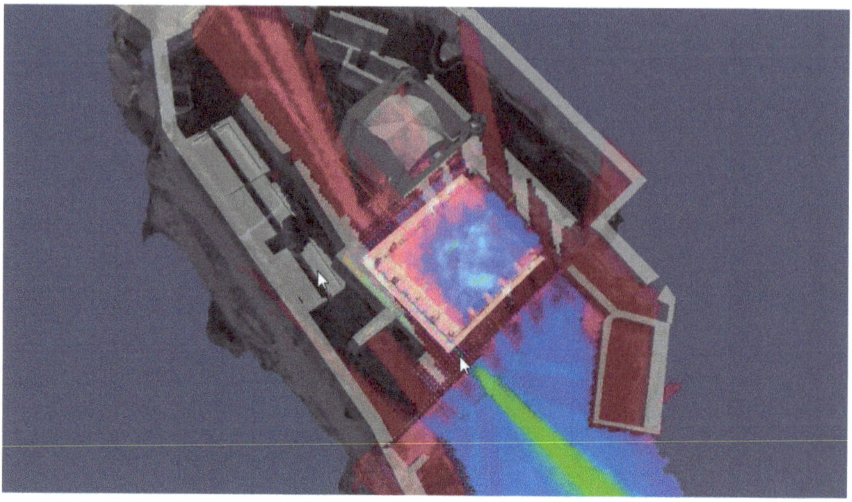

Fig. 18 A Paraview screenshot of the cloud-like viewsheds that emerge from the aggregated data from all of the light meters

Fig. 19 A 3D S Max 2015 render of the cumulative visibility inside the digitally reconstructed cloister at Montesa

is much more difficult to comprehend without viewshed technology. Even more difficult to visualize is the "translucent" quality of the interior volume of the cloister for a person walking past an arcade.

From an aerial perspective (Fig. 20), the voxels of colored visibility in the cloister appear as a cloud of complex, even erratic visibility. It is easy to comprehend that a person walking along a cloister-walk and looking at a particular target inside the cloister garth will periodically see, then briefly not see that target. What is not obvious is which previously obstructed areas of the cloister will become open to vision as one passes next to a column, and then past it. In Fig. 20, the aerial view shows that there is a halo of blue to green colored voxels around a nebulous area of greater visibility in the center of the cloister, but none of these nodes is as deeply in the red (highest visibility) spectrum as the north side of the cloister walk. The simple reason for this is because parts of this vaulted space can be seen by our walker while he is moving inside the lower ward. Figure 19 is a render of the scene from the perspective of a viewer just after they have entered the cloister from the lower ward. The walk to the left is nearly solid red in the direction of the viewer's movement, but the rest of the space is a far more complex distribution of visibility.

Like many data visualizations, this project has only begun to hint at its possible utility, and it is only constrained by the available data. The current system of exporting light-meter CSVs from 3DS Max and re-importing the aggregates is only a three-step process, but it is labor and memory intensive due mostly to the simple issue that the light meters are only sensitive on one side.[4] Given time or changes to the software, it would be helpful to apply a similar grid-like viewer setup to that used by Eleftheria Paliou in her 3D visibility analyses, or add multiple paths simulating typical movements for different groups within Montesa's military-monastic community. A more holistic data capture of visibility at Montesa would also give a greater sense of the degree to which its architects expressed a desire to foster privacy for the clerics by separating them visually from the rest of the community. One could equally identify ways in which the architects strategically designed the fortress so that particular spaces were highly visible for defenders on the walls, or tripped a "visual alarm." The combination of these ideas could begin to re-map the military-monastic spaces at Montesa, and more deeply interrogate the idea of "hybrid" architecture, but as of this publication, this is beyond the scope of the project.

Perhaps most usefully, the process outlined above could easily be repeated for a target-specific analysis in which the target exists in a particular volume of space. For example, it is possible to visualize which parts of a speaker's body were more visible to a crowd standing in a complex architectural space with 3D occlusions. If the size of the volumetric target were increased to be larger than an average human, one could identify a 3D place for the speaker to stand which would maximize their visibility. In the end, the utility of this experiment lies in the power of 3D Studio Max to capture and visualize point data in ways that most data visualization software has been less concerned with.

---

[4]It is possible to clone all the light meters in place and invert their orientation, yet calculating this many light-meters is extremely unstable within 3DS Max and it would require additional programming to replace all zeros with the same XYZ location with a one when the CSV files were combines. The additional problem is rigging the lights so that no matter how glancing the light on a particular node, the node will read as 1 for a single light, 2 for two lights/viewers etc. As the software is currently configured, this is not possible.

Fig. 20 A combination of landscape-scale viewshed (blue terrain) and volumetric viewshed analysis inside the cloister of the Montesa modeling project

The gap between data visualization software and 3D packages with more aesthetic orientations has been partially bridged through 3D Studio Max's ability to export common .csv files and import point clouds. In the past, the main collaboration between data visualization and 3D modeling and animation software has been to create more photo-realistic visualizations of cultural sites or scientific experiments. It appears that a second strong collaboration can occur now that simple spreadsheet data can be attached to a multitude of shader variables, applied to point clouds and rendered spatially.

Multiple-Scale Visibility Analysis

Figure 20 depicts an integration of the ADIMO 2D viewshed analyses with the Montesa modeling project. In the background we can see that the landscape beyond the 3D model of Montesa appears to be painted blue. This transparent blue layer is a 3D polygonal export of the 2D viewshed for Montesa. Beneath it is the faintly visible 3D landscape that was derived from the same ASTER data that drove the viewshed analysis. Shown in 2D, the viewshed for Montesa appears to jitter across the landscape unevenly (Fig. 21). However, when seen from the perspective of a viewer in 3D space just 5 m above the walls of Montesa, this same viewshed seems to thoroughly blanket the valley below. In the foreground, each layer of the modeling project is represented, including the photogrammetric mesh, the transparent reconstruction, and the light-meter derived visibility point-cloud. By

Fig. 21 The viewshed from the fortress-monastery of Montesa exported from ESRI ArcScene and rendered in 3D Studio Max. Note how the viewshed seems sketchy and irregular from this perspective

combining the volumetric and GIS viewshed analyses into a single image in Fig. 20, both projects are given a spatial context that they lacked individually.

The decision to build the new fortress-monastery headquarters for Montesa near the southern boundary of the kingdom of Aragon appears to fit nicely with the narrative of the Military orders as the vanguard and watchmen of the Christian reconquest. By integrating the 3D view, the choice of site becomes far more precise. The rocky outcropping that Montesa was built on juts out perpendicularly from a northern range of mountains that has a parallel range to the south. The valley between is a modern causeway connecting Valencia and Xativa to the northeast with Murcia and Andalucía to the southwest. Given that this fertile valley served the same purpose in the 14th century, Montesa seems particularly well sited to act as a watchtower over the landscape.

2D viewshed analysis allows for a thorough visualization of a highly contested commodity in medieval Iberia's shifting military-religious frontier. Regardless of the degree of architectural survival, or the length of a fortress occupation by a particular order or group, it is possible to visualize what each faction could see from their hilltop locations. The Montesa modeling project takes this visualization a step further by considering not just the tactical decisions evident in the choice of site, but also the intra-site visual relationships that are currently invisible at the largely ruined complex. I have demonstrated that a great deal of the architectural space in this fortress-monastery contained highly controlled, partitioned viewsheds, and that the wide blanket of landscape visibility below the fortress is not a reflection of what could be seen from most locations within the walls. Surveillance was not a passive act, but one that had to be cultivated within architectural spaces that were specifically designed for this purpose. With the assistance of remote sensing technology, this paper has proposed methods for capturing and analyzing multiple scales of visibility at architectural sites where both internal and external visibility were of paramount importance.

References

Anon, *ASTER Global Digital Elevation Map*. Available from: http://asterweb.jpl.nasa.gov/gdem.asp [Accessed April 19, 2015].

Bartlett, R., MacKay, A. & Manuel González Jiménez, M.G., 1989. "Frontier and Settlement in the Kingdom of Castile (1085-1350)." In *Medieval Frontier Societies*. Oxford: Clarendon Press.

Bernad, J., Ruibal, A. & Roca, P.C.i., 1997. *Castillos de España*, Leon: Everest.

Burns, R.I., 1990. "Muslims in the Thirteenth-Century Realms of Aragon: Interaction and Reaction." In *Muslims Under Latin Rule, 1100-1300*. Princeton, N.J: Princeton University Press, pp. 57-102.

CALATRAVA, et al., 1761. *Bullarium Ordinis Militiæ de Calatrava Opus D Ignatij Josephi de Ortega et Cotes directione, D Joannis Francisci Alvarez de Baquedano diligentia, et D Petri de Ortega Zuñiga et Aranda labore completum Cui accessit catalogus Summorum Pontificum ac bullarum, seu indultorum, quæ ab eisdem emanarunt, etc, Matriti.*

Castro, A., 1971. *The Spaniards; an Introduction to their History*, Berkeley: University of California Press.

Creighton, O. & Liddiard, R., 2008. "Fighting Yesterday's Battle: Beyond War or Status in Castle Studies." Medieval Archaeology, 52(1), 161-169.

Dodds, J.D., Menocal, M.R. & Balbale, A.K., 2008. *The Arts of Intimacy: Christians, Jews, and Muslims in the making of Castilian culture*, New Haven: Yale University Press.

Feliciano, M.J. & Rouhi, L., 2006. "Introduction: Interrogating Iberian Frontiers." In *Medieval Encounters*, 12(3), 317-328.

Forey, A.J., 1984. "The Military Orders and the Spanish Reconquest in the Twelfth and Thirteenth Centuries." Traditio, 40, 197-234.

Gerrard, C., 2000. "Opposing Identity: Muslims, Christians and the Military Orders in Rural Aragon." Medieval Archaeology, 43, 143-160.

Glick, T.F., 1995. *From Muslim Fortress to Christian Castle: Social and Cultural Change in Medieval Spain*, Manchester: Manchester University Press.

González Jiménez, M., 1989. "Frontier and Settlement in the Kingdom of Castile (1085-1350)." In *Medieval Frontier Societies*. Oxford: Clarendon Press, pp. 49-74.

Johnson, Matthew. *Behind the Castle Gate: From Medieval to Renaissance*. London: Routledge, 2002.

Moreno, E. M. 1994. "The Christian-Muslim Frontier in Al-Andalus: Idea and Reality." In *The Arab Influence in Medieval Europe*. By Dionisius A. Agius and Richard Hitchcock. Reading, UK: Ithaca, 83-99.

O'Callaghan, J.F., 1986. "The Order of Calatrava: Years of Crisis and Survival, 1158-1212." *Meeting of Two Worlds: Cultural Exchange Between East and West During the Period of the Crusades*. Kalamazoo, Mich: Medieval Institute Publications, Western Michigan University,. 419-430.

O'Callaghan, J.F., 2001. "The Interior Life of the Military Religious Orders of Medieval Spain." *Malta Study Center Lecture Series*.

O'Callaghan, J.F., 2002. The Latin Chronicle of the Kings of Castile, *Tempe, AZ: Arizona Center for Medieval and Renaissance Studies*.

Paliou, E., Wheatley, D. & Earl, G., 2011. "Three-Dimensional Visibility Analysis of Architectural Spaces: Iconography and Visibility of the Wall Paintings of Xeste 3 (Late Bronze Age Akrotiri)." *Journal of Archaeological Science,* 38(2), 375-386.

Pick, L.K., 2004. *Conflict and Coexistence Archbishop Rodrigo and the Muslims and Jews of Medieval Spain* (History, Languages, and Cultures of the Spanish and Portuguese Worlds), New York: University of Michigan Press.

Platt, C., 2007. "Revisionism in Castle Studies: a Caution." *Medieval Archaeology,* 51(1), 83-102.

Rades y Andrada, F., 1980. *Chronica de las Tres Ordenes Militares y Cavallerias de Santiago, Calatrava, y Alcantara* (facsimile of 1572). El Abrir.

Robinson, C., 2011. "Towers, Birds and Divine Light: The Contested Territory of Nasrid and "Mudéjar" Ornament." *Medieval Encounters,* 17(1), 27-79.

Rodriguez-Picavea, E., 2012. "The Military Orders in Medieval Iberia; Image, Propaganda and Legitimacy." *Mirator,* 13(1), 1-35.

Safran, J.M., 2013. *Defining Boundaries in Al-Andalus: Muslims, Christians, and Jews in Islamic Iberia,* Ithaca: Cornell University Press.

Santiago, M.G., 1989. *Documentos de la Orden de Santiago Sobre Castillos Extremeños, Cáceres:* Universidad de Extremadura.

Triplett, Edward. "ADIMO Timeline – Viewshed Analysis on the Frontier." YouTube. Available from: https://youtu.be/P7Z7Ik6_bJU [Accessed May 3, 2016].

Triplett, Edward. "The Resurrection of the Military Order of Calatrava Through the Construction of a New Capital." Thesis. University of Virginia, 2009.

Watt, Kelly L. *Medieval Churches on the Spanish Frontier: How Elite Emulation in Architecture Contributed to the Transformation of a Territorial Expansion into Reconquista.* Diss. U of Louisville, 2011.

Pre- and Proto-Historic Anthropogenic Landscape Modifications in Siem Reap Province (Cambodia) as Seen Through Satellite Imagery

Kasper Hanus and Emilia Smagur

Abstract Since the end of the civil war twenty years ago archaeologists in Cambodia have made a substantial progress in the research on the medieval landscape of the urban complex of Angkor. With modern technologies of archaeological prospection, Air-Sar and LiDAR, the researchers attempted to reconstruct the cultural landscape of the biggest low density urban complex of the preindustrial world. However, on the regional level the sites are still identified based on the colonial-era archaeological reconnaissance. Such a state of affairs is unfavourable for two reasons: it hampers the understanding of settlement patterns in prehistory and history and impedes local heritage protection efforts taken against looting. Therefore a large-scale reconnaissance effort based on satellite imaginary was required to fill the gap in our understanding of the past landscapes in the region. As North-Western Cambodia is mainly an alluvial plain it is only natural for an occupied settlement to leave a recognizable mound of accumulated material. Other recognizable topographic signatures of occupation include one or more moats excavated by the site's occupants to enclose the settlement. While many of the settlements of interest are dated back to the Neolithic period, a remarkable number of them are still inhabited today. These long occupied areas leave features on the landscape that are easily recognizable in satellite images, as well as from the ground level. Furthermore, the structure of past occupation can be deduced from the pattern of rice paddies, especially when it is radial, as contrasted with the modern pattern dominated by right angles. These features related to land use are equally visible in the satellite images. With this paper the authors aim to provide a systematic review

K. Hanus (✉) · E. Smagur
Institute of Archaeology, Jagiellonian University in Kraków, Ul. Gołębia 11, 31-007 Kraków, Poland
e-mail: kasper.hanus@amu.edu.pl

E. Smagur
e-mail: emilia.smagur@gmail.com

K. Hanus
Institute of Prehistory, Adam Mickiewicz University in Poznań, Ul. Św. Marcin 78, 61-007 Poznań, Poland

M. Forte and S. Campana (eds.), *Digital Methods and Remote Sensing in Archaeology*, Quantitative Methods in the Humanities and Social Sciences,
DOI 10.1007/978-3-319-40658-9_10

of research routines applied to identification of anthropogenic landscape modifications.

Introduction—Sensing the Past South-East Asian Landscapes

The landscapes have a vital role in the narratives about the past, despite the fact that the definition of "an archaeological landscape" still rises the controversies. The landscape could be understand or as the stage that thanks to the resources it could offer influencing the cultural systems (Hodder 1985; Shanks and Tilley 1987; Marciniak 2002), or as in the other theoretical approaches it is seen as a social construct (Ingold 1993; Tilley 1994; Rączkowski 2002). Despite the theoretical stance the proper understanding of the physical remains of the past landscapes is crucial in the archaeological investigation procedures. Great deal of work was devoted for developing the methods of ground, airborne and spaceborne remote sensing for prospection and documentation of the landscapes. However most of such research was revolving around the case studies in Europe and the Mediterranea (Rączkowski 2002, 2015).

The authors of this chapter aim to present the realities of applying those methods—mainly satellite images—while creating the narratives about the past of continental South-east Asia. The cultures of this region, as well as the geographical settings are much different than temperate areas of Europe or the basin of Mediterranean sea. The implication is that the methods need to be altered to fit such conditions, what was done by the authors and is going to be presented in this chapter. The chapter's narrative is going to cover the brief history of South-east Asia (focusing on the territories of contemporary Cambodia), the research questions asked by the researchers, methods applied to resolve those problems, and conclusions.

The Settings—Northwest Cambodia Before Angkor

Northwest Cambodia is a land stretching between the rampart of Dângrêk Mountains in the north and shores of Tonlé Sap lake in the south. The heart of this region is constituted by three modern provinces of the Kingdom of Cambodia: Siem Reap, Oddar Meanchey and Banteay Meanchey (Fig. 1). The most striking feature of this landscape is its flatness, as the land was shaped by alluvial activity of Tonlé Sap. The lake, fed by waters of the Mekong, annually floods adjacent plain (Chandler 1991). The Northwest lies in a tropical wet and dry climate zone (McKnigh and Hess 2000) and the annual cycle consist of monsoon and dry period.

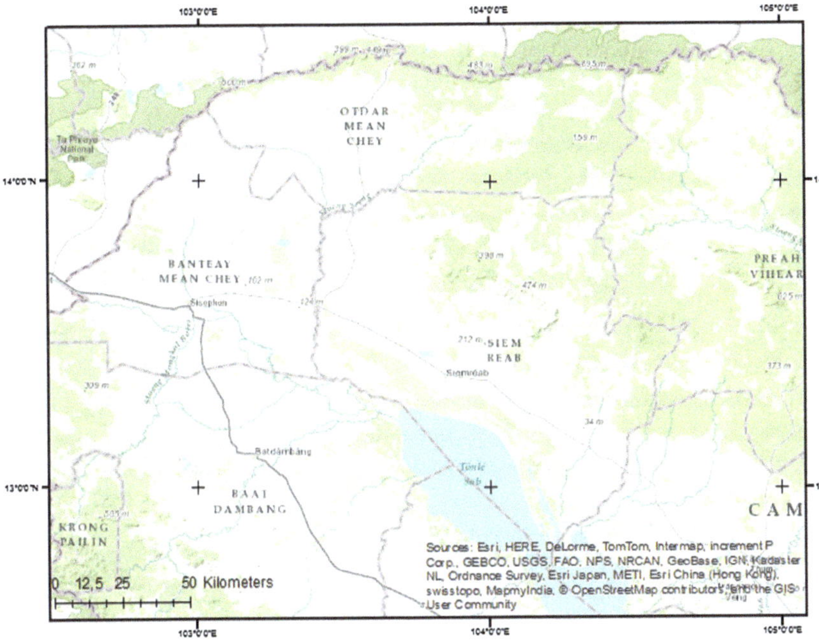

Fig. 1 The topography and administrative division of NW Cambodia. *Data source* ArcGIS Online resources

Modern Siem Reap province was the core zone for a medieval Khmer Empire, that during its peak in 12th century controlled most of mainland Southeast Asia (Coedes 1968). The capital, known to its inhabitant as Yaśodharapura, was one of the biggest pre-industrial low density complex in the world (Fletcher et al. 2008; Evans et al. 2013), covering an area estimated to 1000 km^2 (Evans et al. 2007). Now the capital, known as Angkor, is enlisted on UNESCO World Heritage List. Numerous research projects were devoted to the political history of Angkor (Coedes 1968) as well as functioning and demise of the urban complex (Fletcher et al. 2008; Buckley et al. 2010, 2014). However our understating of the times before 9th century, when Khmer Empire was founded, is not so well established (Higham 2014a).

The main reason, why the state of the arts of the studies on prehistoric and early historic Cambodia is so poor, is undoubtedly a long-lasting military conflict that tore the country apart in the second half of 20th century (Jelonek 2008). While the adjacent regions in Thailand (Higham and Rispoli 2014) or Vietnam (Reinecke and Lê 1998) have well established chronological sequence, the attempt to create a continuous narrative about the early Cambodia is doomed to fail through blind spots in our knowledge.

The introduction of agrarian economy into territories of contemporary Cambodia is a part of bigger process of the Neolithic Revolution in a global scale

(Higham 2014a). It is widely accepted among scholars (e.g. Diamond and Bellwood 2003) that the autochthonic centres of the Neolithisation were located in the Fertile Crescent of Levant and Iraq, Mesoamerica, the Andean region and central China. However the origin of first farmers in Southeast Asia for a long time was subjected to ongoing academic discussion (e.g. Higham 2002; O'Reilly 2006). Recent-most studies in archaeology (e.g. Rispoli 2007), paleolinguistics (Ferlus 2010) and genetics (e.g. Fuller et al. 2010) suggest allochtonic provenance of the first farmers who over generations spread southwards from Yangze Valley. The main two "routes" were indicated by archaeologists (Higham 2002, 2014a): through the coast of South Chinese Sea and over the highlands of Yunnan. However the arrival of new population, most likely speaking a language from Austroasiatic family, has not abruptly ended the existence of hunters-gatherers communities. Moreover evidences from the burial sites of Thailand excavated by Higham (2014a) suggest coexistence and permeation of these populations. Unfortunately just few of the Neolithic sites in Cambodia were excavated. In this context it is worth mentioning the site of Samrong Sen, firstly excavated by the French during colonial period (Mansuy 1902) and recently re-excavated by Khmer team led by Vanna (2002, 2006). The site, which occupation period is estimated to 2300–500 BC, yielded numerous artefacts, including stone adzes and ceramic items, that enriched our knowledge about the material culture of that period. Samrong Sen is one of the limited number of known Neolithic sites, however as it has been accurately pointed by Coe (2003) "Samrong Sen cannot have been unique".

The Bronze Age of mainland Southeast Asia is marked by rapid social changes, that crystallized the ruling elite lineages (Higham 2011; Higham et al. 2011). Another social phenomena, that were a prelude to early states of protohistory, was stratification of a settlement pattern (White 1995) that occurred during that period. Again, only limited number of sites from that period, including praiseworthy example of recent studies at Koh Ta Meas (Pottier 2006), in Cambodia, were excavated. However the well studied adjacent Khorat Plateau Bronze Age can be, with obvious caution, emulated to Northeast Cambodia. The well-studied site of Ban Non Wat (Nakhon Ratchasima Province, Thailand; Higham 2011), especially multiphase cemetery, provided an insight into a Bronze Age society. The close examination of 224 burials revealed that overtime certain percentage of population were inhumated with grave goods that significantly exceeded the average. Assuming that wealth of the burial gifts reflects the position of an individual in the society, it clearly indicates the emerge of the people of high status. However from a landscape perspective, as noticed by O'Really (2014), there is a lack of build environment features, including a defensive structures.

During the late prehistoric and protohistoric period, in some chronological systems demarked as "Iron Age" and subsequently "pre-Angkorian" (Stark 2006), the social and material changes snowballed. The last centuries BC and the beginning of the first millennium AD witness the emerge of first cities and states (Higham 2014b). This period is much better studied in Cambodia, thanks to recent excavations all over the Cambodia, including Angkor Borei and Lower Mekong Region (Stark 2003, 2007; Reinecke et al. 2009); Pouk and Siem Reap rivers basin

(O'Reilly, personal communication) or later site, dated to mid-first millennium AD, of Sambor Prei Kuk (Shimoda et al. 2006). As well the new source of data appeared in that period—the written sources (Bellwood and Glover 2004).

At the dawn of Common Era a proto-states of the Mekong Delta emerged. Conventionally those polities are described by umbrella term "Kingdom of Funan" derived from a Chinese historical sources (Coedes 1968), however the political landscape seems to be much more sophisticated and consisted of numerous mandalas—protostates with concentric sphere of influences around the main city (Maṇḍala मण्डल—circle in Sanskrit; Wolters 1999). Archaeological evidences show an urban civilisation in a Mekong Delta which focal points were sites od Óc Eo (An Giang province, Vietnam) and Angkor Borei (Takeo Province, Cambodia). Óc Eo was discovered during World War II thanks to aerial reconnaissance and subsequently excavated by the French archaeologists led by Malleret (1959a, 1960). The site itself covers the area of 450 ha and it enwalled by complex of ramparts and moats. Malleret's excavations, as well as more recent French-Vietnamese archaeological works (Manguin and Khai 2000) revealed an extensive compact cityscape with monumental architecture, and portable artefacts found on site suggested that craft and commerce was the subsistence of the population. Trade relations linked Óc Eo with trade centres in India and further with Roman Egypt (more on trade relations of Óc Eo: Smagur and Hanus *in press*). Óc Eo was connected by the system of canals with Angkor Borei. The canals are believed to serve both transportation and water management function (Fox and Ledgerwood 1999; Bishop et al. 2004). Angkor Borei, excavated in late 90s (Stark et al. 1999), was another urban complex with monumental brick architecture.

In parallel the Khorat Plateau also was subjected to major changes (O'Reilly 2014). The changes are not only visible in burials (Higham 2002, 2014a, b) that shows inequalities, but, for the first time, also in the large-scale landscape modifications (Moore 1988; Boyd et al. 1999). Such modifications were moats and channels. Moore suggests (1988) the evolutionary explanation, where in the early Iron Age settlements were built on the oxbows or confluences that served as the natural moat and gradually the "artificial" moats, dug by men, started do appear. However as Higham noted (2014b) most of these sites were not subjected to archaeological research, therefore creating both relative and absolute chronology is very challenging.

Method—Anthropogenic Modifications as Seen Through Satellite Imagery

The state of the arts presented above shows a big gap in our understating of the past settlement phenomena in a regional scale. The regions of Khorat Plateau (e.g. Higham 2014a) and Lower Mekong, including the Delta (e.g. Stark 1999) are well studies. In the former region we have well established settlement patterns study, based on aerial images (Moore 1988), consisting of several types of early historic sites (Moore 1988, 1992; Stargardt 1992). The prominent in settlement hierarchy enclosed sites can be were divided by Moore (1992) into three categories: irregular moated sites, characterized by moat or complex of moats that follow the irregular shape of inhabited mound in the centre; territorial sites that alters the landscape on bigger scale than irregular sites that were shaped by local topography; and a rectangular enclosed sites which shape is determined by imposed geometric forms. Those sites were very common in Khorat Plateau, e.g. Moore (1992) counts 101 irregular moated sites in just one province of Thailand—Buriram (10,323 km^2). The moat seems to have a great significance, having military, economical (Moore 1989) and social aspect (O'Reilly 2014). The latter of the regions, lower Mekong and the Delta, served as a venue for a development of an urban civilisation (Stark 1999; Higham 2014a, b).

However little is known about the land that lies in between, the region known today as Northwest Cambodia. The research on the early history there was dimmed by a magnificent medieval capital of Khmer Empire, Yaśodharapura, known today as Angkor. Angkor was subjected to intensive archaeological investigations since late 19th century when Cambodia was incorporated into French Colonial Empire. Last decade resulted in great advance in our understanding of the cityscape of this urban complex, *inter alia* thanks to airborne and space borne survey (Evans et al. 2007, 2013). However our knowledge about the landscapes beyond the borders of Greater Angkor is limited.

Therefore we raised a series of questions about the past landscapes in this region. First of all we were interested in the settlement patterns; especially if we could observe, similar to Khorat plateau, the emerge of moated sites and settlement hierarchy. Likewise it is important to determine if the settlements, both preceding and contemporary to Angkor, had similar disperse characteristic; and, after mid-first millenium AD, rectangular form. Answering such research issues has a potential to resolve if Yaśodharapura—the biggest low density urban complex of the pre-industrial world, was an isolated phenomena or typical urban settlement of medieval Southeast Asia; as well if the sites agglutinated gradually, or at the dawn of Angkor period there was a sudden shift from pre-urban settlements to first urbanised capital of Angkorean Empire—Hariharalaya.

Great contribution to our understanding of the pre-Angkorian landscape has a site of Lovea (Fig. 2), located 17 km west from Angkor Thom. The site is known from the Colonial Period, when incidental discovery of human skulls (during adaptation of Hinduism and Buddhism in early historical period cremation replaced inhumation) and metal objects suggested pre-Angkorian occupation (Malleret 1959b). In recent years the site was excavated by team of Australian archaeologists and final excavation report shall pinpoint the detailed chronology (O'Reilly, personal communication). The first landscape-oriented studies on this site were conducted by Moore (1989), who pointed out the similarities with moated sites of Khorat Plateau. She identified the complex of two moats and ramparts and proposed a theory explaining moats as early water management features (Moore 1989, 1992). Moreover discovery of two similar sites in the vicinity of Lovea

Fig. 2 Satellite image of Lovea. *Data source* ArcGIS Online resources

(Groslier, unpublished materials) made Moore (1989) believe that "Lovea is not unique in Cambodia". The landscape studies of the early sites at the edge of Angkor were carry on by Hawken (2011). He concluded, by painstaking analysing of oxcart tracks and rice field boundaries identified on vertical aerial images that prehistoric and early historic sites were a focal loci for a mesh of tracks and field-boundaries which "radiated" from the settlement mounds (Fig. 3).

In order to extend the understating of pre-Angkorian landscapes of Northwest Cambodia it was decided by the authors to survey this region using satellite imagines. The arbitrary extent of the study area were delineated by the borders of contemporary provinces of Oddar Meanchey, Banteay Meanchey and Siem Reap provinces. As the survey has not been finished yet, the scope of this paper covers the area of western part of Siem Reap province. Another arbitrary choice done, while designing the research, to focus on potentially inhabited areas and features directly related to them. Therefore features related to our investigations were primary occupation mounds with adjacent moats, channels or field systems. Excluded features included mostly rectangular medieval water cisterns. The reason for such discrimination was most likely that irregular in shape, occasionally moated mounds are considered pre-Angkorian (e.g. Moore 1989; Hawken 2011) and straight angled features are connected to medieval, Angkorian period (Hawken 2011). The source of satellite images were as follows: Google Earth and ESRI ArcGIS Base Maps. Considering that overall study area cover the area of over 20,000 km^2, acquiring hi-resolution commercial satellite images would not be feasible for financial reasons. To ensure systematic survey the study area was grided into 10 km × 10 km zones.

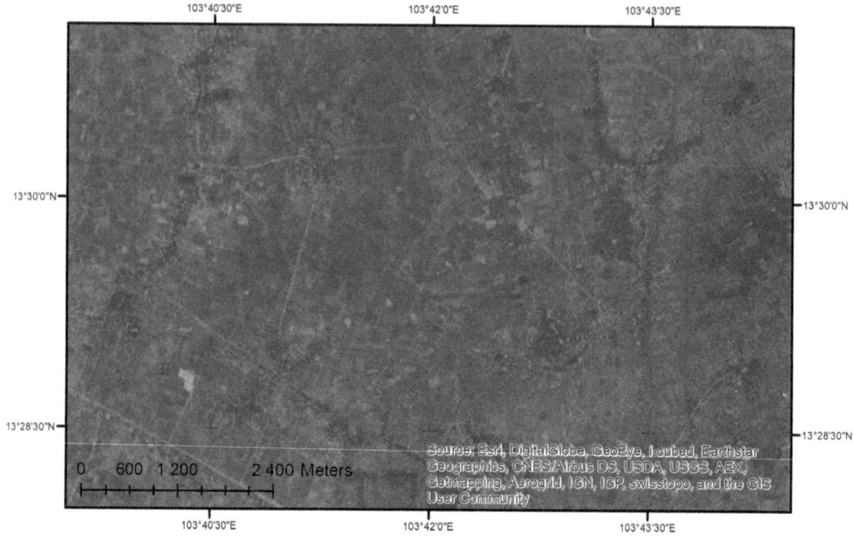

Fig. 3 Satellite image of radial sites in Pouk District. *Data source* ArcGIS Online resources

The biggest challenge was to establish a methodology for site recognition. As there are established routines for such surveys in Europe, incorporating identification based on soilmarks, cropmarks etc. (Palmer and Cowley 2010) mainland Southeast Asia needed a system adjusted to local cultural and environmental settings. The "known to unknown" approach was used, as the authors during their previous work in Cambodia gained necessary "landscape experience" (Rączkowski 2002). Therefore the initial part of the work was both analysis of the methodology of survey done by Moore (1988) in adjacent parts of Thailand, as well close investigation of known sites of our interest, mainly mentioned earlier sites near north-western edge of the Greater Angkor, including Lovea (Evans et al. 2007; Hawken 2011). Furthermore Lovea was subjected to Airborne Laser Scanning (Evans el al. 2013), thereupon the landscape features could be examined for their footprint on satellite image and topographical characteristic.

Based on such premises it was concluded that since prehistoric times the landscape was physically altered and such alternations should be search as indicators of habitation. Those modifications were as follows: raised occupational mounds, similar to the tells known from the Middle East; moats and ramparts; channels; rice field systems. The case study of Lovea as well demonstrated that such mounds can be inhabited even today.

Examination of known sites resulted in development of three research routines, on which identification of prehistoric and historic landscape modifications is possible. The first method is based on geomorphological footprint that as well influences the vegetation. Similar approach was used by Moore (1988) in her Thai survey. The second routine investigates the contemporary settlement patterns, as the Lovea example demonstrates how past landscape features distort the contemporary habitation and land use. Finally, the third form is the implementation of Hawken's (2011) rice field systems studies as the pre-Angkorian fields can be easily distinct from medieval and modern paddies.

The clear example of the site identified using geomorphological features and vegetation is Banteay Plang, site already mentioned by Lunet de Lajonquière (1911) during Colonial Period. The sites morphology is very distinct as it consists of two overlapping moats that makes the plan of the site resembles number "8" or a snowman (Fig. 4). The moats can be identified due to two factors. First of all the soil has clearly lighter shade, suggesting a higher humidity. Another premise lies in the vegetation, as banks of the moat are overgrown with dense bushes, clearly visible as a thin, greenish belt. Clear, oval shape of the entire moated complex advocates for an anthropogenic provenance. On the ground the moat appears as

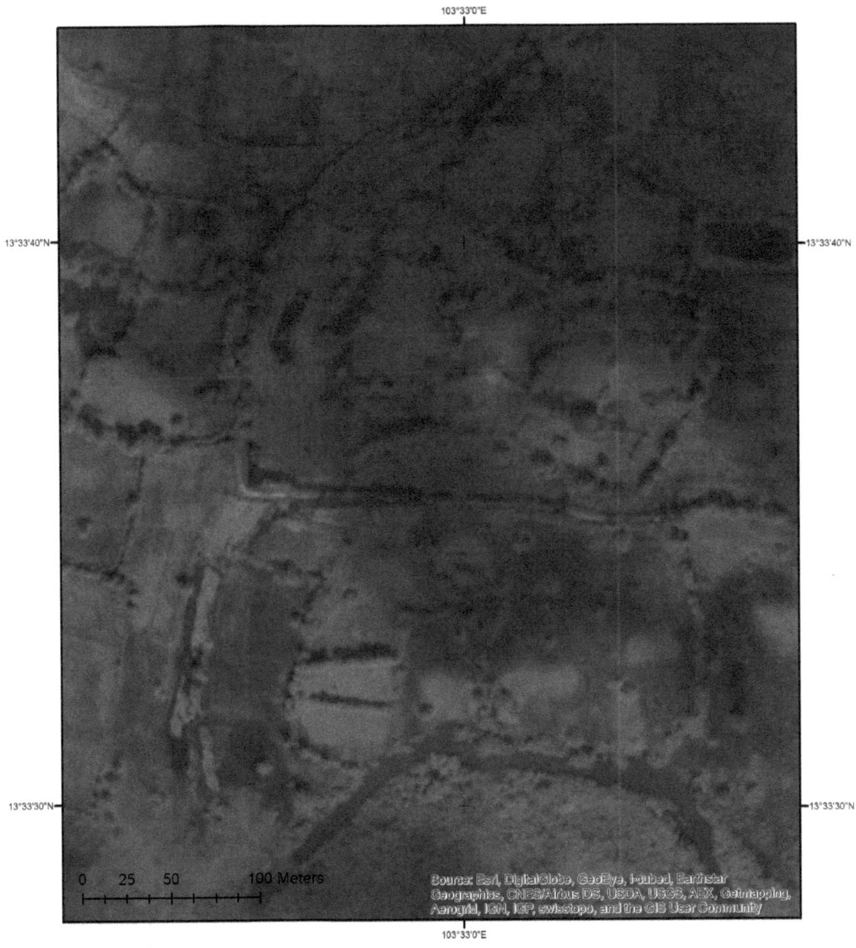

Fig. 4 Satellite image of Banteay Plang. *Data source* ArcGIS Online resources

Fig. 5 Vegetation covering the edge of the moat at Banteay Plang. *Photo* by E. Smagur

pronounced, around 1 m deep negative feature (Fig. 5). As the authors visited the site in January, the peak of dry season, the moat was dry, but mire soil suggested that there is a standing water during more humid parts of the year. The mound, which habitation is confessed by pottery and worked stone shards, stands above surrounding plain. The most contrasting feature of the satellite image is the clump of trees covering the mound, as opposed to bare rice fields surrounding the site. The site's palimpsest is complicated as the satellite image suggests that the site has at least two phases. It can be judged from a fact that the moat of the southern mound cuts through and superpositions the northern moat. The hinterland is all well interesting, as the remnants of the feeding canal or old bed of nearby stream to the north were identified. The stream itself, overgrown by riparian forest, was identified using satellite image as a threat, due to erosion that already has washed away southernmost part of the site. Another two examples, that lack a toponym, can be seen in sparsely populated area east of Srei Snam village. The first, moated site, has a very interesting palimpsest as it is "cut" by a medieval royal road linking Angkor and Phimai in modern Thailand (Hendrickson 2010; Fig. 6), what gives an important clue about sites relative chronology. A double moat is very well visible as parts of it contains a standing water. The rampart differs on the image as a ring of vivid green bushes. The mound itself is, similar to Banteay Plang, overgrown with a clump of high trees. Another site in this area that is presented as the example is again an oval mound, protruding from a flat landscape of rice fields and covered

Fig. 6 Moated site superpositioned by a medieval road. *Data source* Google Earth

with isolated trees. However this site in contrast is not surrounded by the moat. Thus, the close investigation of morphology of the landscape, especially the positive/negative features, soil colours and vegetation, but in contrast to Europe not tones of the cereals, but presence of trees and bushes, can reveal a past alternation of the landscape. Similarly Garrison et al. (2008) pointed out the certain tree-species can help identify the settlements in the Maya lowlands.

In contrast to the sites that are no inhabited anymore, the large body of pre-historic and historic sites are incorporated into modern inhabited parts of the landscape. As a clear example can serve a village of Chres, just 3 km to the east of Banteay Plang. As encountered by ethnographers, such as Szafar (1971), modern Khmer villages are linear settlement following a roads, similar to e.g. medieval European communes called from German Reihendorf—row village. Investigation of satellite images reveals myriad of such villages and small towns build in two rows on two sides of, usually straight, roads. However sometime this settlement pattern is distorted. Village of Chres is a good example of such case. The main body of the village is located on a two sides of the dirt road. However the easternmost part of this village is not following this manner, but rather forms an oval. Therefore it suggests that this part follows a much older settlement pattern (Fig. 7). The site is not moated, however the moat could be filled up during the long occupation of the site. Eg. the authors witnessed the situation at the site of Lovea, where villagers dumped a dirt in the moat in order to create a raised platform for a house. Besides Lovea, which ancient occupation was proved by archaeological excavations (O'Reilly, personal communication) another examples of such oval villages that are inhabited today are Chuo Chakkrei or Reul. Case of Reul is especially interesting, as the oval mound, covered with temporary houses, is cut on its eastern age with a

Fig. 7 Satellite image of Chres village. *Data source* Google Earth

medieval channel, that gives an interesting insight into the relative chronology of the palimpsest. The four examples are verified, it seems that such conservative settlement pattern is more common, but intensive ground-truthing is required for verification. Unfortunately such continuation of habitation rises some serious issues of proper surface survey and excavations, as usually the contemporary housing structures are dense.

The last manner of the site identification is derived from Hawken's (2011) research in rice fields systems. The agricultural practices of rice cultivation in Cambodia requires a standing water to plant the seedlings (Szafar 1971) and such farming technique is knows since late prehistory (Higham 2014a). In order to provide such conditions the artificial basins, called paddies, are created by rising an earthen boundaries, usually around 30–50 cm high. Such fields, usually a squares up to 50 m on a side, creates a whole network of small boundaries. Numerous research endeavours in Europe indicated that the past field systems visible in contemporary landscape (e.g. Baker and Butlin 1973; Hunter 2003), and some of them are two millennial old, like Roman centuration (Romano 2003). Similarly, in modern Khmer landscape it is still possible to reconstruct the spatial settings of old rice fields using aerial and satellite images (Hawken 2011). The field systems are visible from above thanks to raised paddy boundaries, that cast shadows. Hawken (2011) is credited for developing a chronology of the field systems from radial paddies of pre-Angkorian period, though rectangular medieval and modern fields, to enormously big, around 100 m on a side, fields from Khmer Rouge period (1975–1979). As noted in case of Lovea the rice field boundaries were "radiating" from an inhabited mound. The survey done by the authors revealed a vast set of similar sites where the contemporary perpendicular and parallel networks of boundaries were distorted by, on first sight, more "chaotic" paddies. However the closer investigation shows that all the lines converge at one point. Assuming that Hawken is right and such not-gridded field systems have pre-Angkorian provenances and the settlements were the focal points for the past populations, the place where the "lines" of field system come together, can be identified as prehistoric and early historic occupational site. As an example can serve the site of Slaeng Ta Vet (Fig. 8), where more distant hinterland of the site is constituted by common gridded fields, however the fields closer to the site follow the radial pattern. In contrast the nearby village of Kouk Thkov is surrounded by "normal", square fields. The site of Sleang Ta Vet is again an example of the site with contemporary occupation, as well the satellite image does not provide any signs of the moat.

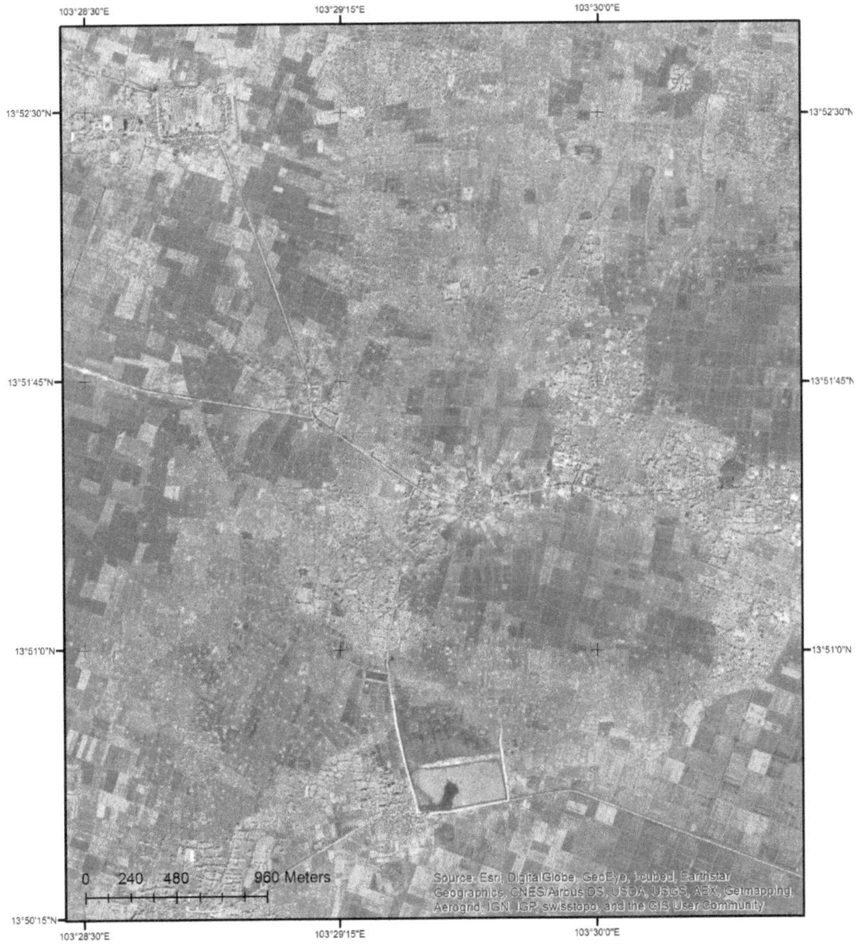

Fig. 8 Rice-fields boundaries pattern as seen using satellite image. *Data source* ArcGIS Online resources

The three identification routines, presented above, usually are applied at the same time, as the site and its hinterland is usually preserved in the landscape. As the routines are based on different principles, when combined, they provide a high level on certainty in the identification of the elements of the past landscapes. However still majority of those sites requires a ground verification, that the authors hope to achieve in following years. Beside standard procedures of surface collection, where the pottery samples can help to determine the period of occupation, the application of UAVs for creation of detailed ortofotomapas and digital elevation models based on photogrammetric elaboration seems to have a great potential. This was proven by the authors at the site of Lovea, where a point clouds derived from both ALS and series of overlapping vertical images done with hexacopter were compared.

Results—Populating the Past Landscapes of Cambodia

As the result of this on-going survey over 300 sites and potential sites, that require further ground truthing, were discovered in just Siem Reap province. This figure underestimates the total count of the pre-Angkorian sites as the limitation of the space borne survey is that areas below the tree canopies, and significant part of Siem Reap province is covered with a tropical forest, are invisible. Therefore, with this survey, the past landscapes of Northeast Cambodia beyond the edges of the Greater Angkor, are repopulated. The results that are under discussion in this chapter have the preliminary character, and therefore yet we are unable to create a historical narratives out of it. For today the most important outcome is subversion of the impression that was created due to the state of research, that areas between the Mekong Delta and Khorat Plateau, were not an occupational void. Since late prehistory it was a living landscape consisting of the settlements and its hinterland, constituted mostly by the rice fields. Most interesting, standing in the contradiction which the results of Moore (1988), the moated sites are not that numerous in the study are of Siem Reap province. The further investigation is needed, however the outcome has a potential to heavily influence the discussion about the importance of the moats in the development of early historic societies in the regional scale of mainland Southeast Asia.

Conclusions—Landscape Studies as a New Approach in Southeast Asian Archaeology

For long time the studies of prehistory and early history of mainland Southeast Asia were very artefact and site oriented (e.g. Higham 2002), however the recent-most studies (e.g. O'Reilly 2014) stress the role of landscape and landscape modifications in our understanding of both materiality and sociality of the past of this region.

The archaeological landscape as shall be define as the past cultural landscape. The cultural landscape is therefore a space that was modified by a man (Spencer—Wood and Baugher 2010) or attributed with definitions, descriptions and meaning (Delle 2002). Therefore presented research can be seen as a proposal for a research routine for creating a datasets for further archaeological interpretation. The aerial archaeology has its roots in Europe (Rączkowski 2002), however we transformed it to match the realities of mainland Southeast Asia. Thus the combination of a vegetation marks and soil marks, as well as contemporary settlement pattern and rice fields systems can help us with the identification and interpretation of past landscape modifications. As well we are aware of the limitations of this method, the one most important, beyond the issue with forested areas, is that this survey is oval mound sites oriented. Therefore habitation sites of different morphology, e.g. Iron Age dated site discovered by looters just east of Banteay Chhmar Cheung village in Banteay Meancheay province (O'Reilly, personal communication) is a cigar-shaped and most likely would be omitted using the proposed research routine. Therefore the survey as well needs a further cross-checking.

Nevertheless the limitations the survey contributed to our understanding of the landscapes beyond the edges of Greater Angkor and clearly helps to fill the data void south of Dângrêk Mountains. Undoubtedly the further research shall focus on interpretation of this dataset, mostly focusing on providing a relievable chronological framework, providing a landscape perspective to archaeological studies of this region and helping the resolve some of the contemporary issues of Southeast Asian archaeology, like formation of first states and cities (Higham 2014a, b) and the processes that led to the creation of Angkor, the biggest low density urban complex of the preindustrial world (Fletcher et al. 2008).

Acknowledgments Funding for this research was generously provided by the Ministry of Science and Higher Education of the Republic of Poland (Grant # DI2012 013542). The authors would like to thank Prof. Roland Fletcher, Prof. Włodzimierz Rączkowski, Dr Rachel Opitz, Dr Dougald O'Reilly, Dr Jarosław Źrałka and Mr Łukasz Banaszek for the discussions and comments on the draft of this paper. As well the author would like to acknowledge the people who provide organizational support: Ms Malay So, Ms Katarzyna Pilipowicz, Ms Paulina Dębicka and Mr Sam Player.

References

Baker, A. R., & Butlin, R. A. (Eds.). (1973). *Studies of field systems in the British Isles*. CUP Archive.

Bellwood, P. S., & Glover, I. (2004). Southeast Asia. Foundations for an archaeological history. *Southeast Asia: from prehistory to history*, 4-20.

Bishop, P., Sanderson, D. C., & Stark, M. T. (2004). OSL and radiocarbon dating of a pre-Angkorian canal in the Mekong delta, southern Cambodia. *Journal of Archaeological Science, 31*(3), 319-336.

Boyd, W. E., McGrath, R., & Higham, C. F. (1999). The geoarchaeology of the prehistoric ditched sites of the upper Mae Nam Mun Valley, Northeast Thailand, II: stratigraphy and morphological sections of the encircling earthworks. *Bulletin of the Indo-Pacific Prehistory Association, 18*, 169-180.

Buckley, B. M., Anchukaitis, K. J., Penny, D., Fletcher, R., Cook, E. R., Sano, M., ... & Hong, T. M. (2010). Climate as a contributing factor in the demise of Angkor, Cambodia. *Proceedings of the National Academy of Sciences, 107*(15), 6748-6752.

Buckley, B. M., Fletcher, R., Wang, S. Y. S., Zottoli, B., & Pottier, C. (2014). Monsoon extremes and society over the past millennium on mainland Southeast Asia. *Quaternary Science Reviews, 95*, 1-19.

Chandler, D. P. (1991). *The land and people of Cambodia*. Harper-Collins Publishers.

Coe, M. D. (2003). *Angkor and the Khmer civilization*. Thames & Hudson.

Coedes, G. (1968). *The Indianized States of South-East Asia*. University of Hawaii Press.

Diamond, J., & Bellwood, P. (2003). Farmers and their languages: the first expansions. *Science, 300*(5619), 597-603.

Delle, J. A. (2002). Power and landscape: spatial dynamics in early nineteenth century Jamaica. *The dynamics of power*. Southern Illinois University, Carbondale, 341 - 361.

Evans, D. H., Fletcher, R. J., Pottier, C., Chevance, J. B., Soutif, D., Tan, B. S., ... & Boornazian, G. (2013). Uncovering archaeological landscapes at Angkor using lidar. *Proceedings of the National Academy of Sciences, 110*(31), 12595-12600.

Evans, D., Pottier, C., Fletcher, R., Hensley, S., Tapley, I., Milne, A., & Barbetti, M. (2007). A comprehensive archaeological map of the world's largest preindustrial settlement complex at Angkor, Cambodia. *Proceedings of the National Academy of Sciences, 104*(36), 14277-14282.

Ferlus, M. (2010). THE AUSTROASIATIC VOCABULARY FOR RICE: ITS ORIGIN AND EXPANSION 47. *JSEALS, 3*, 61-76.

Fletcher, R., Penny, D., Evans, D., Christophe, P., Barbetti, M., Kummu, M., & Lustig, T. (2008). The water management network of Angkor, Cambodia. *Antiquity, 82*(317), 658-670.

Fox, J., & Ledgerwood, J. (1999). Dry-season flood-recession rice in the Mekong Delta: two thousand years of sustainable agriculture? *Asian Perspectives 38 (1),* 37-50.

Fuller, D. Q., Sato, Y. I., Castillo, C., Qin, L., Weisskopf, A. R., Kingwell-Banham, E. J., ... & Van Etten, J. (2010). Consilience of genetics and archaeobotany in the entangled history of rice. *Archaeological and Anthropological Sciences, 2*(2), 115-131.

Garrison, T. G., Houston, S. D., Golden, C., Inomata, T., Nelson, Z., & Munson, J. (2008). Evaluating the use of IKONOS satellite imagery in lowland Maya settlement archaeology. *Journal of Archaeological Science, 35*(10), 2770-2777.

Hawken, S. (2011) *Metropolis of Ricefields. A Topographic Classification of a Dispersed Urban Complex*. Unpublished PhD Dissertation. The University of Sydney

Hendrickson, M. (2010). Historic routes to Angkor: development of the Khmer road system (ninth to thirteenth centuries AD) in mainland Southeast Asia. *Antiquity, 84*(324), 480-496.

Higham, C. (2002). *Early cultures of mainland Southeast Asia*. Chicago: Art Media Resources.

Higham, C. (2011). The bronze age of Southeast Asia: New insight on social change from Ban Non Wat. *Cambridge Archaeological Journal, 21*(03), 365-389.

Higham, C. (2014a) *Early Southeast Asia: From the First Humans to the Civilization of Angkor*. River Books.

Higham, C. (2014b). From the Iron Age to Angkor: new light on the origins of a state. *Antiquity, 88*(341), 822-835.

Higham, C., & Rispoli, F. (2014). The Mun Valley and Central Thailand in prehistory: integrating two cultural sequences. *Open Archaeology 2014; vol. 1*, 2-28.

Higham, C., Higham, T., Ciarla, R., Douka, K., Kijngam, A., & Rispoli, F. (2011). The Origins of the Bronze age of southeast Asia. *Journal of world prehistory, 24*(4), 227-274.

Hodder, I. (1985). Postprocessual archaeology. *Advances in archaeological method and theory, 8*, 1-26.

Hunter, J. M. (2003). *Field systems in Essex*. Essex Society for Archaeology and History.

Ingold, T. (1993). The temporality of the landscape. *World archaeology, 25*(2), 152-174.

Jelonek, A. (2008). *Historia państw świata w XX wieku – Kambodża*. Wydawnictwo Trio.

Lunet de Lajonquière, E. (1911). *Inventaire descriptif des monuments du Cambodge*, vol. 3. Paris: E. Leroux.
Malleret, L. (1959). III. Ouvrages circulaires en terre dans l'Indochine méridionale. *Bulletin de l'École française d'Extrême-Orient*, *49*(2), 409-434.
Malleret, L. (1959). *Làrchéologie du Delta du Mékong: L'exploration archéologique et les fouilles d'Oc-Èo. 2 v* (Vol. 43). Ecole française d'Extrême-Orient.
Malleret, L. (1960). L'archéologie du delta du Mékong, Part 2. *La Civilisation Matérielle d'Oc-Èo, 2*. Ecole française d'Extrême-Orient.
Manguin, P. Y., & Khai, V. S. (2000). Excavations at the Ba The/Oc Eo complex (Viet Nam): a preliminary report on the 1998 campaign. *Southeast Asian Archaeology*, *1998*, 107-121.
Mansuy, H. (1902). Stations Préhistoriques de Samrong-Sen et de Longprao (Cambodge) (Prehistoric sites of Samrong Sen and Longprao (Cambodia). Hanoi: FG Schneider.
Marciniak A. (2012) Paradygmaty badawcze w archeologii [w:] *Przeszłość społeczna. Próba konceptualizacji*, Wydawnictwo Poznańskie, 29–83.
McKnight, T. L., & Hess, D. (2000). Climate zones and types: the Köppen system. *Physical Geography: A Landscape Appreciation. Upper Saddle River, NJ: Prentice Hall*, 235-7.
Moore, E. (1989). Water management in early Cambodia: evidence from aerial photography. *Geographical journal*, 204-214.
Moore, E. (1992). Water enclosed sites: links between Ban Takhong, northeast Thailand and Cambodia. *The gift of water: water management, cosmology and the state in South East Asia*, 26-46.
Moore, E. H. (1988). *Moated Sites in Early North East Thailand* (Vol. 400). British Archaeological Reports Ltd.
O'Reilly, D. J. (2014). Increasing complexity and the political economy model; a consideration of Iron Age moated sites in Thailand. *Journal of Anthropological Archaeology*, *35*, 297-309.
O'Reilly, D. J. (2006). *Early Civilizations of Southeast Asia*. Rowman Altamira.
Palmer, R., & Cowley, D. (2010). Interpreting aerial images—developing best practice. In *Space, time, place. Third international conference on remote sensing in archaeology* (pp. 129-35).
Pottier, C. (2006). Under the western Baray waters. *Uncovering Southeast Asia's Past.* NUS Presse Singapore, 298-309.
Rączkowski, W. (2002). *Archeologia lotnicza: metoda wobec teorii* (No. 47). Wydawnictwo Naukowe.
Rączkowski, W. (2015). Aerial archaeology. In Field Archaeology from Around the World (pp. 19–25). Springer International Publishing.
Reinecke, A., & Lê, D. S. (1998). *Einführung in Die Archäologie Vietnams*. Linden-Soft.
Reinecke, A., Laychour, V., & Sonetra, S. (2009). *The first golden age of Cambodia: excavation at Prohear*. Reinecke.
Rispoli, F. (2007). The Incised & Impressed Pottery Style of Mainland Southeast Asia: Following the Paths of Neolithization. *East and West*, 235-304.
Romano, D. G. (2003). City Planning, Centuriation, and Land Division in Roman Corinth: Colonia Laus Iulia Corinthiensis & Colonia Iulia Flavia Augusta Corinthiensis. *Corinth: Results of Excavations Conducted by the American School of Classical Studies at Athens*, 279-301.
Shanks, M., & Tilley, C. (1987). Social archaeology. *Social Theory and Archaeology, University of New Mexico Press, Albuquerque*, 29-60.
Shimoda, I., Him, S., Seng, K., Chang, V., Chhum, M., So, S., & Nakagawa, T. (2006). Preliminary Report on the Excavation Survey at the Prasat Sambor, Sambor Prei Kuk, 2004 – 2005. *Journal of Southeast Asian Archaeology*, *26*, 117-145.
Smagur, E & Hanus, K. (in press). Kattigara of Claudius Ptolemy and Óc Eo: the issue of trade between the Roman Empire and Funan. *Proceeding of the XIV EurASEAA conference, Dublin 2012*.
Spencer - Wood, S. M., & Baugher, S. (2010). Introduction to the historical archaeology of powered cultural landscapes. *International Journal of Historical Archaeology*, 14 (4), 463 - 474.

Stargardt, J. (1992). 'Muang Fa Daed: from prehistoric moated villages to Mon-Khmer style city; new archaeological studies of aerial images.' in *Ancient Khmer Cities of Lower North-East Thailand*, (eds. Ishizawa, Y. and T.Kano). Tokyo, Sophia University, Institute of Asian Culture, in association with the Fine Arts Department of Thailand, 107-128.

Stark, M. T. (2003). Angkor Borei and the archaeology of Cambodia's Mekong Delta. *Art & Archaeology of Fu Nan: Pre-Khmer Kingdom of the Lower Mekong Valley*, 87-106.

Stark, M. T. (2006). Early mainland Southeast Asian landscapes in the first millennium AD. *Annu. Rev. Anthropol.*, *35*, 407-432.

Stark, M. T. (2007). Pre-Angkorian settlement trends in Cambodia's Mekong Delta and the Lower Mekong Archaeological Project. *Bulletin of the Indo-Pacific Prehistory Association*, *26*, 98-109.

Stark, M. T., Griffin, P., Phoeurn, C., Ledgerwood, J., Dega, M., Mortland, C.& Latinis, K. (1999). Results of the 1995-1996 archaeological field investigations at Angkor Borei, Cambodia. *Asian perspectives 38(1)*, 7-36.

Szafar, T. (1971). *W cieniu świątyń Angkoru. Kambodża wczoraj i dziś*. Warszawa: Nasza Księgarnia.

Tilley, C. (1994). *A phenomenology of landscape: places, paths, and monuments* (p. 10). Oxford: Berg.

Vanna, L. (2002). Rice remains in the prehistoric pottery tempers of the shell midden site of Samrong Sen: Implications for early rice cultivation in central Cambodia. *Aséanie*, *9*(1), 13-34.

Vanna, L. (2006). Living in a Proto-type Prehistoric Alluvial Environment: Water and the Villagers of Samrong Sen in Kampong Chhnang Province, Central Cambodia. *WATER AND CULTURAL DIVERSITY*, 131.

White, J. C. (1995). Incorporating Heterarchy into Theory on Socio-political Development: The Case from Southeast Asia. *Archeological Papers of the American Anthropological Association*, *6*(1), 101-123.

Wolters, O. W. (1999). *History, culture, and region in Southeast Asian perspectives* (No. 26). SEAP Publications.

The Ambivalence of Maps: A Historical Perspective on Sensing and Representing Space in Mesoamerica

John K. Millhauser and Christopher T. Morehart

Abstract Imaging and spatial analysis technologies are revolutionizing archaeological methods and archaeologists' perceptions of space. Rather than view these innovations as inevitable refinements and expansions of the archaeological toolkit, it is useful to critically assess their impacts on theory and practice. In this chapter, we consider what spatial data—data that appear to represent an objective reality—tell us about past and present human experiences of the physical world in terms of abstraction, temporality, and power. We draw on archaeological cases from Mesoamerica to illustrate how these subjective perspectives on space are revealed through technological innovations and how historical and current efforts to map this region play out in the political sphere.

The map is not the territory

—Samuel Hayawaka

Imaging and spatial technologies are revolutionizing archaeological methods and archaeologists' perceptions of space. From high-resolution satellite data to 3-dimensional image processing software, archaeologists have access to spatial resources of greater scope, detail, accuracy, and precision than ever before. The traditional barriers to collecting these kinds of data—high costs and steep learning curves—are falling as instruments produced for the consumer market become smaller, less expensive, and more user-friendly. Greater access to spatial data is facilitated by the internet, the development of free and open-source software for analysis and visualization, and the establishment of digital data curation and sharing repositories. These tools and techniques are becoming customary, and progressively compulsory, in archaeological projects. Rather than view these innovations as no more than

J.K. Millhauser (✉)
North Carolina State University, Raleigh, USA
e-mail: millhauser@ncsu.edu

C.T. Morehart
Arizona State University, Tempe, USA

M. Forte and S. Campana (eds.), *Digital Methods and Remote Sensing in Archaeology*, Quantitative Methods in the Humanities and Social Sciences,
DOI 10.1007/978-3-319-40658-9_11

refinements and expansions of the archaeological toolkit, it is useful to critically assess the impact they are having on the questions archaeologists ask and the ramifications of the choices we make as scholars of the past and, moreover, of space.

We begin this chapter by considering maps as the products of efforts to collect spatial data at a variety of scales and to distill that data into concise, meaningful, and useful representations. Our discussion of maps exposes a tension between the goal of collecting data to produce objective representations of space and the fact that such representations are subjective abstractions. This tension cannot be resolved by appealing to a difference between objective data and subjective information—as if the former were raw, unorganized, and useless until processed and organized into the latter (Kent 1978). Our technological means of observation may improve, but spatial data are still the products of human judgments, imperfections, and histories.

However, we see no inherent contradiction between the contingencies of spatial knowledge and the collection of empirical data to capture and represent reality. Rather, we see opportunities to embrace the inherent social complexity contained within spatial data, which can lead us to understandings of the human experience of space as subjective, dynamic, and historically situated. To show how this can work, we consider maps in terms of their abstractness, their temporality, and their power and then illustrate these points with three sets of archaeological cases from Mesoamerica. While by no means exhausting the possibilities, these three themes bridge the study of the past with the practices of archaeology in the present.

The Limits of Spatial Data: Abstraction, Temporality, and Power

As landscape architect James Corner noted, "mapping is a fantastic cultural project, creating and building the world as much as measuring it and describing it" (1999: 197). This cultural project begins long before we sit down to draw a map; it encompasses our choices of subjects and our means of observation and documentation. We collect and combine spatial data, sift it, mine it, and distill it into representations of the world. These representations support our arguments, allow us to test hypotheses, and help us disseminate knowledge. At every stage, we choose what to collect and what to ignore, what to emphasize and what to sacrifice. This cultural project does not stop after we have made a map; it continues as each map is remembered or forgotten, preserved in the historical record, lost, or destroyed. The persistence of maps in the historical record does not constitute the neutral accumulation of knowledge. Rather, their creation, their mobilization and use, as well as their survival are the direct result of historical processes in which some voices and visions are fostered and celebrated while others are often erased, silenced, and destroyed (e.g., Anderson 1991; Trouillot 1995).

Defining maps as creative and historically contingent products does not necessarily entail rejecting their ability to capture reality via the collection of empirical data. Maps aid navigation, for example, because they are analogous to the physical world that they represent. Maps capture relationships in terms of position, distance, or other criteria that are independent of any viewer's location or perception (Gell 1985). Wherever one is currently located, the statement "New York is 215 miles from Boston" remains true. But maps do not do the same kinds of work in all times and places; nor do they work equally well for everyone. In order to address these limitations and balance the subjective and objective aspects of maps, we suggest that archaeologists working with spatial data consider three areas of concern: (1) abstraction, (2) temporality, and (3) power. We expand on each concern below and then, in subsequent sections, consider each in light of archaeologists' efforts to collect and represent spatial data in meaningful and innovative ways.

Abstraction

On their surface, maps are abstractions. Their relevance and, further, their legibility are tied to the particular cultural and historical contexts in which they were produced (Corner 1999: 199; Ingold 2011: 224). The skills involved in producing and using maps are learned and passed down in particular social groups—from Micronesian mariners navigating by the stars to archaeologists in Mexico with a total station and a GPS unit (Turnbull 1993: 61). To read a map involves at least two modes of translation: first, the translation of symbols and relationships into meaningful spatial information; and second, the translation of a particular way of seeing, being, and moving in the world. Archaeologists who map ancient sites and landscapes perform additional acts of translation: they record space in the system with which they are familiar and then try to find the traces of prior and potentially divergent spatial sensibilities.

Our translations mark the pursuit of meaning and, thus, share operational similarities with hermeneutics, the effort to interpret and understand products of human experience in the contexts of their social, cultural, and cognitive environments. Going beyond the double hermeneutic that Giddens (1982) advocated, Shanks and Tilley (1987: 107–108) suggested archaeologists operate within a fourfold hermeneutic: contemporary life; our discipline; translating an "other;" and interpreting the other in the past. The segmentation of each of these hermeneutics is debatable. But the fact remains that the act of translation requires an effort to move beyond the interpretive fetters of both our contemporary life and of our discipline to access spheres of meaning that are not only temporally removed but also culturally distinctive. Whether or not we succeed in this endeavor, collapsing these acts of translation through mapping into a single hermeneutic would destroy the culturally variable textures of human life across history and space.

Temporality

Maps freeze a moment in time. Their static nature is instrumental. Maps not only freeze a representation of the state of affairs; they also materialize this state of affairs to produce a sense of objective truth. In the contemporary world, the static map is a technology that produces new subjects (Foucault 1977) and empowers emergent geopolitical conditions, such as the nation (Anderson 1991). Yet many maps from other periods of time, cultural contexts, and modes of (knowledge) production were and are created to convey history. For some groups, present landscapes are impregnated with the past (e.g., Basso 1996). Maps of these areas may not depict space and relations in space in ways that coincide with culturally-specific practices; many indigenous maps in Mesoamerica, for example, stress places, physical movements, celestial and seasonal cycles, personages, and spiritual entities (see below).

In many respects, the atemporal nature of the modern map is tyrannical: even sophisticated technological applications, such as GIS, have are cumbersome tools for depicting change. Archaeologists, geographers, and cartographers struggle to overcome technologies built on collapsing time into a single representational view—a struggle likely intensified by advanced capitalism's hyper-production and consumption, which progressively collapses our daily experiences of time and of space (see Harvey 1989). Scott (1998) traced this process historically to European Colonial expansion, which he argues required the top-down imposition of a political, cultural, and economic "grid" across the globe in order to not only extract and conquer but to fundamentally render the illegible legible. But maps are representations, and therefore can be quite flexible if conceived of as less static. Perhaps the temporal conventions to which we are accustomed can bent or broken.

Power

Maps influence what people think and how they act in space. Such influence is not inconsequential—it is a field through which power can be enacted and enforced (Anderson 1991; Mignolo 1995; Scott 1998). The differential value of maps can directly affect the well-being of people whose lives depend upon the spaces that are, or are not, mapped. At the same time, maps can be used to mediate or resist existing powers (Hale 2006; Maya People of Southern Belize 1997; Mundy 1996; Orlove 1991). The act of mapping also influences the world. Collecting and archiving spatial data changes the way the world works and the way people see, experience, and benefit from it.

Even as archaeologists move from more direct and engaged methods (e.g. field walking) to more passive and remote ones (e.g. satellite images and drones), careful attention to ethics, power disparities, informed consent, and community engagement are gaining urgency. As with many technologies used for science, there also exist complicated relationships between governmental (and often military) applications and scientific uses. Such entanglements are deeply rooted in archaeological practice. For example, the Ordnance Survey maps that are so crucial to archaeological surveys in Great Britain are the products of military surveys originally aimed at subduing rebellions (Phillips 1959; Seymour 1980). In a similar fashion, inexpensive and low-to-no-impact imaging tools, such as drones, also represent private companies' development of technology with fundamental military applications. Archaeologists who invest in this technology may be unintentionally financing the creation of instruments of war.

Simultaneously, the hyper-commercialization of mapping technologies can outpace governments' abilities to regulate and control, opening brief windows of democracy. Myers (2010), for example, employed Google Earth to document the infrastructural expansion of the Guantanamo Bay Detention Camp in Cuba during a time when governmental discourse centered on dismantling investment. By factoring the influence that representations of space have in the present day into our choices and actions, including the maps and images that archaeologists produce, we can begin to dismantle what Corner (1999: 199) calls the 'benign neutrality' that we often attribute to the act of mapping and the technology we use to make maps.

Archaeologists in Space: Innovative Approaches to Current Problems

To explore the ways in which the abstraction, temporality, and power of spatial data present challenges and opportunities for archaeologists, we pose three related questions: First, how can sensing and visualization technologies help us to better understand the ways in which past people understood, constructed, and mapped their worlds? Second, how can we use these technologies to better understand dynamic processes in the physical world? Third, what political challenges and opportunities do new mapping technologies present? Our goal is not to offer a Panglossian solution. Rather, by addressing these questions we can expose the limitations of our contemporary gaze. We can also frame objective and subject approaches as, if not complementary, at least non-contradictory. In the process, our questions and methods can make progress toward better understandings of the human experiences of space.

What Can Spatial Data Reveal About Past Spatial Knowledge and Practices?

Mesoamerica provides a good context to begin to answer this question because Mesoamerican societies shared common ideas about space and time, ideas which were often materialized in the locations of sites, the organization of settlements, and the orientation of monuments and other structures (Ashmore 1991; Ashmore and Sabloff 2002; Astor-Aguilera 2010; Aveni 2001; León Portilla 1963). In certain ways, Mesoamericans' views of space were analogous to Cartesian space and to contemporary Western conventions. For instance, Mesoamericans recognized the cardinal directions and the quadripartite division of space among those directions, both of which are intrinsic to grid-based maps. At the same time, the spatial syntax of Mesoamerican sites reflects values linked to the connections of space and time, human and cosmological forces, and the natural and supernatural worlds. The cardinal directions were associated with life and death, creation and destruction, and these meanings appeared in the arrangement of sacred spaces. Space was organized not only by cardinal directions but also by natural features of the landscape as well as solar, lunar, planetary, and other celestial cycles. Furthermore, the Mesoamerican landscape was, and remains, a vital part of religious rituals and pilgrimages (Halperin 2005; Kubler 1984; Palka 2014). These spatial beliefs were organizing principles that shaped all walks of life from the orientation of homes to the layout of cities.

The careful collection of precise spatial data in the present day facilitates our comprehension of Mesoamerican spaces, but not without considerable effort. One must first create an accurate map using modern methods, measurement units, and orientations and then strip away the trappings of the present to reveal the priorities that informed ancient peoples as they modified their world. The ancient city of Teotihuacan (ca. AD 150–750), located in central Mexico, serves as case in point. Teotihuacan is typical of Mesoamerican urban centers for its central location within its local settlement system and the blending of social and political power with religious and ritual spaces at the city's center. However, Teotihuacan stands out among its contemporaries for its size—it may have supported a population of as many as 200,000 people at its peak—and for the grid pattern of its organization (Millon 1973). Although we might be prone to read the grid as a function of scale—a solution to organizational problems endemic to large settlements—there is no single functional cause for orthogonal city plans (Blanton and Fargher 2011: 515; Grant 2001). In the case of Teotihuacan, the grid was established early in the city's growth and therefore was not originally a solution to problems of scale. Given the connection of the grid to the construction of monuments in the city's core, we might consider an alternative explanation of the grid in terms of its symbolic valance.

Saburo Sugiyama (1993, 2004) has combined precise cartography with attention to Mesoamerican cosmology and spatial organization to reveal an alternative mapping of Teotihuacan, one which may reflect the intentions and concerns of its original planners. The pattern is structural in the sense that it incorporates the division of the site into four quadrants, created by a monumental avenue which was bisected by the channelization of the Rio San Juan. The pattern is also numerical. Sugiyama has argued that Teotihuacanos used a basic measurement unit equal to 83 cm—a refinement of previous efforts to find regularity in the dimensions of monumental structures at the site (Almaráz 1865: 212–213; Drewitt 1987; Drucker 1974). When applied to the sizes of monuments or the distances between them, the quantities of units often coincided with cosmologically and calendrically important numbers. For example, the Pyramid of the Moon measured 260 units to a side, a number which corresponds to the number of days in a pan-Mesoamerican ritual calendar. The Pyramid of the Sun measured 520 units across, which is twice 260, as well as three times 173.3 (the days in the eclipse calendar) and ten times 52 (the number of years in an important Mesoamerican calendrical cycle). Sugiyama has argued that these were more than coincidences: they "support the notion that the major monuments were constructed following a master plan with cosmological and calendrical significance" (Sugiyama 2004: 105). In essence, the city became a physical representation of cosmological principles that, in turn, may have been part of a strategy by which leaders legitimized their authority or power.

As precise and rapid three-dimensional mapping technology becomes more readily available, it may provide new opportunities to investigate and recognize the ways in which Mesoamericans observed, experienced, and created space. Sugiyama's efforts have extended to the mapping of earlier structures encased within later construction phases at the major monuments and to subterranean chambers and caves beneath them (Sugiyama et al. 2013; Sugiyama and Cabrera Castro 2007). More broadly, LiDAR mapping of sites and regions is becoming increasingly common in Mesoamerican archaeology, much as in other parts of the world (Carter et al. 2012; Chase et al. 2012; Hare et al. 2014; Prufer et al. 2015; Rosenswig et al. 2013; Zetina-Gutiérrez et al. 2014). In areas with dense ground cover, LiDAR has been especially useful for identifying ancient settlement patterns and human alterations of landscapes. LiDAR data may provide new avenues to investigate Mesoamerican understandings of space because they are collected as a whole, independent of the priorities and intuitions of observers, at least within a survey's boundaries.

When applied at a wide scale to regions and sites, sensing and visualization technologies can help us identify spatial patterns that may reflect culturally-specific understandings of landscapes and built environments. For example, as archaeologists get better at "reading" the topographic maps produced by processed LiDAR data, they develop an eye for unique cultural patterns and conventions—such as the identification of house lots (Hare et al. 2014), ceremonial groups (Rosenswig et al. 2013: 1506), and anthropogenic landscapes (Chase et al. 2012; Fisher, this volume). Importantly, these insights are not self-evident; they emerge from previous understandings and expectations, from the questions that guide our research, and from our qualitative assessments of features on the ground.

How Can We Map Spaces Whose Traces Are All but Erased?

Despite advances in sensing technology, some signs of the past remain beyond the reach of our observational capabilities because they have been buried or obliterated, or because they were dynamic. Nevertheless, earlier observations of ancient sites and landscapes are preserved in historical maps. Spatial analysis and visualization tools can help us to integrate disparate historical sources into a consistent framework, even if standards of precision and cartographic conventions have changed over time. But, when we layer maps made at different times and for different purposes, how and why do we recognize one view as more authoritative than another? To the extent that we encounter gaps or inconsistencies, what significance do we to attach to them? And, more broadly, how can we use disparities constructively in order to represent the dynamic nature of space across time?

In this section, we highlight a single example—an effort to map the lakes that filled the bottom of the Basin of Mexico in Mexico's central highlands. The Basin of Mexico was the seat of Spanish colonial power, as well as the Aztec Empire and Teotihuacan before that. Today it is filled by Mexico City, a megalopolis through which only scant traces of the past are visible. The basin comprises a broad valley that covers an area of approximately 110 km north to south and 80 km east to west. Because the basin lacks an external outlet for water to drain, runoff from precipitation, permanent and seasonal rivers, and snowmelt contributed over millennia to a series of swamps, marshes, and lakes on its floor. These bodies of water stabilized approximately 4000 years ago into a series of five interconnected lakes: Zumpango and Xaltocan in the north, Texcoco in the center, and Xochimilco and Chalco in the south (Caballero and Guerrero 1998; Lozano-García et al. 1993). The lakes underwent cycles of rainy season expansion (when all five were connected) and dry season contraction (when they shrank and separated). The process was dynamic; when Cortés (1971 [1522]: 102) arrived in the Basin of Mexico in 1519, he observed that the lakes' waters rose and fell like ocean tides. Lakes Xaltocan and Zumpango were generally alkaline and brackish. During the rainy season they were replenished by fresh water and drained into Lake Texcoco, but during the dry season they shrank and became more saline. Lakes Chalco and Xochimilco were perennially fresh-water bodies of water that also drained into Lake Texcoco, which was the largest, lowest, and most saline of them all (Sanders et al. 1979: 84–85). The lakes' annual fluctuations produced significant changes in their total surface area which directly affected where and when individuals and communities lived and worked along their margins (Parsons 2006: 17).

Knowing where the lakes' boundaries were at any given time is crucial to our understanding of how local inhabitants adapted to and changed their environment. The lakes and marshes played a pivotal role in social and demographic developments. They supported a form of highly productive agriculture, called *chinampas*, which involved the construction of raised soil beds surrounded by narrow canals (Armillas 1971; Morehart 2010), even in more saline lakes like Xaltocan (Morehart and Frederick 2014). The lakes and marshes also provided wild foods such as fish and amphibians, seasonal and permanent waterfowl, insect larvae, wild plants, and algae as well as other resources, like reeds and salt (Parsons 2001, 2006; Millhauser 2012, 2016). They facilitated the movement of goods and people by canoe and furnished a measure of defense for settlements that lay within their waters. The lakes and marshes therefore provided advantageous resources, but to live among them required a significant amount of ingenuity and persistence relative to the higher and drier piedmont slopes. The conquistadors, for example, marveled at the size and complexity of Tenochtitlan, the island capital of the Aztec empire.

After the conquest, the Spanish built the Ciudad de México on the ruins of the former Aztec capital. Over the next 400 years, however, thousands would work to design and construct a complex system by which to drain the region of its waters and protect the city from flooding—with dramatic social, political, and environmental consequences that are still being realized today. Prehistoric efforts to prevent flooding or the influx of saline water into chinampas included the construction of extensive systems of dikes and the redirection of rivers (Doolittle 1990; Gibson 1964: 236–237; Palerm 1973). Colonial methods expanded upon these efforts to include projects aimed at draining water from the basin (Candiani 2014; Gibson 1964: 303–305; Mathes 1970; Palerm 1973). Yet, despite efforts to control lake levels, Colonial period flooding became an endemic problem that plagued urban centers and rural communities alike (Candiani 2014, Morehart 2016). Upland erosion channeled more runoff into the lakes and sedimentation raised the relative water level and contributed to the progressive desiccation and salinization of Lakes Texcoco, Xaltocan, and Zumpango. By the late eighteenth century, the northern lakes had been drained to the point that they were each divided into two separate lagoons (Humboldt 1836 [1811]: 372–373; Orozco and Berra 1864; Palerm 1973). The drainage of the lakes was completed in the early 20th century, but subsequent improvements and expansions have been necessary in the wake of the growth of Mexico City (Montoya Rivero 1999; Ríos Elizondo 1975).

With this history in mind, identifying the limits of the lakes that filled the bottom of the Basin of Mexico is not—as it might otherwise seem—a simple task. We might ask in what year or what season. Answering these specific questions can help us to better understand how people made use of the lakes, how the lakes shaped peoples' choices, and how water has factored into the political economies of the different states that have dominated the Basin of Mexico. Knowing where the lakes' boundaries were can also help us better design research projects and strategize our efforts to preserve the few vestiges of the past that remain in the region. Answering these questions demand that we produce accurate and legible maps. But why should we expect that such a representation is possible when the boundaries of the lakes changed?

Current technologies of remote sensing provide us with a starting point, but they are limited their coverage and accuracy. A map of the history of the lakes might start with the most precise and comprehensive data available today based on satellite imagery and airborne LiDAR. A digital elevation model (DEM) of the Basin of Mexico is available for public access through the website of the Instituto Nacional de Estadística Geografía e Informática (INEGI) (Fig. 1). The DEM is limited because urbanization over the last century has covered the majority of the

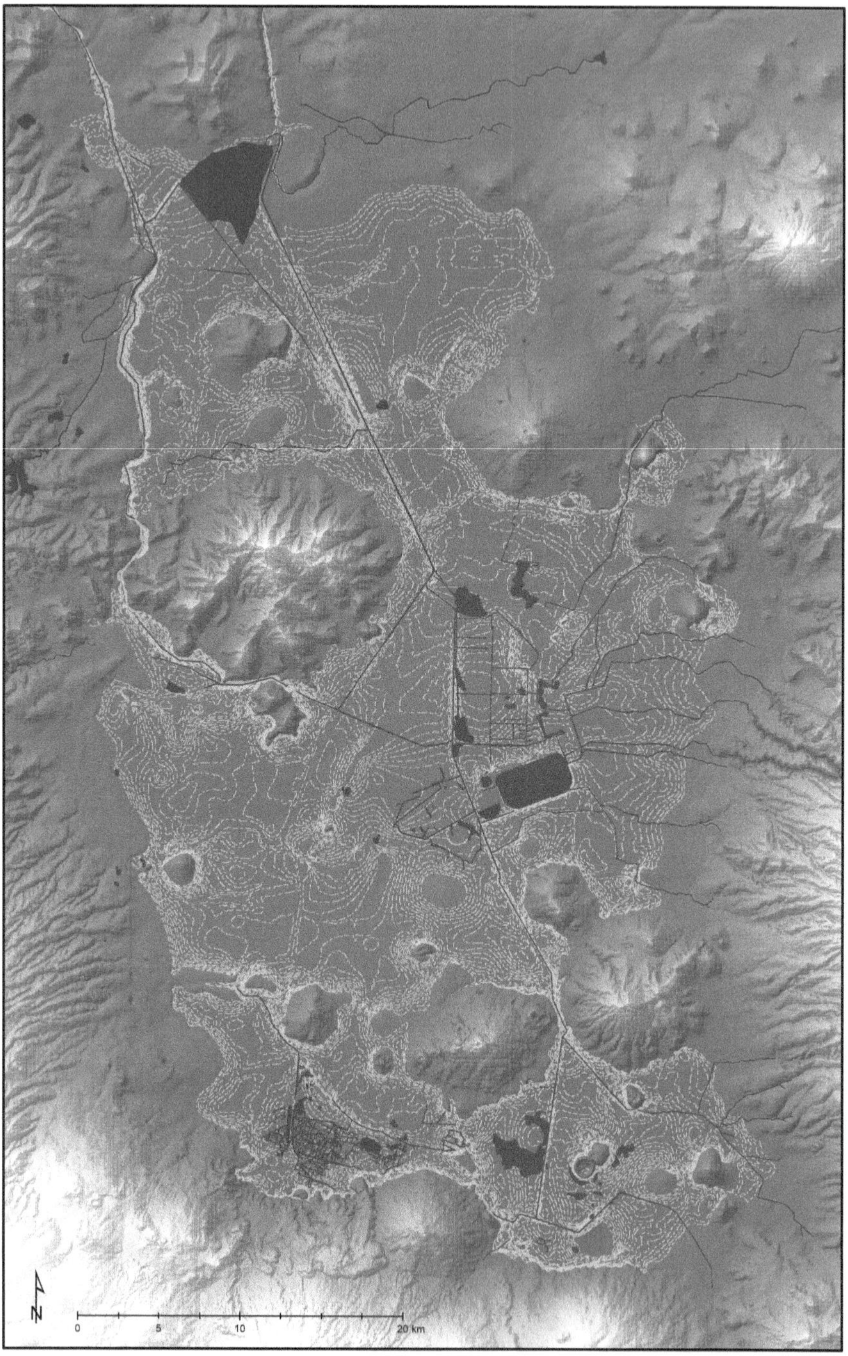
N
0
5
10
20 km

◀ **Fig. 1** Map of the current topography and hydrology of the Basin of Mexico. Digital elevation model (30-m resolution) available for download from the Instituto Nacional de Estadística Geografía e Informática (INEGI). Contour lines range from 2220 to 2250 m above sea level (2-m intervals). Standing water (depicted in *dark blue*) traced from satellite images available through Google Earth, collected ca. 2012

former lakebed. Furthermore, local topography has changed as the lakebeds have dried out—some areas have risen as drying soils expand while others have sunk under the weight of new construction (Ezcurra et al. 1999). Erosion and sedimentation may hide or obscure the furthest extent of the lakes. The DEM, as accurate as it is, reflects the scars of 500 years and freezes a dynamic system at a single moment in time. Even satellite images are limited to a particular season which obscures how bodies of water expand and shrink.

Perhaps, with the right layers and a deft hand, we could fill lakes as if we were focusing a microscope's lens, bringing different depths of field into view. Doing so might make clear the differential impact of rains on the lakes, for example, such as the tendency of the northern lakes to expand and contract to a much greater degree than the southern ones. Layering older maps on top of the new may provide some solutions to the limitations of current sensing technologies. But in layering maps, we must accept a certain degree of uncertainty because each map was produced with different data, priorities, and standards. For example, we can use the DEM to create fine-grained contour lines and map the different possible extents of the lakes depending on the level of water at different times of the year (Fig. 1). We can compare those boundaries to soil maps that demarcate where lacustrine deposits can be found today (Fig. 2a) or to maps historians and archaeologists have produced that account for known archaeological sites which are presumed to have been located on land that was dry for at least part of the year (Fig. 2b–d). Even a cursory examination of the maps in Fig. 2 reveals inconsistencies among these reconstructions, but all is not lost.

The diversity of mappings begs the questions of how and why archaeologists choose to represent the lakes in particular ways, how those representations relate to their assumptions, and how maps shape future archaeological research and interpretations. Interpreting these maps' inconsistencies is difficult. Morehart (2012a), for example, employed historic records and maps from the 16th to 18th century, air photos, Landsat images, and VHR satellite data to reconstruct an ancient raised field (chinampa) and canal system that once existed in Lake Xaltocan. His approach was cumulative, as each new line of data was assessed to evaluate the former. Yet, this approach was not simply an attempt to use each source of visual data to add missing elements from the landscape, thereby compressing the temporal entanglements of different material features. He compared different imagery to untangle change in seemingly statics images, both when the raised field system developed and during the twentieth century. The latter method allowed him to postdate some features as later than the chinampa system and, hence, eliminate them from its reconstruction. In other words, this approach helped him achieve the goal of reconstructing some elements of the landscape's history and understand where their temporality lies.

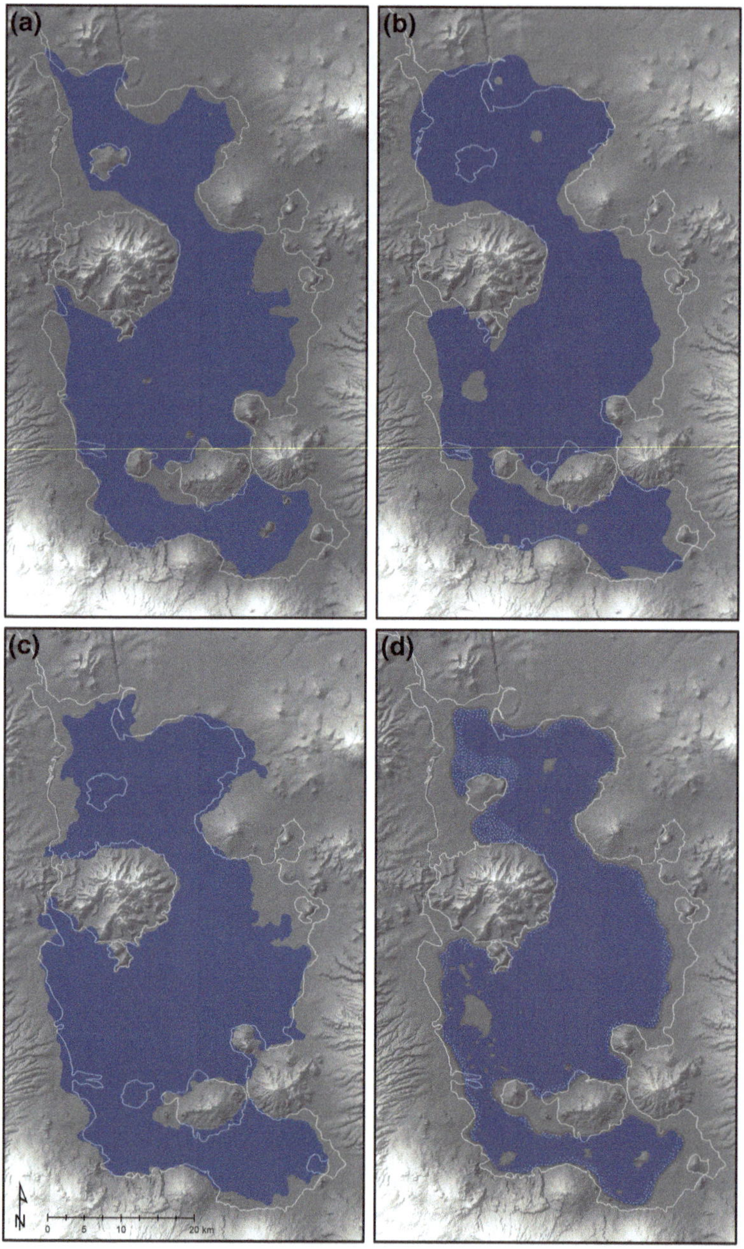

Fig. 2 Maps of the lake system in the Basin of Mexico as reconstructed in several influential twentieth-century geological and archaeological sources. *Dark blue* represents water. *Stippling* represents swampy areas (as presented in original maps). A simplified 2250 m contour is displayed in *white*. Sources include: **a** lacustrine soils (Ríos Elizondo 1975: 1: Plate 1), **b** Gibson (1964: frontispiece), **c** Palerm (1973), and **d** Sanders et al. (1979)

We must expect a certain degree of error because the lakes' boundaries in Fig. 2 have been drawn based on indirect sources rather than direct observations of where they were at any given time. For those kinds of direct observations, we must turn to historical maps produced before the lakes were fully drained. Historical maps have inconsistencies of distance, direction, scale, and/or projection which we can correct —to a certain extent—by using rectification and georeferencing features of GIS software (ArcGIS in our case). Figure 3 presents four nineteenth-century maps that depict the hydrology of the Basin of Mexico. These by no means exhaust all of the possibilities, but they help make clear points of continuity and disjuncture. We have georeferenced and rectified these maps in reference to named settlements and topographic features, such as mountaintops, that are common to them and available in current sources. Some mapmakers acknowledged the dynamic nature of the lakes by distinguishing open water from swamps (Rand McNally and Company 1897) (Fig. 3a). Other mapmakers made choices that are difficult to interpret in the present. For example, the Rand McNally map recapitulates the boundaries of the lakes visible in several maps from as early as 1868 (Ríos Elizondo 1975: 2: Plates 33, 37, and 39). Do we interpret these maps as a document of the lakes' stability over the time elapsed? Alternatively, might they reflect the authority that certain maps took on despite the accumulation of inaccuracies over time? Each disparity leads us to ask whether it was based on an objective difference or a subjective one. In a sense, these maps give a better sense of what was possible than they do of what was actual.

For maps pre-dating the nineteenth century, inconsistencies are too great to rectify them to a grid and coordinate system in any meaningful way. Nevertheless, earlier maps remain informative through the spatial relations and the layered information they represent. Furthermore, they reveal other agendas, interests, and aesthetics than what we are traditionally used to seeing in maps. For example, in the Santa Cruz map, dated to 1560, the lake is depicted as a vibrant part of the daily lives of its past inhabitants, who hunted and fished in its shallow waters (Apenes 1947: Lámina 2; Parsons 2006: Fig. 3.16). Here, behavior is not detached from the map or erased by static and regular spatial units. Yet even these kinds of mappings can be political and are not innocuous. Schematic mappings of Tenochtitlan, the lakes, and the surrounding communities make clear unequal relations between center and periphery, colonizer and colonized, such as the dispropotionately large representation of Tenochtitlan in the Nuremberg Map, produced in 1524 (Mundy 1998: Fig. 1). While these cartographic conventions may seem foreign to us today, we can still discern their content and how maps served as ideological tools.

Even if our layering does not produce a definitive map of the lakes, the layering of maps is good practice in archaeology. Our example of reconstructing the hydrography of the Basin of Mexico makes clear several challenges that may be resolved by technologies of sensing, spatial analysis, and visualization. First, legacy maps produced by earlier archaeologists are crucial to our selection of sites for the application of new sensing technology as well as to our interpretations (Gillespie 2011; Hare et al. 2014; Morehart 2012a, b; Rosenswig et al. 2013). Second, despite their gaps and inconsistencies, historical and legacy data can provide access to views of space that are no longer available first hand. In the Basin of Mexico,

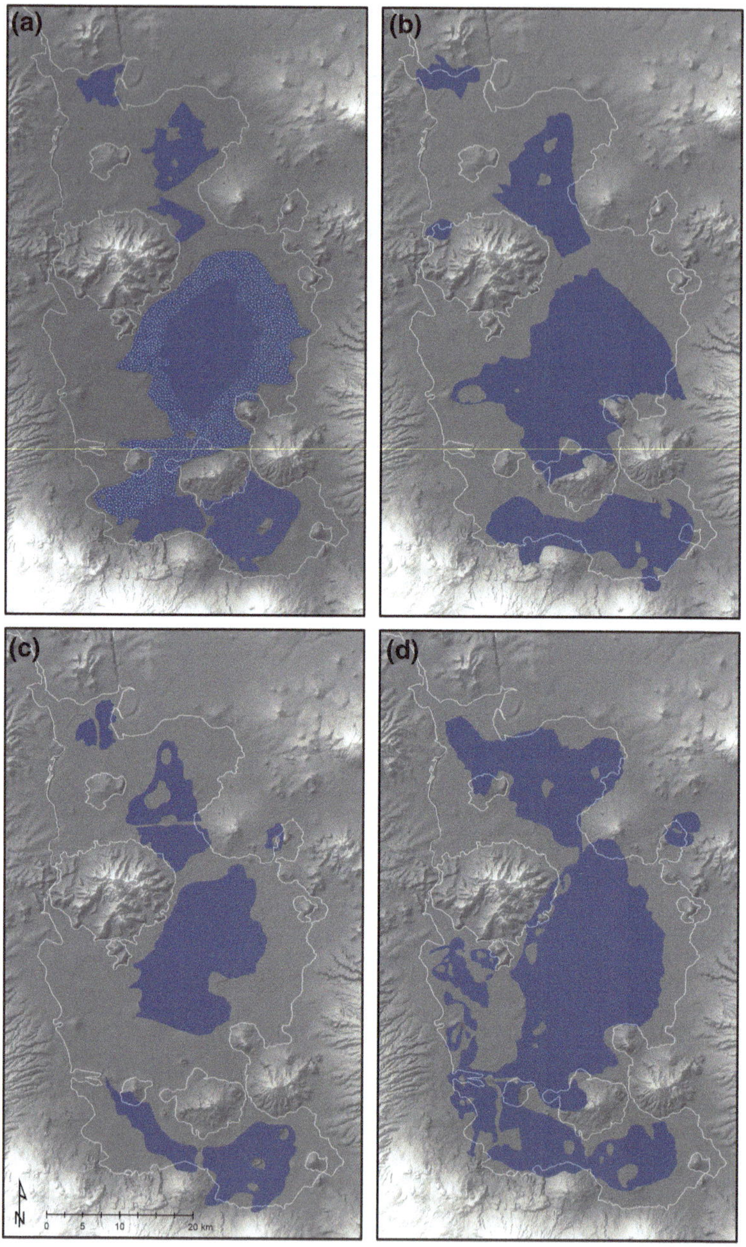
(a)
(b)
(c)
(d)
N
0
5
10
20 km

◀ **Fig. 3** Maps of the lake system in the Basin of Mexico as recorded by nineteenth-century cartographers. *Dark blue* represents water. *Stippling* represents swampy areas (as presented in original maps). A simplified 2250 m contour is displayed in *white*. Sources include: **a** Rand McNally (1897), **b** Mauricio Castro's reduction of Francisco de Garay's 1856 map (Ríos Elizondo 1975: 2: Plate 28), **c** Don Luis Martin's map produced between 1804 and 1807 (Ríos Elizondo 1975: 2: Plate 19), and **d** a map produced circa 1806 with no known cartographer (Ríos Elizondo 1975: 2: Plate 18)

layered maps force us to acknowledge that the solid lines that we prefer to see on maps are actually poor representations of dynamic systems. Moreover, the technologies that we use to layer and visualize can only be applied to the sites and landscapes that we deem valuable in the present day. Even when we incorporate multiple maps, our sample is limited and reflects the historical processes by which some maps are preserved and reproduced while others are lost or forgotten.

What Political Challenges and Opportunities Do New Mapping Technologies Present?

If our previous themes emphasize the multiple understandings of space that have existed in the past, we must also admit that multiple understandings exist in the present. If so, then whose understandings emerge as authoratative? More to the point, who is empowered by these technologies and in what ways? In one sense, new technologies may democratize the process of understanding space by making rich data more broadly available. Accessibility may be measured in terms of access to hardware and infrastructure (computers, internet, GPS) as well as training and competence in the use of instruments and software to visualize, analyze, edit, and enhance spatial data. But access to hardware, infrastructure, and skills—let alone the time and money to use them—is hardly distributed equally among all parties interested in the past. We therefore turn to the structures of power that are embedded in technologies of remote sensing and visualization and how these tools may be useful in efforts to democratize and de-colonize the past.

Archaeology is political at the very least because it involves laws, people, and resources and takes places in politically-discrete jurisdictions. Archaeologists also act politically to the extent that they emphasize or re-write history with particular agendas in mind (Smith and Wobst 2005; Shanks and Tilley 1987; Tilley 1989; Wobst 1989). As a spatial endeavor, archaeological research is directly connected to contemporary (local and global) politics. The spatial nature of both the archaeological record and our interpretations of it is particularly relevant to local politics and ideas related to land ownership and conservation, the significance of archaeological remains, and interpretations of the past (Atalay et al. 2014; McGuire 2008).

In the previous two sections, we showed how the use of sensing and visualization can promote alternative and multiple readings of the past that are less encumbered by present-day subjectivities. In this sense, we could make a strong case that what matters is who wields the tool, or how they weild it. In a practical sense, however, real disparities in who has access to these technologies limit who can wield them or how. In recent years, efforts to address these disparities have emerged in the form of indigenous and post-colonial archaeology (Atalay et al. 2014; Chapin et al. 2005; McGuire 2008; Watkins 2000; Wobst 2005). These need not be limited to areas of former colonial control—they reflect efforts to empower historically marginalized people to speak for, map, identify, and interpret the past.

De-colonizing archaeology forces us to reconsider alternatives to the priorities that archaeologists bring to their study of space (Morehart 2012b). Traditionally, we look for places with high artifact densities, visible remains, and evidence of intense landscape transformation. But these priorities might be limiting because, in practice, we gravitate to places marked by the absence of current use and a disconnection from descendant communities and we tend to shun places that appear empty or unused in the past (Wobst 2005: 20–22). Our gaze is contradictory; to focus on the past we must study a materiality physically in the present. Furthermore, we ignore some of what see, not just in chronological terms but also according to shifting intellectual views of what can and cannot exist in the discursive space of archaeology. For example, the classification of spaces that appear to have little material of value in the traditional archaeological mindset may actually ignore the importance of open, empty, protected, and unmodified spaces in past and present cultures (Basso 1996; Dunnell 1992; Ingold 2011; Robin 2002; Sletto 2009; Watkins 2000). Thus, as part of an intellectual project aimed at better understanding different people's understandings of space, engaged archaeology can help us overcome these oversights by broadening our vision.

As part of archaeological practice, we also must recognize that acts of mapping —which include observing, demarcating, and classifying—are political acts with real and often uneven consequences. For example, the identification of empty lands, so-called 'terra nullis' has its roots in colonial efforts to dispossess native people of their lands—a practice which continues into the present day and in which archaeological investigations can be readily implicated around the world (Denevan 1992; Smith and Gazin-Schwartz 2008; Wobst 2005: 20–21). As we briefly mentioned above, the creation of maps both requires and constitutes technologies that produce subjects. Foucault (1977) and, later, Anderson (1991) and Mignolo (1995) rooted this phenomenon in the uneven emergence of modernity. Modern, post-colonial nation states, for example, required material mechanisms to form and persevere (Anderson 1991): the census recorded their subjects, the museum celebrated their historical identity, and the map inscribed them in space. Given these histories, we might do well to tread carefully into the territory of making maps.

One of the most straightforward steps to de-colonizing archaeology is to collaborate with indigenous people, or any interested publics, in the processes of designing, conducting, analyzing, and presenting spatial data (Colwell-Chanthaphonh and Ferguson 2008; Dongoske et al. 2000; Maya People of Southern Belize 1997; Orlove 1991). Participatory cartographers, for example, take an activist stance by involving communities in every aspect of the process of mapping. Such work detaches investigators from direct control of methods and results but affords more inclusive and productive relationships (Bryan 2011; Fox et al. 2005; McAnany et al. 2015). But, there may be fundamental conflicts of ideologies and epistemologies to overcome. The idea of bounding space can contradict lives lived and understood through connections and open networks that can include people, plants, and animals—especially in terms of the administration of mobile resources like wildlife and herd management (Hale 2006; Nadasdy 2005; Vermeylen et al. 2012). New technologies of mapping and visualization—from handheld GPS to Google Earth—may have a democratizing effect in the collection and dissemination of spatial data, including archaeological research. Such outcomes do not happen, however, without dedicated effort to involve multiple parties in every aspect of the program, provide access to equipment and training, and maintain ongoing connections with local groups (McAnany et al. 2015). When successful, participatory and community mapping can preserve and present multiple mappings and readings of the landscape and history (Maya People of Southern Belize 1997).

Mesoamerican archaeology is often conducted among indigenous communities with long and enduring histories of social, economic, and political marginalization—and for whom mapping has long been a tool of state control (Mignolo 1995). Thus, archaeological mapping in these areas and among these people is a deeply political endeavor. Political lines may exist at many scales; between local groups and state institutions as well as within and among local groups. Because mapping can establish ownership, or at least stewardship of a tract of land, lines drawn can support disempowered communities in their efforts to protect their lands (Mollett 2013). They can also lead to heated discussion between communities or among factions within them (Fox et al. 2005; Morehart 2012b). If we acknowledge that any mapping is political and may have unintended consequences for local and descendent communities, the more ethical approach would be to involve them in the process.

Participatory mapping can empower marginalized groups, but we should not presume that technology or community mapping is a panacea. The act of mapping can be quite powerful to the extent that it serves as a basis for a legal claim to use, administer, or restrict access to land. Communities may use mapping to fix claims to ownership of communal lands against the state, but mapping can also be a precursor to privatization and subdivision (Sletto 2009; Mollett 2013). Moreover, archaeological interpretations of local histories may have little bearing on how those histories are taken up and used in local lives. Such has been the case the community of Xaltocan, in central Mexico, where local historians draw on archaeological evidence to write their own narrative of the community that counters a dominant narrative of marginalization (Brumfiel 2000:189; Morehart 2012b). Thus, as a participatory effort, competition for recognition as well as the resources available through tourism or national and international cultural heritage organizations may inform local decisions in ways that archaeologists do not expect.

Indigenous and community mapping efforts serve as reminder of mapping's inherent political power as well as the responsibility archaeologists take on when they participate in any research tied to space. Archaeological mapping has a tendency to tether people to particular places and to tether particular places and people to particular times (Fabian 1983)—all of which are bound up in relations of power.

By opening the processes of research design, data collection, analysis, and interpretation, archaeologists relinquish control of the products of their work, which can be quite troubling for some. But, doing so makes possible a more collaborative and representative understanding of spaces both past and present. However, while such efforts address important ethical dimension of our own practices, they may have little lasting effect on long-established and pervasive structural inequalities.

Concluding Thoughts: Being Ambiguous and Disruptive Are Not Necessarily Bad Things

Current technologies of remote sensing, spatial analysis, and visualization provide levels of detail, accuracy, and accessibility in the work of archaeological mapping that are greater than have ever been possible. However, this level of precision can lead to a degree of confidence that is stripped of a responsible recognition of human investigators' judgment and imprecision as well as the historical and political context of their investigations. Yet, we have suggested that the objective and subjective nature of spatial data are not in conflict. By approaching our research with both aspects of spatial data at the front of our minds, we can achieve richer, more rewarding results. We focused on three aspects of spatial data that cross the objective/subjective boundary—their abstractness, their temporality, and their power. In the context of Mesoamerica, the study of these aspects have revealed the nested layers of translation involved in spatial research and the practical acts of making and using maps.

The metaphor of translation is apt because it implies communication and social connections among the people who make and use maps, past and present. Archaeologists are members of the social networks facilitated by maps. Hence, we accept some responsibility for our work's political implications and consequences. At the same time, sharing data means accepting that others will likely interpret its meaning and value differently. As new spatial technologies emerge and provide even greater detail and precision, they can, paradoxically, destabilize the authority of a singular interpretation of space. Such disruptions are not necessarily a bad thing. Instead, they are part of the ongoing process of making meaning in space in which we have been, and continue to be, participants.

Acknowledgments We would like to thank the organizers and participants in the Age of Sensing conference for the opportunity to share our work. We also thank Michaela DeSoucey, who read an earlier draft of this paper, and David Jensen, who helped to scan and trace historical maps. All errors are strictly those of the authors.

References

Almaráz, R. (1865). *Apuntes sobre las piramides de San Juan Teotihuacán* (Memorias y trabajos ejecutados por la Comisión Científica de Pachuca). México, D.F.: Ministerio de Fomento.

Anderson, B. (1991). *Imagined communities: reflections on the origin and spread of nationalism*. London: Verso.

Apenes, O. (1947). *Mapas antiguos del Valle de México*. México, D.F.: Universidad Nacional Autónoma de México, Instituto de Historia.

Armillas, P. (1971). Gardens on Swamps. *Science, 175*(4010), 653-661.

Ashmore, W. (1991). Site Planning Principles and Concepts of Directionality among the Ancient Maya. *Latin American Antiquity, 2*, 199-226.

Ashmore, W., & Sabloff, J. A. (2002). Spatial Orders in Maya Civic Plans. *Latin American Antiquity, 13*, 201-215.

Astor-Aguilera, M. A. (2010). *The Maya world of communicating objects: quadripartite crosses, trees, and stones*. Albuquerque: University of New Mexico Press.

Atalay, S., Clauss, L. R., McGuire, R. H., & Welch, J. R. (2014). Transforming Archaeology. In S. Atalay, L. R. Clauss, R. H. McGuire, & J. R. Welch (Eds.), *Transforming archaeology: activist practices and prospects* (pp. 7-29). Walnut Creek, CA: Left Coast Press.

Aveni, A. F. (2001). *Skywatchers*. Austin: University of Texas Press.

Basso, K. H. (1996). *Wisdom sits in places: landscape and language among the Western Apache*. Albuquerque: University of New Mexico Press.

Blanton, R. E., & Fargher, L. F. (2011). The collective logic of pre-modern cities. *World Archaeology, 43*(3), 505-522.

Brumfiel, E. M. (2000). Making History in Xaltocan. In K. E. Dongoske, M. S. Aldenderfer, & K. Doehner (Eds.), *Working together: Native Americans and archaeologists* (pp. 181-190). Washington, D.C.: Society for American Archaeology.

Bryan, J. (2011). Walking the line: Participatory mapping, indigenous rights, and neoliberalism. *Geoforum, 42*(1), 40-50.

Caballero, M., & Guerrero, B. O. (1998). Lake Levels since about 40,000 Years Ago at Lake Chalco, near Mexico City. *Quaternary Research, 50*(1), 69-79.

Candiani, V. S. (2014). *Dreaming of dry land: environmental transformation in colonial Mexico City*. Stanford: Stanford University Press.

Carter, W. E., Shrestha, R. L., Fisher, C., & Leisz, S. (2012). Geodetic imaging: A new tool for Mesoamerican archaeology. *Eos, Transactions American Geophysical Union, 93*(42), 413-415.

Chapin, M., Lamb, Z., & Threlkeld, B. (2005). Mapping indigenous lands. *Annual Review of Anthropology, 34*, 619-638.

Chase, A. F., Chase, D. Z., Fisher, C. T., Leisz, S. J., & Weishampel, J. F. (2012). Geospatial revolution and remote sensing LiDAR in Mesoamerican archaeology. *Proceedings of the National Academy of Sciences, 109*(32), 12916-12921.

Colwell-Chanthaphonh, C., & Ferguson, T. J. (Eds.). (2008). *Collaboration in Archaeological Practice: Engaging Descendant Communities*. Lanham, MD: AltaMira Press.

Corner, J. (1999). The Agency of Mapping: Speculation, Critique and Invention. In D. E. Cosgrove (Ed.), *Mappings* (pp. 231-252). London: Reaktion Books.

Cortés, H. (1971 [1522]). *Letters from Mexico*. New York: Grossman Publishers.

Denevan, W. M. (1992). The Pristine Myth - the Landscape of the America in 1492. *Annals of the Association of American Geographers, 82*(3), 369-385.

Dongoske, K. E., Aldenderfer, M. S., & Doehner, K. (2000). *Working together: Native Americans and archaeologists*. Washington DC: Society for American Archaeology.

Doolittle, W. E. (1990). *Canal irrigation in prehistoric Mexico: the sequence of technological change*. Austin: University of Texas Press.

Drewitt, R. B. (1987). Measurement Units and Building Axes at Teotihuacan. In E. McClung de Tapia, & E. C. Rattray (Eds.), *Teotihuacán: nuevos datos, nuevas síntesis, nuevos problemas* (pp. 79-94). México, D.F.: Instituto de Investigaciones Antropológicas Universidad Nacional Autónoma de México.

Drucker, R. D. (1974). *Renovating a Reconstruction: The Ciudadela at Teotihuacan, Mexico. Construction Sequence, Labour, and Possible Uses of the Structure*. University of Rochester, Rochester, NY.

Dunnell, R. C. (1992). The Notion Site. In J. Rossignol, & L. Wandsnider (Eds.), *Space, time, and archaeological landscapes* (pp. 21-41). New York: Plenum Press.

Ezcurra, E., Mazari-Hiriart, M., Pisanty, I., & Aguilar, A. G. (1999). The Basin of Mexico: critical environmental issues and sustainability. New York: United Nations University Press.

Fabian, J. (1983). *Time and the other: how anthropology makes its object*. New York: Columbia University Press.

Foucault, M. (1977). *Discipline and punish: the birth of the prison*. New York: Pantheon Books.

Fox, J., Suryanata, K., & Hershock, P. (2005). Mapping communities: ethics, values, practice. Honolulu, HI: East-West Center.

Gell, A. (1985). How to Read a Map: Remarks on the Practical Logic of Navigation. *Man, 20*(271-286).

Gibson, C. (1964). *The Aztecs under Spanish rule: A history of the Indians of the Valley of Mexico, 1519-1810*. Stanford: Stanford University Press.

Giddens, A. (1982). *Profiles and critiques in social theory*. Berkeley: University of California Press.

Gillespie, S. (2011). Archaeological Drawings as Re-Presentations: The Maps of Complex A, La Venta, Mexico. Latin American Antiquity, 22(1), 3-36.

Grant, J. (2001). The dark side of the grid: power and urban design. [Article]. *Planning Perspectives, 16*(3), 219-241.

Hale, C. R. (2006). Activist research v. cultural critique: indigenous land rights and the contradictions of politically engaged anthropology. *Cultural Anthropology, 21*(1), 96-120.

Halperin, C. S. (2005). Social power and sacred space at Actun Nak Beh. In K. M. Prufer, & J. E. Brady (Eds.), *Stone houses and earth lords : Maya religion in the cave context* (pp. 71-90). Boulder, CO: University Press of Colorado.

Hare, T. S., Masson, M. A., & Russell, B. (2014). High-Density LiDAR Mapping of the Ancient City of Mayapan. *Remote Sensing, 6*(9), 9064-9085.

Harvey, D. (1989). *The condition of postmodernity: an inquiry into the origins of cultural change*. Oxford: Basil Blackwell.

Humboldt, A. v. (1836 [1811]). *Ensayo Politico sobre Nueva España* (V. G. Arnao, Trans., Vol. 1). Paris: Libreria de Lecointe.

Ingold, T. (2011). *The Perception of the Environment: Essays on Livelihood, Dwelling and Skill*. London: Routledge.

Kent, W. (1978). *Data and reality: basic assumptions in data processing reconsidered*. New York: North-Holland.

Kubler, G. (1984). Pre-Columbian Pilgrimages in Mesoamerica. *Diogenes*(125), 11-23.

León Portilla, M. (1963). *Aztec thought and culture: a study of the ancient Nahuatl mind*. Norman: University of Oklahoma Press.

Lozano-García, M. S., Ortega-Guerrero, B., Caballero-Miranda, M., & Urrutia-Fucugauchi, J. (1993). Late Pleistocene and Holocene Paleoenvironments of Chalco Lake, Central Mexico. *Quaternary Research, 40*(3), 332-342.

Mathes, W. M. (1970). To Save a City: The Desague of Mexico-Huehuetoca, 1607. *The Americas, 26*(4), 419-438.

Maya People of Southern Belize (1997). *Maya Atlas: The Struggle to Preserve Maya Land in Southern Belize*. Toledo, Belize: Toledo Maya Cultural Council.

McAnany, P. A., Rowe, S. M., Cholotio, I. Q., Menchú, E. C., & Quic, J. M. (2015). Mapping Indigenous Self-Determination in Highland Guatemala. *International Journal of Applied Geospatial Research, 6*(1), 1-23.

McGuire, R. H. (2008). *Archaeology as political action*. Berkeley: University of California Press.

Mignolo, W. (1995). *The darker side of the Renaissance: literacy, territoriality, and colonization*. Ann Arbor: University of Michigan Press.

Millhauser, J. K. (2012). *Saltmaking, Craft, and Community at Late Postclassic and Early Colonial San Bartolome Salinas, Mexico*. PhD Doctoral dissertation, Northwestern University, Evanston, IL.
Millhauser, J. K. (2016). Aztec use of lake resources in the Basin of Mexico. In D. Nichols and E. Rodríguez-Alegría (Eds.), The Oxford Handbook of the Aztecs (pp. 301-318). Oxford: Oxford University Press.
Millon, R. (1973). *Urbanization at Teotihuacán, Mexico*. Austin: University of Texas Press.
Mollett, S. (2013). Mapping Deception: The Politics of Mapping Miskito and Garifuna Space in Honduras. *Annals of the Association of American Geographers, 103*(5), 1227-1241.
Montoya Rivero, M. C. (1999). Del desagüe del Valle de México al drenaje profundo. México Desconocido 30. http://www.mexicodesconocido.com.mx. Accessed March 15, 2012.
Morehart, C. T. (2010). *The archaeology of farmscapes: Production, place, and the materiality of landscape at Xaltocan, Mexico*. Unpublished Ph.D. dissertation, Northwestern University, Evanston.
Morehart, C. T. (2012a). Mapping Ancient Chinampa Landscapes in the Basin of Mexico: A Remote Sensing and GIS Approach. *Journal of Archaeological Science, 39*, 2541-2551.
Morehart, C. T. (2012b). What if the Aztec Empire Never Existed? The Prerequisites of Empire and the Politics of Plausible Alternative Histories. American Anthropologist, 114, 267-281.
Morehart, C. T. (2016). Let the Earth Forever Remain! Landscape Legacies and the Materiality of History in the Northern Basin of Mexico. Journal of the Royal Anthropological Institute (incorporating Man), 22, 939-961.
Morehart, C. T., & Frederick, C. (2014). The chronology and collapse of pre-Aztec raised field (chinampa) agriculture in the northern Basin of Mexico. *Antiquity, 88*, 531-548.
Mundy, B. E. (1996). *The mapping of New Spain: indigenous cartography and the maps of the relaciones geográficas*. Chicago: University of Chicago Press.
Mundy, B. E. (1998). Mapping the Aztec Captial: The 1524 Nuremberg Map of Tenochtitlan, Its Sources and Meaning. *Imago Mundi*, 50, 11-33.
Myers, A. (2010). Camp Delta, Google Earth and the ethics of remote sensing in archaeology. *World Archaeology, 42*(3), 455-467.
Nadasdy, P. (2005). The Anti-Politics of TEK: The Institutionalization of Co-Management Discourse and Practice. *Anthropologica, 47*(2), 215-232.
Orlove, B. S. (1991). Mapping Reeds and Reading Maps - the Politics of Representation in Lake Titicaca. *American Ethnologist, 18*(1), 3-38.
Orozco y Berra, M. (1864). *Memória para la Carta Hidrográfica del Valle de México*. México: Sociedad Mexicana de Geografía y Estadística.
Palerm, A. (1973). *Obras hidráulicas prehispánicas en el sistema lacustre del valle de México*. México, DF: Instituto Nacional de Antropología.
Palka, J. W. (2014). *Maya pilgrimage to ritual landscape: insights from archaeology, history, and ethnography*. Albuquerque: University of New Mexico Press.
Parsons, J. R. (2001). *The Last Saltmakers of Nexquipayac, Mexico: An Archaeological Ethnography*. Ann Arbor: Museum of Anthropology, University of Michigan.
Parsons, J. R. (2006). *The Last Pescadores of Chimalhuacán, Mexico: an archeological ethnography*. Ann Arbor: Museum of Anthropology, University of Michigan.
Phillips, C. W. (1959). Archaeology and the Ordnance Survey. *Antiquity, 33*(131), 195-204.
Prufer, K. M., Thompson, A. E., & Kennett, D. J. (2015). Evaluating airborne LiDAR for detecting settlements and modified landscapes in disturbed tropical environments at Uxbenká, Belize. *Journal of Archaeological Science, 57*, 1-13.
Rand McNally and Company (1897). *Indexed Atlas of the World*. Chicago: Rand McNally.
Ríos Elizondo, R. (1975). *Memoria de las obras del sistema de drenaje profundo del Distrito Federal*. México, DF: Departmento de Distrito Federal, Talleres Gráficos de la Nación.
Robin, C. (2002). Outside of houses: The practices of everyday life at Chan Nòohol, Belize. *Journal of Social Archaeology, 2*(2), 245-268.

Rosenswig, R. M., López-Torrijos, R., Antonelli, C. E., & Mendelsohn, R. R. (2013). Lidar mapping and surface survey of the Izapa state on the tropical piedmont of Chiapas, Mexico. *Journal of Archaeological Science, 40*(3), 1493-1507.

Sanders, W. T., Parsons, J. R., & Santley, R. S. (1979). *The Basin of Mexico: Ecological Processes in the Evolution of a Civilization.* New York: Academic Press.

Scott, J. C. (1998). *Seeing like a state: how certain schemes to improve the human condition have failed.* New Haven: Yale University Press.

Seymour, W. A. (Ed.). (1980). *A History of the Ordnance Survey.* Folkestone, UK: Dawson.

Shanks, M., & Tilley, C. Y. (1987). *Re-constructing archaeology: theory and practice.* Cambridge: Cambridge University Press.

Sletto, B. (2009). Special issue: Indigenous cartographies. *Cultural Geographies, 16*(2), 147-152.

Smith, A., & Gazin-Schwartz, A. (2008). *Landscapes of clearance: archaeological and anthropological perspectives.* Walnut Creek, CA: Left Coast Press.

Smith, C., & Wobst, H. M. (2005). Decolonizing archaeological theory and practice. In C. Smith, & H. M. Wobst (Eds.), *Indigenous archaeologies: decolonizing theory and practice* (pp. 5-16). London: Routledge.

Sugiyama, N., Sugiyama, S., & Sarabia G., A. (2013). Inside the Sun Pyramid at Teotihuacan, Mexico: 2008-2011 Excavations and Preliminary Results. *Latin American Antiquity, 24*(4), 403-432.

Sugiyama, S. (1993). Worldview Materialized at Teotihuacan. *Latin American Antiquity, 4*(2), 103.

Sugiyama, S. (2004). Governance and Polity at Classic Teotihuacan. In J. A. Hendon, & R. A. Joyce (Eds.), *Mesoamerican archaeology: theory and practice* (pp. 97-123, Blackwell studies in global archaeology). Oxford: Blackwell Publishing.

Sugiyama, S., & Cabrera Castro, R. (2007). The Moon Pyramid Project and the Teotihuacan State Polity. *Ancient Mesoamerica, 18*(1), 109-125.

Tilley, C. (1989). Archaeology as socio-political action in the present. In V. Pinsky, & A. Wylie (Eds.), *Critical Traditions in Contemporary Archaeology: Essays in the Philosophy, History, and Socio-politics of Archaeology* (pp. 104-116). Cambridge: Cambridge University Press.

Trouillot, M. (1995). *Silencing the past: power and the production of history.* Boston: Beacon Press.

Turnbull, D. (1993). *Maps are territories: science is an atlas, a portfolio of exhibits.* Chicago: University of Chicago Press.

Vermeylen, S., Davies, G., & van der Horst, D. (2012). Deconstructing the Conservancy Map: Hxaro, N!ore, and Rhizomes in the Kalahari. *Cartographica: The International Journal for Geographic Information and Geovisualization, 47*(2), 121-134.

Watkins, J. (2000). *Indigenous archaeology: American Indian values and scientific practice.* Walnut Creek, CA: Alta Mira Press.

Wobst, H. M. (1989). Commentary: a socio-politics of socio-politics in archaeology. In V. Pinsky, & A. Wylie (Eds.), *Critical Traditions in Contemporary Archaeology: Essays in the Philosophy, History, and Socio-politics of Archaeology* (pp. 136-140). Cambridge: Cambridge University Press.

Wobst, H. M. (2005). Power to the (Indigenous) past and present! Or: the theory and method behind archaeological theory and method. In C. Smith, & H. M. Wobst (Eds.), *Archaeologies: Decolonizing Theory and Practice* (pp. 17-32). New York: Routledge.

Zetina-Gutiérrez, G., Castillo-Peña, P., & Iankov, K. (2014). *Archaeological Research Site-based LiDAR in El Tajín, Ver., México.* Paper presented at the 79th Annual Meeting of the Society for American Archaeology, Austin, TX, April 23-27

Part IV
Simulation, Visualization and Computing

Cyber Archaeology: 3D Sensing and Digital Embodiment

Maurizio Forte

> *"In that Empire, the Art of Cartography attained such Perfection that the map of a single Province occupied the entirety of a City, and the map of the Empire, the entirety of a Province. In time, those Unconscionable Maps no longer satisfied, and the Cartographers Guilds struck a Map of the Empire whose size was that of the Empire, and which coincided point for point with it. The following Generations, who were not so fond of the Study of Cartography as their Forebears had been, saw that that vast Map was Useless..."*
> —Borges (1975), Universal History of Infamy

Abstract This chapter is focused on the study of the relations between digital embodiment, close range sensing and three-dimensional information in archaeology. The relations between digital recording and simulation environments generate new research questions and design a new epistemology in cyberarchaeology.

Introduction

This contribution is focused on the study of the relations between digital embodiment and three-dimensional sensing in archaeology. It is aimed at discussing new methods and theories in the domains of virtual reality, 3D modeling, poli sensing environments, digital recording and simulation environments. The hypothesis of the author is that specific kinds of virtual embodiments can accelerate the archaeological interpretation and increase the learning experience. In this scenario, the discussion should be focused on poli sensing environments, because of the involvement of multiple sensors and multimodal digital information.

Perception, recording, interpretation depend on specific codes able to trigger the information. Is the informational process different in the cyber-world? Digital triggers are recognizable in the interaction between human minds and virtual environments. This is the main task in cyber archaeology: the core of the informative process is in the multiple interactions/feedback and not in the model itself

M. Forte (✉)
Art, Art History, and Visual Studies, Department of Classical Studies, Duke University (USA), Durham, USA
e-mail: maurizio.forte@duke.edu

M. Forte and S. Campana (eds.), *Digital Methods and Remote Sensing in Archaeology*, Quantitative Methods in the Humanities and Social Sciences,
DOI 10.1007/978-3-319-40658-9_12

(Forte 2009, 2010). The simulation and not the reconstruction of the past is the final goal. The interaction stimulates and triggers our interpretation and makes the model/environment understandable, explorable, and sometimes also reproducible. The interpretation is designed by context and affordances (Gibson 1979): "An important fact about the affordances of the environment is that they are in a sense objective, real, and physical, unlike values and meanings, which are often supposed to be subjective, phenomenal, and mental. But, actually, an affordance is neither an objective property nor a subjective property; or it is both if you like. An affordance cuts across the dichotomy of subjective-objective and helps us to understand its inadequacy. It is equally a fact of the environment and a fact of behavior. It is both physical and psychical, yet neither. An affordance points both ways, to the environment and to the observer" (Gibson 1979, p. 3).

According to Malgraves "our brains, bodies, and environments (material and cultural) are no longer seen as entities to be independently investigated, but as highly dynamic and interacting systems connected with each other biologically, ecologically, and socially" (Malgrave 2015).

In archaeology, the affordances design the relationships mind/body/object/environment across space and time. Any artifact, object, site, has to be interpreted according to its potential meaning/action and not to its taxonomy.

Cyber archaeology, because it boosts digital interaction and feedback, actually multiplies the affordances/potential relationships in virtual environments. In particular, it deals mainly with:

- *Avatarizing* the present (perception by observation)—By managing so called "empirical models".
- *Avatarizing* the past (perception by imagination/simulation/context)—By mediated experience.

In the first case an empirical observation is actually "filtered" by the human perception and physical embodiment. It is somehow mediated by our senses. We experience the environment through our senses and we consider this information usually empirical. In the second action, the reconstruction/simulation of the past entails a mediated experience in between human imagination and digital environment. From a phenomenological perspective the "empirical" domain should be objective, while the mediated experience, subjective and depending on the digital tools. A common bias is to consider individual senses as objective, while any mediated experience should be totally subjective.

I argue that any mediated or non-mediated experience is in any case subjective and it entails the analysis and evaluation of models and embodied simulation (Forte 2015). Therefore a correct methodological approach should consider the validation and consistency of the experiential activity as we do in cyber archaeology. Moreover the embodied simulation works according to genetic and cultural triggers: the genetic ones are related with our perceptual-motor skills, the cultural ones with our capacity to translate this information in an understandable cultural context.

In other terms, genetic behaviors co-evolve very slowly, while the cultural ones change more rapidly according to different societies, time and space.

Vittorio Gallese states that "Neuroscience allows us to understand how the line between what we call reality and the imaginary and imagined worlds of fiction is much less sharp and clear than one might think. Indeed, experiencing an emotion and imagining it are both underpinned by the activation of partly identical brain circuits, although differently connected in these two different cognitive and phenomenal situations" (Gallese 2016, 21). Does this happen even in a virtual environment? The line between imaginary/evocative/reconstructed and reality is hybrid, smooth and undefined. Cyber worlds in archaeology comprehend data recorded, reconstructed and simulated (Forte 2010, 2014a, 2015): the interaction between these domains is the booster of the interpretation. Therefore, cyber archaeology deals mainly with the simulation and not the reconstruction of the past (Fig. 1): 3D integration of digital data in the same cyber world, 3D embodiment and virtual reality applications in archaeology, and real time performances and interaction with digital data. The focus of the discipline is in the cybernetic feedback/enactment/embodiment and not just in the virtual content. The interpretation is provided by interaction and it is malleable.

The integration of different ontologies of data in the same environment opens new research perspectives in data simulation and visualization. In particular it is possible to recognize three main domains:

- Digital/empirical" (digital raw data, data collected in the field by different sensors)

Fig. 1 Comparative analysis of virtual (on the left the cover of the first book on this topic) and cyber archaeology (on the right). The picture on the right recalls the importance of virtual interactive environments in a cyber perspective

- Processed (data elaborated in lab)
- Simulated/reconstructed (by interpretation)

The third category requires a kinesthetic approach: a digital embodiment in the cyber-environment. This can be immersive, semi-immersive, haptic, or in other categories.

In this new digital-archaeological hermeneutic cycle, what happens to the interpretation? Is it at the intersection of every single step of the simulation and/or at the end of the process? Or in both? How can we evaluate the effects of this virtual hermeneutics? In cybernetic terms the feedback augments in the loop of the hermeneutic circle: 1st observation of digital raw data (for example in the field); 2nd observation of processed data (in the field and in lab); 3rd observation by simulation environments (by virtual reality applications).

The integration of the entire cycle (1–3) in a single virtual environment can possibly increase the level of interpretation in multiple scenarios. This, unfortunately, is quite difficult because it requires computing and visualization power and as well the capacity to visualize multiple data sources on and off site. However, the integration of bottom-up information (data recording) with top-down models (data interpretation) is fundamental for the consistency and validation of the entire process (Fig. 2). Data interpretation, in fact, requires comparative studies, visual analysis and the access to multiple digital libraries and archives.

In the near future virtual environments will be designed very differently and they will look as frameless realities or frameless environments. For frameless I intend tools or applications not embedded in a screen, CAVE or any dispositive with a frame describing the location of the digital performance (inside the frame or through the frame). Like in a picture, the frame delineates the target, the visual narrative, and the action point of the performance. Frameless will mean also any 3D

Fig. 2 the DiVE-CAVE (Duke Immersive Virtual Environment)

projection or floating models projected in the real world: at that point real/empirical and virtual will coexist in the same space and with the same ontology. It will be an hybrid world, hyper-real, “more real than real” in the Baudrillar’s sense (Baudrillard 1994). In that world, digital and empirical realities will coexists in the same space but without a clear distinction between ontologies: what perceived as real will be real. The frame is the cybernetic place where the virtual performance acts and recalls the interaction. The frame around a picture, if we consider this frame as a message intended to order or to organize the perception of the viewer, says “Attend to what is within and do not attend to what is outside. “Figure and ground, as these terms are used by gestalt psychologists, are not symmetrically related as are the set and non set of set theory. Perception of the ground must be positively inhibited and perception of the figure (in this case the picture) must be positively enhanced ...” (Bateson 1972, 187). In other terms:

- Any message, which either explicitly or implicitly defines a frame, ipso facto gives the receiver instructions or aids in his attempt to understand the messages included within the frame. (Bateson 1972)
- The frame in cybernetic terms represents the interface between ontologies: in the picture the frame distinguishes the interpretation, the visual difference between what is observing and what is observed, the focal direction of the context: “... the human animal, because it interposes a semiotic screen between mind and external environment ... can drive from inside the perception, getting free from the direct influence of the external environment” (Bateson 1972).
- In the relationship between interaction and virtual contexts the frame is typically constituted from a display, which ontologically separates the perception or informative feedback that is generated by the differences between real and virtual, map and territory. The crossing of the frame constitutes the metaphor of the passage from the territory to the map and of the consequent feedback.

This theoretical approach helps us to understand that the recognition of the digital world passes through the frame: the difference between user and frame represents the feedback deriving from the observation of the virtual world through the interface of the system (frame). Therefore, difference between user and interaction is given from the feedback of each action within the virtual world. It is in this domain of virtuality that one seeks to in a correct spatial connection between observer and observed world, between place and space.

Sensing and Poli sensing

All the contributions of this book show clearly that the definition of “remote sensing” has to be reviewed and rediscussed in the light of new applications, methods and digital products.

"Remote Sensing" means literally perceiving not by direct perception. It deals with any distant classification, analysis, visualization and interpretation of physical data. These datasets come usually from so called "mediated" tools, satellite, airplanes, geophysical instruments, Lidars, laser scanners, digital cameras and so on. Many of these data are digitally born in 3D: point clouds, meshes or geometrical reconstructions, but they are usually split in different domains.

For several decades archaeology dealt with satellite imagery and aerial photos, more recently with LIDARS, drones, laser scanners, new and more accurate geophysical instruments, optical scanners, 3D cameras and different active sensors with a high level of accuracy and resolution (Fig. 3). The 70s–90s was the age of satellite-remote sensing-archaeology; the first decade of the third millennium the age of laser scanner; how can we name the current period? Are we still in the category of "remote sensing"? How remote? How accurate?

Remote sensing archaeology changed dramatically in the last decade. The performance of laser scanners, the standardization of image modeling techniques (Forte et al. 2012; Forte 2014, 2014a), the systematic use of very high resolution imagery by UAVs and satellites, increased exponentially the production of digital datasets. For example, in 2001 a time of flight scanner was able to record a thousand points per second, while now it goes over a million points per second.

Nowadays archaeology as many other research fields, generates big data and this is a new and unexpected challenge. Applications and digital devices in archaeology produce likely hundreds or thousands times the amount of data we were generating a decade ago: in short, we produce big data as many other scientific fields. This involves new methodological problems such as digital curation, preservation,

Fig. 3 3D visualization by point clouds of a Neolithic building in the archaeological site of Catalhoyuk (Turkey). The 3D model is made by terrestrial laser scanners

archiving, standardization, transmission and collaborative 3D interaction. In short, we face different qualitative and quantitative analyses.

There is still an evident discrepancy between raw and processed data: the power of personal computers is not adequate to the complexity of models recorded by laser scanners and image modeling for example. Actually, there is an impressive and growing use of sensors controlled by human interaction and they are changing the relation between humans and environments. In these terms we should define a new ecology of sensing based on human interaction with digital/virtual environments. In the future, sensors will be an extension of our body, recombining genetics, kinesthetic and advanced technologies. We assume that the human body will be a polisensing maker system, close to a cyborg.

Poli sensing will include big data, since sensors and devices produce a very large amount of datasets, some raw, some processed, some usable and accessible. The current bottleneck of this workflow is that just a low percentage of datasets produced by multiple instruments and tools are really accessible, open and usable. This depends on the fact that there are no standard (or too many) for big data, by the unsuitableness of computing power and virtual environments where to host models and datasets. The contribution of Bill Seaman in this book designs quite well this polisensing scenario, but I think it is important to highlight that the multiplication of sensors is related to a different multi sensorial and poli sensing archaeological activity we should expect in the near future. For instance, the combination of drones, terrestrial lidar and laser scanners, environmental sensors, digital cameras and image modeling will transform the archaeological survey in a poli sensing (by multiples sensors) and multi sensorial (by VR) environment. Multi sensorial because of the digital/immersive embodiment, which is not anymore purely visual; poli sensing because of the involvement of multiple sensors detecting the environment.

Some of the data produced by sensors are not completely discretizable in the environment since they need additional mediated tools for being classified and interpreted. I would suggest replacing the traditional category of "remote sensing" with polisensing, because this represents much better the state of the art and the multiplication of close range sensing applications simultaneously. Cyber-archaeology nowadays entails multiple interactive and poli sensing activities in the same space and frequently at the same time. Therefore, the interpretation process in the field is different because it is related with a multiplicity of embodied actions, technological affordances and tools.

Virtual Reality and Embodiment

The massive use of digital technologies and 3D data recording are having an outstanding impact on archeological research. This impact is slightly different when these data interact in real time in virtual reality systems. At Duke University we have developed a VR platform in Middle VR and Unity 3D able to operate in different virtual environments (Figs. 4, 5, 6). This is a very strategic approach because the same application can run, simultaneously, on different immersive and

Fig. 4 3D immersive visualization of the Roman Forum of Regium Lepidi (today Reggio Emilia, Italy) by Oculus Rift

Fig. 5 Duke Immersive Virtual Environment: the virtual excavation of the Neolithic site of Catalhoyuk (Turkey)

Fig. 6 The Virtual Museum of Regium Lepidi (Reggio Emilia, Italy): interaction by holographic display (zSpace)

semi-immersive systems. This approach mitigates the issue of models optimization and data migration which are very common in the case of real time performance.

The digital workflow involves data recording (by computer vision, scanners or remote sensing), optimization/decimation, 3D modeling and data simulation/visualization. The standard format for 3D models is .obj, the standardization of data is made by Middle VR, the final simulation engine is Unity 3D. This digital standardization allows a successful digital preservation of raw data and as well 3D models in obj, which is an "old", standardized and very stable format.

In terms of virtual environments, we are testing different levels of immersion and embodiment: Z-Space (proprioception), Oculus Rift (spatial embodiment, Fig. 4) and the DiVE-CAVE (Duke Immersive Virtual Environment, Fig. 2). Z-space (Fig. 7) is a holographic virtual reality collaborative platform managed by a 3D stylus. Here the users can explore the potential of proprioception (a sense of how our bodies are positioned) and eye-tracking in the virtual exploration of archaeological artifacts. This interaction is collaborative, since the interaction of the user with tracking glasses is displayed in an external monitor by a video camera. This monitor can show in augmented reality real people and virtual objects in the same frame.

Oculus Rift (DK2, Fig. 4) is the well-known immersive helmet (now produced by Facebook) with an extraordinary capacity to reproduce the model in scale. The

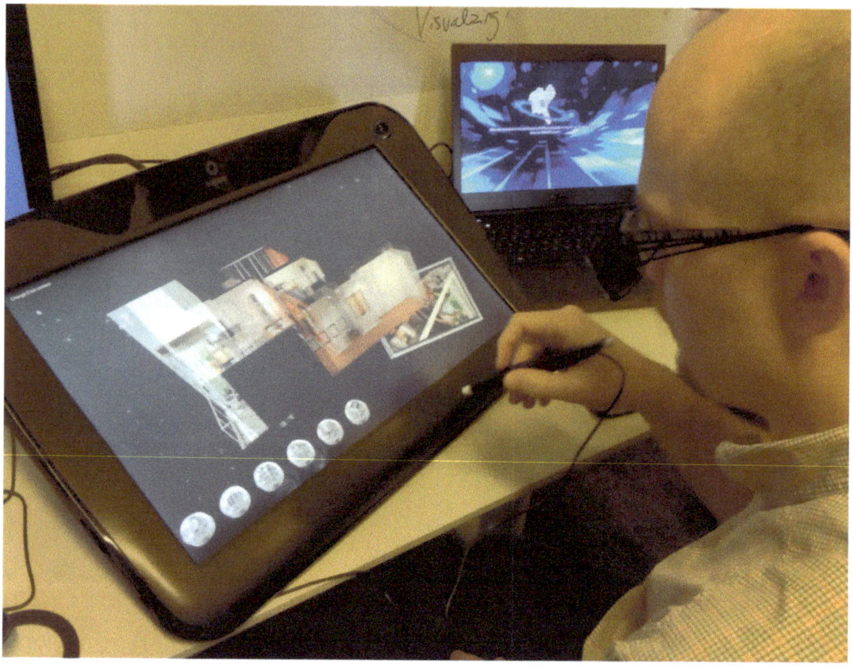

Fig. 7 Holographic virtual reality collaborative platform managed by a 3D stylus. Here the users can explore the potential of proprioception (a sense of how our bodies are positioned) and eye-tracking in the virtual exploration of archaeological artifacts

level of visual embodiment with Oculus is impressive and it recalls the potential of the above-mentioned frameless reality. The tracking system and the Gyroscope automatically give the perception of space and scale. In our experiments we have use the DK2, below the data sheet.

Sensors	Accelerometer, Gyroscope, Magnetometer
Inertial update rate	1000 Hz
Positional update rate	60 Hz
Field of view	100°
Video in	HDMI

Finally, the DiVE (Figs. 2, 5) is a 3 × 3 × 3 m stereoscopic rear projected room with head and hand tracking and real time computer graphics. All six surfaces—the four walls, the ceiling and the floor—are used as screens onto which computer graphics are displayed. For one, the newly installed system generates 1920 × 1920 pixels on each wall (versus the original resolution of 1050 × 1050 pixels). In order to generate a higher resolution image, each wall has two projectors working in unison. These projectors are simultaneously generating the same image, blending the overlap zone between them as a means to increase the output resolution. The DiVE supports multiple platforms—including *C*, *C++* (through CCG), *MATLAB*, *AVIZO*, and *Unity*—which minimizes future compatibility issues.

The inferential model generated by and with virtual environments uses different kinds of embodiments, which require more accurate studies and investigations. We assume that interaction design, interface and information layers influence deeply the user feedback and the qualities of embodiment. In a recent experiment (Heimann et al. 2014) demonstrated that the use of steady cams in movies increases meaningfully the involvement of the watchers. In short, the participants perceived the movement of the steady cam as being the most natural and most resembling the movements of an approaching observer, thus eliciting the feeling that the observer would walk towards the scene (Heimann et al. 2014). This might suggest that the 1st person view of a virtual reality application (for example a game or a walk-through dynamics) would increase as well the level of embodiment of the user.

At theoretical level it is appropriate to recall the metaphor of puppets and strings discussed by Deleuze and Guattari (1987) and Forte 2015). Puppet strings (Fig. 8), as a rhizome or multiplicity, are tied not to the supposed will of an artist or puppeteer but to a multiplicity of nerve fibers, which form another puppet in other dimensions connected to the first. In a virtual environment we know the content, the modality of interaction, but not the cybernetic feedback, which is subjective and

Fig. 8 The metaphor of puppets and rhizomes (Deleuze and Guattari, Capitalism and Schizophrenia (1972–1980) project)

individual. In this case the puppets represent the virtual content, the strings the interaction, the puppeteer the software/design. This complex interaction among puppeteer, strings and puppet changes all the time and it is not predicable since it is designed according to a bodily-kinesthetic approach and a digital narrative. The strings can be considered as affordances since their action is managed by the environmental context. The way we approach a 3D model is controlled by the relationship/affordance user/environment; if we change the affordance, we modify the interpretation or we add multiple interpretations. In short, the interpretation is performed, acted through feedback and interaction.

So, where are we now? Virtual reality was a missed revolution in the last 20 years: a lot of expectations, a very limited social impact. The situation is radically changed because of the massive use of virtual devices and by the customization of portable and low costs immersive systems. The games for example, contributed substantially to this revolution and to improve a collaborative access to virtual content.

In this scenario, the digital archeological work is located in between empiricism and simulation and it is characterized by multi vocality, collaborative interpretation, networking, and open digital narratives. This kind of (cyber) archeology is based on the capacity to interpret and learn through interaction in a specific sociocultural context: an interaction that involves relations in the past and from the past. Digital data are at a distance; in fact the human interpretation is in between the real datum (on site, by artifacts, etc.) and its virtual ontology, the simulation.

The Embodied Archaeologist

"Embodied simulation is a functional mechanism in social cognition, not confined to the domain of action, but encompassing other aspects of intersubjectivity, such as emotions and sensations" (Gallese 2005). Because of a shared representational format, we map the actions of others onto our own motor representations, as well as others' emotions and sensations onto our own sensory-motor representations (Gallese 2000, 2001). The bio-cultural evolution of embodiment reflects the human minds' relationships with different social contexts and environments. The study of the ancient mind is thus one of the most challenging and fascinating research activities regarding *Homo sapiens* and human society. This kind of study requires a multidisciplinary approach to involve different disciplines and research backgrounds relevant to understanding how ancient minds thought about the world. This can be possible through new archeological study of traces and cultural inscriptions of brain processing, and in particular connecting the study of art and material culture with cultural, visual models, cultural patterns and the evolution of the brain.

Human cognition cannot occur without the interaction of the lived body in the unfolding of experience and learning: "knowledge begins with enaction" (Bruner 1962). In the Embodied Cognition Theory, every learning process occurs when a human being enacts in an environment. (Maturana and Varela 1980; Varela 1997).

In fact, sensorimotor perception and action are interdependent, "coupled" or linked by a relation of co-determination and of the contextualized experience of a body-environment system. In other terms, the cognition is embodied and embedded in its context of affordances it experiences (Varela 1997). What happens in virtual environments? Without a context, there is no communication, no information exchange (Bateson 1972). This exchange requires the perception of a difference, between humans and environment (Ingold 2000). Meaning is hence a dynamic, situational and dialogical process. As we have seen, meaning emerges as something that is contextualized, as some kind of reflection emerging from being involved and in action. (Gibson 1979). There are different modalities of being coupled with the environment, but "through *embodying* the world around us and through *interiorizing* the environment, things become *meaningful*; and this happens in an expressive form or mode which seems to be right for the subject in the here and now of the situation" (Stelter 2005; Sheet-Johnstone 1999). The tacit, implicit and embodied knowledge, which is pattern-based and builds on intuitive expertise, is actualized in the present moment of action (Forte and Bonini 2010).

Archaeology of embodiment, embodied archaeology and related terms stimulated a very broad cross-disciplinary debate in the last decades in archaeology (e.g. in Csordas 1990, Meskell 2004; Lesure 2005; Joyce 2005; Pearson and Meskell 2014; Hodder 2012; Forte and Gallese 2015). This kind of conversation was mainly centered on material culture, embodied experience, spatial models, ethnographic studies but not on digital or cyber archaeology.

The idea of embodied (digital) archaeologist is inspired by the fact that the IT impact on archaeology is changing the experience of the archaeologist in relation to all the information is surrounded by. The embodied simulation generates new cycles of iterative interpretations. This has somehow generated a new digital positivism with an exaggerate emphasis on quantitative data rather than an adequate attention to the semantics of qualitative data, the cognitive impact, the meaning of the methodological approach and the effects of simulation. In short, a sort of new "digital processualism". For example, the discussion about interpretation in virtual archaeology was mainly focused on evocative imagination of a single past and not on the evaluation of the inferential method for interpreting it (multiple past). In other terms, the computer rendering was the target of this debate instead of computer interaction and human embodiment.

The evolution of virtual archaeology in cyber-archaeology reflects the relevance/dynamics of interaction-simulation-performance versus the pre-determined static reconstructive process. The current problem is that we continue to classify, study and interpret 3D interactive models and virtual environments with the same methodological approach we used for 2D data, maps, metadata and so on. Is that sufficient and what's the next step? The computer simulation creates feedback, affordances and interactions otherwise not achievable and this might be recognizable as "digital mind", a combination of ancient (simulation) and modern mind (interaction).

At this stage of research how can we study the digital mind? How can we study cognitive universals in the ancient mind? A new methodological approach is necessary and it should be at the intersection of brain sciences, humanities, archaeology, anthropology, art, philosophy, aesthetics and visual studies. Cross-disciplinary contributions concerning embodiment and enaction in visual models, the interpretation of visual patterns and memes in cultural transmission should entangle a neuroscientific approach: we need to know what happens in the brain during a virtual simulation of ancient worlds.

I quoted above the metaphor of puppets and strings in Deleuze and Guattari (1987): in this scenario there are independent and dependent factors. The dependent factors are the hands moved by the puppeteer, the independent factors are the strings that, by mutual interactions, generate uncontrolled and unpredictable engagements. This metaphor recalls what happens in a simulation environment where different users collaborate, operate, and interact with models and data; the result is unpredictable and the whole is more than the sum of its parts (Forte 2000). More important, during the simulation, the exchange of information increases and consequently the interpretation becomes an open and dynamic process.

Mediated/remote tools of investigation operate almost in real time: it means that for example when a laser scanner records millions of point clouds or a stratigraphy is generated by computer vision applications in a tablet pc, the operator, the archaeologist, is still there, comparing digital data on a screen and "real" data on site. In short, it is nowadays often unnecessary to wait until the lab post processing (of course important) for interpreting and evaluating data and models. The digital performance acts on site, in different operative contexts (inter-sites, intra-site).

In "Simulacra and Simulation", Baudrillard (1994) introduces the opposition between real and hyper-real, where the hyper-real is the simulacrum of the reality. For Baudrillard a simulacrum is not a replica of the real but become truth in "its own right": the hyper real. The hyper reality is the core condition of the simulacrum, "more real than real", with the ability to enrich, transform and transfigure the reality in something else. "In a network post-modern society the hyper real "avatarizes" the information, disseminating digital replicas, models and virtual collaborative experiences. In its immateriality the hyper real is like a virus in the sense that it multiplies its code in order to transmit the content in endless forms.

In Baudrillard (1994) there is a remarkable distinction among simulation, representation and simulacrum. Whereas representation attempts to absorb simulation by interpreting it as a false representation, simulation envelops the whole edifice of representation itself as a simulacrum (p. 4). There is a sort of syllogism among these terms: if representation interprets simulation as false representation, the representation itself is a simulacrum. The Baudrillard's idea of simulacrum applied to an archaeological context might suggest that digital data are "signs", markers of the past, new and lost cybernetic codes re-processable during a simulation. If one day we could synchronize these "signs" or messages with our mind (bio-cultural) universals, we could digitally reconstruct the past.

Conclusions and Future Perspectives in Cyber archaeology

This contribution focused on the new research perspectives generated by polisensing and 3D embodiment in cyber archaeology. The revolutionary impact of massive digital datasets in the phase of real time data recording and virtual interactions generates new research questions but also advocates a different approach in remote sensing archaeology. Sensing is not anymore remote: drones, lidars, computer vision, phone, mobile devices and new generations of geophysical tools are changing our approach to the traditional "remote sensing". Speed, accuracy, quality and quantitative of data, and new digital semantics require a different methodological approach (Fig. 9).

The author proposes to replace the terms "remote sensing" with "poli sensing", since this reflects actually the variety of technologies and ontologies of multimodal data recorded in the same space by multiple sensors. Poli and not remote, anymore. The first effect of this revolution is in the digital performance, "data in action", which starts already during the acquisition: real time interaction is possible by the simultaneous use of sensors. Therefore the first archaeological interpretation is already "mediated" and it is challenging to contextualize of the digital datasets in a searchable and classifiable domain.

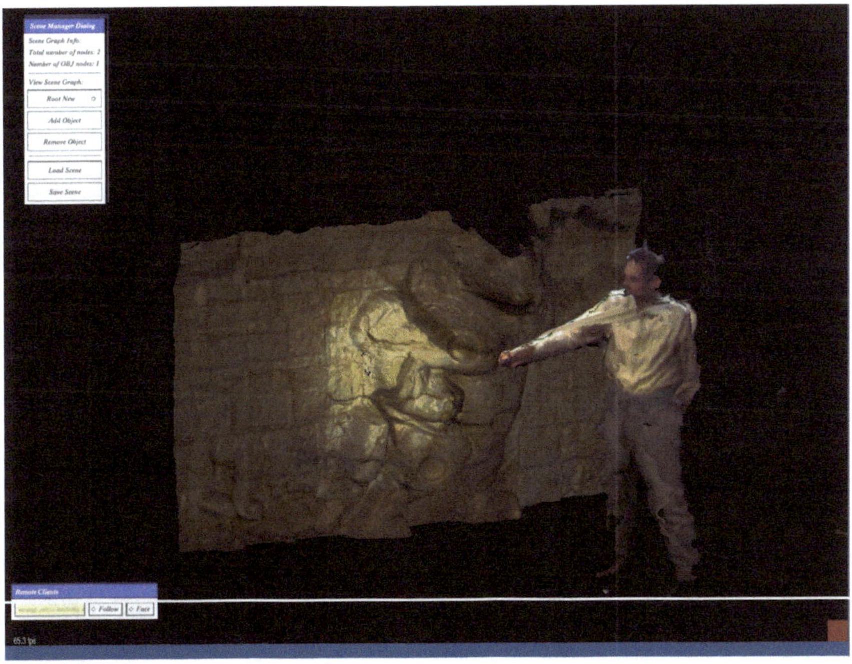

Fig. 9 Teleimmersive archaeology (UC Berkeley, UC Merced, in collaboration with G. Kurillo): multiple remote users can interact in the same collaborative cyberspace

Nevertheless, there is clearly still a gap between data entry, digital recording and processing and a successful evaluation of this digital hermeneutics. We don't have yet adequate instruments of evaluation and validation and the risk of hyper positivism is quite tangible.

Some scholars have recently debated about the "objective positivism" of cyber archaeology in relation to heritage interpretation and possible risks of discontinuity/mismatching between digital methods, archaeological practice and heritage interpretation. I think this misunderstanding comes from an old processualist idea of cyber archaeology as heavy-quantitative research field. This kind of criticism reminds me a similar debate at the very beginning of virtual archaeology (1993ss). At that time most part of the academic positions was born by the misconception that virtual archaeology could somehow replace other methods of investigation or could transmit the idea of a fake/artificial/visual past. At that time the core of the debate was on the peremptoriness of the reconstructions and not on the potential of the simulation (Forte 1997, 2000, 2015; Fig. 10). In short, virtual archaeology was discussed in terms of aesthetics of computer graphic models and not in terms of potential interaction and cybernetic feedback inside the virtual worlds. This is in part understandable because of the low-tech impact of virtual reality in the users' experience at the very beginning of this research field.

There is a general misunderstanding on considering the massive and immersive digital recording the core of the cyber archaeological approach, while the main goal in cyber archaeology is the simulation, the feedback, and the interaction through

Fig. 10 Computer graphic rendering of the Imperial Villa of Livia (I-II cent. AD) in Rome

virtual environments: this generates an intersubjective interpretation. Cyber archaeology doesn't look for objectivity, since it is focused on the simulation and performance of a multiple and open past. Semantics and not quantity is the keyword for designing future simulation environment according to a cyber archaeological approach.

Archaeologists can digitally "perform" the past if the information is correctly recorded and embodied in the human experience. The mediated knowledge is by nature multi vocal because it allows different categories of users to get full access to the digital simulation. In this way users and content makers become all content providers because of the cybernetic feedback.

This debate recalls somehow the opposition between empiricism and rationalism. Rationalists believed that knowledge can be generated independently of sense experience and that the ultimate starting point for all knowledge is not the senses but reason. In that sense, without prior categories and principles supplied by reason, it wouldn't be possible to organize and interpret our sense experience. Empiricists argued that sense experience is on the contrary the core of human knowledge. Where does human knowledge ultimately come from?

The poli sensing data acquisition is the necessary starting point in order to act with data, to reformulate hypotheses, compare inferential models. So, here we have to collect more, not less data. The bottleneck is that we continue to collect digital data in the same domain (geometrical) and not in other non-geometric domains (auditory, environmental, chemical, physical) and this creates an unbalanced result in our way to interpret the world. In addition, the capacity to classify certain categories of data (for example lidar, laser scanning) are still limited.

GISs and most of the remote sensing applications are oriented to geometrical and spatial data, because they are all easily geo-referable and geo-located in the same space (Fig. 11). This kind of discretization uses relatively simple codes with limited processing capabilities for 3D data and related fields.

Virtual environments, on the contrary, recall the importance of multimodal and poli sensing data integration. In fact, future development of digital technologies

Fig. 11 3D procedural modeling and virtual reconstruction of the Roman city of Regium Lepidi (Reggio Emilia, Italy). Immersive interaction in virtual reality

should involve a poli sensing vision in data recording and in data simulation. This second aspect sounds more difficult because of the constrains of virtual reality systems, actually merely visual. In fact, it is necessary to project multisensory virtual environments in order to increase the level of embodiment, engagement and simulation. This won't happen soon, but the mass use of portable VR systems will change radically our access and embodiment to real time data and models before any possible revolution in multisensory domains. Mass VR by multiple devices will accelerate and expand human capacities of virtual interaction and cybernetic feedback in a range of scale never experimented before.

The hyper reality discussed in this paper suggests also that in the near future the category of the "Virtual" might loose part of its immateriality: it could be part of the human material culture, like an archaeological artifact. Why? How? The future of frameless reality, holograms, holographic projection, floating models, mass virtual reality will produce so sophisticated models that we won't recognize them anymore as virtual/fake/artificial, but we will perceive them as core elements of empirical reality.

A neuroscientific experiment based on the effect of camera movements (steadycam) on motor cortex activation during action observation (Heimann et al. 2014) implies that a similar effect might be possible during a 1st person interaction in a virtual environment. This recalls the importance of situated presence, embodiment and participated interaction in a visual/virtual context (in this case narrative). Shall we expect the same in a virtual reality simulation?

References

Bateson G., 1972, Steps to an Ecology of Mind: Collected Essays in Anthropology, Psychiatry, Evolution, and Epistemology. Chicago, Illinois: University of Chicago Press.

Bateson, G., 1979, Mind and Nature: A Necessary Unity (Advances in Systems Theory, Complexity, and the Human Sciences). Hampton Press.

Baudrillard J. 1994, *Simulacra and Simulation*, Ann Arbor: The University of Michigan Press.

Bateson, G., 1979, Mind and Nature. A Necessary Unity. New York: E.P. Dutton

Borges, J.L., 1975, Universal History of Infamy, Penguin Books, 1975.

Bruner, J.S. 1962 The conditions of creativity. In: Gruber, H., Terrell, G., Wertheimer, M. (Eds.), Contemporary Approaches to Creative Thinking. Atherton, New York (NY), pp. 1–30.

Csordas, T. (1990) Embodiment as a paradigm for anthropology. Ethos 18: 5–47.

Deleuze G. and Guattari F. 1987, *A Thousand Plateaus*. Capitalism and Schizophrenia. Trans. By Brain Massumi. Minneapolis: The University of Minnesota Press.

Forte M., 1997 (ed.), *Virtual Archaeology*, Re-creating Ancient Worlds. London: Thames and Hudson, 1997.

Forte, M., 2000, *About virtual archaeology: disorders, cognitive interactions and virtuality*, in *Virtual Reality in Archaeology*, (ed. by J. Barcelo, M. Forte, D. Sanders), "Computer Applications in Archaeology" Oxford, 2000.

Forte M., 2009, *Participatory Research in Cyber archaeology*, in CAA 2009. Proceedings of the 37th Annual Computer Applications and Quantitative Methods in Archaeology Conference, March 22-26, Williamsburg, Virginia, USA, 76-85.

Forte M., 2010, *Introduction to Cyber archaeology*, in "Cyber archaeology" (ed. by Maurizio Forte), BAR International Series 2177, Oxford 2010, pp. 9-13.
Forte, M., 2014, Çatalhöyük: a Digital approach for the Study of a Neolithic Town. In: R. Tamborrino, ed. *Digital Urban History: Telling the History of the City at the age of the ICT Revolution*, Roma: CROMA, pp. 1-10.
Forte, M., 2014a, *3D Archaeology. New Perspectives and Challenges. The example of Çatalhöyük*, in Journal of Eastern Mediterranean Archaeology, Vol.2, n.1, 2014.
Forte, M. 2015, Cyber archaeology: a Post-Virtual Perspective, in D. T. Goldberg - P. Svensson (eds.), Humanities and the Digital. A Visioning Statement, Boston, MIT Press, 2015, 295-309.
Forte et al. 2012, Forte, M., Dell'Unto, N., Issavi, J., Onsurez, L. & Lercari, N. 2012. 3D Archaeology at Çatalhöyük. *International Journal of Heritage in the Digital Era*, 1: 351–378.
Forte, M. Bonini, E., 2010, *Embodiment and Enaction: a Theoretical Overview for Cybercommunities,* in "Heritage in the Digital Era" (eds.M.Ioannides, A.Addison, A. Ltd, 2010, 45-56. Georgopoulos, L.Kalisperis, A.Brown, D.Pitzalis), Multi-Science Publishing Co.
M.Forte, V.Gallese, *Embodiment and 3D Archaeology: a Neolithic House at Çatalhöyük*, in "Breaking Barriers", Proceedings of the 47th Annual Chacmool Archaeological Conference, November 7-9, 2014, edited by Robyn Crook, Kim Edwards and Colleen Hughes Chacmool 2015, The Chacmool Archaeological Association of the University of Calgary, pp. 35-55.
Gallese, V. (2000) The inner sense of action: agency and motor representations. Journal of Consciousness Studies: 7; 23-40.
Gallese, V. (2001) The "Shared Manifold" Hypothesis: from mirror neurons to empathy. Journal of Consciousness Studies: 8(5–7): 33–50.
Gibson, J.J.,1979, *The Ecological Approach to Visual Perception*, Houghton Mifflin, Boston.
Gallese V., 2005, Embodied simulation: from neurons to phenomenal experience. Phenomenology and the Cognitive Sciences, 2005, 4:23–48.
Gallese V., 2016, Neoteny and social cognition: A Neuroscientific Perspective on Embodiment. In: Durt, Fuchs, Tewes (Eds.). Embodiment, Enaction and Culture. MIT Press, 2016
Hodder I., 2012, Entangled: An Archaeology of the Relationships between Humans and Things, Wiley-Blackwell, 2012.
K. Heimann, M.A.Umilta', M.Guerra, V.Gallese, Moving mirrors: a high-density EEG study investigating *the effect of camera movements on motor cortex activation during action observation*, in "Journal of Cognitive Neuroscience", 2014, pp. 1-15.
Ingold T. (2000) The Perception of the Environment. Essays on Livelihood, Dwelling and Skill. London and New York: Routledge.
Joyce R.A., 2005, *Archaeology of the body*, in Anny. Rev. Anthropol. 2005, 34, pp. 139-158.
Lesure R. G., 2005, *Linking Theory and Evidence in an Archaeology of Human Agency: Iconography, Style, and Theories of Embodiment* in "Journal of Archaeological Method and Theory", Vol. 12, No. 3 (Sep., 2005), Springer, pp. 237-255
Malgrave H.F., 2015, *Enculturation, Sociality and the Built Environment*, in "Architecture and Empathy"(eds. J. Pallasmaa, H.F. Mallgrave, S. Robinson, V.Gallese), pp. 20-41.
Maturana H.,Varela. F., 1980.Autopoiesis and Cognition:The Realization of the Living. In: *Boston Studies in the Philosophy of Science,* ed. by Robert S. Cohen and Marx W.Wartofsky, vol. 42, Dordecht (Holland): D. Reidel Publishing Co..
Meskell, L. (2004) *Object Worlds in Ancient Egypt: Material Biographies Past and Present.* London: Berg
Pearson, J.A. and L.M. Meskell (2014) Biographical bodies: esh and food at Çatalhöyük. *Proceedings of the British Academy* 198: 233–50.
Sheet-Johnstone, M., 1999. *Thinking in movement.* John Benjamins, Amsterdam.
Stelter, R. 2005. Embodiment and Learning. *Agora* nr. 6/2005, Copenhagen.
Varela, F.J. 1997. Neurophenomenology: a methodological remedy for the hard problem. *Journal of Consciousness Studies*, June.

Emergent Relationality System/The Insight Engine

Bill Seaman

Abstract The Emergent Relationality System is a work in progress that enfolds multiple research agendas. The system seeks to define a new wholistic approach to CyberArcheology including new forms of multi-modal sensor hardware to work in conjunction with current sensor systems; a new software/search paradigm; and a novel generative virtual environment; to empower the user to bring media materials into "intelligent" proximity/juxtaposition. This system seeks to include the visualization of multiple kinds of sensor data relevant to CyberArcheology as it is brought into dynamic relation with media elements that might not normally exist in the same associational virtual space. Here polysensing systems (parallel multimodal sensing over time) enable the creation of a form of media object that can be given additional meta-data, and can be explored via state of the art search algorithms, new meta mark-up methodologies, and virtual visualization. The system is modular and non-hierarchical and hence a potentially combinatoric modular format is explored in the following text.

When one attends contemporary conferences related to CyberArcheology, one notices on the one hand the incredible exploration and proliferation of new sensing technologies and alternately one gets the feeling that a form of wholism related to the digital-archeological project is missing. Of course the concept of artifacts, locations and cultural processes, and how they can be addressed to form a proposed "picture" of a historical time/space is central to this research. The capture of multiple forms of digital sensing processes enables new access to the relational, in the spirit of ongoing understanding, in particular via the intelligent mining of relational databases through linguistic and pattern related means. Here we seek to form a deep accretive notion of context via computational means. Let me here give a simple example of media that might be brought into relation to help extend our understanding of context. One person might be focused on the photography of a

B. Seaman (✉)
Media Arts + Sciences, Department of Art, Art History & Visual Studies,
Duke University, Durham, N.C., USA
e-mail: bill.seaman@duke.edu

M. Forte and S. Campana (eds.), *Digital Methods and Remote Sensing in Archaeology*, Quantitative Methods in the Humanities and Social Sciences,
DOI 10.1007/978-3-319-40658-9_13

particular site as drawn from historical archives; another might be interested in contemporary oral histories that have been passed down; a third might be looking deeply into the archives of past research; and a forth foci might be using contemporary digital tools to scan and map the site—the sensing technologies that are central to this text. New forms of multi-modal visualizations become an essential new form of media-element. Of course the project is never finished but is always unfolding, adding new layers of meaning as an ongoing process over time. Often different areas of research are kept separate. My contention is that if they can be brought into relation, where sensing data is dynamically juxtaposed with other forms of media objects, via new computational methodologies, this relationality might bring insight to a more holistic historical understanding.

This suggests the necessity of authoring an extensible intelligent network of multi-perspective knowledge production technologies. These could be explored through the enfolding of multiple forms of mapping, multi-modal data collection and subsequent visualization (polysensing), storage, annotation, text analysis, pattern analysis, AI, and insight production systems.

My goal is to articulate a multi-modal sensing environment that is linked to a networked virtual space, related database and 'intelligent' media connectivity systems. If we draw from human sensing as a model, we can employ multiple senses to provide for us a multi-perspective understanding of an environment over time. How can we re-apply this methodology to enhance our understanding of new forms of relational technological artifacts? One foci is to collect a stream of bio-mimetic multi-modal pattern flows of sensed data as derived from a series of different linked machinic sensing systems including multiple sensors on a chip [employing as many chips as needed for different sites]. This kind of complex chip or hardware unit including multiple sensors could function in conjunction with other sensing/scanning technologies that are being employed at a given site via new forms of database mining and subsequent relational visualization. Here, the time-based layering of data from differing sensors begins to present coherent data-objects in a similar fashion to that of the how the human builds up knowledge of context over time. Here a site-based virtual timeline becomes one organizing methodology. One will also be able link and/or intermingle "associated" "informational" data of relevance via differing linguistic data-mining methodologies—e.g. coherent textual mark-ups here give an extended voice to differing digital objects in the database, key-words, meta-tags, etc—sensing data in particular will also be visualized (and/or sonified) in new ways that make it easy to juxtapose with other media elements. The potential is to link an Internet of Things and new forms of sensing, with particular spatial environments and distributed researchers from differing, yet related fields creating a more extensive operative holistic CyberArcheology.

Additionally in the future one seeks to extend these multi-modal sensing environments with new forms of light sensors enabling intelligent transduction of multi-modal sensor data. The potentials of new forms of bio-abstracted robots that might have mobile Polysensing environments linked to particular "learning" systems are also of long-term value. One also seeks to create a flexible "Grammar of

Attention" for the sensors that is easily coded by a remote observer via an "Emergent Intention Matrix", an object-based coding system. Imagine that one could scroll though a list of functionalities, and easily choose (through a menu system) to have a particular sensor "pay attention" to the site in a specific way, e.g. track the light, temperature, oxygen level, etc. as these factors change over a period of time. It could then communicate with the smart sensor system and undertake this particular functionality. Such a system can be applied to many different kinds of environment—scientific, artistic, and humanities related. One can easily see the specific relevance of such a project to CyberArcheology. Additional new technologies might be brought into the network including Seaman's concept of the R-map (Room/Region/Relation Mapping) which explores the mapping of a space via multiple video zooms, fanning out in 180° or 360°, that can later be navigated (moving in and out of the zoom) providing detailed images of surfaces as well as wider angle images collected in the zoomed-out mode. This system creates a video virtual space. Hypothetically this space could also be annotated at specific parts of the zoom, and searched in relation to semantic, syntactic and statistical textual information as well as pattern-oriented information.

The Relational World Generator, a new iteration of the Insight Engine, is also in its initial production stages. This system would link a database of sense-oriented media elements with more traditional media-objects like texts and images that are embued with textual meta-data, within a generative virtual environment. One could make queries that bring different media elements into proximity in the virtual space e.g. the media elements described above might be brought into dynamic proximity and shed new light on a particular site and aspects of the cultural relationality of particular artifacts combining the most contemporary of sensing processes with older forms of representation.

Seaman's Insight Engine project enables searches across disciplines to bring textual and media materials into proximity via the linguistic analysis of meta-tags, stored and scraped texts. By bringing these different kinds of functionalities together in a holistic system, the new technology might enable exploratory relational approaches to differing forms of contemporary and historical data. Consider for example old maps, images and photos; oral histories; along with media objects (visualizations) produced via contemporary sensing technologies, all brought into dynamic proximity via an intelligent query system. The goal of the overarching system is to enable intelligent meta-tag as well as corpus searches via statistical, semantic, and syntactic analysis to enable new forms of intellectual relationality to be defined by individual observers and/or groups of researchers, and/or via computational means—ai employment. This will also in time include intelligent pattern matching as well as new forms of visual and/or multi-modal intelligence. This later investigation can be considered an outgrowth of Seaman and Rössler's Neosentient research exploring new approaches to artificial intelligence and learning systems as coupled with multi-modal sensing systems.

As I unpack the system described above, I seek to do this in a different way to that of a traditional book chapter. I am interested in what I call recombinant informatics. Of course we know that computers can easily bring into juxtaposition different kinds of data and I could author a program that would allow me to combine and recombine these paragraphs. Historically this methodology has been explored more via combinatoric artworks, novels, poems and games. Yet here we explore in the service of creating new approaches to cyberarcheology through recombinant informatics.

For my Chapter, I will ask you as a reader to take an active role. Take any two paragraphs that are to follow and consider how they suggest a new research paradigm through their bridging. They will be intentionally short and modular so you can jump though them in a non-hierarchical manner. Of course this is antithetical to most book chapter publication but I think you will begin to see a parallel between this methodology and the system that the chapter itself is describing.

The goal of potential new technological system is to create an extensible intelligent network of multi-perspective knowledge production technologies, to be explored through the enfolding of multiple forms of mapping, polysensing (Bonin 2002; Seaman and Verbauwhede 2002), multi-modal data collection, storage, annotation, and insight production systems. The system will be designed to be accessed from multiple kinds of viewing systems, for differing kinds of viewing experience. I will discuss this below.

This New Field

As a media researcher one of the things I notice about the field of CyberArcheology is that it might better progress in the goal of creating new knowledge, by cross-pollinating differing modes of research. What kinds of system can we best author to accomplish this goal, mixing sensing system data with other historical forms? In particular this could be enhanced by articulating a network of intelligent relationalities (Hodder 2012) between actual sites, the media objects and technologies that represent them in differing ways, as well as through differing representations of objects that belong to earlier time periods. One can imagine a generative associational network that could function as a spatial visualization system dynamically linked to a relational database. Such a system could be explored within a generative virtual environment. This environment would ostensibly enable the creation of intelligent juxtapositions between disparate media materials, thus potentially providing new insight into the “reading” of a particular media object, site, artifact, or historical notation. New notations and/or observations could be added in an accretive manner. One imagines a virtual space housing maps from differing time periods, elaborate 3d Models, still photos, video, text documents, oral histories, GIS data (Geographic Information System), and GPR data (Geological Penetrating Radar) etc. All of this material will be explored in a rhizomatic manner enabling any media-object to be juxtaposed with any other in a dynamic intelligent manner.

Relational Database

A relational database will function as a central storage system, housing these differing forms of sensed data, visualizations, media-elements, media annotations, as well as specific forms of metadata. Yet, the connectivity of the system via the internet will enable differing researchers around the world to access it, adding in, in an accretive manner new data as well as new interpretations of that data. It will potentially be connected to multiple networked viewing spaces, i.e. a single screen presenting 3d data e.g. laptop or monitor; VR presented for 3D stereo goggles; and/or 3D immersive environment—e.g. The Duke Dive (DIG Lab Duke University 2015), or semi-immersive environments in the DIG lab (DIVE, Duke Immersive Virtual Environment 2015); and/or the VR Dome as in SAT, the Society for Technological Arts, Montreal, Canada (SAT, Society for Technological Arts 2015).

Polysensing System

The polysensing system bundles together multiple sensors. Its goal is to explore bio-mimetic and bio-abstraction in terms of sensing over time, taking clues from human sensing. Here, we begin to be able to define computational time-based media code "objects" that have "consistent" attributes. One imagines this kind of sensing system at the CyberArcheological site in the future, and/or being employed by robotic learning systems that also function as part of the larger networked learning system. One idea about developing a deep knowledge of the context related to the site, is to collect parallel streams of multi-modal sense data over time, and to begin to build up knowledge of a site much in the way a human does—employing multiple senses over the course of a lifetime (or over a chosen duration of time). The programming of the system could potentially be modular in nature exploring "object-based code." Here the user of the system focuses the "Grammar of Attention" for the sensors that is easily set in motion by a remote observer via an "Emergent Intention Matrix,"—a dynamic menu system of differing options to help in defining the functionality of this multi-modal sensing system at any given time. Here the user selects interactively, differing kinds of functionality related to focusing the attention of the sensors (Seaman and Rössler 2011; Seaman 2005, 2009). One asks how would this be beneficial to extend the study of already existing artifacts? Here, new insights may be provided via new data where the interpretation and reinterpretation of that data becomes central. What can a series of machinic sensors tell us that we don't already know—new kinds of information about site that include temperature over time; qualities of light over time; qualities of sound over time; qualities of chemical reaction over time; qualities of spatial movement over time; qualities of larger flows of individual movements over time; and patterns of potential use of artifacts leading to new cultural understandings etc. These are just some of the initial ideas. Yet, given a particular time code as another stream of data, one can make correlations that link all of this different data at specific moments in time. This might contribute to the construction of potentially more holistic multi-perspective understandings.

Relational World Generator

One function in the system would enable a user to explore a Relational World Generator (Seaman 2010). Such a system becomes a visual database query and perusal system, enabling users to select differing kinds of media objects to be brought into proximity either through interaction with the system via menu or via an entered, automated search query. This might bring into proximity images, maps, scans, 3d models and/or entire virtual environments, texts, oral histories, videos, key words, specific annotations and meta-data—essentially any form of media-object or media-process, to be juxtaposed with other forms of visualized sensed data. The central aspect of this apparatus would be to function as a combinatoric associative network. Here users defining queries could call together objects using textual and image recognition potentials. In Seaman's, (along with Todd Berreth, and Olivier Perriquet) Insight Engine, currently users can make a query and the system will do a search for relevant media objects based on semantic and statistical data, looking for other relevant media-objects. This draws on Arthur Koestler's notion of bisociation as discussed in 'The Act of Creation.' "I have coined the term 'bisociation' in order to make a distinction between the routine skills of thinking on a single 'plane', as it were, and the creative act, which, as I shall try to show, always operates on more than one plane …" "We learn by assimilating experiences and grouping them into ordered schemata, into stable patterns of unity in variety. They enable us to come to grips with events and situations by applying the rules of the game appropriate to them. The matrices which pattern our perceptions, thoughts, and activities are condensations of learning into habit. The bisociative act connects previously unconnected matrices of experience …" (Koestler 1964). I imagine expanding on this notion exploring polysociation, where one could choose multiple media objects and have the system look for other relevant media objects to be brought into proximity. It is this notion of virtual proximity and real-time perusal that may potentially enable new insights to be formed by users of the system. Using this system one will be able to articulate new forms of relationality that might not otherwise surface. I will discuss later in the paper.

R-map—Room|Region|Relation Mapping

One potential is in developing a new technology for high resolution coordinated video of environments from many different chosen perspectives. Such a system would then provide a (navigable sequenced stills) video-virtual environment. The idea is that any detail in at the site might have an annotation made using this sub-system. The person doing the mapping of the site sets up the camera system, zooms in, focuses, and does a very slow power zoom with a consistent speed, recording many frames from different parts of the zoom. Such a zoom could be made at any angle from a central location. At any "virtual" location along the zoom, one could stop and add an annotation or link. These annotations would be searchable and would enable other media elements to be called into the Relational World Generator for intelligent juxtaposition, again, becoming part of the semantic and statistical analysis of text that has been entered across the entire system (Koestler 1964). One could also imagine such a system being mounted onto a robotic system to enable the mapping of hard to reach environments deep inside fragile and/or dangerous architectural spaces; and/or extreme environments e.g. radioactive or on other planets; or underwater sites, etc. Additionally new forms of 3D video recoding systems are now in production. The key here is in searchable textual mark-up systems that work in conjunction with such recording mechanisms.

The Existing Tools of CyberArcheology

There are many types of CyberArcheological sensing tools that can be leveraged in the study of a particular site. Maurizio Forte at Duke's DIG lab in the program of the Media Arts + Sciences at Duke University, provides a description of his working environments:

> The project "3D-Digging at Çatalhöyük" started in 2009–10 as an on site digital experiment with the scope to record every phase of excavation in 3D using different technologies: laser scanning, computer vision and photogrammetry (image modeling). The final goal is to make the excavation process virtually reversible in a simulation environment at different levels, from laptop computers to virtual immersive systems.
>
> In the last two years the project was focused on two main documentation levels: macro scale (all the excavations in the East mound, North and South shelters) by laser scanning; micro-scale (the Neolithic house B89) by computer vision. In addition all the burials are systematically documented in 3D using image modeling techniques (Photoscan and Meshlab) and integrating all the data in GIS format (ArcGIS, QGIS).
>
> Actually, the entire archaeological excavations on the East mound are totally recorded in 3D, paperless and georeferenced in the same virtual space. The project has also introduced 3D stereo visualization systems during the archaeological fieldwork in order to interact with 3D models along the excavation period. In 2012 the introduction of tablet PCs has actually reset all the digital drawing, field notes, mapping and 2D documentation in a complete digital format compatible with GIS and databases. More recently, a

multidisciplinary team of Duke University has implemented all the virtual stratigraphy of a Neolithic house (B89) for the DiVE (Digital Immersive Virtual Environment) a fully immersive system of visualization and interaction. In the DiVE it is possible to virtually dig the house by browsing layers and archaeological datasets (Seaman 2003; Forte's Group at the Dive 2015).

The concept here is develop a next generation system that can include both metadata, and media—location-based textual annotations that can be "scraped" and put into the relational database such that one can do a statistical, syntactic and/or semantic analysis, and call up appropriate relational materials.

The Insight Engine

A related project that explores the use of statistical and semantic analysis is The Insight Engine (Seaman 2014). This functioning computational system seeks to draw on my long history as a media researcher designing new forms of interface and qualities of interactivity, and to expand this via a strong interdisciplinary collaboration that bridges Neuroscience, Computer Science, the Arts and Humanities at Duke as well as through international collaboration. The Insight Engine is now expanding its potentials into the CyberArcheological arena. Such a project presents a multi-perspective approach to knowledge navigation and subsequent knowledge production. This research has been in the creation of a tool to empower insight production, distributed interdisciplinary team-based research and to potentially enable bisociational processes as discussed by Arthur Koestler in The Act of Creation (mentioned above). We are now in the process of making a 3d version of the Insight Engine to work in concert with the Relational World Generator.

If we reverse engineer differing research communities across multiple disciplines we can assume that many researchers undertake similar practices—reading papers, viewing diagrams, observing media-objects, exploring data sets, creating and viewing visualizations, annotating research materials, watching and annotating videos, as well as partaking in discussions among other activities. Interdisciplinary, crossdisciplinary, and transdisciplinary research also means crossing "linguistic" domains framing that research. Here the generation of shared language (developing bridging languages) is essential. Could we make a new system that heightens the potential for insight and creative juxtaposition of essential ideas that cut across multiple research communities/domains that are central to CyberArcheology—including the Digital Humanities, Computer Science, Computational Linguistics, Imaging Science, Visualization creation, Historical research, the employment of differing viewing technologies, the internet of things, the distributed internet-based potentials of a unified system, differing potential media storage systems, as well as relational database set-up and management etc.

Searchable Mark-up for All Media Objects including Sensing System Data

If we could scrape textual data in conjunction with metadata in terms of each media-object input into the database i.e. if each media object was associated with an abstract, we could begin to be able to analyze and articulate how each media object could fall in relation to the others in the database. In terms of research, this means that a piece of text in an oral history might be brought into juxtaposition with an ancient map and/or a contemporary scan of the environment etc. Each media object carries a field of potential meaning, and use of the system would provide an ongoing meaning-summing through focused juxtaposition (Seaman 2010). Alternately, pattern recognition systems could enable other forms of relational search.

Semantic, Syntactic, Statistical Analysis, as well as AI and Pattern Recognition

The potential is to scrape the text and use computational linguistic techniques including Semantic, Syntactic, and Statistical Analysis in order to find relevant media objects for juxtaposition as part of an automated query. Here the development and employment of a contemporary VR and/or sensing data mark-up language is essential, one that can be searched and accessed easily. One might also visually choose materials via the menu system. New forms of pattern recognition and/or vision systems might also be employed (Kpalma and Ronsin 2007; Tomasi and Duke 2015). Otthein Herzog explores automatic content analysis and annotation of still images, videos and sound for content-driven multimedia archiving, retrieval, and video abstracting (Herzog 2015). This is also where artificial intelligence might become involved in CyberArcheology as part of a holistic networked system enabling researchers access anywhere connectivity is available.

A Compendium of Relationalities as Applied to Learning Systems

I am very interested in defining a "compendium of relationalities." This could be helpful in the long run in terms of articulating relationalities of differing varieties of media-object in an automated manner. In terms of AI, an interesting initial attempt at parsing important categories of relationality was discussed by Joscha Bach in his book Principles of Synthetic Intelligence in terms of causal relations (Bach 2009). This could also be something that is done organically by having the system ask the user to explain the relationality between the media-objects they have chosen to explore, and add this into the database in a focused manner. This would also become part of the searchable/analyzable text. Mentioned earlier, Ian Hodder's book *Entangled—An Archeology of Relationships between Humans and Things*, presents many "relationalities" across his entire text (Hodder 2012). Here one could begin to employ computational learning theory, neural networks, and machine learning (a subfield of AI) as areas that might become part of the Emergent Relationality System. Here data-mining becomes central to emergent aspects that arise through the functionality of the system (Izenman 2008; Greiner et al. 1997). Can we use learning systems to help us define relationalities we wouldn't normally come across? Can we also use learning systems to help us define categories on the fly? This becomes part of the goal set for the operational system.

Neosentience

Seaman and Rössler have been interested in ideas surrounding learning systems for over a decade as explored in their book *Neosentience|The Benevolence Engine* (Seaman and Rössler 2011). The term "Neosentience" was coined by Seaman. It describes a new branch of scientific inquiry related to artificial intelligence. Seaman and Rössler's book explores the potential of creating an intelligent robotic entity in possession of a form of sentience similar to that of a human being. The juxtaposition and re-juxtaposition of many micro-chapters in the Neosentience book enables one to reflect on new research possibilities and approaches—what Seaman calls Recombinant Informatics. Seaman and Rössler suggested a pragmatic set of benchmarks to articulate Neosentience—We consider a Neosentient robotic entity to be a learning system that could exhibit well-defined functionalities: It learns; it intelligently navigates; it interacts via natural language; it generates simulations of behavior (it "thinks" about potential behaviors) before acting in physical space; it is creative in some manner; it comes to have a deep situated knowledge of context through multimodal sensing; it exhibits a sense of play; it will be mirror competent and will in this sense show self-awareness. In the long run such a robotic learning system could be networked with the other technologies discussed above. It could actively employ a polysensing environment (discussed above) as part of its own synthetic sensing system. One could imagine implementing aspects of the Neosentient paradigm in our Emergent Relationality System, adding new functionality over time.

Future Approach—The Light Data Domain

Seaman has been in discussion with Tuan Vo-Dinh, PhD, Director, Fitzpatrick Institute for Photonics, R. Eugene and Susie E. Goodson Professor of Biomedical Engineering, and Professor of Chemistry at Duke University, about a hypothetical technological approach to networked communication systems. As one surveys the field of contemporary sensing, communication and computational research, one sees a number of different research areas that each explore the use of light as a process-related vehicle. If we begin to articulate connectivity between differing research areas, does this suggest a new holistic paradigm shift toward the use of light as a vehicle for multiple functionalities to be housed in our holistic system? Here different modalities of light-oriented technologies may potentially function in concert to achieve a cybernetic cycle flowing from multi-modal sensing system to computational system to memory and database systems to visualization systems, then back to the observer, who programs, designs, and interacts with the systems within a dynamic distributed network of intra-actions. Where some have felt that we have reached the end of Moore's Law, might the use of light in the light of these new technologies, function as a means to transcend current electron-based methodologies? Does this suggest a radical shift from the electron to the photon as a contemporary physical propagation mechanism for data? Why would this be of interest? What are the unique qualities of light that play into this paradigm shift? Vo-Dinh and Seaman call this holistic network of technologies and processes the *light-data domain*.

Conclusion

I have discussed at length the creation of an extensible intelligent network of multi-perspective knowledge production technologies, to be explored through the enfolding of multiple forms of mapping, multi-modal data collection (polysensing), storage, annotation, text analysis, and insight production systems. Of course this is still just a hypothetical network, but I believe this to be a promising area of future investigation for the field of CyberArcheology. Here the dynamic employment of new sensing systems as they are brought together with database search methodologies, and new viewing mechanisms, may contribute to new knowledge and insight production. I will here focus the multiple ways sensing will be employed in this holistic system.

(1) Differing forms of machinic sensing will be employed on the site, this includes currently used systems for scanning and mapping archeological sites.
(2) A video VR mapping system Seaman calls R(map)—room/region/relation mapping that includes dynamic annotation will be developed, with the potential of scrapable text and/or pattern recognition capabilities.
(3) A poly-sensing system which encodes multiple streams of sense-oriented data over time, in parallel, will be developed. This is to draw on the abstraction of human sensing as it is implemented via a multi-modal machinic sensing system. The notion is to develop "code" objects that help the system develop new forms of deeper context awareness over time. This may also enable new forms of dynamic correlation of data from differing sensed perspectives.

(4) Seaman and Rössler's Neosentient research presents a learning system that integrates poly-sensing data with new approaches to AI and learning systems.
(5) The focus is on an integrated holistic system that enables the various forms of sensed data to be housed in a relational database and called up in an intelligent manner such that it can be juxtaposed with other "sensed" data (or other data in general) though differing linguistic search methodologies (semantic, syntactic, and statistical) as well as pattern-oriented approaches to discerning and categorizing the data.
(6) The sensed data will be made available for viewing/listening in a interactive generative virtual environment and/or via other forms of viewing apparatus (laptop/ipad) etc. Thus a cybernetic loop will be entertained that moves from integrated sensing, to storage, to search, to categorization, to dynamic juxtaposition in viewing, to additional intellectual input/analysis and or the creation of additional dynamic links that future users and collaborative researchers can explore.
(7) In this holistic system, many different kinds of researchers can work together juxtaposing differing forms of "sensed" media objects that are housed in the system within an accretive distributed learning environment.

This holistic Emergent Relationality System integrates many technologies that are currently part of CyberArcheology, with new integrative technologies, in an intelligent manner that may potentially contribute in the long run to new forms of deep contextual historical understanding.

References

Bonin, V. (2002) Bill Seaman and Ingrid Verbauwhede, Poly-sensing Environment, Toward the Development of an Integrated Distributed Technology Exploring Poetic/ Informational Grammars of Attention and Functionality. http://www.fondation-langlois.org/html/e/page.php?NumPage=49 (accessed 4 April, 2015)

Seaman, B., Verbauwhede, I. (2002) The Poly-sensing Environment and Object Based Emergent Intention Matrix: Toward an integrated Physical/Augmented Reality Space http://students.dma.ucla.edu/~fwinkler/PSE/ (accessed 4 April, 2015)

Hodder, I. (2012) Entangled, An Archeology of the Relationships Between Humans and Things, Wiley/Blackwell, West Sussex, UK

DIG Lab Duke University, https://duke.academia.edu/MaurizioForte/Posts (accessed 7 April, 2015)

DIVE, Duke Immersive Virtual Environment, http://virtualreality.duke.edu (accessed 7 April , 2015)

SAT, Society for Technological Arts, http://sat.qc.ca (accessed 7 April, 2015)

Seaman, B. and Rössler, O. E. (2011). Neosentience | The Benevolence Engine. London: Intellect Press

Seaman, B. (2005) "Pattern Flows | Hybrid Accretive Processes Informing Identity Construction." Convergence Magazin 7.2, special Issue on Intelligent Environments.

Seaman, B. (2009), "(Re)Thinking – the body, generative tools and computational articulation", Technoetic Arts: A Journal of Speculative Research 7: 3, pp. 209–230, doi: 10.1386/tear.7.3.209/1

Seaman, B (2010). Recombinant Poetics/Emergent Meaning as Examined and Explored Within a Specific Generative Virtual Environment. VDM Press, Germany

The Insight Engine – See William Seaman, (2014) "A multi-perspective approach to knowledge production", Kybernetes, Vol. 43 Iss: 9/10, pp.1412 – 1424, see also documentation video http://athanasius.trinity.duke.edu/projects/I_E/i_e.html (accessed 8 April, 2015)

Koestler, A. (1964).The Act of Creation. New York: Macmillian Co.

Seaman, Bill (2003) Sentient R-map - Toward the Creation of a Specific Computer-based Memory Theatre: An Exploration into a Poetics of Machinic Sensing, Multi-modal Searching and Thought Augmentation Approaches. http://projects.visualstudies.duke.edu/billseaman/pdf/sentientRMap.pdf (accessed 4 April, 2015)

Forte's Group at the Dive http://vis.duke.edu/projects/catalhoyuk/ (Accessed 7 April, 2015)

Seaman, B. (2010), Recombinant Poetics: Emergent Meaning Examined and Explored Within a Specific Virtual Environment, Berlin: VDM Press.

Kpalma, K. and Ronsin, J. (2007) An Overview of Advances of Pattern Recognition Systems in Computer Vision, http://www.cubs.buffalo.edu/govind/CSE705_2011-resources.pdf (accessed 7 April, 2015)

See Carlo Tomasi's, Duke CS, research, https://www.cs.duke.edu/~tomasi/papers/ (accessed 7 April, 2015)

Otthein Herzog, http://ai.uni-bremen.de/team/otthein_herzog (accessed 7 April 2015)

Bach, J. (2009) Principles of Synthetic Intelligence, Oxford Press, New York (p 89)

Hodder, I. (2012) Entangled, An Archeology of the Relationships Between Humans and Things, Wiley/Blackwell, West Sussex, UK

Izenman, A. J. (2008) Modern Multivarite Statistical Techniques – Regression, Classification and Manifold Learning. Springer, New York, p.3

Greiner, R. ; Petsche, T. ; Hanson, S. (1997) Scaling Up. Computational Learning Theory and Natural Learning Systems: Making Learning Systems Practical, MIT Press, Cambridge, see also http://ieeexplore.ieee.org/xpl/mostRecentIssue.jsp?punumber=6267512 (accessed April 8, 2015)

Seaman, W. & Rössler, O. E. (2011). *Neosentience: The Benevolence Engine*. Intellect Press

Using 3D GIS Platforms to Analyse and Interpret the Past

Nicoló Dell'Unto

Abstract The diffusion of digital technologies has strongly affected the way scholars and researchers use and perceive the archaeological information detected during the field investigation process. Archaeologists are still only beginning to realize the full potential of these applications beyond the attraction of providing visually engaging documentation and focus on the analytical and interpretive power. Today, digital instruments are used in archaeology at any level, and their employment increases the possibilities to document and visualize the information detected during investigation campaigns. In particular, the recent development of powerful visualization platforms, such as virtual reality or the three-dimensional Geographic Information System (3DGIS), and the introduction and diffusion of digital acquisition tools have provided the opportunity to fully visualize and study in three dimensions (3D) the spatial and temporal relations between the fragmented information detected on-site. The combination of these technologies and the construction of more and more functional field workflows of data acquisition allow for defining new solutions to manage and analyse large three-dimensional data sets of archaeological information, opening new discussions concerning the theoretical and methodological implications connected with the introduction of these new approaches in the field, and highlighting archaeological information previously impossible to detect. These new and non-conventional field documentation strategies give new possibilities and dimensions on how to approach the material and inevitably provide archaeologists with the opportunity to formulate new research questions. This chapter will discuss how the development and use of such new simulation systems are affecting the way archaeologists retrieve and analyse material detected in the field in support of more accurate archaeological interpretations.

N. Dell'Unto (✉)
Department of Archaeology and Ancient History, Lund University, Lund, Sweden
e-mail: nicolo.dellunto@ark.lu.se

M. Forte and S. Campana (eds.), *Digital Methods and Remote Sensing in Archaeology*, Quantitative Methods in the Humanities and Social Sciences,
DOI 10.1007/978-3-319-40658-9_14

Introduction

Technological development has always represented an important part of archaeological best practices, and with the recent introduction of new instruments of documentation, it has been possible to experiment and introduce new data acquisition workflows for the implementation and use of 3D information in aid of archaeological investigation.

Since its appearance, digital technology has changed archaeological documentation on multiple levels by providing a new dimension for visualizing and communicating the past. The significant possibilities for retrieving, analysing and interpreting different types of archaeological information, all performed within the digital environment, have only recently come into focus in the greater archaeological community. This text will discuss this development and the current research directions on a global level while highlighting some cases of best practice.

Only recently, experiments to test the efficiency of 3D models in support of field activities have been performed (Doneus and Neubauer 2005; Dellepiane et al. 2013; Opitz and Nowlin 2012; Forte et al. 2012; De Reu et al. 2013) and questions concerning the impact that 3D models have on the process of field interpretation have been formulated (Losier et al. 2007).

Three dimensional models, as a result of field 3D acquisition campaigns, proved to be a robust tool for generating highly accurate bi-dimensional maps and sections of the site (De Reu et al. 2013; Quartermaine et al. 2014) and for monitoring, with high accuracy of details and in three dimensions, the actions performed by the archaeologists during field investigations (Callieri et al. 2011; Forte et al. 2012). These typologies of data, if implemented into 3D visualization platforms, allow for virtually visualizing and reviewing the complex sequence of archaeological information detected in the field.

3D models, when used in relation with other typologies of data, facilitate a deeper understanding of the details that characterize the investigated context, allowing for accurate, visual, and spatial comparisons among the different elements retrieved during investigations.

The introduction of 3D surface models in support of field activities represents an important achievement in archaeological practice. This specific type of 3D data has the unique capacity to display aspects of the material culture, which are usually difficult to visualize with more traditional methods (Fig. 1). As presented in Fig. 2a, 3D models (as well as ortho-images) are capable of displaying virtual representations of the geometric characteristics and colours of contexts or features exposed at the specific moment of their documentation. In contrast, drawings performed by the excavator (Fig. 2b) have the capacity to present the interpretation performed in the field as a result of a direct observation. For this reason, it is important to consider 3D models (as a result of field acquisition) not as a substitute but rather as complementary information to use together with the rest of the visual documentation produced on-site (Fig. 2).

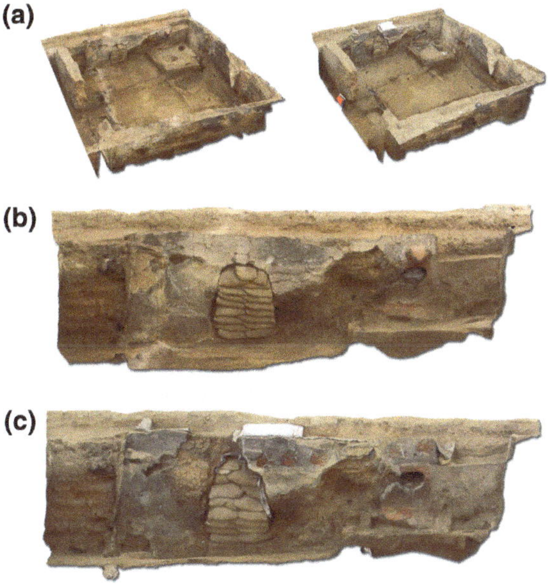

Fig. 1 Different excavation phases of a Neolithic building at the archaeological site of Catalhöyuk, Turkey. The building has been excavated by the Catalhöyuk research project team and documented in three dimensions during different phases of the investigation. **a** On the *left side* the building at the end of the excavation season 2011, on the *right side* the same building at the end of the investigation season 2012. **b** Detail of the building at the end of the investigation season 2011. **c** Detail of the building at the end of the investigation season 2012

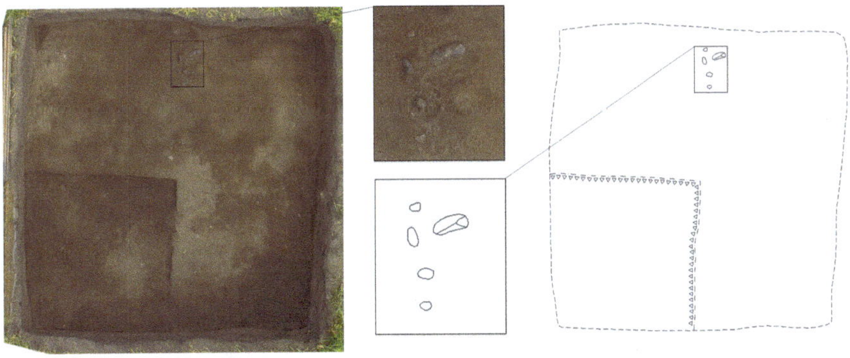

Fig. 2 Comparison between a digital drawing as a result of a field interpretation (*right side*) and a 3D model of the same context realized by means of image-based 3D reconstruction techniques (*left side*)

Moreover, it is important to consider that 3D models are themselves the result of an interpretation process, which starts with their acquisition in the field and ends with their visualization. Operations such as: points cloud optimization, mesh generation and texture projection are, in fact, performed by the archaeologists with a specific visualization goal in mind. For this reason, the result of this process must be considered as purely subjective information, realized to primarily describe the information as perceived in the field. 3D models, if not properly realized, can negatively affect the final site interpretation, and their generation should always be the result of a strategy focused on obtaining data capable of answering research questions formulated before or during the field practice.

A large production and employment of 3D data in the frame of any site investigation activity implies the use of new visualization systems with the capacity to (i) display, in three dimensions, the information recorded on-site, (ii) simulate dynamically the spatial relations among the different materials retrieved during the investigation, and (iii) provide new tools for the detection and analysis of existing archaeological information otherwise undetected by more traditional methods.

To increase the impact that this new data type could have on the investigation campaign, it is imperative to develop new and efficient visualization systems capable of hosting, managing, and visualizing, in full 3D/4D, the multitude of digital data recorded in the frame of the field campaign.

The recent development of 3D visualization systems, such as game engines or virtual reality platforms (so far mainly employed in the domain of Virtual Museums and public communication), has allowed for experimenting with the use of these instruments to visualize archaeological data in support of field investigations and to retrieve new and more accurate archaeological site interpretations.

A ground-breaking example in this direction is represented by MayaArch3D (http://mayaarch3d.unm.edu/). This project is focused on investigating innovative methods to combine and use 3D-Models partially with functions that are usually implemented in more traditional Geographical Information Systems (GIS). In the framework of this research activity, a web-based system called QueryArch3D has been developed. This tool consists of an online repository that provides access to 3D models, GIS maps, and virtual environments with the aim of simulating landscape and architecture (http://mayaarch3d.unm.edu/). This platform allows the user to display and connect a multitude of data (2D/3D) concerning the eighth-century Maya kingdom of Copan, Honduras, and through this application, it is possible to link together 3D data at different levels of detail, providing the user with full access to a heterogeneous dataset of information (Agugiaro et al. 2011; Agugiaro and Remondino 2014).

Another interesting experimentation currently under development is represented by the project *Gabii goes digital* where a team of archaeologists are experimenting with the use of the game engine Unity 3D for creating more complex archaeological virtual scenes capable of presenting 3D models of multiple contexts on-site. This project aims to investigate the use of 3D models as an essential basis for analysis and represents an important reference for the study of the delicate relation between 3D models and archaeological practice (Opitz and Johnson 2016).

Geographic Information Systems and 3D Models in Archaeology

Another important indication in this direction comes from the development of more traditional Geographic Information Systems, where a community of developers has recently invested considerable resources in the creation of 3D interfaces. These systems allow visualizing 3D high resolute surface models, providing new instruments for measuring and mapping (directly in the 3D space) the relations among the different elements uploaded in the system.

Geographic Information Systems are currently widely acknowledged as one of the most influential instruments for the management and analysis of archaeological data, and these are considered the standard for archaeological documentation in many countries (Allen et al. 1990; Lock and Stancic 1995; Weathley and Gillings 2002; Chapman 2006; Connolly and Lake 2006). The importance and potential of using 3D data inside GIS have been extensively discussed and theorized (Conolly and Lake 2006; Lock 2001:161, 2003:182). However, due to the technical limitations of previous versions of GIS in managing surface textured 3D models, most of the early case studies presented in the literature (concerning the employment of the third dimension for the use of 3D GIS platforms in support of field investigation activities) have been mainly characterized by the implementation of Digital Elevation Models (DEM), such as the Triangulated Irregular Network (TIN) or Raster (GRID), or by the use of 3D models as a result of computer-based visualization (Katsianis et al. 2008).

So far, these limitations have prevented GIS from being used as an instrument for the visualization and management of complex 3D geometries as a result of field acquisition campaigns performed by means of a laser scanner or image-based 3D reconstruction techniques.

The possibility to implement and visualize these types of data within a 3D GIS platform and in the frame of an archaeological excavation would allow for experiencing a completely new approach to the dataset collected in the field. This would provide archaeologists with a powerful simulation environment in which it would be possible to retrieve a more accurate and holistic overview of the complex information retrieved on-site. To achieve this goal, the simple visualization of 3D surface models into a GIS platform is not sufficient. It is, in fact, necessary to define new data structures capable of interlinking, as much as possible, the information retrieved and documented on-site. Such systems would provide researchers and scholars with the opportunity to visualize data in a non-linear way, making the excavation process more reflexive and contextual (Berggren et al. 2015) and, at the same time, providing the opportunity to combine, in the same visualization system, the results retrieved by both specialists and excavators in the field and/or post-excavation.

One of the first archaeological projects developed in this direction is the prehistoric site of Paliambela Kolindros, Greece, where a data structure for archaeological documentation was designed to merge together aspects of a GIS platform with an extended use of 3D vectors and 2.5D raster surfaces (Katsianis et al. 2008). Despite the technological limitations at that time in managing more complex textured 3D models within GIS, this work proved how the combination of 3D data and a three dimensional GIS platform could allow researchers to efficiently use the system in support of archaeological investigation. Today, more recent developments of GIS software include new, powerful 3D visualization tools capable of importing and managing texturized 3D complex geometries. This important improvement has led to the initiation of a number of projects based on the use of a full three dimensional geo reference platform to function as a simulation environment in which all the information retrieved and documented on-site could be merged. Archaeological projects such as: "The Swedish Pompeii Project" (Leander Touati 2010; Dell'Unto et al. 2015), "3D Digging at Çatalhöyük" project (Forte 2014) "Gabii goes digital" project (Opitz and Johnson 2016), and the "Uppåkra project" (Dell'Unto 2014a) are examples of this. In the frame of these research activities, 3D surface models (as a result of laser scanning and/or image-based 3D reconstruction techniques) systematically produced in the framework of field investigations have been imported into a 3D GIS system and used in support of the investigation activity. As previously stated, the discussion thus far concerning the use of these new types of 3D data has been mainly focused on investigating the sustainability in using these new acquisition techniques in support of site documentation. In the future, we can expect a more diffuse and systematic use of these new typologies of 3D visualization systems to integrate the complete datasets of information produced and used by archaeologists and specialists to investigate the site.

3D GIS in Practice

The term 3D Geographic Information Systems refers to specific visualization platforms capable of displaying geo-references spatial data (such as raster or vector information) by means of a three dimensional visualization package. These systems allow for operations such as visualization, editing, and analysis. For this reason, when referring to 3D GIS, it necessary to clarify that these platforms do not represent a novelty per se.

Instead, what must be considered as an important development is the recent possibility of these systems to import and manage high resolute, textured, surfaced 3D models resulting from a field acquisition campaign performed by means of instruments such as laser scanning and image-based 3D reconstruction techniques. The majority of the projects currently presented in the literature concerning the use of this new version of 3D GIS platforms for the visualization and analysis of archaeological sites have mainly advocated using ESRI ArcGIS. The development of the 3D Analyst extension makes it possible to use this product to directly import, visualize, and analyse 3D surface models together with more traditional data sets. This new tool enables taking advantage of a powerful, fast rendering speed three-dimensional platform (ESRI 2010) to explore, in full 3D, all of the data acquired in the field.

Another important aspect is the fact that ArcGIS can be easily implemented and used in the field without any IT-support due to a relatively user friendly interface. To be visualized, 3D models resulting from field surveys are transformed in multipatch files (a format specifically designed for visualization of the boundary representation of 3D objects) (ESRI 2012). Once imported into the geodatabase, the multipatch files can be visualized in spatial relation with all of the rest of the data. Despite the multiple possibilities currently available for managing and visualizing 3D data in user friendly platforms, traditional GIS visualization systems (due to their large diffusion and use in archaeology) rely on a very large community of developers and can benefit from tools already developed and largely employed in archaeology. Moreover, it is important to highlight that open source GIS platforms, such as GVSIG (http://www.gvsig.com/en), are currently developing tools for the management and visualization of 3D textured surface models and that their use has, to a minor extent, been tested in the field.

In the last few years, the Digital Arkeology Laboratory (DARK Lab) at Lund University has been actively participating in several field investigation projects focused on exploring how the systematic use of this new visualization approach could impact archaeological practice in a variety of different archaeological contexts. After the initial experimentation phase, where 3D models were generated and used in support of a more traditional documentation method, it was clear that these new type of data, if used in spatial relation with larger and different typologies of digital information, could have been used to simulate, in three dimensions, the temporal sequence of actions performed by the archaeologists during the entire investigation process, creating the conditions for reviewing every step of the investigation and allowing the detection of archaeological evidence that would have been impossible to notice otherwise. The case studies presented in the following paragraph have been selected to describe our experience in using 3D GIS in support of archaeological practice. Specifically, a discussion concerning how this new typology of data has been used to enhance the field experience and to retrieve new archaeological evidence will be presented.

Experiences in Contextualizing 3D Models

One of the first tests that realized contextualizing 3D models was conducted in the archaeological site of Uppåkra, Sweden. Since 2009, an international team of archaeologists has been testing the use of these methods in field investigation activities organized by the Department of Archaeology and Ancient History at Lund University. The archaeological site of Uppåkra is considered one of the most important Iron Age central places in Sweden, with a temporal continuity spanning

1000 years. The site is located 5 km south of Lund and has an extension of approximately 100 acres. The site, which was discovered in 1934, has been the subject of archaeological investigations since 1996 (Larsson 2007).

After an initial phase of experiments based on the employment of different instruments and methods to test the use of 3D resolute surveyed 3D models (Dellepiane et al. 2013), the team began systematically employing image-based 3D reconstruction techniques to document the on-going field investigation. The results of the first experimentation allowed for testing the use of 3D surface models as a reference for reviewing the different steps conducted during the investigation activity. The possibility to use these data in the field proved the capacity of 3D models to enhance the archaeological field experience, positively effecting the decisions made by archaeologists in the field (Callieri et al. 2011). Despite the important results achieved in the first phase of this work, the difficulties in visualizing and using this new typology of data in spatial relation with the rest of the field documentation represented a strong limitation. The most "obvious" solution in this case would have been the direct implementation of the 3D models into the Geographic Information System (GIS) that was used to manage the rest of the data detected and recorded on-site. This opportunity did not come until the release of ArcGIS 10.0, when this system was used to import and manage all of the information previously recorded as shapefiles together with the geo-referenced 3D textured models realized in the field.

Since 2012, textured 3D surface models have been systematically imported into a 3D Geographic Information System (GIS) and used to study the relations among the different strata detected during the investigation process in Uppåkra (Dell'Unto 2014a, b). In the frame of this case study, it was also possible to explore the potential of using the 3D editing tool available in ArcGIS to realize graphic documentation by using the 3D models as a geometrical reference. The use of this system allowed for generating 3D drawings to map the different contexts retrieved on-site. With this method, it has been possible to visualize the graphic description of any trench superimposed with the virtual replica of the site in the same visualization system (Fig. 3).

Another project where a similar methodology has been tested is the "3D-Digging at Çatalhöyük" research activity, which was developed in the frame of the Çatalhöyük research project (Forte 2014). The project started in 2009 at the archaeological site of Çatalhöyük, Turkey and can be considered as an on-site digital experiment focused on studying every phase of field archaeological excavation by means of 3D technology (from the acquisition to visualization) (Forte et al. 2012; Forte 2014). Among other visualization systems tested in the frame of this work, since 2012, a 3D GIS platform has been used by the team to manage all of the information retrieved on-site (Fig. 4).

These two projects provided clear indications concerning the way of merging this typology of data with more traditional archaeological datasets of information and eventually led to defining attributes capable of describing and classifying the different elements displayed in the 3D models. Once imported into the geodatabase, the 3D models can be connected to an attribute table and used as a standard

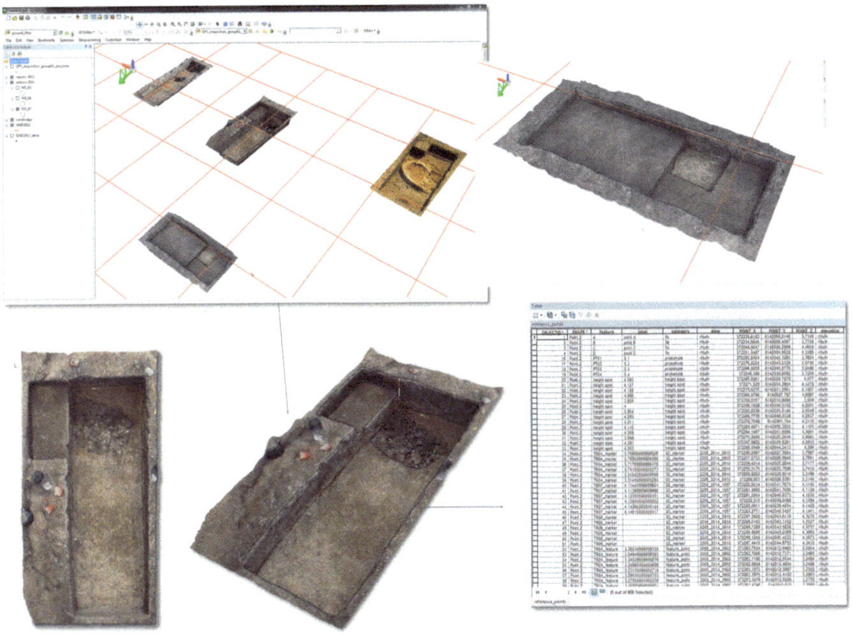

Fig. 3 Example of 3D graphic drawing, realized in ArcScene, using a 3D georeferenced surface model as a geometrical reference. The archaeological feature has been detected in the archaeological site of Uppåkra, Sweden, during the 2014 field investigation season at the time of the excavation seminar led by the Department of Archaeology and Ancient History, Lund University

shapefile to perform spatial analysis or to build 3D representations of the site as result of a specific query. The correct definition of specific attributes to describe the elements displayed for each model could, in the future, provide archaeologists with the possibility to query the 3D GIS to generate virtual simulations of the ancient space as a result of specific archaeological research questions.

The Use of 3D Geographic Information Systems in Support of Archaeological Interpretation

The combination of 3D GIS platforms and 3D surface models represents a powerful tool for analysing and detecting new data to use in support of archaeological interpretations. For example, the possibility to visualize, in the same three dimensional environment, data representing contexts and features retrieved and removed during different investigation campaigns allows for reviewing aspects of the site that would have been impossible to discover otherwise.

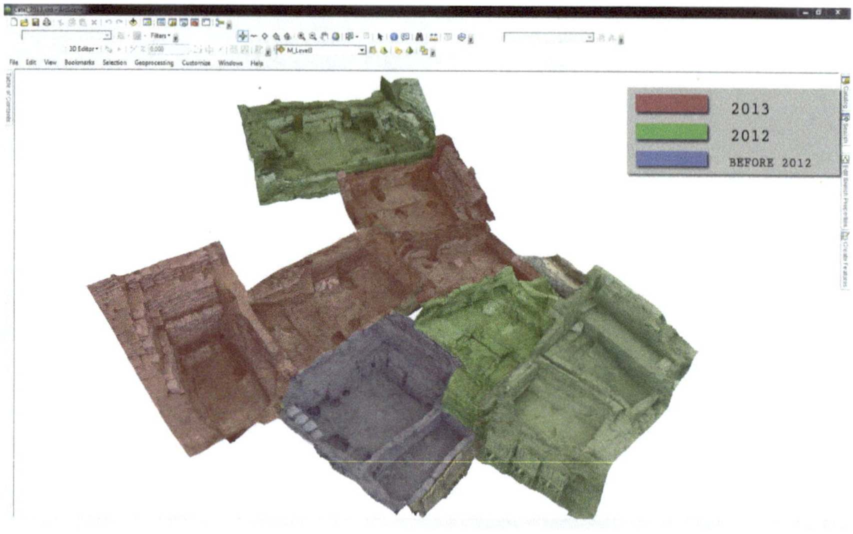

Fig. 4 3D buildings acquired by means of image-based 3D reconstruction techniques recorded at the archaeological site of Catalhöyuk, Turkey, in occasion of different field campaigns. Once generated, the 3D models were implemented and visualized into the 3D GIS platform (ArcScene)

An interesting example can be seen in the Swedish Pompeii project. This research was initiated in 2000 at the Swedish Institute in Rome and is currently directed by Professor Anne Marie Leander Touati, Lund University. This work has been focused on the documentation and study of the insula V, 1 in Pompeii, Italy. The overall aim of the project is to investigate the relation between the different types of entities that characterize Pompeian domestic architecture to comprehend the development of Pompeian building and social history (Leander Touati 2010; Staub 2009). Since 2011, a project focused on the digital acquisition of the structures of insula V 1 by means of integrated 3D spatial technologies has been developed (Dell'Unto et al. 2013a, b). The results of this work have been used to generate a high-resolution model of the entire city block for use in support of the investigation activities in accomplishing the project aims.

Recently, part of the 3D buildings, acquired in the frame of this work, have been georeferenced and imported into a 3D GIS platform and then connected, through a data model specifically designed for this project, with all of the documentation so far retrieved during the site investigation. The use of this platform resulted in the very efficient retrieval of new information concerning the relation among the different structures that composed the site (Fig. 5).

Specifically, the possibility of using any type of device (desktop computer or tablet PC) to draw, directly on the models, the features detected on the structures (using the 3D textured models as a geometrical reference) made it possible to experience a completely new approach. For example, once realized, the 3D polygons and polylines can be used as instruments to assess the geometrical correspondence between features belonging to different sides of the same wall. Another important issue is represented by the 3D measuring tools provided by the software, the use of which makes it possible to easily retrieve linear distances between different surfaces, to measure areas and the length of any surface, and to obtain the global elevation of any element visualized into the system (Fig. 6) (Dell'Unto et al. 2015).

The recent improvements in the visibility analysis toolkit in ArcGIS made it possible to perform quantitative and qualitative analysis based on the interpolation of different types of 3D information to estimate the level of visibility of different categories of objects (i.e., alphabetic inscriptions and wall paintings) retrieved in the house of Caecilius Iucundus (Fig. 7). This approach (although currently under development) provides important information concerning the activities that took place in the house and the way the space was used in ancient Pompeian houses (Landeschi et al. 2015). Despite only part of the insula having been implemented in the 3D GIS, the results obtained so far have encouraged the team to complete the implementation and adopt this system as the main visualization platform.

Another important aspect in using 3D GIS platforms in support of archaeological investigation is the possibility to merge and visualize, in three dimensions, data retrieved by the specialists during post-excavation analysis together with the virtual contexts documented on-site and visualized in the 3D system. An experiment in this direction has been conducted on the Baltic island of Öland, Sweden at the archaeological site of Sandby Borg (http://www.sandbyborg.se), which comprises a

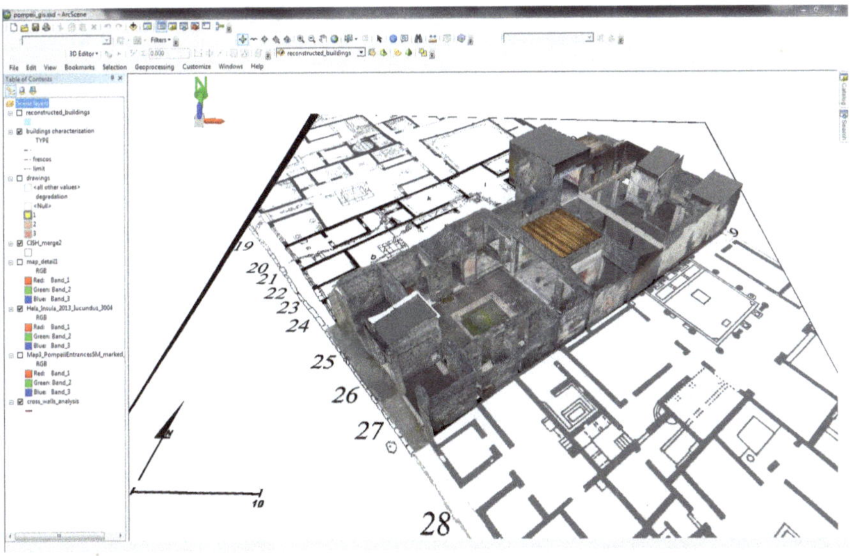

Fig. 5 3D model of the southern part of the house of Caecilius Iucundus, insula V 1, imported and visualized with ArcScene and superimpose to the map of the insula realized in 2005 by Ezequiel Pinto-Guillaume

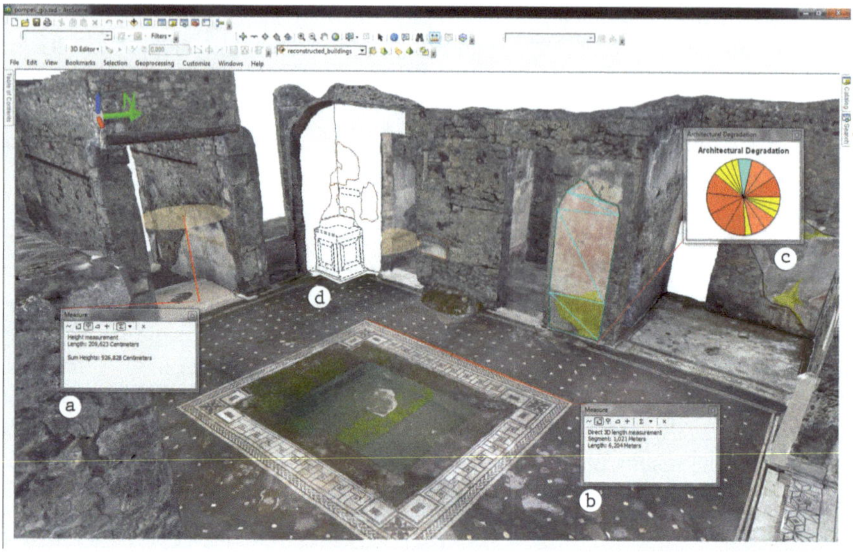

Fig. 6 Example of the different tools available in Arcscene for3D editing and measuring. *a* Height measurement tool, *b* direct 3D length measurement tool, *c* generation of graphs for the quantitative assessment of the architectural degradation of the structures, *d* 3D drawing performed using the 3D models as geometrical reference

ring fort located on the east coast of the island and dated to the late fifth century (460–490 AD) (Viberg et al. 2012). The site has been under investigation by a team led by the Department of Museum Archaeology at Kalmar County Museum, Sweden since 2011. During the investigation campaigns in 2012 and 2013, an experiment was developed by Lund University to test and study the impact a field documentation method based on the combination of image-based 3D reconstruction and 3D GIS platforms could have on the archaeological interpretation of the site. This approach was specifically directed at simulating and investigating the spatial relations among human skeletal remains retrieved on-site and their integration with the archaeological context. A particular focus was placed on the possibilities to reconstruct in three dimensions the field investigation process and study two individuals identified with a very close spatial relation but who were excavated and documented during different field campaigns. To achieve a deeper understanding of the sequence of events forming this unusual context with two individuals lying in a house, it was crucial to "re-compose" the scene by virtually relocating the two bodies in the building by merging, into the 3D system, the data retrieved in laboratory during the post-excavation activities (Fig. 8).

To implement these data into the system, a Geo database management system (Geo-DBMS) was designed by using ESRI ArcGIS. The specific observations detected by the osteologist during the field and lab analysis have been edited

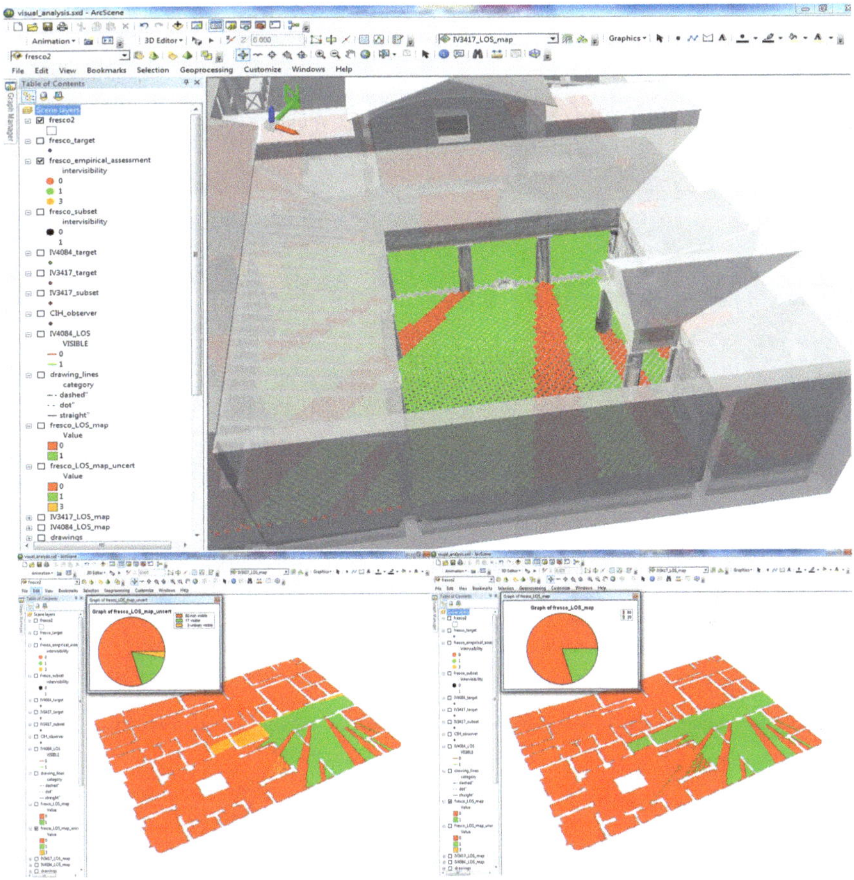

Fig. 7 Use of the visual analysis tool to calculate the visibility impact of the electoral and alphabetic inscriptions retrieved in the house of Caecilius Iucundus

directly on the three dimensional models by using 3D polylines and 3D polygons. These were linked to an attribute table specifically designed to host the information retrieved in the laboratory, including the fracture patterns and anatomical representation (Fig. 9) (Wilhelmson and Dell'Unto 2015).

The possibility to integrate and re-interpret the data directly into the 3D GIS considerably facilitated the opportunity to view the results integrated in the physical archaeological context and to use the system to retrieve new data as a result of specific queries. The results of the 3D spatial analysis performed with the 3D GIS indicated clear connections among the fractures, spatial location where the bodies were found and archaeological structures (elements of the house). The visualization of these relations and the possibility to highlight specific information as a result of a spatial queries provided new ways of viewing the osteological material, both adding

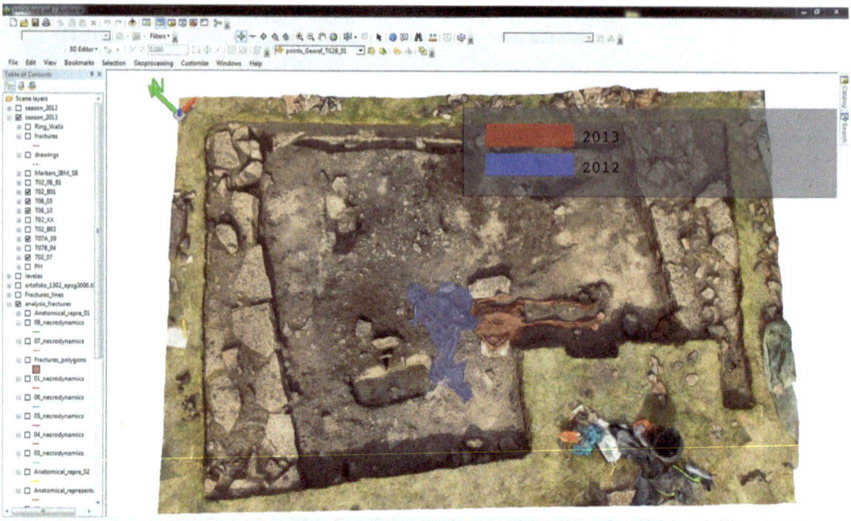

Fig. 8 3D simulation of the archaeological site of Sandby Borg, Sweden realized in ArcScene. In the image the two bodies, excavated during seasons (2012 and 2013), are visualized together in situ inside the house. 3D models realized by Lund University. Excavation led by the Department of Museum Archaeology (Kalmar County Museum)

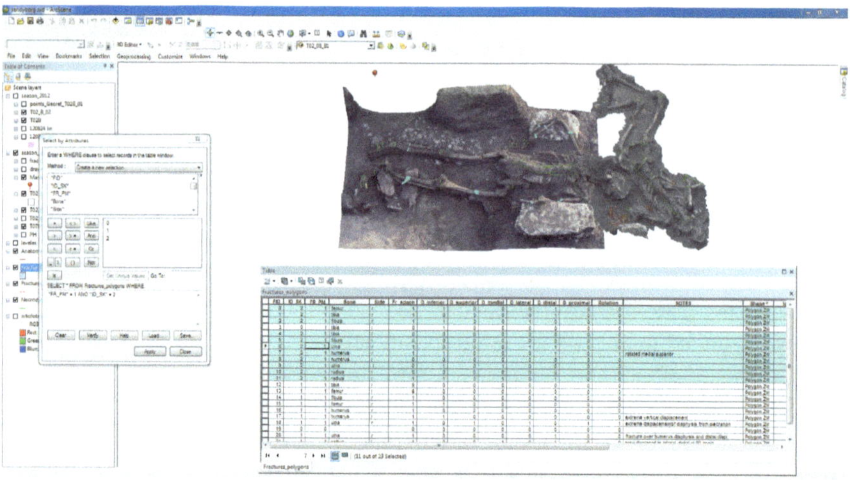

Fig. 9 Integrated GIS database and a query based on an attribute table related to osteological observations made both in the excavation and post-processing of the bones. For example, it is possible to, via the query function, highlight, and show the spatial distribution, in situ in the archaeological context in 3D, of specific types of fractures present on the bones of one or both individuals. This approach makes it possible to execute multiple independent and highly detailed spatial analyses directly in the 3D models. Three-dimensional models and osteological analysis realized by Lund University. Excavation led by the Department of Museum Archaeology (Kalmar County Museum)

great value to the analysis, yielding previously unattainable information, and playing a crucial role in the interpretation of the bones, as well as the archaeological context (Wilhelmson and Dell'Unto 2015).

Conclusions

Three-dimensional Geographic Information Systems represent an interesting option for managing on-going investigation activities, supporting researchers in building more accurate archaeological interpretations both in the field and in post-processing. These systems do not require any specific interaction nor visualization device. However by using a tablet PC, 3D GISs can be employed during the field activities to collect and manage (in real time) the entire documentation process, or to build on site 3D simulations capable of combining and visualizing contexts retrieved at different stages of the field investigation. By using tablet PCs to produce and manage the three dimensional model it is possible to keep a direct contact with the contexts retrieved on site. This approach provides archaeologists with the opportunity to validate the 3D documentation directly in the field, using the original layer as main source of reference. From a methodological perspective, this represents an important achievement because it allows keeping a high control on the quality of 3D information produced on site.

These systems, with no doubt, are highly affecting archaeological practice, in particular by providing archaeologists with the possibility of being more reflexive and by adopting more dynamic approaches in terms of data management.

The extent of their use allows for achieving (i) the construction of a more accurate and detailed documentation of the archaeological process, (ii) visualization of the results of analysis coming from different specialists working on site, and (iii) the opportunity to record and visualize all of the data detected and removed during different sequences of the excavation.

These systems can be customized to host and cross match data coming from new acquisition instruments, providing archaeologists with the opportunity to gain a more complete overview of the dataset of information retrieved. Three-dimensional GIS proved so far to be a very functional simulation platform. However, the development of more customized tools for 3D visualization and analysis needs to be undertaken. In specific, more efficient and friendly tools of mesh editing could greatly support archaeologists in simulating and testing different interpretations. Moreover, the future implementation of these systems as web platforms will possibly allow gathering more complete and accurate representations of the past; by connecting a larger spectrum of archaeological data coming from different typologies of analysis. As of now, the work has highlighted the importance and urgency to start theorizing and testing new typologies of investigation approaches capable of defining new and more efficient ways to collect, combine and analyse information as a result of more accurate and descriptive typologies of data acquisition workflows.

Acknowledgments Thanks to the Swedish Pompeii project, particularly project director Prof. Anne Marie Leander Touati, Dr. Giacomo Landeschi and Arch. Danilo Marco Campanaro from Lund University, the Department of Museum Archaeology at Kalmar County Museum, and, specifically, project director Dr. Helena Victor (from Kalmar County Museum) for allowing the author to participate in the excavation of Sandby Borg and to study and collect the material during the field campaigns of 2012–2013 and to Helene Wilhelmson, for the feedback concerning the osteological material. Thanks to the Uppåkra project team, the Catalhöyuk research project and to the 3D Digging at Çatalhöyük project for allowing the author to participate in the excavation and to study and collect the material. This research activity was generously supported by the Birgit och Sven Håkan Ohlssons Foundation, the Swedish Research Council Grant (340-2012-5751) Archaeological information in the digital society (ARKDIS), the Humanities Lab, Lund University and the DARKLab, Lund University.

References

ESRI. 2010. What's new in ArcGIS 3D Analyst 10,. Resource document. http://help.arcgis.com/en/arcgisdesktop/10.0/help/index.html#//00qp0000000z000000.htm Accessed 20 February 2014

ESRI (2012). The Multipatch geometry type—an ESRI white paper. December 2008. Resource document. http://www.esri.com/library/whitepapers/pdfs/multipatch-geometry-type.pdf. Accessed 20 March 2014.

Allen, Kathleen M. S. Green, Stanton W, and Zubrow, Ezra B. W. 1990. *Interpreting space: GIS and archaeology*. London: Taylor and Francis

Agugiaro Giorgio, Remondino Fabio, Girardi, Gabrio., Von Schwerin Jennifer., Richards-Rissetto Heather, De Amicis, Raffaele. 2011. QueryArch3D: querying and visualising three-dimensional archaeological models in a web-based interface. Geoinformatics, Faculty of Civil Engineering, Czech Technical University. Resource document. http://3dom.fbk.eu/sites/3dom.fbk.eu/files/pdf/Agugiaro_etal_CIPA2011.pdf. Accessed 20 March 2014.

Agugiaro Giorgio and Remondino Fabio. 2014. 3D GIS for Cultural Heritage Sites: The QueryArch3D Prototype. *In 3D surveying and modeling in archaeology and cultural heritage Theory and best practice*s, ed. Campana Stefano. and Remondino Fabio, 145-150. BAR international series, Oxford

Berggren, Åsa, Dell'Unto Nicoló, Forte Maurizio, Haddow Scott, Hodder Ian, Issavi Justine, Lercari Nicola, Mazzuccato Camilla, Mickel Allison and Taylor James. 2015. Revisiting reflexive archaeology at Çatalhöyük: integrating digital and 3D technologies at the trowel's edge. *Antiquity*, 89:433-448.

Callieri Marco, Dell Unto Nicoló, Dellepiane Matteo, Scopigno Roberto, Soderberg Bengt and Larsson, Lars. 2011. Documentation and interpretation of an archeological excavation: an experience with dense stereo reconstruction tools. In VAST The 11th International Symposium on Virtual Reality Archaeology and Cultural Heritage, 33–40. Eurographics.

Chapman Henry. 2006. *Landscape archaeology and GIS*. Stroud: Tempus.

Conolly James and Lake Mark. 2006. *Geographical information systems in archaeology (Cambridge manuals in archaeology)*. United Kingdom: Cambridge University Press.

Dellepiane Matteo, Dell'Unto Nicolo, Callieri Marco, Lindgren Stefan. and Scopigno Roberto. 2013. Archeological excavation monitoring using dense stereo matching techniques. *Journal of Cultural Heritage. Elsevier*

Dell'Unto Nicoló, Callieri Marco, Dellepiane Matteo, Leander Touati Anne Marie, Lindgren Stefan & Larsson Carolina. (2013a). Pompeii revived: Scanning Mission- insula V 1. I Earl, G., Sly, T., Chrysanthi, A., Murrieta-Flores, P., Papadopoulos, C., Romanowska, I. & Wheatley, D. (Red.) Archaeology in the Digital Era. Papers from the 40th Annual Conference of

Computer Applications and Quantitative Methods in Archaeology (CAA), Southampton, 26-29 March 2012 (pp. 199-207). Amsterdam University Press.

Dell'Unto Nicoló, Ferdani Daniele, Leander Touati, Anne Marie. Dellepiane Matteo, Callieri Marco. and Lindgren Stefan. 2013b. Digital reconstruction and visualization in archaeology Case-study drawn from the work of the Swedish Pompeii Project, ed.Addison, A., Guidi, G., De Luca, L. and Pescarin, S. Digital Heritage International Congress 621–628

Dell'Unto Nicoló. 2014a. The Use of 3D Models for Intra-Site Investigation in Archaeology. *In 3D surveying and modeling in archaeology and cultural heritage Theory and best practices*, ed. Campana Stefano. and Remondino Fabio, 151-158. BAR international series, Oxford

Dell'Unto, Nicoló. 2014b. 3D models and archaeological investigation. *In Perspectives to archaeological information in the digital society, ed.* Huvilla Isto, 55-71. Uppsala University, Department of ALM.

Dell'Unto Nicoló, Landeschi Giacomo, Leander Touati Anne Marie, Dellepiane Matteo, Callieri Marco. and Ferdani Daniele. 2015. Experiencing Ancient Buildings from a 3D GIS Perspective: a Case Drawn from the Swedish Pompeii Project. *Journal of archaeological method and theory*. DOI 10.1007/s10816-014-9226-7

De Reu Jeroen., Plets Gertjan., Verhoeven Geert, De Smedt Philippe, Bats Michel., Cherretté Bart., De Maeyer Wouter., Deconynck Jasper., Herremans Davy., Laloo Pieter., Van Meirvenne Marc. and De Clercq Wim., 2013. Towards a three-dimensional cost effective registration of the archaeological heritage. *Journal of Archaeological Science*. 40: 1108–1121

Doneus Michael and Neubauer Wolfgang. 2005. Laser scanners for 3D documentation of stratigraphic excavations. International workshop on recording, modeling and visualization of cultural heritage, Ascona, May 22–27, Taylor and Francis 2005.

Forte Maurizio, Dell'Unto Nicoló, Issavi Justine., Onsurez, L. and Lercari, Nicola. 2012. 3D Archaeology at Çatalhöyük. *International Journal of Heritage in the Digital Era*, 1: 351–377.

Forte Maurizio. 2014. 3D Archaeology: New Perspectives and Challenges—The Example of Çatalhöyük. *Journal of Eastern Mediterranean Archaeology and Heritage Studies*.1:1-29

Forte, M., Dell'Unto, N., Jonsson, K. & Lercari, N. 2015. Interpretation Process at Çatalhöyük using 3D. I Hodder, I. & Marciniak, A. (Red.) Assembling Çatalhöyük, 1. European Association of Archaeologists Contemporary Themes in Archaeology.

Landeschi Giacomo, Dell'Unto Nicoló, Ferdani Daniele, Leander Touati Anne Marie and Lindgren Stefan. 2015. Enhanced 3D-GIS: Documenting Insula V 1 in Pompeii.in Pompeii, Proceedings of the 42nd Conference on Computer Applications and Quantitative Methods in Archaeology – Paris 22nd to 25th April

Larsson Lars. 2007. The iron age ritual building at Uppåkra, southern Sweden, *Antiquity* 81: 11-25.

Lock Gary and Stancic Zoran.1995. *The impact of GIS on archaeology: a European perspective*. New York,Taylor and Francis

Lock Gary. 2001. Theorising the practice or practising the theory: archaeology and GIS. *Archaeologia Polona*, 39, 153–64.

Lock Gary. 2003. *Using computers in archaeology: towards virtual pasts*. Routledge

Losier, L.,M., Pouliot, J., and Fortin, M. 2007. 3D geometrical modeling of excavation units at the archaeological site of Tell 'Acharneh (Syria). Journal of Archaeological Science 34, (pp. 272-288).

Katsianis Markos, Spyros Tsipidis, Kotsakisa Kostas and Kousoulakoub Alexandra. 2008. A 3D digital workflow for archaeological intra-site research using GIS. *Journal of Archaeological* Science, 35: 655–667.

Leander Touati Anne Marie. 2010. Water, well-being and social complexity in insula V 1 a pompeian city block revisited. *Opuscula*, 3: 105–161.

Opitz Rachel and Nowlin Jessica. 2012. Photogrammetric modeling + GIS. Better methods for working with meshdata. ArcUser. Resource document. http://www.esri.com/news/arcuser/0312/files/archaeology-inventory. Accessed 20 March 2014.

Opitz Rachel, Johnson Tyler D., 2016. Interpretation at the Controller's Edge: Designing Graphical User Interfaces for the Digital Publication of the Excavations at Gabii (Italy). Open Archaeology. 2: 1–17

Staub Thomas. 2009. Decorative effects and room functions. Evidence of thresholds studied in the residential quarters of insula V 1, Pompeii. *Opuscula* 2: 205–218.

Quartermaine Jamie, Olson Brandon R. and. Killebrew Ann E. 2014. Image-Based Modeling Approaches to 2D and 3D Digital Drafting in Archaeology at Tel Akko and Qasrin: Two Case Studies. *Journal of Eastern Mediterranean Archaeology and Heritage Studies.* 2:110-127

Viberg Andreas, Victor Helena, Fischer Svante, Liden Krestin, Andren Andreas. 2012. A room with a view: archaeological geophysical prospection and excavations at Sandby ringfort, Öland, Sweden. *In Remnant echoes of the past: archaeological geophysical prospection in Sweden, ed.* Viberg Andreas, 1-10. Stockholm University.

Wheatley David and Gillings Mark. 2002. Spatial technology and archaeology: the archaeological applications of GIS. London: Taylor and Francis.

Wilhelmson Helene and Dell'Unto Nicoló. 2015. Virtual Taphonomy: a New Method Integrating Excavation and Post-processing of Human Remains. *American Journal of Physical Anthropology*. Doi:10.1002/ajpa.22715

Archaeology in the Age of Supercomputing

Devin A. White

Abstract As a discipline, archaeology is poised to fully embrace both the power and the peril of big data analysis. Our datasets are growing ever larger, especially those generated via remote sensing and geospatial processing activities, as is the computational complexity of algorithms designed to exploit them. Analyses are quickly outpacing what can be done using a single processing core on a desktop computer, leveraging off-the-shelf commercial and open source software. Our research needs are becoming increasingly sophisticated, to the point where relying wholly on outside experts in computer science and related fields is untenable. While the above statements could be viewed primarily as challenges, it is better to think of them as opportunities for archaeology to grow technologically and retain more ownership of our hardest problems. High performance computing, i.e., supercomputing, is already having an impact on the field, but we are moving into an era that promises to put the power of the world's largest and fastest computers at archaeologists' fingertips. What does the state of the art look like? How could we use the coming power? What lines of inquiry and analysis could we pursue once long-standing technical limitations have been removed? This chapter will focus on the present and the future of archaeological high performance computing, using several ongoing projects across a broad swath of the discipline as examples of where we are now and signposts for where we are heading.

Introduction

As a discipline, archaeology is poised to fully embrace both the power and the peril of big data analysis. Our datasets are growing ever larger, especially those generated via remote sensing and geospatial processing activities, as is the computational

D.A. White (✉)
Geographic Data Sciences, Geographic Information Science and Technology Group, Oak Ridge National Laboratory, Oak Ridge, TN, USA
e-mail: whiteda1@ornl.gov

D.A. White
Department of Anthropology, University of Tennessee, Knoxville, TN, USA

M. Forte and S. Campana (eds.), *Digital Methods and Remote Sensing in Archaeology*, Quantitative Methods in the Humanities and Social Sciences,
DOI 10.1007/978-3-319-40658-9_15

complexity of algorithms designed to exploit them. Analyses are quickly outpacing what can be done using a single processing core on a desktop computer, leveraging off-the-shelf commercial and open source software. Our research needs are becoming increasingly sophisticated, to the point where relying wholly on outside experts in computer science and related fields is untenable. While the above statements could be viewed primarily as challenges, it is better to think of them as opportunities for archaeology to grow technologically and retain more ownership of our hardest problems. High performance computing, i.e., supercomputing, is already having an impact on the field, but we are moving into an era that promises to put the power of the world's largest and fastest computers at archaeologists' fingertips. What does the state of the art look like? How could we use the power already available, much less what is coming next? What lines of inquiry and analysis could we pursue once long-standing technical limitations have been removed? How will fieldwork be transformed?

This chapter, after an introduction to the world of high performance computing, will focus on the present and the future of archaeological supercomputing, using several ongoing projects across a broad swath of the discipline as examples of where we are now and signposts for where we are heading, concluding with some thoughts on the art of the possible, given current and emerging technological trends. It is hoped that the reader will come away feeling less intimidated by the idea of using supercomputing to solve archaeological problems, and knowing that they can and should take full advantage of the computing power available today as well as help drive how the systems of tomorrow are designed.

The Current State of the Art for High Performance Computing

"Supercomputing" is the widely-recognized name for what has now become a much more varied computational landscape, driven largely by the commoditization, democratization, and more recently, miniaturization of the necessary hardware and software. The three trends have created, and will continue to create, opportunities for archaeologists to generate, process, analyze, visualize, and contextualize data more quickly, accurately, and creatively. To do so most effectively, and to better imagine what could come next, it is important to appreciate the current state of the art and to understand that there are resources available to scientists at multiple levels, some of which are relatively straightforward to access and leverage. This section will serve as a relatively non-technical primer on high performance computing (the more proper term), focusing on the definition and function of a broad set of technologies archaeologists are already using, would like to use, or may not even know exist. To that end, it lays the foundation and establishes a frame of reference for the illustrative use cases highlighted below and sets the stage for the subsequent discussion on the art of the possible.

The most fundamental concept associated with high performance computing (HPC) is that of *parallelism.* In its simplest form, parallelism is the process by which a large problem is broken up into smaller, more manageable pieces and distributed to a large number of computers (hereafter referred to as nodes) that can each work on their assigned piece independently. The two most common problems encountered are (1) single case, where one larger set of calculations is run to produce a single result, and (2) ensemble case, where multiple smaller sets of calculations are run to produce multiple results. The latter is generally used when exploratory research is required to understand the full behavior of an algorithm or model, so many variations on inputs are supplied and the output is analyzed statistically to derive meaningful trends. When all of the pieces have been worked on, the individual results are brought together to form the final output. This is the ideal case, where little to no communication between nodes is required. There are far more complicated cases, which is more of the norm, where nodes must update each other on what they are doing and exchange data. There are several different ways, with respect to both hardware and software, that parallelism is "expressed" in HPC systems, and they are not always mutually exclusive. Each one, discussed below, was born of a different computational need and thus will differentially apply to a problem of interest—and may even be combined when it is advantageous to do so. A scientist working within the HPC domain is primarily trying to figure out how to break up a problem so that it can run most efficiently on the hardware and software they have available (square peg in round hole) or trying to seek out the right hardware and software to meet the requirements of their current solution (square peg in square hole). Both avenues create challenges and opportunities. It is best to begin with an overview of hardware, after which software will be discussed.

HPC Hardware

Modern computers, with rare exceptions, have a central processing unit (CPU) that is made up of multiple cores. Each core is capable of executing a computational task independently of the others, but can do so within the same workspace (system memory), enabling each of them to easily talk to one another and exchange data while they are working on a common problem. In this tightly-coupled arrangement, sometimes referred to as *strong parallelism*, each core can be thought of as a very efficient node. Sadly, most of today's desktop software—especially the applications focused on image processing and geospatial data analysis—is not written to take advantage of multiple cores despite their widespread availability, but the situation is slowly improving and archaeologists are beginning to benefit from the more rapid data analysis available through tools they are already using. The size of strongly parallel solutions, which tend to work best for problems that require a significant amount of communication between nodes and/or for inputted data to remain whole throughout the process, are most often limited by the number of cores and the amount of available memory on the system. While these numbers are always increasing (for example, this manuscript was written on a system with 32 cores and 512 GB of RAM, which is close to the high end at present), they cannot keep pace with the exponentially-growing size of big data problems. Larger problems are handled one of two ways: (1) build a larger tightly-coupled system or (2) find a way to use a more loosely-coupled approach.

Larger systems, known as Symmetric Multiprocessing (SMP) solutions, can be thought of as gigantic workstations. They generally have hundreds or thousands of cores that all have access to a large pool of shared memory. SMPs are purpose-built and not very common, but for certain classes of problems that simply cannot be solved in a loosely-coupled way, or where doing so would be prohibitively expensive, they are the best option. For scientists who have access to such systems, they are also a great way to test the *scalability* of an existing tightly-coupled approach, i.e., the efficiency of it when a much larger amount of resources are available and used. In some cases, simply throwing a bigger system at the problem does not help, which is when the overall approach has to be rethought.

A special case of strong parallelism is *massive parallelism*, which is more commonly referred to as hardware acceleration. Massively parallel arrangements are the domain of General Purpose Graphics Processing Units (GPGPUs) and more specialized devices called coprocessors. Both, but primarily GPGPUs, have revolutionized HPC over the past decade. Originally designed to rapidly process data for display on individual computers, in particular to support the video game industry, they have been repurposed, or even purpose-built, to instead execute mathematical operations of interest to scientists. The highest end GPGPU available today, the NVIDIA Tesla K80, has the equivalent of 4992 cores, which is on par with what one finds in SMP solutions, but the required hardware is the size of a small book (Fig. 1). The smaller size, and its design legacy, imposes some very significant restrictions on the type of problems it can solve. First, it is not a computer in its own right—it has to be connected to a traditional computer, but more than one can generally be connected to the same computer (an important distinction that will be revisited below). Second, it is designed to execute a massive number of very simple calculations that don't depend on one another, which is what is needed for data display. As communication requirements gradually increase, the utility of a GPGPU, or even a coprocessor, rapidly decreases. Developing software that can effectively run on this kind of hardware is often a laborious, expensive, frustrating, and counter-intuitive process, but when done right, the resulting speedups are extremely impressive. One area where archaeologists are already benefiting from GPGPUs is in viewshed analysis, exactly the kind of problem the hardware was designed to solve. You are in essence creating a three-dimensional environment and determining what can be seen from a particular vantage point, which is a crucial element of modern, immersive video games.

A loosely-coupled arrangement, sometimes referred to as *weak parallelism*, is traditionally associated with the notion of a *cluster*, which many readers are likely familiar with. A cluster is a collection of individual nodes that can act as a greater whole through software-based orchestration, but the connections between them are very tenuous and computationally expensive to use, and resources like memory are

Fig. 1 GPGPUs, and related hardware known as coprocessors, are small devices capable of quickly executing massive numbers of calculations, but they have to be connected to a regular computer to function. The NVIDIA Tesla K80 GPGPU, currently the fastest in the world, is pictured here. Image courtesy of NVIDIA Corporation

not shared. As such, they work best when communication is kept to a minimum. It might be surprising to the reader, but all of the leading supercomputers in the world, including Titan at Oak Ridge National Laboratory, employ this arrangement (Fig. 2). Titan is composed of 18,688 nodes (16 cores and 32 GB of RAM per node), each of which is capable of operating as an individual workstation. The modern science (and art) of HPC is centered on how to use a system like that to its fullest potential. Apart from ease of maintenance (nodes can be easily replaced), one of the main advantages of loose coupling is that the solution can more easily scale in a physical sense, i.e., nodes can be quickly added or removed as needed to meet the requirements of a specific problem. There are two common types of loose coupling arrangements, most often differentiated by whether or not the nodes exist in the same physical location. When that is the case, and every effort is made to accelerate communication between nodes through specialized hardware and software, the arrangement is *scientific*. When off-the-shelf hardware and software are used, or more importantly, the nodes are scattered across a wide geographic area and talk to one another over the internet, the arrangement is *commercial*, what we generally think of as "the cloud." The former (think Titan) is built for absolute speed, the latter (think Google) is built for more general purpose computing tasks like sending email and streaming movies. An apt comparison would be that of a sports car to a minivan: They each excel at certain tasks, and can each do what the other is good at, but not necessarily very well. They also have very different price tags, due to how they are designed and built. Examples of archaeologists using large clusters are few and far between, but that is beginning to change as they begin to explore large-scale modeling and simulation frameworks (scientific) and/or need to process massive quantities of data using off-the-shelf software (commercial).

An increasingly common approach to problem-solving in the HPC domain is to combine the different types of parallelism, leveraging each one for its strengths, an arrangement known as *hybrid parallelism*. For example, a single case problem can

Fig. 2 ORNL's Titan supercomputer, which employs weak parallelism to connect a large number of individual computers for complex problem-solving. This is the most common arrangement for today's high-end scientific systems. Only a small part of Titan, currently the second-fastest computer in the world, is pictured here

be broken down into several large chunks, each of which is passed to a strongly parallel node on a weakly parallel system, which then further breaks down the chunk so each of its CPU cores is working on a different sub-chunk, and some of those cores might be passing part of their sub-chunk to one or more locally-connected massively parallel GPGPUs for rapid analysis (because at this point the calculations are simple). A scientist that is able to successfully harness the available parallelism at all of these levels is akin to a symphony conductor. It should be noted that it is not necessary, or even advantageous, to do this for every problem. Archaeologists reconstructing three-dimensional scenes from large volumes of imagery have probably taken advantage of hybrid parallelism and did not even know it. Most desktop commercial photogrammetry software (e.g., PhotoScan and Pix4D) can take advantage of multiple CPU cores and multiple GPGPUs, if they are present, but the advantage does not extend beyond a single machine.

There are two additional hardware trends worth noting here, as they will be particularly useful for archaeologists: miniaturization and specialization. Within the past few years, innovative hardware manufacturers have found ways to shrink the size of the components required to make a functional computer, resulting in affordable units that are the size of candy bars (and smaller). The two most well-known examples at present are the Raspberry Pi and the NVIDIA Jetson. Both are smaller than a GPGPU and are full-fledged computers with no moving parts, running regular operating system software, capable of completing a wide range of tasks. It did not take long for curious scientists to find ways to link several of them together to create what is essentially a desktop supercomputer that runs exactly the same software as larger machines and, with some creativity, can harness hybrid

Fig. 3 ORNL's Tiny Titan supercomputer, built from nine Raspberry Pi computers, pictured in front of Titan for comparison. It can run many of the same applications as its larger cousin, albeit more slowly and at a smaller scale

parallelism to solve problems. One of these, dubbed Tiny Titan (Fig. 3), is used at ORNL as an instructional aid during visits by local schoolchildren, but it is capable of doing much more and cost very little to make. Imagine being able to take a small, lightweight, durable, cheap supercomputer with you to the field. At the other end of the spectrum are increasingly specialized machines that are designed to solve one type of problem extremely well, which in some ways brings HPC full circle to its early days, when custom-built computers were the norm. The most notable example at present is the "graph discovery" appliance built by Cray, which at every level, from basic design and construction to the software it runs, is focused on quickly analyzing extremely large networks, which has been a very difficult problem for traditional HPC systems to solve due to the amount of shared memory and communication required. Given that archaeologists are beginning to more seriously focus on social network analysis, and are rapidly running into limits with respect to the amount of data they can process, a specialized system like this could be very helpful.

HPC Software

Hardware serves as the physical foundation layer for HPC, but there is an equally important virtual foundation layer that is important to briefly review here: The software that governs how the various components of the system are exposed to the scientist, communicate with one another, and exchange data. As with the hardware discussion above, the review will begin with what can run on a single computer (node) and build upwards and outwards from there.

For strongly parallel systems, where CPU cores and shared memory are the main units of currency, there are several software frameworks available that make it relatively easy for a scientist to develop a parallel application. The most popular of them is OpenMP (http://openmp.org/wp/), but an alternative is Intel's Threading Building Blocks (https://www.threadingbuildingblocks.org/), also called TBB. They are both written in C/C++, work across multiple operating systems, and enjoy widespread support within the scientific HPC community. C/C++ is not often considered to be an approachable language, though, so it is important to note that similar functionality is available in many of the cross-platform languages that are popular with archaeologists, specifically Python, R, MATLAB, IDL, Java, and JavaScript. In many cases, all one has to do is add a small amount of new code to an existing application to get an instantaneous boost in performance. It may not be the largest possible boost, but it generally helps and allows you to better understand what your application is capable of. Some languages, like MATLAB and IDL, will automatically detect and use all available cores for certain tasks.

For massively parallel systems, where GPGPUs and coprocessors are the main units of currency, there are two main software frameworks available, both of which are becoming easier for scientists to use, but generally speaking require significantly more experience to leverage to their fullest potential. The most popular one by a large margin is CUDA (http://www.nvidia.com/object/cuda_home_new.html), which only works with GPGPUs manufactured by NVIDIA. The alternative, which works with all types of hardware accelerators, is OpenCL (https://www.khronos.org/opencl/). Both are written in C/C++ (CUDA also offers a Fortran version) and are relatively difficult to master. They each, like OpenMP and Threading Building Blocks, enjoy widespread use within the scientific HPC community—especially since many of the world's largest and fastest computers have at least one GPGPU or coprocessor attached to each node, as do miniaturized systems like Raspberry Pi and NVIDIA Jetson. They are also widely used in the computer vision and computational photogrammetry communities due to their unique ability to quickly process imagery. As a result, many desktop image processing packages used by archaeologists are already taking advantage of CUDA and/or OpenCL. It should be noted that many of the archaeologist-friendly languages mentioned above do provide methods for communicating with CUDA and OpenCL, but they are primarily limited to offering "faster" versions of specific, popular computational tasks (e.g., multiplying matrices, filtering images) if acceptable hardware is available on the computer. In other words, one gives up all control, but some results may come back faster than if no hardware acceleration was available.

For weakly parallel systems, where nodes are the main units of currency, there are also two main software frameworks available. One is specific to scientific systems and the other to commercial systems. Message Passing Interface, or MPI, is the standard way nodes communicate with one another on a scientific system, and there are several flavors available, including MPICH, MVAPICH, and Open MPI. They are all written in C/C++ and have moderate learning curves with respect to picking up the basics. MapReduce, specifically the free and open source implementation of it called Hadoop (https://hadoop.apache.org/), is hugely popular within the commercial HPC space. The paradigm was originally developed by Google, but quickly caught on with non-scientists because it is written in Java, runs on almost anything, and has a relatively shallow learning curve. The core concept behind the framework is that a single problem can be broken down into smaller pieces and distributed to several nodes (mapping), where the pieces are analyzed to generate a small number of meaningful values (reducing), which are then aggregated to create output. A very large community has built up around it, even on the academic side, but interestingly enough, Google has already moved on to something else. MPI and MapReduce each have strengths and weaknesses, and there are passionate supporters on both sides, but at a higher level, while they share the basic concept of passing information between nodes, MapReduce performs only a subset of what MPI is capable of—but it is a very useful subset for a wide variety of relatively simple problems that require quickly processing a large amount of data and for which it is possible for each node to work independently. It is possible to be very creative in using Hadoop to solve far more complex problems, but there is a lot of overhead, and some risk, involved in doing so. Unlike MPI, which is tied to a specific system, it is also capable of using networked nodes that are scattered across a wide geographic region (i.e., the cloud) and most commercial cloud vendors already support it. It is at a distinct disadvantage compared to MPI when it comes to hybrid parallelism, though. For example, a single node with multiple CPU cores is generally treated as multiple single-core nodes in MapReduce, greatly limiting the potential to break down a large problem even further or leverage the strengths of all available hardware. Again, archaeologist-friendly languages like Python and R provide ways for scientists to quickly and easily tackle big data problems on weakly parallel systems, especially in the area of statistical analysis, by leveraging MPI or Hadoop behind the scenes if either framework is present, but one loses a great deal of control over how that happens. Lower-level access, where one can more fully control how nodes communicate with one another, is also available, but it does require the programmer to do a lot more work to manage the entire process.

As a footnote to the above discussions on hardware and software, the line between weakly parallel scientific and commercial HPC systems has started to blur in the past few years. Scientific HPC researchers are finding ways to run Hadoop efficiently on their systems so that they do not have to rewrite useful software and commercial HPC providers are teaming up with those researchers to find ways to run MPI-based scientific applications in the cloud. There is a price to pay in terms of performance when going either of those directions, but the end results are that (1) software that is easier to write is becoming easier to run on large supercomputers that already exist, lowering the barrier to entry for archaeologists who might already have access to those kinds of systems through their home institutions, and (2) it is becoming possible to temporarily and inexpensively build a traditional scientific supercomputer whose nodes could exist all over the world, when the need arises.

At this point in the discussion, the reader's head is most likely spinning, given the wide array of available hardware and software options that vary substantially with respect to availability and usability. Table 1 provides a summary of the software paradigms mentioned above, organized by hardware paradigm, and also indicates which ones can be accessed through several archaeologist-friendly languages (and at what level). The take home message for the reader should be that there are several levels at which HPC can be applied to archaeological problems, and those levels can be initially explored easily through languages like Python and R. When more sophisticated control is required, or a problem outgrows locally-available computing resources, more complicated frameworks do exist that can help, but finding the right one(s) to use and exploiting them to their fullest extent may require teaming up with a computer scientist.

Available Resources Beyond the Desktop

There is a wide array of HPC hardware and software resources available to archaeologists, some of which might be free or very low cost to use—if one knows who to talk to and what questions to ask.

On the scientific side, the prime example is the National Science Foundation's Extreme Science and Engineering Discovery Environment (https://www.xsede.org/), also known as XSEDE. Scientists can apply for computing grants, ranging from small seed projects that test out ideas to large projects that might consume a significant amount of resources, and the program has several machines distributed around the country that specialize in different aspects of HPC. Before a large project grant is awarded, the scientists submitting it must document their application's performance on several systems of increasing size and complexity and thoroughly justify why one of the NSF's largest machines is required. No direct funding is awarded at any level. Instead, grantees are given time allocations that roughly equate to the number of hours a single CPU core can be used. So, for example, if a grant is awarded for 100,000 h on a system with 100,000 CPU cores (Titan has almost 300,000), a scientist could theoretically use up their entire allotment in an hour if they ran an application such that it requested and used all available resources. Generally speaking, multiple projects are running on large systems at the same time, so it is very rare to have access to all of it, but it is something that should be kept in mind.

Other countries with robust scientific research programs have similar grant initiatives, as do most universities. In almost all cases, both during the application process and after a grant is awarded, the scientists are paired with one or more technical support liaisons whose job may include translating algorithms so that they can run on the requested system—especially when the scientists come from a non-traditional computing discipline like archaeology. It is important to note that as of the time this chapter is being written, social science is still considered a novel focus area for scientific HPC, so there is great interest from that community and many opportunities to find support for little or no cost.

Table 1 A summary of HPC software paradigms, organized by hardware paradigm, with indicators for the level of support available through a selection of archaeologist-friendly languages

	Domain	Popularity	Language	Sophistication	Learning curve	Python	R	Java	JavaScript
Strong parallelism									
OpenMP	Scientific	High	C/C++	High	Moderate	High	Medium	High	Medium
TBB	Scientific	Low	C/C++	High	Steep	High	Medium	High	Medium
Massive parallelism									
CUDA	Scientific	High	C/C++	High	Steep	Low	Low	Low	Low
OpenCL	Scientific	Low	C/C++	High	Steep	Low	Low	Low	Low
Weak parallelism									
MPI	Scientific	High	C/C++	High	Moderate	Medium	Medium	Low	Low
Hadoop	Commercial	High	Java	Low	Shallow	Medium	Medium	High	High

The situation on the commercial side is more heterogeneous, given that the suppliers of available computing resources are focused on making a profit. The three largest are Amazon Elastic Compute Cloud (http://aws.amazon.com/ec2/), Microsoft Azure (http://azure.microsoft.com/en-us/), and Google Compute Engine (https://cloud.google.com/compute/). All of them, as noted above, support Hadoop, but support for MPI is minimal to nonexistent. For scientists working with satellite imagery and derived products, there are two additional options worth considering: Google Earth Engine (https://earthengine.google.org/), which is focused on analyzing coarse-grained data produced by sensors like Landsat and MODIS, and DigitalGlobe's Geospatial Big Data platform (https://www.digitalglobe.com/), which at present is focused on analyzing the fine-grained data produced by their own sensors. In both cases, the user can access data stored in the cloud and analyze it in place, which can be very fast and convenient. The simplest usage option in this for-profit environment is to pay for the compute time you need, but that can quickly deplete project funds, even when the expense is built into a traditional research grant—especially if you underestimate your project's resource requirements. Most companies have nonprofit arms that award time grants that are similar to those available on the scientific side of HPC, but the biggest differences between the two are that the awards tend to be a lot smaller and, more importantly, you are largely on your own to figure out how to use those resources to solve your problem.

Archaeological Supercomputing: Illustrative Use Cases

The goal of this section is to highlight several active research areas within archaeology that have benefitted from, or could definitely benefit from, the use of HPC hardware and software as described above. Specific projects will be discussed for each one to give the reader an idea of what has been attempted, where successes were achieved, and where challenges still remain. In other words, this is a tour of the state of art in archaeological supercomputing. To that end, the emphasis will be on what the researchers did, not what they specifically discovered. As a discipline, we are only beginning to take advantage of available resources, altering our thinking about the scale and scope of the questions we can ask and the problems we can solve. Once the tour is complete, this chapter will conclude with a discussion of the art of the possible, i.e., thoughts on where we can go from here.

Four areas will be highlighted here: landscape recording and reconstruction, terrain analysis, social network analysis, and complex adaptive systems. The examples of each discussed below do not, and cannot, cover the entire breadth and depth of archaeological supercomputing. They do, however, represent a reasonable cross-section of computationally intense problem solving and touch upon many topics of current interest to the discipline. One thread that ties all of them together is their geospatial focus. While that is likely not surprising to the reader, it is important to point out.

Landscape Recording and Reconstruction

The processing of capturing accurate, detailed three-dimensional information at the feature, site, and regional levels has undergone a revolution over the past decade. While traditional survey methods are still widely used, active scanning technologies like LIDAR and passive methods like photogrammetry-based Structure from Motion (SfM) are starting to play integral roles in most field projects (Opitz and Cowley 2013). Each is capable of producing massive numbers of precise point measurements, on the order of billions and trillions, that can be used to create models of the environment that can be analyzed and visualized in a wide variety of ways, a process that often involves fusing imagery, often collected simultaneously, to make the results more realistic-looking and to provide more quantitative depth. While the analytical and visual techniques employed are generally not novel, working with that much data is definitely a new frontier for archaeology, but has been explored at great lengths within the established HPC community—especially on the scientific side. Desktop software designed to exploit point clouds, as they are called, quickly breaks down at that scale. To make the situation even more challenging, archaeologists are working at multiple levels, from documenting excavations at individual sites to scanning huge swaths of jungle.

The two most noteworthy large-scale landscape recording LIDAR projects in recent years, ones that have pushed the boundaries of the discipline, are centered on the site of Caracol in Belize and Angkor Wat in Cambodia. Over two collection campaigns spanning several years, Arlen and Diane Chase have collected LIDAR data for more than 1200 km^2 of the triple-canopied Belizean rainforest, a region that they have been painstakingly, and slowly, exploring via traditional pedestrian survey for three decades (Chase et al. 2012, 2014). The resulting point cloud, consisting of trillions of points, required a great deal of trial-and-error processing via specialized software in order to remove the trees and underlying vegetation that was obscuring features of archaeological interest, ranging from agricultural terraces to roads and household groups. The result, even though far from perfect, was an incredibly detailed bare earth digital elevation model that could be run through traditional GIS software to generate standard products like shaded relief maps and hydrological flow models. They have also, through a hybrid parallelized application provided by the author of this chapter, created a Sky-View Factor map for the entire region, which is a substantial improvement over shaded relief due to its omni-directionality (Zakšek et al. 2011), as can be seen in Fig. 4. Not satisfied with stopping there, the Chases are now embarking on the creation of an automated framework for classifying features of interest in the terrain data to make map creation, and subsequent interpretation, possible for the entire region. Doing so manually would require many years and a substantial amount of funding, but the machine-learning-based automated approach carries a cost as well: required computing power. It simply cannot be done within a desktop environment. Damian Evans has faced similar survey challenges in Cambodia and was able to reveal a much larger, better organized, and more varied landscape surrounding Angkor Wat

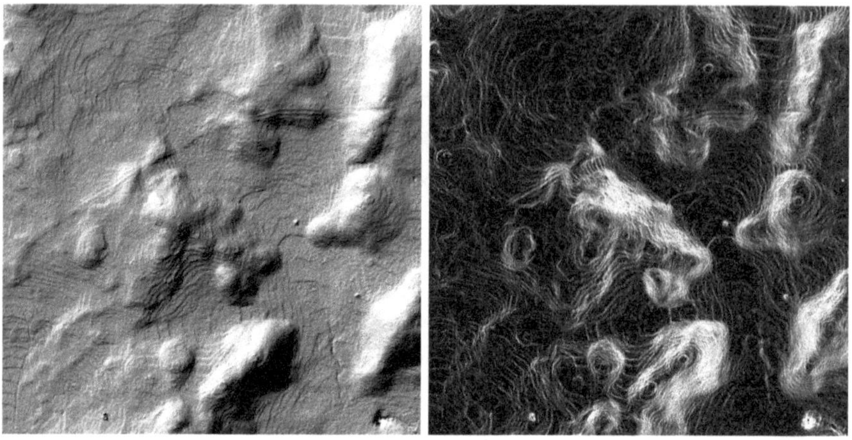

Fig. 4 High spatial resolution LIDAR data from Caracol, Belize, visualized using a traditional shaded relief method (*left*) and an inverted version of a more computationally intensive method known as Sky-View Factor (*right*). Note how many more potential features of interest are visible in the latter image

than anyone anticipated (Evans et al. 2013). The data collected for that project included next-generation full waveform LIDAR, but the computational requirements for fully exploiting it far outstrip the capacities of all but the largest computers in the world, so only a traditional point cloud was provided by the vendor. Full waveform, already shown to be valuable to archaeologists who are attempting to produce more accurate bare earth models in challenging terrain (Doneus et al. 2008; Lasaponara et al. 2011), will become more common in the next decade. What both projects clearly demonstrate is that this is just the beginning. LIDAR technology, deployed via manned aircraft and drones, is quickly dropping in price and it is only a matter of time before it is a mainstay for any field project.

Landscapes come in many sizes, though, and the challenges being faced by those working at the regional level are faced by those working at the individual site level, too. LIDAR data can be collected using a sensor mounted on a tripod, and can produce point counts of a similar volume, but what is far more practical for most projects on a budget is to use SfM (Structure from Motion) to record individual site features and, when possible, every aspect of an excavation (Green et al. 2014; Opitz and Cowley 2013; Remondino 2011). SfM uses specialized photogrammetry software and a large number of images of an object, taken from different perspectives, to reconstruct that object's three-dimensional characteristics and even build immersive environments. It requires a great deal of computing power if one wants the results for a small area quickly, or to do a large area at all. For an individual site, running the software on a desktop computer, which is common for archaeologists working with true color or thermal imagery collected by a low-flying drone to create digital elevation models and orthophotos, is usually sufficient (Casana et al. 2014). However, what happens when your area of interest is much larger? The Center for Advanced Spatial Technologies at the University of Arkansas has experimented with running PhotoScan on their institutional scientific cluster in an ad hoc hybrid parallel framework, with mixed results, but the approach shows a lot of promise. As with the LIDAR examples discussed above, the amount of three-dimensional data being collected, and associated imagery, is only going to increase. In light of how inexpensive it is to produce mass quantities of photos, and how affordable handheld active scanners like Google's Project Tango (https://www.google.com/atap/project-tango/) are quickly coming to market, we will all be drowning in useful, multidimensional data. We have mastered many of the analytical and visual techniques we want to use through existing software, but there is a huge gap between where that software and expertise ends and where we want to (and need to) be as a discipline.

A special case of recording and reconstruction worth mentioning here is the CORONA Atlas of the Middle East (http://corona.cast.uark.edu/), a multi-year project that was originally designed to recreate the landscape of the Fertile Crescent during the time when now-declassified spy satellites collected imagery over the entire region (Casana and Cothren 2013). In many cases, sites visible in those images no longer exist, destroyed through processes as diverse as agricultural expansion, urban sprawl, and armed conflict. Working with the imagery is extremely difficult, but in partnership with a photogrammetrist, archaeologists were able to properly geolocate a large number of historical stereo pairs, which can now be publicly accessed and used to produce orthophotos and terrain models using the techniques mentioned above. Creating those usable images required a hybrid parallel computing framework that leveraged GPGPUs, where possible. There is a great deal more the authors want to do with the Atlas, including an expansion into new regions and creating an automated process for detecting unrecorded sites, but recording damage at known sites has become a more pressing focus (Casana 2014; Casana and Panahipour 2014), at least in the short term.

Terrain Analysis

Terrain analysis, specifically the extraction of meaningful information from analyzing digital elevation models, has been a staple of archaeological GIS for at least two decades. The most common products generated are slope and shaded relief (mentioned above), both of which are relatively fast to calculate and are relatively easy to move into HPC environments because the mathematical operations required are *embarrassingly parallel*, as in they can map well into strong, weak, or even hybrid parallel frameworks because each one is completely independent and not very complex. This is also the case for Sky-View Factor, but that approach does require more time to compute than the others. There are three types of analysis that are far from embarrassingly parallel, though, and they are growing in significance for archaeologists who are interested in how things (people, water, ideas, social connections, goods, etc.) flow across landscapes. They are focused on watershed, peoplesheds, and viewsheds. It should be noted that all three are of interest to the researchers recording and reconstructing landscapes in the ways discussed in the previous section, in particular those struggling with how to effectively process large datasets without sacrificing fidelity.

The first is hydrological flow modeling, which has been used by archaeologists for many years, but requires a significant amount of computing power to execute and has been limited to the desktop environment, so regions of interest have remained somewhat small. A given terrain model is analyzed to find all high points from which will always flow downhill. Water is placed in those locations and the flow to all possible low points is modeled, after which a virtual stream network is extracted and its components are classified by rate of flow, creating a "realistic" representation of local watersheds along the way. While this can be done in traditional GIS software, that software cannot handle the size of landscapes that are now of interest to archaeologists, where "size" refers to how much data one has, not its physical geographic extents. Work has been done in the HPC domain to solve this problem, though, so archaeologists are encouraged to use tools like TauDEM (http://hydrology.usu.edu/taudem/taudem5/index.html), which leverages MPI and can be accessed through an ArcGIS extension, if needed.

The second is least cost analysis. What if one is interested in modeling how people, not water, flow across landscapes? Traditional GIS software allows for the generation of a small number of least cost paths across relatively small landscapes (it suffers the same technical limitations as hydrological analysis), but that is rarely the scale at which archaeologists are thinking about connections between, and travel to, locations of interest within a region. What if you are not sure where travelers are coming from or going to, but instead want a more general sense of how a landscape might channel movement, akin to water flow? The only feasible way to answer either of those questions, especially when one is working with a very large landscape, is to use some form of HPC. The From Everywhere To Everywhere (FETE) project, initially focused on the state of Oaxaca in Mexico, is doing just that (White and Barber 2012). The software written for FETE is capable of using strong parallelism on a desktop or hybrid parallelism on a cluster to quickly generate tens to hundreds of millions of theoretical travel routes across a region, which are then aggregated into a map that indicates the rate of people-flow, creating something akin to a "peopleshed." If locations of interest are known, they can be used, but it starts with the assumption of no a priori knowledge and instead samples terrain to build up an understanding of how it directs movement. The more samples requested, the more complex the overall set of calculations. Figure 5, where all of Mesoamerica has been analyzed at relatively high spatial resolution to highlight potential pedestrian trade routes across the entire region, demonstrates some of what can be done with the approach. It is somewhat reminiscent of a circulatory system, which makes sense.

The third is viewshed analysis, which like the previous two, can and is often done to a limited extent with GIS software packages. The traditional approach is to pick a small number of points of interest, specify a visibility extent, and generate a map of what can be seen from which points. Also like the previous two, calculating

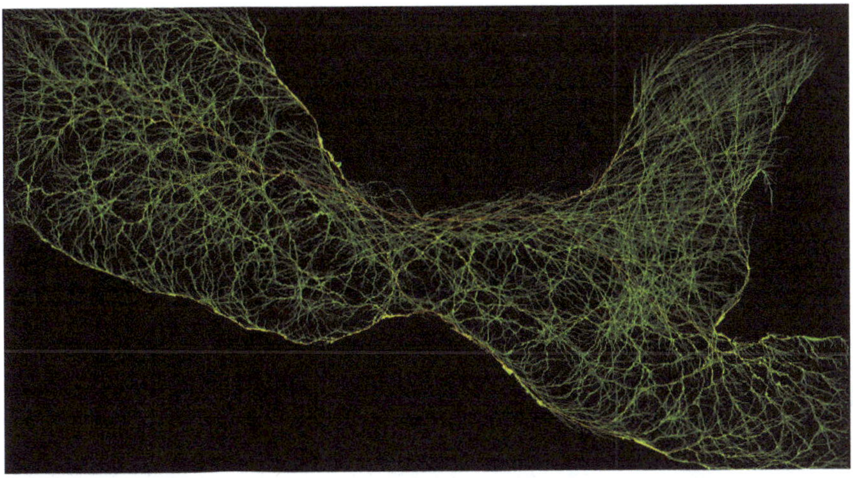

Fig. 5 Theoretical, terrain-based pedestrian trade routes throughout all of Mesoamerica, generated at relatively high spatial resolution using the FETE HPC application. Hundreds of millions of least cost routes were required to create the map. *Green* routes are high traffic, *yellow* are higher traffic, and *red* are highest traffic

a viewshed is computationally expensive, but the end result is extremely useful. As early as 2003, archaeologists began to hit a substantial computational wall with respect to viewshed analysis, expressing an interest in producing two types of products: *aggregate viewsheds*, where results are built up in a fashion similar to FETE through directed or systematic sampling, and *total viewsheds*, where results are built up for every single cell in the supplied elevation model (Llobera 2003). Figure 6 shows an example of an aggregate viewshed, created for a large region in the North American Southwest using a massively parallelized algorithm running on a GPGPU. In either approach, the main goal is to highlight prominent features on a landscape through sheer computational brute force, which requires HPC. To date, that brute force has been expressed most elegantly through the visual prominence research of Bernardini, who was interested in finding out which communities on a landscape could see the same prominent features and might then be considered part of the same "sight communities" (Bernardini et al. 2013; Bernardini and Peeples 2015), which could possibly share other things in common as well—despite great distances between them. Creating the baseline visual prominence map, derived from skylines extracted from viewsheds, involved deploying traditional GIS software in a weak parallel arrangement in a commercial-style cloud, a process that required a month to complete due to the inefficient, single-core nature of the software. Work is ongoing to translate the algorithms so that they can run within a hybrid parallel framework and leverage GPGPUs for the viewshed calculations, leveraging concepts developed by the broader Geographic Information Science community (Zhao et al. 2013). When complete, the anticipated speedup, and the overall extent of a region that can be analyzed, will be significant.

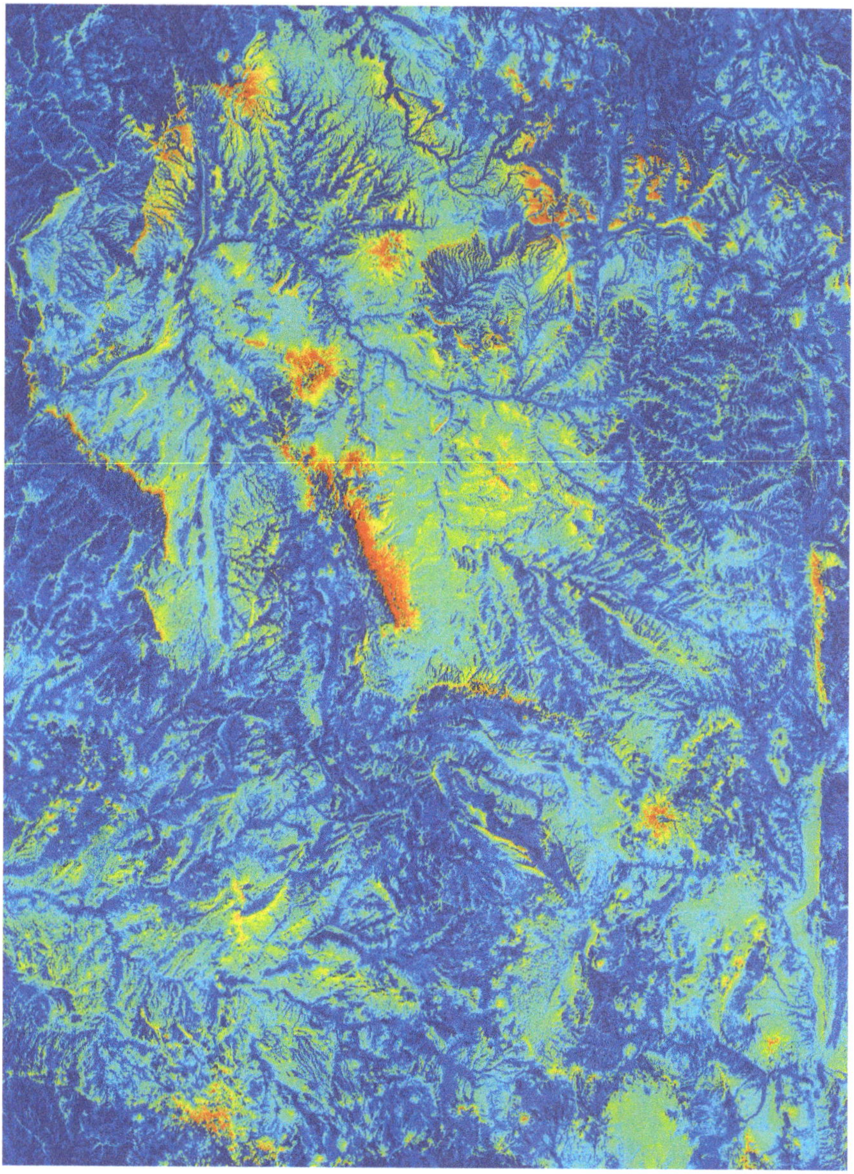

Fig. 6 An aggregate viewshed model for the entire Chaco regional system, which spans 158,000 km^2 in the North American Southwest. *Blue* areas are the least visible, *red* areas the most visible. Hundreds of thousands of viewshed analyses were executed using a GPGPU at regular spatial intervals, and then consolidated, to produce the output

Social Network Analysis

Social network analysis (SNA) has only recently been embraced by archaeologists, but the broader quantitative social science community has been exploring its utility for almost fifteen years (Borgatti et al. 2009) and it is an integral underpinning of Silicon-Valley-based social media outlets like Facebook and Twitter. Archaeologists have been able to benefit from those earlier explorations, as well as more recent research within the commercial sector, translating some of the core methods so that they have meaning within our discipline. At its heart, SNA is graph theory, which for the purposes of this discussion is a branch of mathematics and computer science concerned with how entities are connected to one another and how information flows between them. Given that archaeologists are interested in the flow of ideas, goods, and people between discrete locations, for example, using SNA would appear to be a natural fit for the discipline. The most substantial and ambitious efforts to date have come out of the Southwest Social Networks Project, which is examining community interactions (via a large standardized ceramics and architecture database) across space and time in the North American Southwest (Mills et al. 2013, 2015). Where their research, and ultimately all SNA projects, run into issues is when their graphs get so large that they cannot be processed on a single workstation or, if they can be processed, doing so requires a great deal of time. That is becoming increasingly common. As mentioned above, graph computing is a challenging problem, even for HPC. Many advances have been made, including the development of specialized computers and open source software packages like GraphLab and GoldenOrb, the latter of which is based on another Google standard named Pregel, but none of these are particularly easy for archaeologists to use at present. Where the situation becomes even more interesting is when SNA is combined with one or more of the terrain analysis techniques discussed above (Bernardini and Peeples 2015). As the reader has seen, each one is challenging on its own, but the results are highly complementary because they each speak to a different aspect of flow across a landscape. The use of SNA, in particular the combining of it with other more well-established quantitative methods, will continue to grow in the coming years and eventually become a common approach within the discipline. Making it a practical one will take time, though.

Complex Adaptive Systems

Modeling and simulating complex adaptive systems, like SNA, has quickly moved from a niche research space to one that is being more fully embraced by archaeologists as established projects have published their findings and clearly demonstrated its utility. The software required has also become much easier to use, with two packages being the most prominent at the time of writing: NetLogo (https://ccl.northwestern.edu/netlogo/) and Repast Simphony (http://repast.sourceforge.net/).

Both make it relatively easy to create what are known as agent-based models (ABMs), where complex (and nonlinear) interactions between people can be simulated over varying amounts of time and space. The most well-known ABM effort in archaeology to date is the Village Ecodynamics Project (VEP), a multi-year effort focused on understanding the formation and eventual abandonment of villages in the Mesa Verde Region of the North American Southwest (Kohler and van der Leeuw 2007; Kohler and Varien 2012). ABMs are by far the most computationally demanding entities discussed in this chapter. Communication between all of the elements of a model is required, so as the model scales to larger numbers of people and/or larger or more fine-grained space-time contexts, desktop software solutions quickly break down. NetLogo, by its own admission, is designed to be an educational tool, not a production-level solution, so models must be kept small. Repast Simphony can operate at the production level, but even it runs into technical limitations as a model grows in size. VEP has run into several issues related to computing capacity, which has constrained their ability to look at more regions at great levels of spatial and temporal detail. This is a common problem for archaeologists working with ABMs, who are generally interested in exploring the interactions of many people in very detailed ways. One solution is available at present, which is to employ an HPC-enabled version of Repast, unsurprisingly called Repast HPC (http://repast.sourceforge.net/repast_hpc.html). One archaeologist has already attempted to use Repast HPC to explore an entire Hohokam irrigation system in southern Arizona and the results to date are promising (Murphy 2012). As with many of the technologies discussed above, there is a steep learning curve associated with the framework, but the hope is that it will become more accessible in the coming years because the disciplinary need is definitely present.

Concluding Thoughts: The Art of the Possible

Roughly speaking, there has been a trillion-fold increase in computing power since 1956 (for more details, see http://pages.experts-exchange.com/processing-power-compared/). To put that increase in perspective, the newly-released Apple Watch is on par with a Cray-2 from 1985, which means we can now walk around with a thirty-year-old supercomputer on our wrist, one that can connect to a much deeper pool of computing power via the internet. Smartphones are even more powerful (fifteen to twenty times greater). Where will computing power be in thirty years? Will we be able to walk around with the equivalent of Titan on our wrists? What can archaeologists possibly do with that much power, given that we take so little advantage of what is already available to us on the scientific and commercial sides? Granted, it is not necessarily easy to use what is available today, but it is hoped that the survey provided above will open new doors for archaeologists and help them connect with the right technical resources who can help them, with the ultimate goal being empowered archaeologists, fluent in the languages of computer science, helping themselves. We should, in the end, completely own our problems and their solutions.

A more important question to ask is this: If unlimited computing power was available, how would archaeologists interact with that technology, what questions could they ask, and what problems could they solve? As archaeologists broaden their regions of interest and/or examine smaller areas at increasingly finer levels of detail, with respect to both space and time, the amount, variety, and complexity of the data they must work with grows exponentially, which means that HPC will inevitably become a deeply integrated and transformative element of the discipline, much in the same way as radiocarbon dating and LIDAR/SfM. What would that look like?

A natural place to start would be the "grand challenge" topics proposed by Kintigh et al. in PNAS (2014), topics that until very recently could not be realistically explored at a global scale due to data sparsity, lack of sufficient technical expertise, and lack of available computing power:

- Emergence, communities, and complexity
- Resilience, persistence, transformation, and collapse
- Movement, mobility, and migration
- Cognition, behavior, and identity
- Human-environment interactions

Beneath each of these general topics are multiple questions and the reader is encouraged to consult the article for more detailed information (Kintigh et al. 2014). Archaeologists have attempted to address these topics in relatively small, focused ways over the past several decades, but momentum has built up recently to address them in a cross-cultural way that incorporates as much space and time as the extant archaeological record will allow. The recent work by Kohler on the spatially variable Neolithic Demographic Transition in the North American Southwest is an excellent example of this trend (Kohler and Reese 2014). That is not just a big data problem, it is a massive data problem, one that is fraught with peril due to the fragmented and inconsistent nature of global archaeological datasets. Two initiatives are taking the first steps towards creating a consolidated archaeological database for the world: the Digital Archaeological Record (http://core.tdar.org/), also known as tDAR, which is focused on archiving all types of critical archaeological data, from site reports to datasets, and the Digital Index of North American Archaeology (http://ux.opencontext.org/blog/archaeology-site-data/), also known as DINAA, which is focused on creating interoperability models between archaeological site databases, thus enabling analysis at larger scales and finer spatiotemporal resolutions. Between the two, assuming enough compute power is present, archaeologists can now ask regional and even continental-level questions in a way that was not possible previously. Whether tDAR and DINAA specifically persist is not really the point here: consolidated archaeological databases are the future and will enable researchers to finally, after centuries of trying, ask the really big questions in a quantitatively defensible way (Kintigh 2006). It is important to note, however, that those topics are just a subset of a much richer tapestry of "big questions" that HPC and the data analytics it supports, in the hands of archaeologists, can address.

Putting aside grand challenges, there are practical areas where increasingly-available and increasingly-accessible HPC can and should transform the discipline. They are summarized here as a series of desires, something all archaeologists should want: everything global, everything detailed, everything mobile, everything fast, and everything smart.

What is meant by everything global is the desire to have the world's archaeological knowledge base available at one's fingertips for analysis, visualization, and contextualization, no matter the location, with the ability to contribute back to it in real time. Repositories like tDAR and DINAA are an excellent start, but the broader knowledge base should include the entire breadth and depth of the discipline. Given how compute power and storage capacities continue to rapidly increase, this is an achievable goal.

What is meant by everything detailed is the desire to have fine-grained spatiotemporal (four-dimensional) models of archaeological features, sites, regions, cultures, continents, and even the entire world. The standard can and should be the digital capture of archaeological information in multiple dimensions and the ability to immerse oneself in it at multiple scales, from virtually exploring the intricacies of an individual artifact to experience a reconstruction of an ancient city, complete with people. More nuanced reconstructions based on existing data is one path, but another is to more fully embrace the wide array of options now available for recording data in the field, not the least of which are smartphones, drones, Microsoft Kinect, and laser scanners. A logical extension of this desire would be the ability to test out hypotheses in virtual environments, seeing how events might play out under varying circumstances over large expanses of space and/or time. Many of the foundational technologies, including compute power, are already available to reach this goal, but the discipline (like so many others) currently lacks the technical expertise and resources to take full advantage of it. As costs continue to drop, and technological barriers continue to fall, the situation will greatly improve.

What is meant by everything mobile and everything fast is the desire to have HPC resources available at one's fingertips, regardless of location, as quickly as possible (at the speed of research). That means being able to collect and analyze mass quantities of data in the field, perhaps in a disconnected fashion (no internet access), including accurate real-time recording and analysis of excavations by multiple sensors (terrestrial and airborne) and being able to use augmented reality displays while one works. This mobility and speed should extend to the lab environment as well.

Lastly, what is meant by everything smart is the desire to have the equivalent of IBM's Watson for archaeology. Artificial intelligence research is currently undergoing a renaissance and human-trained systems like Watson are now able, in fields like medicine, to digest vast storehouses of information, find non-obvious connections between elements, and make suggestions to researchers, in close to real time. By connecting a system with this kind of potential to others that address the previous four desires, archaeologists will be able to explore the past in ways that we cannot even possibly imagine.

At the center of this new type of exploration are archaeologists who are not just passive recipients of technologies and methods developed by others. We can, and should, harness the potential of HPC for ourselves. This chapter, only the latest step in that direction, has introduced the reader to the current landscape of HPC, how to take advantage of it, and some of what archaeologists have already attempted. It is very exciting to think of what could happen next.

Acknowledgments This submission was written by the author acting in his own independent capacity and not on behalf of UT-Battelle, LLC, or its affiliates or successors.

References

Bernardini, Wesley, Alicia Barnash, Mark Kumler and Martin Wong 2013 Quantifying visual prominence in social landscapes. *Journal of Archaeological Science* 40:3946–3954.

Bernardini, Wesley and Matthew A. Peeples 2015 Sight Communities: The Social Significance of Shared Visual Landmarks. *American Antiquity* 80(2):215–235.

Borgatti, Stephen P., Ajay Mehra, Daniel J. Brass and Giuseppe Labianca 2009 Network Analysis in the Social Sciences. *Science* 323(5916):892–895.

Casana, Jesse 2014 Regional-Scale Archaeological Remote Sensing in the Age of Big Data: Automated Site Discovery vs. Brute Force Methods. *Advances in Archaeological Practice* 2 (3):222–233.

Casana, Jesse and Jackson Cothren 2013 The CORONA Atlas Project: Orthorectification of CORONA Imagery and Regional-Scale Archaeological Exploration in the Near East. In *Mapping Archaeological Landscapes from Space*, edited by D. C. Comer and M. J. Harrower. Springer, New York.

Casana, Jesse, John Kantner, Adam Wiewel and Jackson Cothren 2014 Archaeological aerial thermography: a case study at the Chaco-era Blue J community, New Mexico. *Journal of Archaeological Science* 45:207–219.

Casana, Jesse and Mitra Panahipour 2014 Satellite-Based Monitoring of Looting and Damage to Archaeological Sites in Syria. *Journal of Eastern Mediterranean Archaeology and Heritage Studies* 2(2):128–151.

Chase, Arlen F., Diane Z. Chase, Jaime Awe, John F. Weishampel, Gyles Iannone, Holley Moyes, Jason Yaeger and M. Kathryn Brown 2014 The Use of LiDAR in Understanding the Ancient Maya Landscape. *Advances in Archaeological Practice* 2(3):208–221.

Chase, Arlen F., Diane Z. Chase, Chistopher T. Fisher, Stephen J. Leisz and John F. Weishampel 2012 Geospatial revolution and remote sensing LiDAR in Mesoamerican archaeology. *Proceedings of the National Academy of Sciences of the United States of America* 109 (32):12916–12921.

Doneus, Michael, Christian Briese, Martin Fera and Martin Janner 2008 Archaeological prospection of forested areas using full-waveform airborne laser scanning. *Journal of Archaeological Science* 35:882–893.

Evans, Damian H., Roland J. Fletcher, Christophe Pottier, Jean-Baptiste Chevance, Dominique Soutif, Boun Suy Tan, Sokrithy Im, Darith Ea, Tina Tin, Samnang Kim, Christopher Cromarty, Stéphane De Greef, Kasper Hanus, Pierre Bâty, Robert Kuszinger, Ichita Shimoda and Glenn Boornazian 2013 Uncovering archaeological landscapes at Angkor using lidar. *Proceedings of the National Academy of Sciences of the United States of America* 110(31):12595–12600.

Green, Susie, Andrew Bevan and Michael Shapland 2014 A comparative assessment of structure from motion methods for archaeological research. *Journal of Archaeological Science* 46: 173–181.

Kintigh, Keith 2006 The Promise and Challenge of Archaeological Data Integration. *American Antiquity* 71(3):567–578.

Kintigh, Keith W., Jeffrey H. Altschul, Mary C. Beaudry, Robert D. Drennan, Ann P. Kinzig, Timothy A. Kohler, W. Fredrick Limp, Herbert D. G. Maschner, William K. Michener, Timothy R. Pauketat, Peter Peregrine, Jeremy A. Sabloff, Tony J. Wilkinson, Henry T. Wright and Melinda A. Zeder 2014 Grand challenges for archaeology. *Proceedings of the National Academy of Sciences of the United States of America* 111(3):879–880.

Kohler, Timothy A. and Kelsey M. Reese 2014 Long and spatially variable Neolithic Demographic Transition in the North American Southwest. *Proceedings of the National Academy of Sciences of the United States of America* 111(28):10101–10106.

Kohler, Timothy A. and Sander E. van der Leeuw (editors) 2007 *The Model-Based Archaeology of Socionatural Systems*. SAR Press, Santa Fe, NM.

Kohler, Timothy A. and Mark D. Varien (editors) 2012 *Emergence and Collapse of Early Villages: Models of Central Mesa Verde Archaeology*. University of California Press, Berkeley, CA.

Lasaponara, R., R. Coluzzi and N. Masini 2011 Flights into the past: full-waveform airborne laser scanning data for archaeological investigation. *Journal of Archaeological Science* 38:2061–2070.

Llobera, M. 2003 Extending GIS-based visual analysis: the concept of visualscapes. *International Journal of Geographical Information Science* 17(1):25–48.

Mills, Barbara J., Jeffery J. Clark, Matthew A. Peeples, W. Randall Haas, Jr., John M. Roberts, Jr., J. Brett Hill, Deborah L. Huntley, Lewis Borck, Ronald L. Breiger, Aaron Clauset and M. Steven Shackley 2013 Transformation of social networks in the late pre-Hispanic US Southwest. *Proceedings of the National Academy of Sciences of the United States of America* 110(15):5785–5790.

Mills, Barbara J., Matthew A. Peeples, W. Randall Haas, Jr., Lewis Borck, Jeffery J. Clark and John M. Roberts, Jr. 2015 Multiscalar Perspectives on Social Networks in the Late Prehispanic Southwest. *American Antiquity* 80(1):3–24.

Murphy, John 2012 A High-Performance Computing Simulation of an Irrigation Management System: The Hohokam Water Management Simulation II. In *Archaeology in the Digital Era Volume II: e-Papers from the 40th Conference on Computer Applications and Quantitative Methods in Archaeology*, edited by G. Earl, T. Sly, A. Chrysanthi, P. Murrieta-Flores, C. Papadopoulos, I. Romanowska and D. Wheatley, pp. 225–230. Amsterdam University Press, Amsterdam.

Opitz, Rachel S. and David C. Cowley (editors) 2013 *Interpreting Archaeological Topography: 3D Data, Visualisation and Observation*. Oxbow Books, Oxford, United Kingdom.

Remondino, Fabio 2011 Heritage Recording and 3D Modeling with Photogrammetry and 3D Scanning. *Remote Sensing* 3:1104–1138.

White, Devin A. and Sarah B. Barber 2012 Geospatial modeling of pedestrian transportation networks: a case study from precolumbian Oaxaca, Mexico. *Journal of Archaeological Science* 39:2684–2696.

Zakšek, Klemen, Kristof Oštir and Žiga Kokalj 2011 Sky-View Factor as a Relief Visualization Technique. *Remote Sensing* 3:398–415.

Zhao, Yanli, Anand Padmanabhan and Shaowen Wang 2013 A parallel computing approach to viewshed analysis of large terrain data using graphics processing units. *International Journal of Geographical Information Science* 27(2):363–384.

Part V
Interpretation and Discussion

Measuring the Face of the Past and Facing the Measurement

William Fred Limp

Abstract We are in the midst of major changes in the practice of archaeology and heritage studies. Technologies for high-precision, high-density measurement of objects, sites and landscapes coupled with computer-based methods for the visualization of these data and other (re)creations of the past are growing in use. These approaches have and will continue to fundamentally alter our field. The term "high density survey and measurement" (HDSM) covers methods such as airborne lidar, real time kinematic GNSS/GPS survey, robotic total stations, terrestrial "laser" scanning, structured light scanning and close range photogrammetry (CRP, also known as structure from motion—SfM) and UAV-based SfM/CRP and scanning. The analytical process that characterized our field before HDSM can be (over-) simplified to a sequence of **observe, interpret/abstract**, **measure**, **record**, and **analyze**. HDSM breaks us out of this process in that it pushes us toward a recursive and reflexive engagement with the data, in which we **observe**, **record**, **measure**, **analyze**, **and abstract/interpret** repeatedly, and in various orders. The growth in the use of HDSM methods is paralleled by increasing applications of computer-based visualization. Effective use of both requires attention to a scholarly digital ecosystem that addresses the archive and reuse of these digital objects and includes strategies to reuse these digital objects in other scholarly representations along with the tools for citation and other aspects of scholarly discourse.

The Dimensions of the Past

In this paper I will examine some of the implications of the new approaches to the measurement and representation of space and form in archaeology and heritage studies that are emerging early in the 21st century. I suggest that these new approaches are not simply evolutionary developments but, instead, represent an inflection point in our approaches to the past. The word revolutionary has been

W.F. Limp (✉)
Department of Geosciences, University of Arkansas, Fayetteville, USA
e-mail: flimp@uark.edu

M. Forte and S. Campana (eds.), *Digital Methods and Remote Sensing in Archaeology*, Quantitative Methods in the Humanities and Social Sciences,
DOI 10.1007/978-3-319-40658-9_16

overused in archaeology and I will not use it here but I firmly believe that future historians of our field will clearly be able to define the first decade of the 21st century as a point of major change. Up to this point in time investigations of the past can be (over-) simplified to a sequence of **observe**, **interpret/abstract**, **measure**, **record**, **analyze**. The new tools for measurement break us out of this process as it moves us toward a recursive and reflexive engagement with the data, in which we **observe**, **record**, **measure**, **analyze**, **and abstract/interpret** repeatedly, and in various orders.

The reason that the use of new methods of measurement are changing our field in fundamental ways is simple. A central focus in archaeology from the very beginning has been the measurement and analysis of the properties of objects and sites and their constituent features and (more recently) the landscapes around them. More than half a century ago Gordon Willey stated that the objectives of archaeology are "approached by the study and manipulation of three basic factors: form, space and time" (Willey 1953, 361). Later Albert Spaulding identified what he termed the "dimensions" of archaeology. He defined dimension as "an aspect or property of the subject matter which requires its own special measuring device" (Spaulding 1960, 438). Over the last five decades we have seen the dramatic improvement in the measurement of the dimension of time—using "instruments" such as radiocarbon, archaeo-magnetism, thermo-luminescence and others. Prior to these measuring "devices" time was commonly inferred (not measured) using stylistic or classificatory methods. It was a dependent variable in the process of analysis. Once the proper instruments were available it became, properly, an independent measure. This change was (and continues to be) profound. We no longer infer time from other properties—we measure it directly and, thus, can use these independent observations in our analyses. As an aside, it can be argued that the late 20th century's "New (aka processural) Archaeology" was simply the inevitable consequence of time becoming an independent measurement.

High Density Measurement

While we have applied many techniques to the measurement of space and form it is only recently that we have been able to apply measurements that are simultaneously intensive and extensive, both precise and comprehensive. In another recent paper Rachel Opitz and I (Opitz and Limp 2015) have coined the term "high density survey and measurement" (or the natural acronym HDSM) for the suite of new methods that allows form and space to become, properly, independent variables in our studies.

> "HDSM is the collective term for a range of new technologies that give us the ability to measure, record and analyze the spatial, locational and morphological properties of objects, sites, structures and landscapes with better resolution and more precision than ever before. The particular technologies considered include airborne lidar, real time kinematic GNSS/GPS survey, robotic total stations, terrestrial "laser" scanning, structured light scanning and

> close range photogrammetry (CRP, also known as structure from motion—SfM) and UAV-based SfM/CRP and scanning" (Opitz and Limp 2015)

HDSM is both precise and dense; it is also three-dimensional. For objects these measurements are typically in the millimeter to sub-millimeter range while for structures they are typically sub-decimeter to millimeter and for sites and landscapes are typically decimeter to sub-meter in density and precision (and hopefully accuracy).

These dense and precise measurements have the ability to move our consideration of form and space from their past roles as ***dependent observations*** to **independent**. By dependent observations I mean the use of classificatory terms that were (and are) used to communicate form such as "excurvate," "jar," "beaker," "Clovis point," "Mesa Verde black on white bowl." These classificatory terms were meant to encompass simultaneous considerations of shape, size, surface and embedded within were many other properties. At the structure or landscape scale we applied similar logic. Descriptors such as "village," "palace," "terrace," "mound," "house" were dependent classificatory terms that encoded multiple characteristics and implications. Since time immemorial archaeologists have believed that statements such as "there were four Clovis points in Level 4" or "the pit had a two-handled lidded chytra" or "there were four houses in the village" are statements of data. They are not. If we use Ackoff's (1989) classification—which is among the foundational statements in the field of knowledge management—as elaborated by Bellinger et al. (2004):

> Data… data is raw. It simply exists and has no significance beyond its existence (in and of itself). It can exist in any form, usable or not. It does not have meaning of itself.
>
> Information… information is data that has been given meaning by way of relational connection. This "meaning" can be useful, but does not have to be (Bellinger et al. 2004: 1)

The typical type of statements of archaeological "data" that we constantly encounter are in fact "information"—a second step up (or down?) the abstraction (interpretation) stairs—though it often appears as a slide rather than a stairs. The data are the measurements, surfaces, and other physio-chemical properties of the object (or feature, or site or landscape). The reason that this is an important distinction is because of the extensive, unconscious encoding and assignment of meaning, interpretation, that is embedded in even the simplest classificatory statement. Classifications move (often without thought) from data to information. There is a perfectly good reason for our classifications. If we did not use these information-laden terms even a short descriptive report would be of an impossible length. It would appear absurd to always provide a paragraph length listing of the metric and physical properties of each object when a simple term will do. There is logic to that. What we lose, however, is that once we are provided information and not data we can never follow the chain of logic back to the origin. We cannot reproduce the results—and the ability to reproduce results is essential. Kansa emphasizes that, "[s]cholarship is better served if claims about the past can be evaluated in terms of appropriate use of evidence to support arguments and interpretations." (Kansa 2005, 100) and the most effective approach is to be able to provide the original data.

This concern has been around a long time in archaeology. For a discussion that is foundational but still timely see (Read 1974) or, more recently, (Niccolucci et al. 2015) and it is far beyond the remit of this article to do the topic justice. While no panacea HDSM provides a strategy to approach this problem with more nuance.

> "HDSM allows us to decouple measurement from interpretation—data from information, leading to a fundamental alteration of the abstractive process in archaeology. Consider the common task of profiling a wall or mapping a floor in an excavation. The first step is to visually process the "data" and abstract from it relevant "objects" of interest to be recorded–a pit, post mold, layer or sediment change, and so on. These abstracted elements are then recorded, usually by selecting relevant two-dimensional points and recording them on graph paper. The traditional process can be intentionally (over-) simplified to a sequence of **observe**, **interpret/abstract**, **measure**, **record**, **analyze**. HDSM breaks us out of this process in two ways. First, it pushes us toward a recursive and reflexive engagement with the data, in which we **observe**, **record**, **measure**, **analyze**, **and abstract/interpret** repeatedly, and in various orders" (Opitz and Limp 2015).

Representation

HDSM is not the only significant aspect of the "great inflection" of the 21st century. The second is representation. The increasingly wider and wider adoption of new methods of representation of the past are equally "inflectionary." These methods include a range of computer-based approaches to first (re)create and then present realistic 3D representations of archeologically derived/investigated objects, structures and landscapes. Increasingly these approaches include virtual realities and explorations via gaming technologies.

The co-occurrence of HDSM and new methods of representation is not coincidental but has, at their base, the common development of computational tools and techniques. I would argue that beyond their common computational foundations the two are inextricably linked—any representation should have as it's foundation high quality measurement—but this is not as frequently the case as it should be.

Abstracting the Past

As we consider these arguments it should be quite clear that the ways in which we describe the past(s) are based on processes of extensive abstraction and essentially reductive. "Science" is reductive—that's why it is so powerful but it is also clear that much of human behavior must be understood as "expressive" built on direct experience. Perhaps the "opposite" of reductive is "artistic"? Such a view is not new, for example, in 1938 "Dewey saw reality as fundamentally an ongoing process whose parts could not exist independently of the interactions in which they participated" (Rockwell 2005, ix).

We are all quite familiar with the argument by C. P. Snow that there are "two cultures" separating the sciences and humanities. Whatever we think about the specifics of his argument it is clear that there was and is a divergence between these two ways of knowing. Within the last few years this gap has increasingly been reduced, however, as reflected in the growth of the "digital humanities."

What Would Leonardo Do?

Even before the divergence of the sciences and the humanities, we can see an earlier one where art and science diverged. While pinpointing the time of the divergence is an exercise for others I would think it very clear that both science and art are seamlessly merged in the work of many Renaissance "artists," with no more powerful example that Leonardo da Vinci. We can see in his work that his art was grounded in observation and recording of the natural world and the discovery of basic principles. His position is clear when he states.

> "[t]hus, you who observe rely not on authors who have merely by their imagination wished to be interpreters between nature and man, but on those alone who have applied their minds not to the hints of nature but to the results of their experience" (da Vinci 1906, 147).

Without entering into the debate it seems to me [and others (Cochrane and Russell 2007)] that the divergence of art and science accelerated in the later 19th century and the 20th century because "art" (particularly after the development of photography) moved increasingly towards an abstract forms of representation. Fortunately, with the development of computational visualization, this distance in now again closing.

The separation of art and science has lead to the separation of observation and representation. It usually comes as a serious criticism of a modern artist to suggest that her representation has photographic realism. Since we are no longer encouraged to "draw what we see" our process of representation has become one of conversion of the observed "data" into symbols that encode the content not the data. This encoding is a process of mental abstraction (but not necessarily abstraction as in abstract art) where the complexity of the "real world" is encoded into a much reduced set of symbols. The ways in which this process is deeply embedded in all our thinking can be illustrated in a simple experiment. Consider the photograph shown in Fig. 1. If the great majority of us were asked to "draw" the image we would very likely come up with something like Fig. 2. I believe we would all agree that the drawing looks nothing like the photo but is, instead, a symbolization of the content. Our brains "see" the house or people and reduce them to the "standard" abstractions of the drawing. I was first exposed to this issue in the late 1970s by a popular book *Drawing on the Right Side of the Brain* published by Edwards (

Fig. 1 An artistic diversion. *Image source* Flickr CC 2.5 license

1979). The premise of the author is encapsulated in this quote from the neurosurgeon Richard Bergland.

> "You have two brains: a left and a right. Modern brain scientists now know that your left brain is your verbal and rational brain; it thinks serially and reduces its thoughts to numbers, letters and words… Your right brain is your nonverbal and intuitive brain; it thinks in patterns, or pictures, composed of 'whole things,' and does not comprehend reductions, either numbers, letters, or words." (Bergland 1986, 1).

As some one who could not draw Edward's book and ideas attracted me. The first exercise in the text is to take a complex drawing and "copy" it. Unless you are already an "artist" the results are always embarrassing, similar to the results in Fig. 2 above. Edwards then says that you should rotate the original drawing until it is upside down and try again. The results are astounding. Your drawing will look so very much better. Edwards proposition is that the inversion "turns off" the abstracting pathways of the brain. The pathways that convert people to stick figures are blocked and your hand simply draws what the eye sees. It works. The point of all this is to argue that for many (most?) archaeologists the past is deeply embedded in abstracted representations and that the great majority of us are deeply challenged to understand the past in any formulation that would begin to approximate a "lived experience," as we are always abstracting (reducing) and then symbolizing the complexities that are world as it is experienced.

Not all of us have this limitation, of course. We know that archaeologists who have been the "great synthesizers" of our field process massive amounts of visual

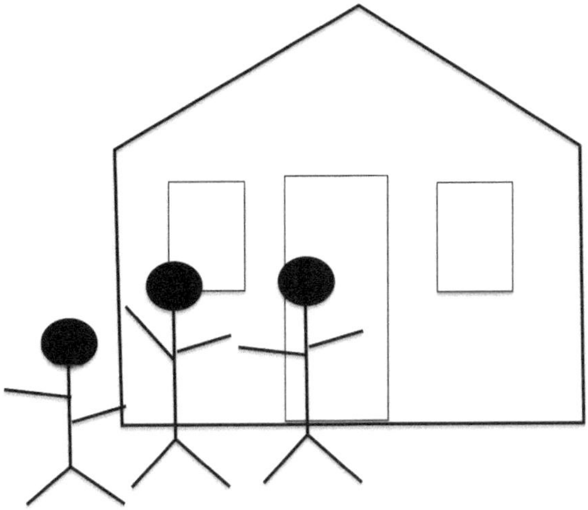

Fig. 2 A typical artistic rendering of Fig. 1

data to derive "grand patterns" and they have strong visual memories. They do not abstract and symbolize the data—they just store it. Some examples:

> "[V. Gordon Child's] already extensive command of European literature and an acute visual memory enabled him to visit and assemble data from museums and excavations across the whole of Europe" (Trigger 1989, 169).
>
> "A good visual memory … allowed him [John Griffin] to recall the shape, decoration, and composition of any artifact he had ever seen" (Griffin and Griffin 1996, 11).
>
> "The extent of this visual memory [in archaeology] is never realized until one meets with some who are so unlucky not to possess it" (Petrie 1904, 19).
>
> "[Joffre Coe's] ability to extract meaning from a handful of pottery sherds or the shape of an arrowhead was legendary. This ability was rooted in his broad interests and a keen visual memory" (Anon 2001, 2).

By way of example of the value of the visual representational strategy lets compare two strategies for displaying the "information" about the nature of an archaeological site. The traditional abstraction-based representational approach is shown in Fig. 3. This is a classic example of the "Cartesian gaze." Such a representation is so common to us that we have all developed complex decoding processes that allows us to "understand" what these symbolic representations "mean" but it is only through this reverse decoding that we can create any sense of an understanding. This understanding is not clearly encoded in this abstraction but follows traditional disciplinary encoding protocols. This encoding process, when learned, not only allows us to reduce the content to a minimum (keeping publishers happy) it also excludes the uninitiated from participating. Compare the information content in Fig. 3 with the "same" location represented in Figs. 4 or 5.

Notwithstanding the visual memories of many of our prominent archaeologists, vision as a "way of knowing" can seem alien—particular to scholars.

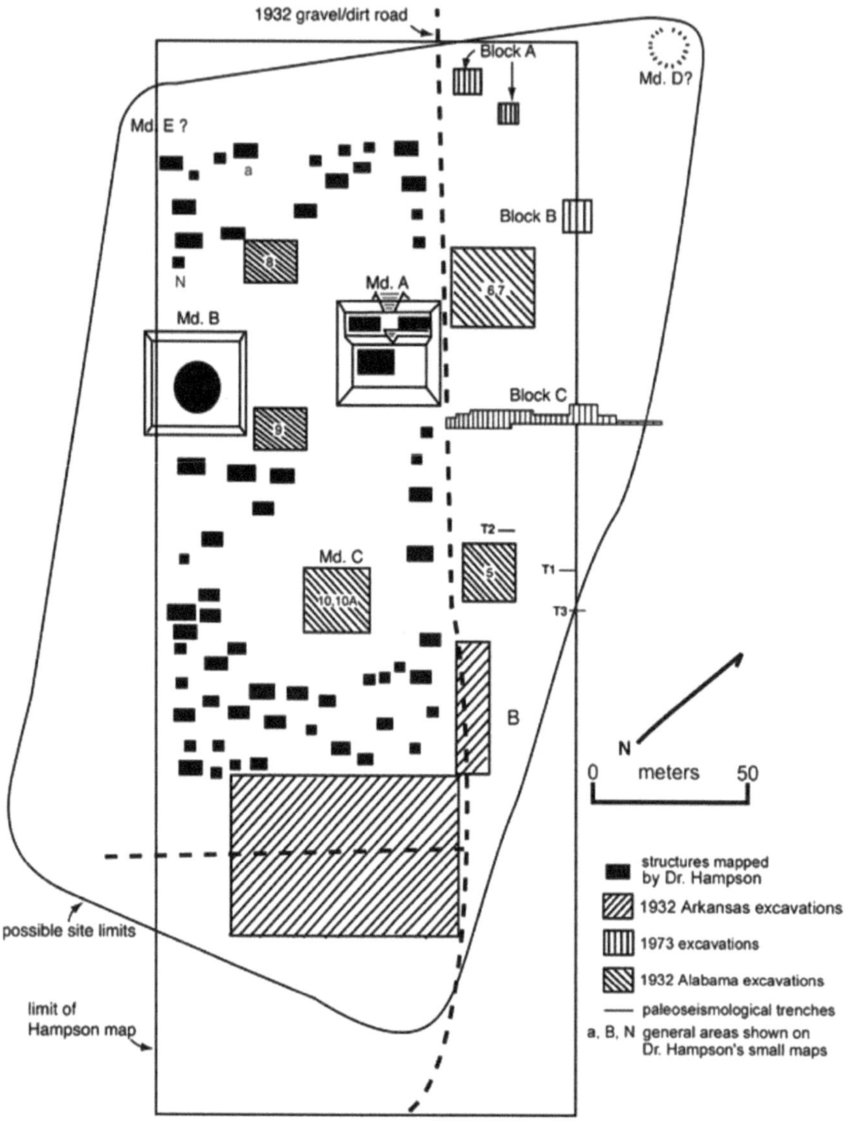

Fig. 3 The traditional site map for the Nodena Site (Payne and Limp 2008)

"Some people are so deeply accustomed to a linear view of intelligence and potential that they find it almost impossible to believe that certain persons may find advanced subject matter quite easy while they find some elementary subject matter quite difficult. …They may have had difficulty learning from books and lectures, .. [but are] better adapted to learning from simulations of reality. For these people, it is sometimes far easier to learn firsthand from nature than it is to learn secondhand from books" (West 2009, 24–25)

Fig. 4 The Nodena "site map" as a visualization (Payne and Limp 2008)

Fig. 5 An alternative Nodena "site map" representation (Payne and Limp 2008)

It seems clear that a full understanding of the world (especially of the past) must be

- Abstracted **and** Contextualized
- Artistic **and** Scientific
- Analyzed **and** Experienced.

Visualizations and Phenomenology

Many authors in this volume have discussed the implications and values of virtual reality and of representations of the past that have high verisimilitude and there is little need to repeat these. In addition to the other rationales for visualization, however, I would also emphasize that such representations have a role to play in the issues that underlie the debate around the value of phenomenological approaches in archaeology. Again there is a massive body of literature on the topic but to briefly (and relatively superficially) to summarize archaeologically-relevant phenomenology would be to emphasize that it is, at its core, an appreciation that experiential (e.g. "sensuous") elements are key to understanding human life and (very specifically) the past and that space is lived and meaningful constituted. It is generally understood as having entered archaeological visibility with Tilly's work but also Thomas and others. The use of phenomenology in archaeology has been one of the most provocative theoretical developments in the discipline in recent years (Brück 2005). Computer visualization representations of the past serve as an effective transition between the "scientific" way of knowing and the "sensuous" one. Much of the argument to date has seen these two ways as conflicting or competitive but with the growing use of computerized tools, visual reality and immersive experiences the boundaries between scientific and phenomenological ways of knowing are blurring.

Linking Measurement and Representation

It would initially appear reasonable that measurement and representation are fundamentally linked but this is not always the case. We could imaging a process where (perhaps) laser scanned data of a structure was "brought into" a visualization package and used as the foundation for a representation. Until relatively recently, however, this has not been the case. There are a number of reasons for this but one technical reason has been the difficulty in bringing the massive and extensive measurement data (scanned, photogrammetry, etc.) into the software used to create the representations. Developments in both the visualization software and data acquisition communities will certainly reduce, if not eliminate, this constraint in the near future.

Much of this measurement data and our representations are "floating" in space—at best using a local grid system. In order to make comparisons and to place these in proper context it is necessary to convert these "local grid" measurements to recognized coordinate systems (Limp and Barnes 2014). Again there are often technical constraints in that many systems that record and process these data and to develop the representations do not support map system coordinates. This too will change as the systems evolve and practitioners should be attentive and adopt these tools as they emerge.

A Digital Ecosystem

I believe it is fair to suggest that substantial portions of the efforts towards both measurement and representation in archaeology are the result of enthusiastic efforts by single or small groups of scholars. Such as it ever was in new approaches in our field. While such efforts are nimble and creative they are not, inherently, sustainable. For these to continue they must be an embedded part of a comprehensive digital ecosystem (Limp et al. 2011). Key elements of this ecosystem are

- Access via search
- Re-use by others
- Lineage of idea, representation, measurement and authorship
- Elemental decomposability
- Scholarly—"cite-ability"—authorship
- Archive
- Sustainability.

Like the biological ecosystem the digital one requires that **all** aspects be "healthy"—in the following I consider some strategies.

A central aspect to any sustained scholarly endeavor is that the results can be accessed by others and "reused"—in this case made part of the next cycle of scholarly product. This goal is accomplished by placing all the "digital objects"—measurements, representations etc. in an accessible digital archives well populated with "finding aids"—metadata.

There are extensive materials on strategies and proper metadata elements for many of archeology's measurement systems at the Archaeology Data Service/Digital Antiquity's Guides to Good Practice. Some particularly relevant ones are:

- Scanning (Payne 2011)
- Close range photogrammetry (Barnes 2009)
- Aerial photography (Bewley et al. 1998)
- Virtual reality (Fernie and Richards 2002).

Other essential recent reading on the topic includes Corsi et al. (2013), Bentkowska-Kafel et al. (2012) and Remondino and Campana (2014) to select a key few. The requirements of quality metadata are substantial. For example Table 1 is the project level metadata for scanning while Table 2 is the metadata needed for each scan. Given the finite amount of time/money in any field or lab project there is always tension—one can scan another object or record the metadata for the ones already done—but not both. Just as we would not pull an object from a pit wall and not record its provenance and its other "metadata" we need to be fierce in our certainty that digital metadata is essential for digital objects. Without metadata they will have been "digitally looted" from their contexts.

While comprehensive metadata is essential for the evaluation of data by future scholars it is also essential that the digital objects are accessible and can be found through search and can be readily cited when reused by others. The requirements

Table 1 Project level laser scanning metadata. *Source* Payne (2011; Table 3.2.1)

Element	Description
Project name	Name of the project
Name of monument, survey area, or object	Name of object, monument, or area scanned
Monument/object number	The ID number or code, if applicable, of the object or monument
Monument/object description	Brief description of the monument/object being scanned
Survey location	Exact location of survey with complete address and/or coordinates
Survey date(s)	Dates(s) of survey
Survey conditions	The overall weather trend during survey (sunny, overcast, indoors, etc.)
Scanner details	Details of the instrument(s) with serial number(s) and scan units
Company/operator Name	Details of company and scan operator name
Control data collected?	If yes, then list *control_file_name.txt*
Turntable used?	Yes/No
RGB data capture?	If yes, then specify whether: – Internal or external? – Was an additional lighting system used? If yes, then provide a brief description of the lighting system
Estimated data resolution	The estimated data resolution across the monument or object
Total number of scans in project	Total number of scans
Description of final datasets for archive	What datasets will be archived (include file names if possible)
Planimetric map of scan coverage areas	If applicable, then provide the image name.
Additional project notes	Additional notes
Images from survey	Optional, if yes, then provide the image file names

Table 2 Laser scan level metadata. *Source* Payne (2011; Table 3.2.2)

Element	Description
Scan filename	The name of the scan. A suggested filename for original raw scans for archiving is in this format: *ProjectName_scan1.txt*.
Scan transformation matrix	The name of the transformation matrix used in Global Registration. Suggested file name: *ProjectName_scan1_mtrx.txt*
Matrix applied to scan?	Y/N Has the matrix been applied to the archived scan?
Name of monument/object	Name of monument or object being scanned
Survey date	Date of scan

(continued)

Table 2 (continued)

Element	Description
Data resolution	Fixed resolution or data resolution at specific range.
Number of points in scan	Total number of points in the scan file
Additional scan notes	Additional notes
Scanner technology	TOF/phase/triangulation
First or last return (TOF scans only)	Indicated whether first, last, or both returns were specified
Camera exposure settings (If RGB acquired) (TOF and phase scans)	Exposure settings set on scanner
Frequency settings (Phase scans only)	Frequency settings set on scanner
Noise settings (Phase scans only)	Noise settings set on scanner
Lense or FOV details (Triangulation scans only)	Indicate which lense or FOV was used during scan

and issues associated with creation of capable digital archives have been well document (Chan 2004). I would only highlight two key ideas. The first is the critical role of persistent uniform resource locators (PURL). For digital objects created through visualization or measurement the PURL is the citation that follows the object. The second is the use of Context Objects in Span (COinS—ocoins.org) in the web pages that provide access to the digital objects. The presence of COinS code means that scholars using any capable citation software (e.g. Zotero, EndNote, etc.) can immediately obtain the appropriate citation data and utilize it in their future publications.

Figure 6 visually illustrates the importance of COinS to effective scholarly re-use of digital objects. The Zotero screen displays the citation data that is automatically populated by a single mouse click when the user access the digital object.

Decomposition of Representation

If we are to have robust and defensible representations we must move from a single "take it or leave it" image, video, game or other visualization to one that is constructed from multiple elements. Our scholarly articles are composed of individual paragraphs (each usually an idea) and those who would wish to use, attack or build on our work can cite a specific section or extract text and embed it in their publications. These are the essential parts of the scholarly armature. Many of our visualizations, however, are "of a whole" and difficult, if not impossible to decompose. What I would urge is that each element in a representation should have the same status as a paragraph or a sentence in a text. Each post, each object, each surface should have its own digital existence so that the next scholar may "reuse" or cite the specific content and critique each element. By way of illustration Fig. 7 shows a Hampson digital object that has been placed in the (then) Google 3D

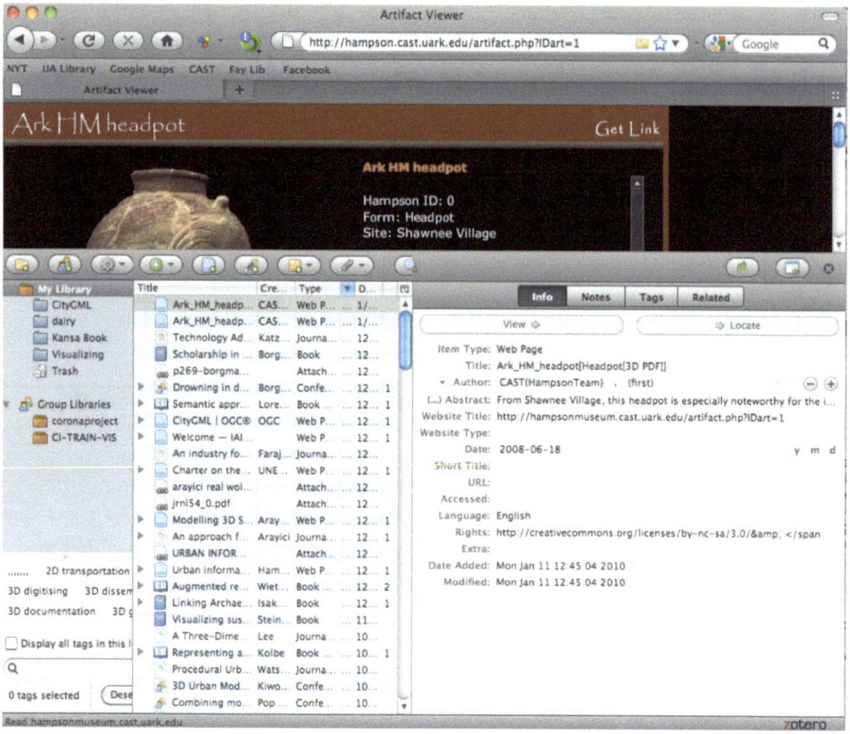

Fig. 6 CoINs produced output for digital objects from the Hampson Virtual Museum (hampson.cast.uark.edu)

warehouse. The following illustration (Fig. 8) shows how that specific digital object can be easily added (simply drop and drag) to an existing representation and (via the metadata and COinS) it is then possible that each individual element can be referenced and discussed. Until we have a structure in which we can readily decompose and "cite" specific elements in a representation, as we do a paragraph in a text, our representations will never have a robust place in scholarly discourse.

Reuse—The Ultimate Objective

All of these steps are designed to provide the lineage and accessibility for digital objects such that the objects themselves, and, more fundamentally, the measurements and/or ideas that underlie them can be re-used and/or critiqued. The Hampson Museum, accessible on the web since 2008 under a CC 3.0 license (Limp et al. 2011), provides an illustration of the potential for re-use. Since its creation the digital objects themselves have served as analytical elements in five different research projects by others with objectives completely different from the ones that

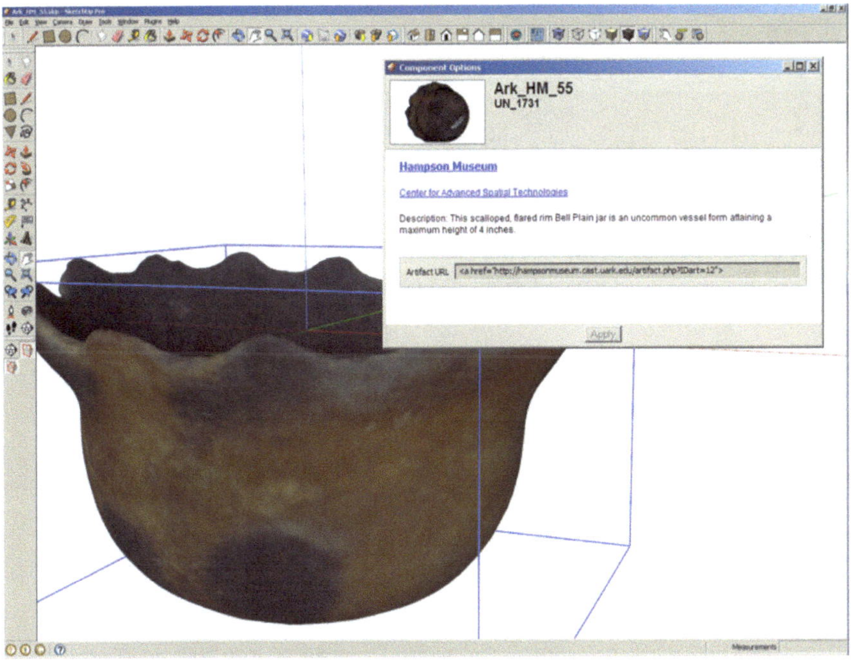

Fig. 7 Digital object from the Hampson Museum in the Google (now Trimble) 3D Warehouse (3dwarehouse.sketchup.com)

served as the basis for the initial project (Gilboa et al. 2013; Koutsoudis and Chamzas 2010, 2011; Sfikas et al. 2013; Savelonas et al. 2015). In an unexpected reuse the digital objects were used as the basis for a 2013 art installation in Toronto (Hanes 2013).

Published HDSM data allows others to produce 3D printed versions of the object and Olson et al. (2014) and Reilly and Beale (2014) provide examples of the value of 3D printing. Figure 9 shows 3D printed versions of a digital object from the Hampson project discussed above. We are very early on in the use of 3D printing but it is clear that the explosive growth of the "maker culture" (c.f. Tanenbaum et al. 2013) and the tools and strategies that it enables will impact archaeology and heritage studies even more over the next years.

Sustainability

All our extraordinary measurement and representation efforts will be for naught unless we can insure that they will be available for future scholars and not just in the next few years but over decades and centuries. Archaeology's digital record, so far, is not encouraging. I was fortunate enough to part of one of the earlier

Fig. 8 Introducing scholarly content (a digital object) into a new narrative

Fig. 9 Two 3D printed versions of one of the Hampson digital objects. Photo courtesy Hanna Ford

summaries of the state-of-the-art in computer data bases (then called "databanks") in the late 1970s published as *Databank Applications in Archaeology* (Gaines 1981). The edited volume described 10 different "databanks" then in use (see also Loewen 2015). As best as I can determine today, some 35 years later, only two are still "alive." In general those "systems" that were not part of an institutional infrastructure (as opposed to individual researcher or even groups of researchers) did not survive. The two that did, AZSITE and AMASDA have had the ongoing commitments of the Arizona State Museum and the Arkansas Archaeological Survey, respectively. Even an institutional home does not insure continuity. The Oracle System (oh how I wished we had trademarked that name) developed in the late 1970s at the Glenn Black Lab of Archaeology at Indiana University (Limp 1978a, b) died a slow death more than 20 years ago. Both AZSITE and AMASDA are examples of Washington's axe—the one he used to cut down the cherry tree. It is still his axe but it has had three new handles and two new axe heads. IN AZSITE and AMASDA it is only the data that endures. AMASDA, with which I am more familiar, has gone though multiple software incarnations. Originally developed in GRIPHOS (Scholtz and Chenhall 1976) it has since had versions in INFORMIX and MySQL across a number of operating systems. Each time the existing data was migrated. It is clear to me, at least, that no matter the capabilities, work and enthusiasm of any individual or group of individuals data, especially digital data, cannot long survive unless there is institutional support. Institutions are complex and often slow moving and it's much easier for an research team to move ahead on their own. We must address the long-term sustainability of a project's digital products. Not recording metadata is digital looting—not placing the digital objects in a sustainable archive will mean that, very shortly, they will be lost and inaccessible.

Digital Publication Issues

The growing use of HDSM and computer based representation are straining the traditional structures of publication. We have seen a new journal *Digital Applications in Archaeology and Cultural Heritage* that has adopted the publication of 3D models as a central focus. The ongoing projects at Catalhöyük (Forte 2014, Forte et al. 2012, 2015) and Gabii (https://sites.lsa.umich.edu/gabiiproject/gabii digital and http://m-gabii.adsroot.itcs.umich.edu/gabiigoesdigital/) have both extensive HDSM efforts and are developing strategies for 3D publication of the results. We are early in this process and many conceptual and practical matters still need resolution. But they will be resolved and our communications about the past will be much richer.

Conclusions

We must always be aware that our excitement surrounding new methods in our field should be mediated by our failures (and successes) in the past. Our attitude towards new methods in research, like serial polygamy, can often be seen as the triumph of hope over experience. Nonetheless I do think that what we are seeing with HDSM and new directions in representation do imply significant changes in our study of the past. A key impact of HDSM is the potential to avoid or at least reduce innattentional blindness.

> "[A]nyone who has followed the ground breaking work of Simons and Chabris (1999) on "inattentional blindness" is well aware that we [truly] only see what we are prepared to see, a point discussed in an archaeological context by Barceló (2010, 93). Semi-automatic and metrics-driven analyses, promoted by HDSM simply by dint of the scale of the data, provide a parallel abstractive process and the results of these analyses may point us at aspects of the data or information sitting in our inattentional blindspots" (Opitz and Limp 2015).

At the same time we must also recognize that these methods come with a cost. They add complexity to our field activities, developing practitioners is demanding and the equipment can be expensive. Our "budgets" for projects, whether calculated in actual money or in the time and effort involved, are likely fixed. Thus "something must give." There is the hope that some methods can actually speed up field recording (see the Catalhöyük and Gabbi projects) but it seems likely that all in all these new approaches will add to the total "budget." What is it that we don't do when we do HDSM or computational visualizations?

As I hope I have made clear robust institutional structures are necessary, to field these methods in the first place, but more critically to insure that the results will be around for future scholars to use. Recent efforts by NSF and NEH to require data management plans send a strong signal to the field that storage, maintenance and access to these digital products are an essential part of research and most universities are moving to set up institutional repositories to insure that they meet NSF requirements. But the great majority of these efforts do not anticipate the complex digital objects created by HDSM or the digital products created by visualizations. The success of the Archaeology Data Service in Britain and the Digital Archaeological Record in the US provide a basis for optimism that we can address this need but also give us a clear sense of the complexity and cost that sustainable digital archives require.

References

Ackoff, Russell L. 1989. "From Data to Wisdom: Presidential Address to ISGSR, June 1988." *Journal of Applied Systems Analysis* 16 (1): 3–9.

Anon. 2001. "In Memoriam: Dr. Joffre Lanning Coe." *Common Ground: A Publication of the Coe Foundation for Archaeological Reserch* 13 (1): 1–3.

Barceló, Juan A. 2010. "Computational Intelligence in Archaeology." In *EUROGRAPHICS Workshops on Graphics and Cultural Heritage*, preprint. BAR International Series S2079. Archaeopress.

Barnes, Adam. 2009. "Close-Range Photogrammetry A Guide to Good Practice#." *ADS Guides to Good Practice*. http://guides.archaeologydataservice.ac.uk/g2gp/Photogram_Toc.

Bellinger, Gene, Durval Castro, and Anthony Mills. 2004. *Data, Information, Knowledge, and Wisdom*. http://geoffreyanderson.net/capstone/export/37/trunk/research/ackoffDiscussion.pdf.

Bentkowska-Kafel, Anna, Hugh Denard, and Drew Baker, eds. 2012. *Paradata and Transparency in Virtual Heritage*. Farnham, UK: Ashgate Publishing, Ltd.

Bergland, Richard. 1986. *The Fabric of Mind*. Viking.

Bewley, Robert, Danny O'Donoghue, Vince Gaffney, Martijn van Leusen, and Alicia Wise. 1998. "Archiving Aerial Photography and Remote Sensing Data: A Guide to Good Practice." http://www.narcis.nl/publication/RecordID/oai:pure.rug.nl:publications%2Ff2e07245-a095-4014-ac9c-93679668098c.

Brück, Joanna. 2005. "Experiencing the Past? The Development of a Phenomenological Archaeology in British Prehistory." *Archaeological Dialogues* 12 (01): 45–72. doi:10.1017/S1380203805001583.

Chan, Leslie. 2004. "Supporting and Enhancing Scholarship in the Digital Age: The Role of Open Access Institutional Repository." *Canadian Journal of Communication* 29 (3). http://cjc-online.ca/index.php/journal/article/view/1455.

Cochrane, Andrew, and Ian Russell. 2007. "Visualizing Archaeologies: A Manifesto." *Cambridge Archaeological Journal* 17 (01): 3–19. doi:10.1017/S0959774307000029.

Corsi, Cristina, Božidar Slapšak, and Frank Vermeulen, eds. 2013. *Good Practice in Archaeological Diagnostics*. Natural Science in Archaeology. Cham: Springer International Publishing. http://link.springer.com/10.1007/978-3-319-01784-6.

da Vinci, Leonard. 1906. *Thoughts on Art and Life*. Translated by Maurice Baring. Whitefish (Mont.): Kessinger Publishing.

Edwards, Betty. 1979. *Drawing on the Right Side of the Brain*. Fontana/Collins.

Fernie, Kate, and Julian Richards. 2002. "Creating and Using Virtual Reality: A Guide for the Arts and Humaities." http://guides.archaeologydataservice.ac.uk/g2gp/Vr_Toc.

Forte, M., 2014, 3D Archaeology. New Perspectives and Challenges. The example of Çatalhöyük, in *Journal of Eastern Mediterranean Archaeology*, Vol. 2, n. 1, 2014.

M. Forte, N. Dell'Unto, K. N. Lercari, K.Jonsson, 2015 "3D and Archaeological Interpretation at Catalhuyuk", in *Assembling Çatalhöyük*, ed. By I. Hodder and A. Marciniak, Maney Publishing,

Forte, M., Dell'Unto, N., Issavi, J., Onsurez, L. & Lercari, N. 2012. "3D Archaeology at Çatalhöyük." *International Journal of Heritage in the Digital Era*, 1: 351–378.

Gaines, Sylvia W., ed. 1981. *Data Bank Applications in Archaeology*. Tuscon, AZ: Univ of Arizona Pr.

Gilboa, Ayelet, Ayellet Tal, Ilan Shimshoni, and Michael Kolomenkin. 2013. "Computer-Based, Automatic Recording and Illustration of Complex Archaeological Artifacts." *Journal of Archaeological Science* 40 (2): 1329–39. doi:10.1016/j.jas.2012.09.018.

Griffin, John W., and Patricia C. Griffin. 1996. *Fifty Years of Southeastern Archaeology: Selected Works of John W. Griffin*. University Press of Florida.

Hanes, David. 2013. "Review of 'An Illuminated iDentity' an Art Exhibit at the Ontario College of Art and Design." http://xpace.info/wp-content/uploads/2013/04/Main-Space_iDentity_2013.pdf.

Kansa, Eric. 2005. "A Community Approach to Data Integration: Authorship and Building Meaningful Links across Diverse Archaeological Data Sets." *Geosphere* 1 (2): 97–109.

Koutsoudis, Anestis, and Christodoulos Chamzas. 2010. "3D Pottery Shape Similarity Matching Based on Digital Signatures." In *Making History Interactive: Computer Applications and Quantitative Methods in Archaeology (CAA) ; Proceedings of the 37th International Conference, Williamsburg, Virginia, United States of America, March 22–26, 2009*, edited by Bernard Frischer, 2079:166–71. BAR International Series. Oxford. Archaeopress.

Koutsoudis, Anestis, and Christodoulos Chamzas. 2011. "3D Pottery Shape Matching Using Depth Map Images." *Journal of Cultural Heritage* 12 (2): 128–33. doi:10.1016/j.culher.2010.12.003.

Limp, Fred, Angelia Payne, Katie Simon, Snow Winters, and Jackson Cothren. 2011. "Developing a 3-D Digital Heritage Ecosystem: From Object to Representation and the Role of a Virtual Museum in the 21st Century." *Internet Archaeology*, no. 30. intarch.ac.uk/journal/issue30/limp_toc.html.

Limp, Fred. 1978a. *The Oracle System Survey Manual*. Research Report 3. Bloomington IN: Glenn Black Laboratory of Archaeology.

Limp, Fred. 1978b. *The Oracle Systems User's Manual*. Research Report 2. Bloomington, IN: Glenn Black Laboratory of Archaeology.

Limp, Fred, and Adam Barnes. 2014. "Solving the Grid-to-Ground Problem When Using High Precision GNSS in Archaeological Mapping." *Advances in Archaeological Practice: A Journal of the Society for American Archaeology* 2 (2): 138–43. doi:10.7183/2326-3768.2.2.138.

Loewen, G. V. 2015. "Thinking the Past Anew: The Prehistory of Non-Linear Digital Systems in Archaeology: 1960–1990." Accessed June 5. http://irssh.com/yahoo_site_admin/assets/docs/6_IRSSH-55-V2N1.51195229.pdf.

Niccolucci, Franco, Sorin Hermon, and Doerr. 2015. "The Formal Logical Foundations of Archaeological Ontologies." In *Mathematics and Archaeology*, edited by Juan A. Barcelo and Igor Bogdanovic, 86–99. Boca Raton, FL: CRC Press Taylor & Francis Group.

Olson, Brandon R., Jody M. Gordon, Curtis Runnels, and Steve Chomyszak. 2014. "Experimental Three-Dimensional Printing of a Lower Palaeolithic Handaxe: An Assessment of the Technology and Analytical Value." *Lithic Technology* 39 (3): 162–72.

Opitz, Rachel, and Fred Limp. 2015. "Recent Developments in HDSM for Archaeology: Implications for Practice and Theory." *Annual Review of Anthropology* 44 (1). doi:10.1146/annurev-anthro-102214-013845.

Payne, Angelia. 2011. "Laser Scanning for Archaeology: A Guide to Good Practice." York University: Archeological Data Service. http://guides.archaeologydataservice.ac.uk/g2gp/LaserScan_Toc.

Payne, Angelia, and Fred Limp. 2008. "3D Visualizations of the Upper Nodena Site." http://hampson.cast.uark.edu/nodena_3D.htm.

Petrie, Sir William Matthew Flinders. 1904. *Methods & Aims in Archaeology*. Macmillan.

Read, Dwight W. 1974. "Some Comments on Typologies in Archaeology and an Outline of a Methodology." *American Antiquity* 39 (2): 216–42. doi:10.2307/279584.

Reilly, Paul, and Gareth Beale. 2014. "Additive Archaeology: Towards a Virtual Archaeology Reprinted?" *CAA-GR Conference 2014 (CAA-GR 2014), Rythymnon, GR, 07–08 Mar 2014.* 19 pp. http://eprints.soton.ac.uk/364455/.

Remondino, Fabio, and Stefano Campana, eds. 2014. *3D Recording and Modelling in Archaeology and Cultural Heritage: Theory and Best Practices*. BAR International Series 2598. Oxford England: Archaeopress.

Rockwell, W. Teed. 2005. *Neither Brain Nor Ghost: A Nondualist Alternative to the Mind-Brain Identity Theory*. MIT Press.

Savelonas, Michalis A., Ioannis Pratikakis, and Konstantinos Sfikas. 2015. "Partial 3D Object Retrieval Combining Local Shape Descriptors with Global Fisher Vectors." In *Eurographics Workshop on 3D Object Retrieval*, edited by I. Pratikakis, M. Spagnuolo, T. Theoharis, L. Van Gool, and R. Veltkamp. The Eurographics Association. doi:10.2312/3dor.20151051.

Scholtz, Sandra, and Robert G. Chenhall. 1976. "Archaeological Data Banks in Theory and Practice." *American Antiquity* 41 (1): 89–96.

Sfikas, Konstantinos, Ioannis Pratikakis, Anestis Koutsoudis, Michalis Savelonas, and Theoharis Theoharis. 2013. "3D Object Partial Matching Using Panoramic Views." In *New Trends in Image Analysis and Processing – ICIAP 2013*, edited by Alfredo Petrosino, Lucia Maddalena, and Pietro Pala, 169–78. Lecture Notes in Computer Science 8158. Springer Berlin Heidelberg. http://link.springer.com/chapter/10.1007/978-3-642-41190-8_19.

Simons, Daniel J., and Christopher F. Chabris. 1999. "Gorillas in Our Midst: Sustained Inattentional Blindness for Dynamic Events." *Perception-London* 28 (9): 1059–74.

Spaulding, Albert C. 1960. "The Dimensions of Archaeology." In *Essays in the Science of Culture*, 437–56. New York City, New York: Cromwell.

Tanenbaum, Joshua G., Amanda M. Williams, Audrey Desjardins, and Karen Tanenbaum. 2013. "Democratizing Technology: Pleasure, Utility and Expressiveness in DIY and Maker Practice." In *Proceedings of the SIGCHI Conference on Human Factors in Computing Systems*, 2603–12. CHI '13. New York, NY, USA: ACM. doi:10.1145/2470654.2481360.

Trigger, Bruce G. 1989. *A History of Archaeological Thought*. Cambridge University Press.

West, Thomas G. 2009. *In the Mind's Eye: Creative Visual Thinkers, Gifted Dyslexics, and the Rise of Visual Technologies*. Prometheus Books.

Willey, Gordon R. 1953. "Archaeological Theories and Interpretation: New World." *Anthropology Today, University of Chicago Press, Chicago*, 361–85.

An Integrated Archaeological Prospection and Excavation Approach at a Middle Neolithic Circular Ditch Enclosure in Austria

Jakob Kainz

Abstract The aim of this paper is to present an approach combining archaeological excavation with geophysical prospection. This is achieved by a combination of magnetometry, magnetic susceptibility, ground penetrating radar (GPR) and pXRF measurements, on archaeological features before and during excavation. Soil properties, such as soil colour, organic content, pH, magnetic susceptibility, chemistry and composition are influenced by natural and human activities and these changes can be identified by various prospection methods. The data was collected at the Middle Neolithic circular ditched enclosure (Kreisgrabenanlage, KGA) at Hornsburg, Austria in the Kreuttal area, which is a case study area of the Ludwig Boltzmann Institute for Archaeological Prospection and Virtual Archaeology (LBI Arch Pro, http://archpro.lbg.ac.at/). Archaeological prospection in the majority of cases is carried out prior to excavation, in order to map the archaeology for the excavation or for planning procedures. The full archaeological potential of the various prospection methods therefore is not attained; as these measurements can help corroborate excavation results as well as providing further archaeological data that cannot be seen by the excavator's eye. Furthermore, excavations provide an opportunity to investigate specific anomalies allowing for an examination of the processes, whether human or natural, influencing the prospectability or non-prospectability of these features. This can provide a link between past human actions and specific anomaly signatures, adding further archaeological interpretation to the prospection data as well as providing a greater archaeological insight during and after the excavation. Data analysis is still ongoing, so the paper will mainly focus on preliminary results obtained from the magnetometry and magnetic susceptibility measurements. Additionally smaller contributions from GPR, aerial photographs and orthophotos are presented here whilst future publications will integrate these alongside ultraviolet and infrared photographs and pXRF measurements.

J. Kainz (✉)
Initiative College for Archaeological Prospection, Vienna Institute for Archaeological Science, University of Vienna, Vienna, Austria
e-mail: jakob.kainz@univie.ac.at

J. Kainz
Vienna Insitute for Archaeological Science, University of Vienna, Vienna, Austria

M. Forte and S. Campana (eds.), *Digital Methods and Remote Sensing in Archaeology*, Quantitative Methods in the Humanities and Social Sciences,
DOI 10.1007/978-3-319-40658-9_17

Introduction

Buried archaeological sites are a rare and precious commodity of our cultural heritage, often under the threat of destruction. Their investigation and management should involve minimal damage or destruction. Due to this, archaeology has a need for fast and accurate mapping of such buried anthropogenic traces, in order to preserve and manage them or in other cases to plan the most cost-effective form of excavation or investigation. Therefore, various non-invasive archaeological prospection methods, such as geophysical prospection and remote-sensing techniques, are generally used to map archaeological structures in the subsurface.

Archaeological prospection methods in general work on the principle that archaeological structures possess physical or chemical differences from their surrounding matrix; and these therefore these are in contrast to the soil or sediment in which they are imbedded (Scollar et al. 1990). These differences are called anomalies and can be of anthropogenic (pits, ditches etc.), geological (bedrock), geomorphological (palaeochannels) and biological natures (animal dens). Each different method responds to certain physical properties. Geophysical methods such as magnetometry relate to soil magnetism whereas ground penetrating radar relates to the relative electrical permittivity properties of the soil. In remote sensing, such as aerial photography, archaeological structures are visible due to their physical and chemical properties seen as soil, moisture or crop marks (Hejcman et al. 2013). Gaining knowledge of the physical properties causing such anomalies will help significantly, as often geophysical anomalies are not easy to understand and therefore need a better knowledge of the anomalies' soils and sediments physical properties (Fassbinder and Stanjek 1993; Fassbinder 2015). Not only will this improve the understanding of why some features are detected by some methods and not by others, but it also can provide new insights into their archaeological interpretation.

Recently there have been two major incentives within archaeological prospection. The first is changing the focal point from the local to a landscape approach. This has partially been due to the need of understanding the archaeological features in terms of the 'site' they form, as well as the 'site' in terms of its landscape. Advances in technology have also allowed extensive surveys to be undertaken ranging from complete cities to landscapes (Gaffney et al. 2000, 2012; Powlesland et al. 2006). The second is the application of multi-method approaches (Neubauer & Eder-Hinterleitner 1997; Gaffney et al. 2000; Neubauer 2001a, b; Clay 2001; Kvamme 2007; Keay et al. 2009) for the reason that additional information can be gained by comparing data from different techniques. Different archaeological structures manifest themselves differently in the various techniques, due to the variations in their physical properties, and hence such an approach offers confirmatory, complementary and new information on archaeological and non-archaeological anomalies.

In order to relate the archaeological anomalies from the various prospection methods with their physical properties and the structures themselves, such a multi-method approach prior to excavation is ideally complemented by measurements undertaken during excavation. The latter can answer questions relating to the basic archaeological materials, soil and sediments and therefore provides an insight into the dynamics of the formation and deformation of archaeological structures and terrains.

The circular ditched enclosures (*Kreisgrabenanlagen* in German, henceforth KGA) of Central Europe date to the Middle Neolithic period (4800–4500 BC). Among the oldest monumental structures known in Central Europe, they represent one of the earliest form of large transcultural ritual monuments in Europe (Neubauer 2012), which can be seen as a reinterpretation of social organisation caused by an increase in social complexity (Ridky et al. 2014). These monuments can be found throughout Central Europe; in Germany, Poland, Czech Republic, Slovakia, Austria and west Hungary (Melichar and Neubauer 2010; Ridky et al. 2014). KGA's consist of one to four concentric V-shaped ditches, ranging from 45 to 180 m in diameter, with at least two entrances. The interior is encircled by one or two wooden palisades creating a defined space for communal events. These could serve various functions, ranging from meeting-places to ritual centres, possible astronomical characteristics indicating some form of calendrical use (Petrasch 1990; Trnka 1991; Becker 1996; Podborský 1999; Neubauer 2012). No obvious defensive function has been identified. Often these monuments have related settlements in close proximity but show no evidence of being used for habitation themselves. Nonetheless, the KGA phenomenon has not been explained fully, with function still open to debate, and it disappears as rapidly as it starts, lasting only 200–300 years.

Location

The Hornsburg I KGA lies on a ridge to the west of the village of Hornsburg in the Kreuttal area, Lower Austria, Austria which is a case study of the LBI Arch Pro, approximately 20 km to the north of Vienna (Fig. 1). The Kreuttal valley lies between the Bay of Korneuburg and the Vienna Basin and cuts east-west through a north-south running forested ridge. The area is dominated by loess deposits on which a multi-period landscape, dating from the Neolithic to modern times, is preserved. Sites include settlements, enclosures, cemeteries, hillforts, KGA's and a motte dating from the early Neolithic to medieval times as well as modern elements. About 500 m to the west, on the opposite slope above the village of Hornsburg, lays another KGA, Hornsburg II. Further to the west of Hornsburg lies one of the largest Early Neolithic enclosures, Grossrussbach-Weinsteig, and other nearby Middle Neolithic KGA's are known at Würnitz and Kleinrötz. Further Neolithic to Iron Age occupation has been identified through extensive magnetometry surveys to the north at Oberkreuzstetten (LBI Ach Pro). Two possible Bronze Age enclosures are situated in close proximity, one to the north-west of Hornsburg on the Ochsenberg and the other enclosure in the forest to the south. An Iron Age hillfort to the south-west of Hornsburg lies directly above the Kreuttal valley on the northern slope. A Medieval motte is also situated in the southern part of the village of Hornsburg (http://lbi-archpro.org/cs/kreuttal/motivation.html).

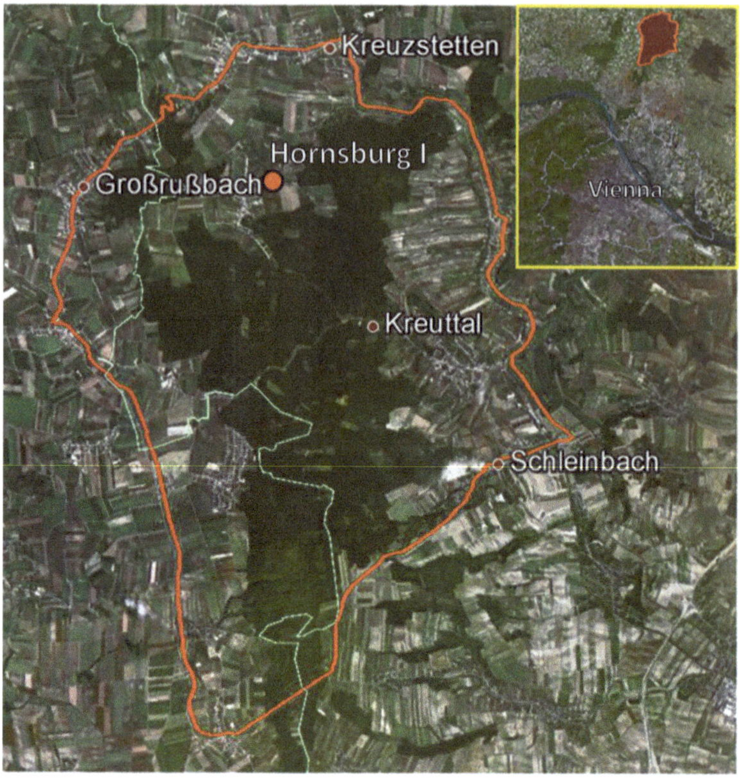

Fig. 1 Location of Hornsburg I and the LBI Arch Pro Kreuttal case study area (Google Earth 2012)

The Site

Hornsburg I (Melichar and Neubauer 2010) lies on a ridge to the east of the village of Hornsburg on a plateau situated between the watershed of the Haselauer and Hautzendorfer streams. These flow into the Rußbach stream which runs through the Kreuttal valley to the south. Hornsburg II, situated on a ridge to the west of Hornsburg and has an adjoining settlement to the north. Both the sites are visible from each other. Hornsburg I was discovered through aerial photography in 1982 and subsequent magnetometry surveys were carried out in 1989, 2003 and 2014 by Archaeo Prospections (http://www.zamg.ac.at/cms/en/products/geophysics/archeo-prospections-r) and the LBI Arch Pro.

The magnetogram (Fig. 2) shows various dipole anomalies caused by iron objects within the ploughzone and a wayside cross with adjoining track between the innermost and middle ditch in the north-eastern part of the KGA. An old track can be seen crossing over the north-eastern entrance from east to west and a present day track runs slightly to the east of the KGA's centre from north to south. The KGA

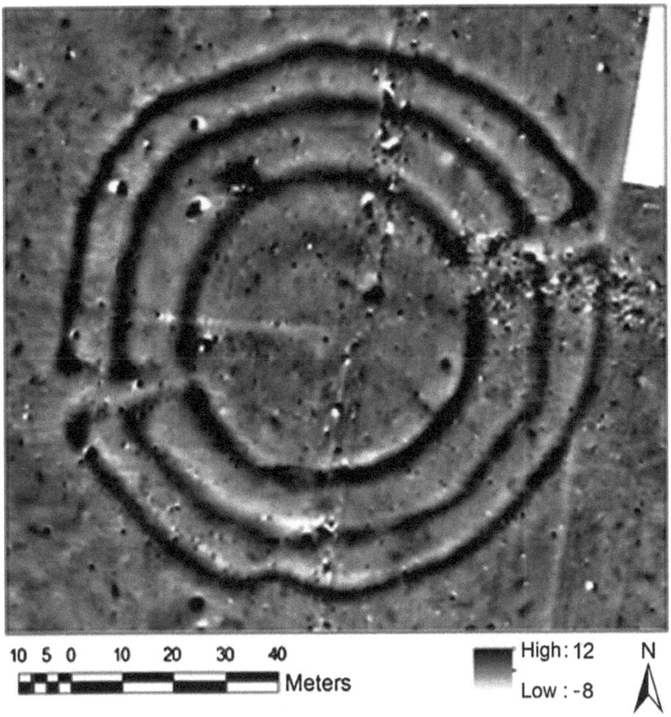

Fig. 2 Magnetogram of Hornsburg I (Melichar and Neubauer 2010, Archaeo prospections)

consists of three concentric V-shaped ditches with two entrance passages roughly aligned southwest to northeast. The outermost ditch measures 106 m in diameter and varies between 4 and 5 m in width, the middle ditch has a diameter of 85 m and is 2.8–5.5 m wide while the innermost ditch is 58 m in diameter and varies between 3–5 m in width. At the entrances, the innermost and outermost ditch ends bend inwards along the earth bridges of the entrance passages whereas the innermost ditch ends are rounded. A palisade measuring 46 m in diameter encircles the central area. The magnetogram suggests another possible palisade around it; although this was not verified in excavation. Two pits were identified in the central area, while various pit complexes, pits and postholes can be recognised in the vicinity of the KGA indicating possible related structures.

An excavation was carried out in 1987 by the Institute of Prehistoric and Historical Archaeology, University of Vienna (uha.univie.ac.at) on the ditches in the south-eastern part of the KGA (Trnka 1991). This consisted of a trench 35 m long and 2 m wide. Further excavations were carried out by the same department in cooperation with the LBI Arch Pro in 2013 and 2014 with a trench covering the south-western entrance passage, palisade and part of the interior during which the measurements below were taken.

Methods

Multi-method Approach

A multi-method approach can improve the understanding, not only of the various prospection data themselves (Neubauer 2001b; Schmidt 2001; Linford 2005) but more importantly of the archaeology itself, which is the prime reason for applying the surveys. The various methods, due to their different nature, complement each other. This allows for the comparison of different archaeological features obtained from a variety of techniques, in which the features may manifest themselves differently due to the physical properties of the features themselves as well as the properties the different techniques measure, thus providing a rich source of data from which a research framework can be made. The focus of this paper is the establishment of a magnetic susceptibility framework for the excavation to aid the interpretation of the archaeology as well as gaining a better insight of the magnetic anomalies.

A comparison of the multi-method archaeological prospection data (magnetic, GPR, aerial photography) allows identification of where the data complement each other. This can be achieved through the analysis of the data via GIS or through the use of data fusion approaches (Neubauer & Eder-Hinterleitner 1997; Schmidt 2001; Kvamme 2007; Keay et al. 2009). Such allows for an analysis of the different archaeological anomalies which are visible/invisible in the various prospection data. The data was combined in ArcGIS and ArcScene for this purpose.

As archaeological prospection anomalies are the product of numerous interacting formation processes, not all of which are possible to control for or identify, the focus will be on formation processes and properties most relevant to the archaeological prospection methods described in the paper, namely magnetic susceptibility for the magnetometry results. The measurements enabled the identification of important sources of variation in the archaeological record and permitted their association with specific anomalies, thus allowing an assessment of the relationship of physical properties, via magnetic susceptibility, to certain anomaly signatures. This provides a qualitative and quantitative insight into the relationships between formation processes and prospectability of archaeological features.

Magnetic Measurements

Magnetic susceptibility is an active method and is a measure of the degree to which a material can be magnetised (Dearing 1999; Evans and Heller 2003; Dalan 2006). Magnetic susceptibility allows for the examination of the induced magnetic component of a material. In comparison, magnetometer surveys, a passive technique, measures the variations in the earth's magnetic field. The values measured make no differentiation between magnetic changes due to induced or remnant magnetisation

(Dalan 2006), and therefore characterise only the net effect in the earth's magnetic field. Induced magnetization is the magnetization of a sample in the presence of an inducing magnetic field, i.e. the earth's magnetic field, whereas remnant and permanent magnetization is the magnetization in the absence of a magnetic field (Neubauer 2001a; Evans and Heller 2003; Fassbinder 2015). The magnetisation of a material is possible and dependent on the inherent forces or energies generated by electrons that form atoms (Dearing 1999). The orbiting electrons generate magnetic fields and all atoms react to magnetic fields and therefore have a magnetic susceptibility (Clark 1996; Neubauer 2001a). Thus, magnetic susceptibility is the ratio of the intensity of the induced magnetization to that of the inducing magnetic field and values measured quantify the response of a material to an external magnetic field (Dalan and Banerjee 1998).

Magnetic Properties

Archaeological features become detectable in magnetic surveys due to the enrichment of soil and sediments with ferrimagnetic minerals, as maghaemite, magnetite, titanomagnetite and greigite, caused by pedogenic, anthropogenic and bacterial action (Neubauer 2001a; Fassbinder 2015). As soils and sediments are often comprised from various sources, these will therefore also have a mixture of minerals varying in concentration, composition and crystal size, shape and domain (Dearing 1999; Neubauer 2001a). The sum of which will affect a soil's, sediment's and material's magnetic properties and therefore its susceptibility.

The magnetic properties of a material can be used to identify, differentiate and categorise different materials (Dearing 1999; Dalan and Bevan 2002). This can be used in a similar way to the description of contexts in archaeology, in which different contexts are categorised by their colour, texture and inclusions. Understanding the magnetic mineralogy, the composition, concentration and grain size and shape, can help to distinguish between different contexts and provides an opportunity to understand their varied histories (Evans & Heller 2003). This is possible as magnetic susceptibility is environmentally sensitive and offers an access to soil forming factors (Dalan 2008) as soil colour, structure and fabric can partially be affected by iron oxides (Maher 1986; Fassbinder 2015). The magnetic enhancement of a material is conditioned not only to the degree of such but more importantly dependent on the original quantity and type of iron available for transformation, which depends on the geological and pedological background of a soil or sediment (Tite 1972; Maher 1986; Evans and Heller 2003; Crowther 2003). These factors can be used to gain an understanding of the soil formation processes, past environments and the nature of human impact (Dearing 1999).

Different physical properties affect the magnetic susceptibility of soils, sediments or materials and therefore their prospectability within magnetometry surveys. Sometimes anomalies detected via magnetic measurements can appear to be invisible during excavations (Dalan 2008; Simon et al. 2012) as only a minute change in the amount of ferrimagnetic iron oxides can alter the measured magnetic susceptibility (Dunlop and Özedemir 1997). The crystal structure of a magnetic mineral influence to what extent the magnetic effects of the electrons reinforce or balance each other out (Clark 1996). Therefore materials react differently to magnetism thus magnetic susceptibility varies from material to material due to different categories of magnetic behaviour which allows for differentiation between them.

The weakest magnetic behaviour is known as diamagnetism (Neubauer 2001a). This is common in materials that do not contain iron molecules, such as organic matter and water. In materials with such behaviour the magnetic field interacts with the orbital motion of the electrons to produce a very weak or negative magnetic susceptibility (Dearing 1999). In materials that display paramagnetism the magnetic moments of atoms or molecules align only when exposed to a magnetic field but disappears when the field is removed (Clark 1996; Neubauer 2001a). These have weak positive magnetic susceptibility values. Paramagnetic behaviour is often seen in materials containing iron and these minerals are often present in rocks and soils. The crystal structure in materials with antiferromagnetic behaviour allows for well-aligned but opposing magnetic moments (Dearing 1999; Neubauer 2001a). These virtually balance each other out, although with a slight misalignment (Clark 1996). Therefore these materials have a moderate positive magnetic susceptibility similar or higher to materials displaying paramagnetism. Only a few materials fall into this category, such as haematite. The most important category of magnetic behaviour is ferrimagnetism. The crystal structure in these minerals allows for the magnetic moments to be strongly aligned but in opposing sets with unequal forces (Dearing 1999; Neubauer 2001a). These minerals display a strong magnetic susceptibility and their behaviour is similar to ferromagnetic minerals but with a diminished magnetic moment. This is due to the iron atoms being in opposition to the remaining atoms. If present, these will dominate the magnetic susceptibility and include iron-bearing minerals such as magnetite and maghaemite (Clark 1996). Minerals, e.g. pure iron, are classified as ferromagnetic. In these, the magnetic moments are highly ordered and aligned in the same direction (Dearing 1999).

Minerals that fall into the ferro-, ferri- or antiferromagnetic category are able to remain magnetised in the absence of a magnetic field and therefore may be identified using remanence measurements (Dearing 1999), such as a magnetometer. The magnetic susceptibility of a material depends on the combined total of the magnetic susceptibility of all the ferri-, antiferro-, para- and diamagnetic minerals and components. The most important of these are the minerals in the ferrimagnetic category, as these have the highest magnetic susceptibility and contribute the most towards magnetic susceptibility, even if only present in a very small quantity (Dearing 1999). Diamagnetic components are usually very weak and often negative; these components can be ignored as they only have little influence in the overall magnetic susceptibility of a material.

The magnetic behaviour of a material does not only depend on the kind of magnetism it displays but also on the crystal size and domain. Magnetic grains are divided up into different cells of magnetisation and these are known as domains (Neubauer 2001a; Evans and Heller 2003). Adjacent domains have contrasting directions of magnetisation and are separated by walls, c.0.1 μm thick. Varying domain states and sizes are the results of different processes and therefore can be used as a further indicator. There are four main domain states (Dalan and Banerjee 1998). Magnetic grains (Dearing 1999) above diameters of ~110 μm are referred to as multidomain (MD) as energetically it is favourable to have more than one domain. Single domain (SD) grains are <0.2 μm and due to their small size only allow one domain. Pseudo-single domain (PSD) grains, 0.2–110 μm in size, are large enough to have more than one domain but have the properties of single domain grains. Superparamagnetic (SP) grains are ultrafine grains and are <0.03 μm in size. These are SD and have a strong magnetisation. However their magnetisation is unstable due to thermal energies counteracting induced magnetisation very quickly after the removal of a magnetic field. Similar behaviour is seen in paramagnetism except that with SP which has a much higher magnetic susceptibility.

Magnetic Susceptibility Measurements

Magnetic susceptibility can either be measured in the field or in the laboratory. Field instruments measure in volume magnetic susceptibility and laboratory measurements can be made in mass magnetic susceptibility. The laboratory allows for further magnetic susceptibility measurements to investigate the magnetic properties of different sediments and strata for a more methodical analysis (Maher 1986; Dalan and Banerjee 1998).

When interpreting and discussing magnetic susceptibility measurements, however, one should bring to attention that such measurements of soils and sediments are made of a large set and mixture of minerals with different values in a volume (Fassbinder 2015). Therefore measurements based on susceptibility exclusively, as Low Field Mass Susceptibility (X_{LF}), cannot account for the magnetic minerals in a sample. Such measurements are not absolute but rather a guidance or semi-quantitative assessment of the properties of the magnetic minerals within a soil or sediment. Further laboratory measurements (Maher 1986; Dalan and Banerjee 1998; Evans and Heller 2003) may involve: Frequency Dependence of X (X_{fd}), Anhysteretic Remanent Magnetization (ARM), Saturation magnetization (Js), Saturation Remanent Magnetization (Saturation Isothermal Remanent Magnetization) (Jrs (SIRM), Coercive Force (Hc), Coercivity of Remanence (Hcr), "S" Ratio or HIRM. These can provide a more accurate understanding of the composition, size and concentration of magnetic minerals; however some of these measurements can irreversibly change the magnetic minerals thus destroying the samples original state and prevent further measurements (Dalan and Banerjee 1998).

The paper focuses on the more basic magnetic susceptibility measurements of X_{LF} and X_{fd}, which are nonetheless very useful (Evans and Heller 2003). X_{LF} is the ratio of the strength of magnetisation, created by a weak magnetic field from an alternating current, to a magnetic field (Dalan and Banerjee 1998; Dearing 1999; Evans and Heller 2003). The other measurement carried out was X_{fd}, which is the measurement of a sample at a low frequency (X_{LF}), the standard measurement, with the Bartington MS2/3, at 0.46 kHz and a high frequency (X_{HF}) measurement, at 4.6 kHz (Dearing 1999; Evans and Heller 2003). This is helpful in indicating the presence of SP crystals, as X_{LF} measurements permit SP crystals near the border to SSD grains to contribute completely to the susceptibility (Dearing 1999). The domain boundary between SP and SSD crystals shifts towards smaller crystals when measuring X_{HF}. Therefore crystals with SP domains close to this boundary behave as SSD grains but lowering the susceptibility. The measurements in this paper are repeatable and less time consuming and especially more applicable during the time constrain of an excavation when compared to the highly informative but elaborate laboratory measurements. X_{LF} and X_{FD} can be done during an excavation, even in the field with the right setup; then the results can immediately flow into the feedback and management of an excavation. Whilst the more elaborate laboratory measurements, due to their time consuming nature, would only take place after an excavation is finished.

Magnetometry

This is the most common and developed archaeological geophysical investigation technique (Neubauer 1990; Scollar et al. 1990; Becker et al. 1996; Clark 1996; Gaffney and Gater 2003; Gaffney 2008; Fassbinder 2015) as the method dates back to 1958 (Aitken 1958) with continuous development until present (Neubauer 2001b; Fassbinder 2015). Its status results from the rapidity of data collection, the general applicability of the technique for shallow investigation, and the high data resolution which can be achieved (Neubauer 2001b; Gaffney 2008).

Magnetometry relies on the condition that topsoils display an enriched magnetic susceptibility (Le Borgne 1955, 1965; Mullins 1977; Maher and Taylor 1988; Fassbinder et al. 1990; Fassbinder and Stanjek 1993; Dalan 2008). The passive measurement of the earth's magnetic field therefore allows the identification of anomalies caused by archaeological structures. Features cut into the substrate, such as pits, postholes and ditches, can measure between 0.1–10nT while anomalies caused by heat, such as fireplaces, hearths and kilns, can measure up to 100nT (Neubauer 2001b).

The use of magnetics for archaeological prospection was largely influenced by Le Borgne (1955, 1965) who suggested two processes that may cause magnetic enhancement. One of these is a fermentation mechanism. This involves oxidation and reduction cycles that occur during pedogenesis, which can be caused by wet and dry periods and affect the magnetic minerals of a material. The other mechanism is affected by burning, caused either by natural or human means, and plays a major role in the detection of archaeology. If a material is exposed to high enough temperatures in an anaerobic atmosphere, as for example directly under a fire, and is afterwards exposed to aerobic conditions, it can significantly change the magnetic mineralogy, thus the magnetic enhancement (remanent magnetization). These causes explain the higher magnetic enhancement (magnetic susceptibility) of topsoil (Le Borgne 1955, 1965), which consequently fills archaeological structures cut into the subsoil through natural and human means.

Another reason for magnetic enhancement was investigated by Fassbinder (1994), Fassbinder et al. (1990), Fassbinder and Stanjek (1993) in the form of magnetic bacteria, containing magnetite, in normal meadow soil but also in high quantities in postholes on archaeological sites. This means archaeological structures are detectable by means of magnetic surveys as they often have a greater magnetic enhancement than their surroundings. However, not all archaeology is more magnetically enhanced, as can been seen by stone structures such as buildings or roads, which consist of material that has low magnetic susceptibility.

The current trend within magnetic surveys is a movement away from single hand held sensor surveys based on grids, to multi-sensor, cart-based systems with either caesium vapour or fluxgate sensors relying on GPS for positioning (Neubauer 2001; Gaffney 2008; Gaffney et al. 2012) as used by the LBI-Arch Pro. These systems are capable of high resolution data collection at 0.25 m × 0.125 m up to 0.005 nanotesla sensitivities, depending on the sensor. However these may not be applicable in places where one cannot drive on smaller sites, as Alpine meadows of less than 1 ha, rock shelters and caves, or very undulating surfaces as well as excavations as discussed in this paper.

Ground Penetrating Radar (GPR)

Much like magnetic surveys, GPR surveys are tending to go from single antenna surveys to motorised multi-channel GPR arrays. GPR is based on the transmission of high-frequency (100 MHz–1 GHz) electromagnetic radio pulses into the ground (Neubauer 2001b; Conyers 2004; Gaffney 2008). As radar pulses are transmitted through various materials in the subsoil, their velocity changes depending on the physical and chemical properties of the material through which they travel. Consequently some of the electromagnetic energy is absorbed and reflected by the interfaces of layers or objects with different physical properties. The travel time of the reflected radar pulses are measured by the receiver antenna and when the velocity of these is known, depth can be accurately measured and three dimensional images of the various interfaces and objects in the subsoil can be made.

Currently the LBI Arch Pro uses a the motorized 16 channel 400 MHz MALÅ Imaging Radar Array (MIRA) in combination with GPS, permitting a reliable high-definition survey with eight centimetres GPR trace spacing, both inline and cross-line (Trinks et al. 2012). However surveys can be carried out with single or multi-channel carts on smaller areas or areas which do not permit driving as done for the excavation.

Aerial Photography

Aerial photography has been in use since the 1920s and has proven to be one of the most cost effective tools for locating and mapping archaeological remains. Existing aerial photographs can be used to map archaeological structures or produce three-dimensional models from rectified oblique photographs (orthophotos), through the application of analytical and digital photogrammetry (Scollar et al. 1990; Doneus 1996; Neubauer 2001a, b). The archaeological features become visible due to their physical properties seen as soil, moisture or crop marks as well as their thermal properties such as frost or snow marks.

Measurements

Magnetometry

A number of magnetometry surveys were carried out before and during the course of the excavation. A Scintrex CS-3 caesium magnetometer was used in a gradiometer setup, consisting of six sensors arranged in two rows above each other, with the lower row approximately being 0.3 m of the ground surface and the higher row 1 m above. The sensors were arranged with a cross spacing of 0.5 m with an inline sampling of circa 0.1 m. For the positioning of the measurements a GPS was used in conjunction with the LBI Arch Pro Logger Vis system to track and record the measurements. The magnetometer was placed on a rail system in order to allow for an easier survey especially when the depth of the ditches increased due to the removal of stratigraphic layers (Fig. 3).

The primary survey, measuring an area of 20 m × 63 m, was carried out on the top soil over the planned trench prior to excavation. After the removal of the topsoil, the measurements were repeated within the trench over an area of 15 m × 42 m. The following measurements were then conducted over the area of the southern ends of the ditches, covering the area of the main excavation activity. The first of these was carried out after the removal of the first stratigraphic layers of all three ditches over an area of 6 m × 42 m. This was then continued only for the inner ditch after the removal of the next four subsequent stratigraphic layers covering an area of 6 m × 12 m.

Fig. 3 Magnetic measurements during excavation (Matthias Kucera)

Magnetic Susceptibility

In Situ Measurements

The Bartington MS3 magnetic susceptibility meter with sensors MS2D and MS2F and the ZH Instruments MS 30 were used for the in situ measurements of horizontal and vertical areas. These varied from larger area surveys, i.e. the whole trench after topsoil removal, with sampling intervals at 0.5 m × 0.25 m, to medium area surveys, as over features and their immediate vicinity at 0.25 m × 0.25 m, to small area surveys, of a features section or horizontal distribution, providing high resolution surveys at 0.05 m × 0.0 m. Furthermore the profiles of the ditches were surveyed at 0.2 m × 0.05 m with the ZH Instruments MS 30. A part of the project was to investigate a range of sampling rates to establish a best practice for surveys with different and specific questions.

Laboratory Measurements

In total 299 samples were collected from all three ditch sections and from a variety of features as the palisade, pits and natural subsoil. These were air dried and measured with the Bartington MS2 and MS3 magnetic susceptibility meters using the MS2B sensor for X_{LF} and X_{HF}. These measurements were used for X_{FD}

calculations. The measurements were repeated with different measuring times. The MS3 measuring times were 1 s, 2 s and 1 min. The MS2 measurement time was 1 s. Each sample was measured three times for each of the different measuring times, in order to test the repeatability of the measurements. The low and high frequency measurements were used to calculate the frequency dependency (%) of each sample, after Dearing (1999):

$$X_{FD\%} = 100(X_{LF} - X_{HF})/X_{LF}$$

The $X_{FD\%}$ was then used to semi-quantitatively assess the presence of ultrafine superparamagnetic ferrimagnetic minerals occurring in the samples. $X_{FD\%}$ was plotted against the low frequency mass specific susceptibility (X_{LF}) and benchmarks, from Dearing (1999), for different magnetic domains and sources were used as a comparison.

Samples (ID_1–18) from the inner ditch of the 2013 excavation season were sent to the University of Toronto Department of Physics, Canada for measurements with the University of Toronto Electro Magnetic Induction Spectrometer (UTEMIS II) which measures the real and imaginary components of magnetic susceptibility of volume susceptibility over a frequency range of 150–63 000 Hz (West and Holladay 2014). This is similar to the above mentioned measurements except that the UTEMIS II results provide a wide spectrum of magnetic susceptibility measurements, compared to only two measurements with the MS2 and MS3. This can provide a better understanding of the magnetic susceptibility properties of the samples due to the wider spectrum measured.

GPR

The GPR survey was carried out in cooperation with Guglielmo Strappazon (University of Padua) with a Sensors & Software Noggin system using a 500 MHz antenna. The surveys were located over a section of the topsoil, to the north and south of the trench (2 m × 49 m; 35 m profile), as well as in the trench after the removal of the topsoil over the unexcavated northern ditch ends (7 m × 40 m). All surveys were collected at a line spacing of 0.1 m with an inline spacing of 0.01 m.

Aerial Photography

The aerial imagery used in this reconstruction was acquired at Hornsburg around 1700 h using a radio-controlled Microdrone md4-1000 quadrocopter by Geert Verhoeven (LBI Arch Pro). Using such an Unmanned Aerial System (UAS) for image acquisition has become quite popular in archaeological remote sensing over the past years. Since the beginning of aerial photography, researchers used all kinds of devices (from pigeons, kites, poles, and balloons to rockets—see Verhoeven 2009) to take cameras aloft and remotely gather aerial imagery. To date, many of these traditional unmanned devices are still used, but radio-controlled (multi-)copter platforms have recently added a new aspect to archaeological low-altitude unmanned aerial imaging. Besides the flexibility they offer in imaging from different positions, they also allow for the exploitation of new imaging techniques, as it is often only a matter of lifting the appropriate device.

The md4-1000 is a high-end quad(ro)copter of the German firm microdrones GmbH. This electrical multi-copter uses four propellers for its propulsion and can be remotely controlled as well as programmed to fly a specified route. Its maximum payload is 1.2 kg, which means it can accommodate a camera gimbal and smaller camera. Due to this payload constraint, a full-format digital reflex camera cannot be taken aloft (at least not when one wants to stay within the warranty rules). As long as there is minimal wind (i.e. not surpassing 5 m/s), the md4-1000 is very suited for detailed mapping of relatively large areas. In this case, a common Sony NEX-5 N (a 16.1 megapixel APS-C format mirrorless camera) equipped with a Sigma 19 mm f/2.8 EX DN lens were taken aloft by the UAS.

Results

Aerial Photography

The aerial photographs of the trench after the topsoil removal reveal soil marks of various features (Figs. 4 and 5). All the ditch ends can be identified, pits outside the

Fig. 4 Aerial photograph of trench after topsoil removal (Geert Verhoeven)

Fig. 5 Aerial photograph interpretation

entrance passageway, in the entrance passageway, in the space between the ditches and pits at the entrance into the palisade. Furthermore soil marks within the central area were left from the topsoil layer, which had not been sufficiently excavated. The aerial photograph enables the identification of most of the features within the trench. However such identification is limited due to the exposure of the soil marks, to air and sunlight, and limits their visibility, depending on their intrinsic properties, as these become invisible with the loss of moisture.

Ground Penetrating Radar

The GPR surveys provided mixed results (Fig. 6), as the subsoil proved not to be suitable for such a technique. These were collected over the topsoil and after the removal of the topsoil. The topsoil surveys however could not penetrate the subsoil enough to identify the ditches. The survey carried out, over the northern ends of the ditches in the trench after the removal of the topsoil enabled to collect usable data, although only to a depth of c. 1–1.5 m.

The data from the trench reveals all three ditch ends as low to non-reflecting anomalies whereas the subsoil, i.e. loess, in which the ditches were cut can be identified by strong reflections in the depth slice (Fig. 6). This is the reason why the topsoil surveys did not penetrate very deep, through which the radar signal could not penetrate whereas the ditch fills absorbed such. The features are not easily discernible when viewed as a depth slice and therefore the data was visualised as ISO surfaces made by Guglielmo Strapazzon. Usually ISO surfaces are used to visualise hard reflections however in this dataset the interest lay more with features

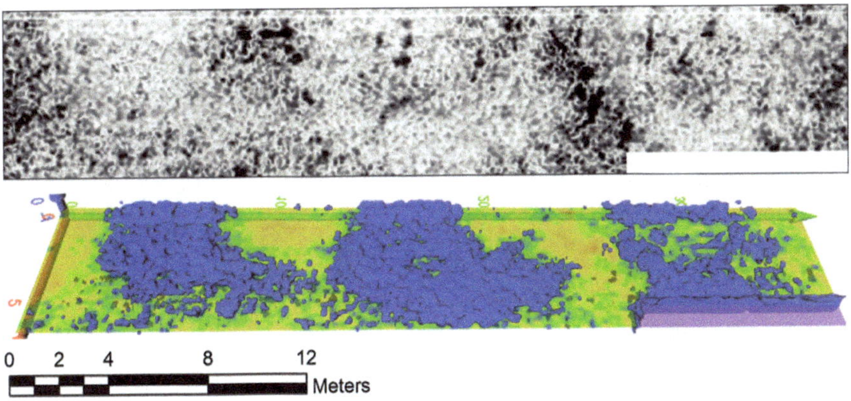

Fig. 6 GPR depth slice and isosurfaces (Guglielmo Strappazon)

with weak reflections, ones that rather absorbed the electromagnetic waves. This allowed for the visualisation of the ditches as well as the pit with postholes to the south of the outermost ditch. Furthermore it also highlighted areas of stronger reflections within the top of the middle and innermost ditches which coinside with the weaker magnetic susceptibilty measurements of the ditches. Such could be interpreted as a different infill and therefore visible as higher relfections in the GPR and lower magnetic susceptibilty.

Magnetometry

Topsoil Magnetometry

The first survey was carried out on the topsoil over the excavation trench prior to the stripping of such (Fig. 7). Similar to the magnetometer survey from 2003, numerous dipole anomalies caused by iron objects close to the surface are visible. Subtle stripping can also be seen in the western part of the survey area which correlates to the ploughing of the field. Nonetheless the three ditches, in total six ditch ends with the outermost and middle ditch ends turning towards the centre alongside the entrance passageway can be identified. However no trace of the palisade can be seen compared to the 2003 measurements.

After Topsoil Removal Magnetometry

The topsoil was removed for a depth of c.0.4 m after which the trench was cleaned with trowels. The magnetometer survey (Figs. 8 and 9) was then carried out over

Fig. 7 Magnetogram of trench over topsoil

Fig. 8 Magnetogram of trench after topsoil removal

the cleaned surface within the trench. The survey basically shows the same ditch ends and passageway as the topsoil magnetometer survey. However the anomalies are crisper and better defined and further features, not visible in the topsoil survey, can be identified. These are pit features, to the west of the beginning of the entrance passage; between the outermost ditch ends; to the east of the northern ditch end; between the southern middle and innermost ditch ends as well as in the central area. Furthermore the palisade can be identified, although very weak, with pit settings at either end in line with the entrance passageway.

Fig. 9 Magnetogram interpretation after topsoil removal

The after topsoil survey enhances the prospectability of the archaeological features within the excavation trench when compared with the topsoil survey. This is to be expected as firstly the topsoil, as can be seen from the topsoil survey, is littered with iron objects causing dipole anomalies that far exceed those of the archaeological anomalies. Secondly the topsoil although only 0.4 m deep is enough to mask the response of the smaller features, as the pits, postholes and palisade, which either means that the layer removed has either a higher magnetic signal then most of the features or that the features themselves are only very weakly magnetic when compared with the topsoil. This can be seen in the magnetic susceptibility survey of the same area, especially within the centre where not all of the topsoil was removed as well as in vertical in situ magnetic susceptibility surveys which show a vast increase in susceptibility for the topsoil compared to the natural and the fill of the archaeological features. Though similar to the magnetic susceptibility the magnetometer measurement of the trench after the topsoil removal shed new light on the archaeological features present within the trench prior to such becoming invisible after their exposure. However the magnetic susceptibility survey shows these features with more detail.

Magnetometer Measurements During Excavation

Several magnetometer surveys were carried during the course of the excavation, especially focusing on the southern end of the innermost ditch. This was done in

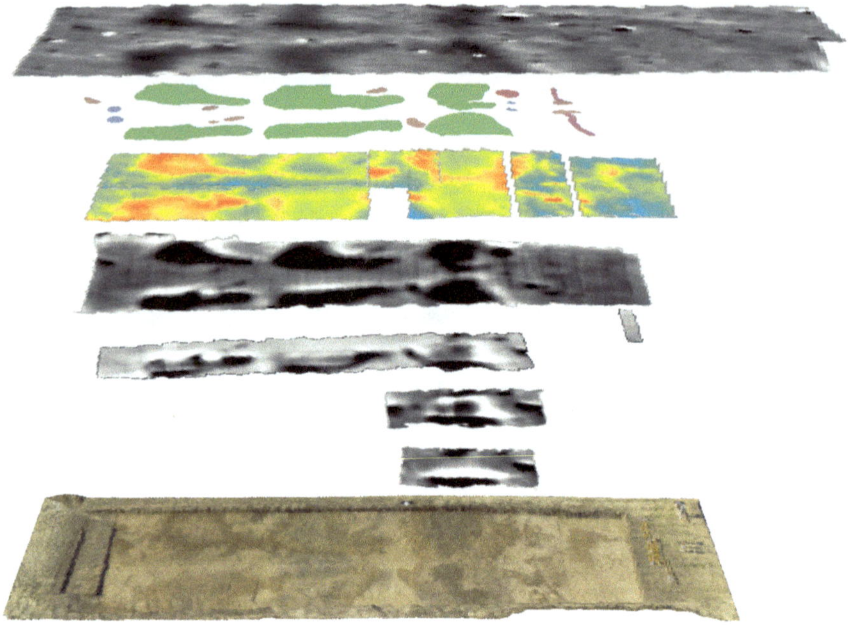

Fig. 10 Magnetograms of different stratigraphic layers

order to measure the stratigraphic layers, which were being excavated, to better understand their contribution to the overall magnetic measurements (Fig. 10).

After the removal of the first stratigraphic layer of all three southern ditch ends a magnetometer survey was carried out after which this was only repeated for the innermost ditch for the following four stratigraphic layers. These measurements show the decrease of the magnetic anomaly of the ditch with depth until no or only a very small magnetic signal is detected. This showed that a layer in the upper fill of the ditch contributed the most to the magnetic signal in the magnetometer measurements. This can be seen as the magnetic signal from the excavated layers diminishes with depth. Such is confirmed by the magnetic susceptibility measurements of the ditches section. However the results could have been affected by two other parameters. Firstly the contribution of layers from deeper strata will contribute less to the magnetic signal due to the decrease of these contributing to the overall signal measured by the magnetometer with distance. Secondly the magnetic signal from the layers within the ditch, due to lack of distance from the ditch section, are overshadowed by the magnetic response of the ditch section. However such would also occur when measured over the non-excavated ditch in the previous surveys.

Magnetic Susceptibility

Field Magnetic Susceptibility

In total forty-two individual magnetic susceptibility surveys were made covering a variety of features over horizontal and vertical areas. These were undertaken in order to improve the understanding of the magnetometer results but as well in their own rights to help investigate the exposed archaeology. The three surveys discussed here are from the trench after the topsoil removal, the profile of the inner ditch and an area of the palisade. These highlight the need for different sampling strategies to fit specific needs.

Trench After Topsoil

The magnetic susceptibility survey of the trench after the topsoil removal used a sampling interval of 0.5 m line and 0.25 m inline spacing (Figs. 11 and 12). Not only did the survey provide a very good comparative dataset to the magnetometry data, due to the similar measuring properties but also as a comparison between an active and a passive method. Another benefit was the help the survey provided in identifying archaeological features in the loess, as these tended to dry out and turn virtually invisible once exposed to the air and sunlight.

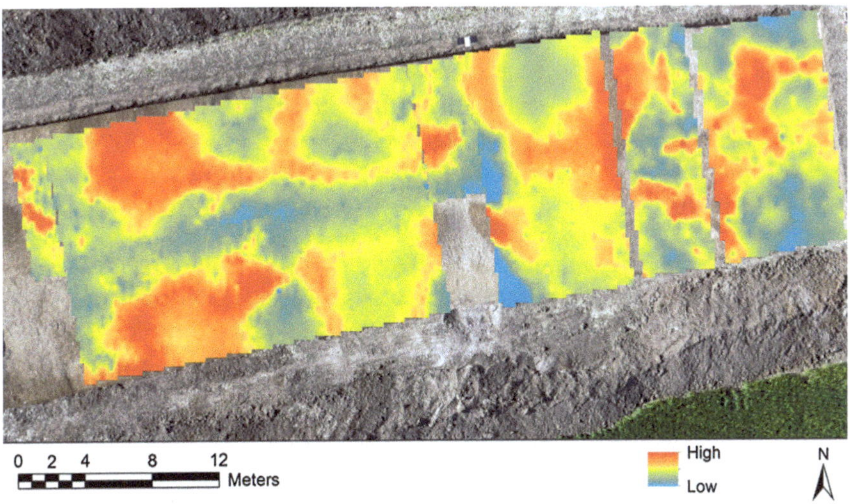

Fig. 11 Magnetic susceptibility measurement of trench after topsoil removal

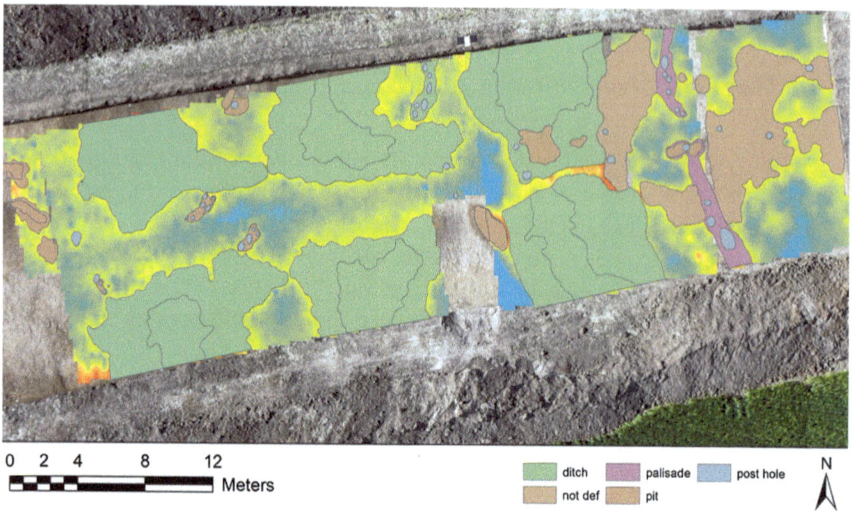

Fig. 12 Magnetic susceptibility interpretation of trench after topsoil removal

In the survey one can clearly see the six ditch ends along the entrance passageway. The measurements also indicate different strengths of magnetic susceptibility of the ditches infill and show the horizontal change over the ditches. Furthermore the survey also reveals a variety of shallower features such as postholes, pits, pits with postholes, the palisade and especially boundaries of soils and sediments with different magnetic susceptibility. The main difference to the magnetometer survey is the increased detail of the individual features, as one can identify different fills of these due to their magnetic susceptibility whereas the magnetometer data does not differentiate such. Therefore possible postholes can be made out in the pit features in the western part of the entrance passage between the outermost ditches. Also the magnetic susceptibility data more readily identifies postholes within the passageway and the palisade trench, whereas these are harder or not at all identifiable in the magnetometer survey. This is partially due to the smaller volume of soil measured by the MSD sensor when compared with the magnetometer measurements as well as the advantage of an active method over a passive one. This allowed the identification of these slight anomalies which were even barely identifiable with one's bare eye. Not only archaeological features can be identified in the survey but changes in soil and sediments across the trench. Such can be seen in the central area to both the east and west of the palisade. This higher readings are due to a red-brownish layer which formed under the ploughsoil and exhibits higher susceptibility values then the other soils and sediments and had not been sufficiently excavated prior to the survey.

Inner Ditch Magnetic Susceptibility Profile

The profile (Figs. 13 and 14) was measured with a line spacing of 0.2 m and measurements were taken every 0.05 m along these. The magnetic susceptibility profile shows a variety of infills which can be identified due to their change in susceptibility. Different phases of infilling as well as recutting of the ditch can be identified in the measurements suggesting a varied life history of the ditch. Thus reflecting periods of quicker and slower backfilling highlighting stages were soils were allowed to form and when sediments or soils were deposited faster by either natural or cultural events.

The susceptibility of the ditch can generally be split into three categories: high, medium and low. The bottom third shows a very low to low susceptibility, from the bottom of the ditch to a V-shaped area with high susceptibility above it. The high susceptibility area is itself cut through by a V-shaped band of medium susceptibility followed by another V-shaped band with high susceptibility at the top. These V-shaped bands can be identified as the major phases of the ditch however finer phases can also be identified within these more general areas indicating various cleaning, recutting and infilling phases of the ditch. Furthermore these can also reflect different formation processes of the soils and sediments and hence various processes leading to the backfilling of the ditch, as can be distinguished from the different susceptibility readings.

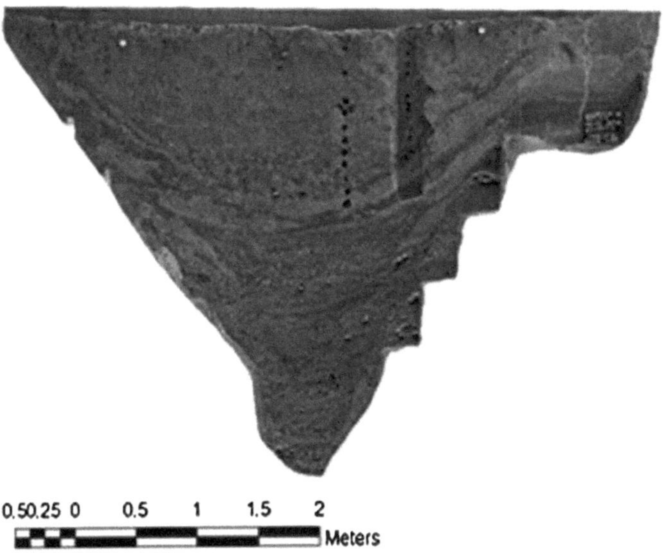

Fig. 13 Photograph of inner ditch profile (Matthias Kucera)

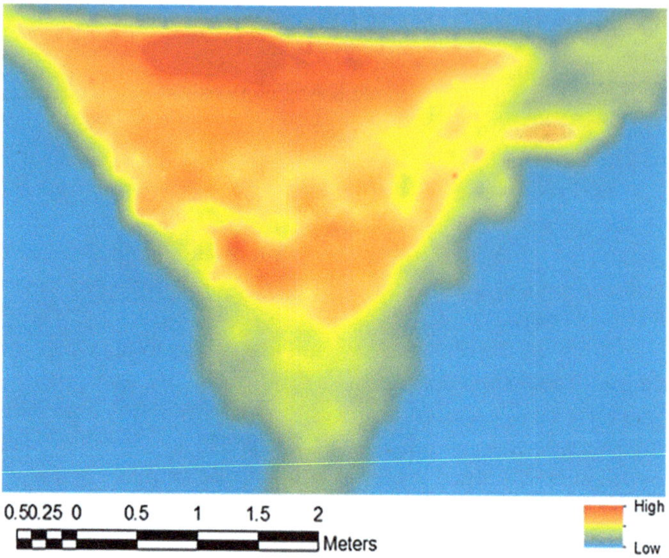

Fig. 14 Magnetic susceptibility measurement of inner ditch profile

Palisade Trench

Various surveys were carried out over the palisade trench in order to measure profiles as well as horizontal areas of the palisade. All of these surveys were measured at a 0.05 m line and 0.05 m inline spacing, as such allowed for a very detailed survey of the feature (Figs. 15 and 16).

The survey was carried out due to the difficulty of identifying the posthole settings within the palisade. This was reflected in the very low magnetometer readings as well as from the excavation during which the postholes could hardly be identified by eye. Therefore a part, in the southern area of the palisade, was excavated in 10 cm spits to allow for detailed horizontal measurements. This was repeated six times covering a depth of 0.6 m until natural was reached. The measurements not only revealed the cut of the palisade trench itself and various infills which helped identify possible posthole settings within the palisade trench. This showed postholes set at approximately 0.8 m apart and indicates that the last phase of the palisade did not consist of continuous postholes but rather ones spaced apart.

Laboratory Magnetic Susceptibility

In the laboratory, the samples from the inner ditch were measured with the MS3 and MS2 susceptibility meters, using the MS2D sensor, with different measuring times

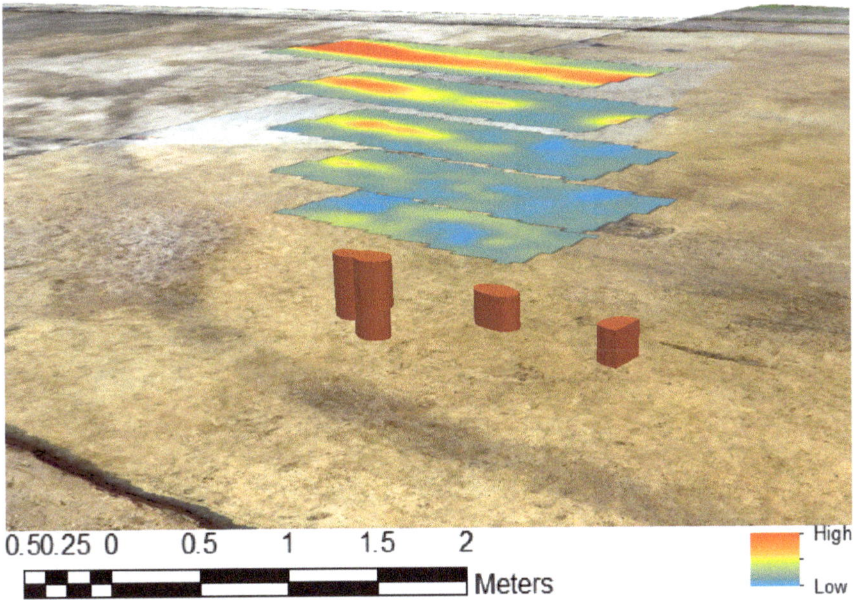

Fig. 15 Magnetic susceptibility measurements of the palisade at different depths and the postholes indicated from these measurements shown below

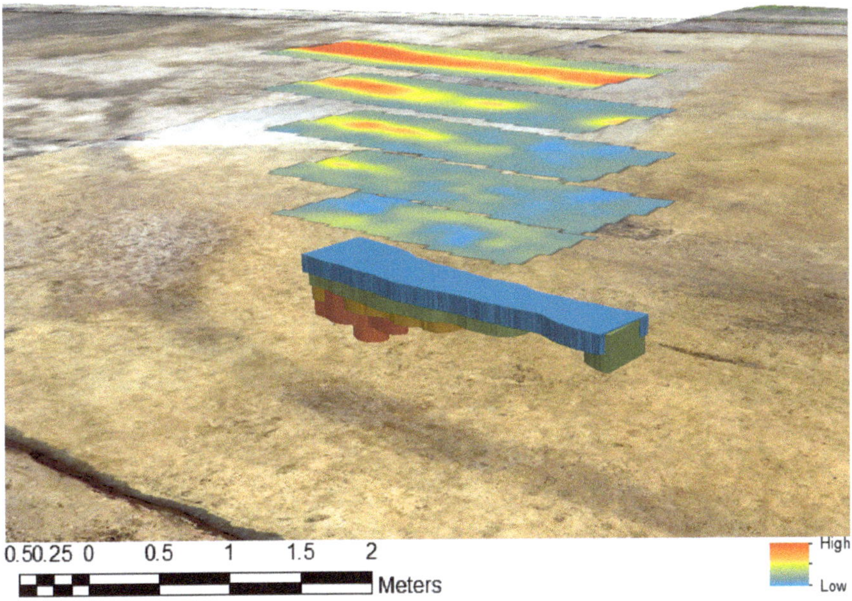

Fig. 16 Magnetic susceptibility measurements of the palisade at different depths and the resulting palisade ditch from the measurements shown below

at 1 s, 2 s and 1 min. Only the MS3 measurements for a 2 s period are discussed below, however differences were observed for the MS2 and MS3 magnetic susceptibility readings of a sample dependent on the length of the measurement. Especially when comparing the results of the MS2 and MS3 for 1 s.

Measurements were taken at low (0.46 kHz) and high (4.6 kHz) frequencies. The X_{LF} measurement allows SP crystals close to the boundary with SSD grains to contribute fully to the susceptibility, whereas the X_{HF} measurement does not. Therefore the superparamagnetic crystals close to the boundary behave like SSD grains and lower the susceptibility value. The forty-three samples discussed below are all from the inner ditch with the remainder still awaiting publication in the near future.

The magnetic susceptibility results (Fig. 17) display a wide variety, as X_{LF} measurements range from 8.28 to 52.40 SI × 10^{-8} m^3/kg and the $X_{FD\%}$ calculated between 0 and 27 %. The general trend is samples with a low X_{LF} also correlate with a low $X_{FD\%}$ and vice versa but this does not always have to be the case. Of the forty-three samples taken from the inner ditch profile, when compared to Dearing's (1999) interpretation of frequency dependent susceptibility values, ten (ID_13, 14, 34, 36, 37, 39, 40, 41, 42 and 43) fall into the low $X_{FD\%}$ range (<2 $X_{FD\%}$) which relates to virtually no SP grains (<10 %). While thirty (ID_1–7, 9–12, 15–19, 21–29, 31–33, 35 and 38) are in the medium range (2–10 $X_{FD\%}$), which is an admixture of SP and coarser non-SP grains, or SP grains which are <0.005 μm. Only two (ID_8 and 20) are in the high category (10–14 $X_{FD\%}$) which equates to virtually all SP grains (>75 %). One sample, ID_30, relates to the very high range (>14 $X_{FD\%}$) which are rare values, erroneous measurement, anisotropy, weak sample or

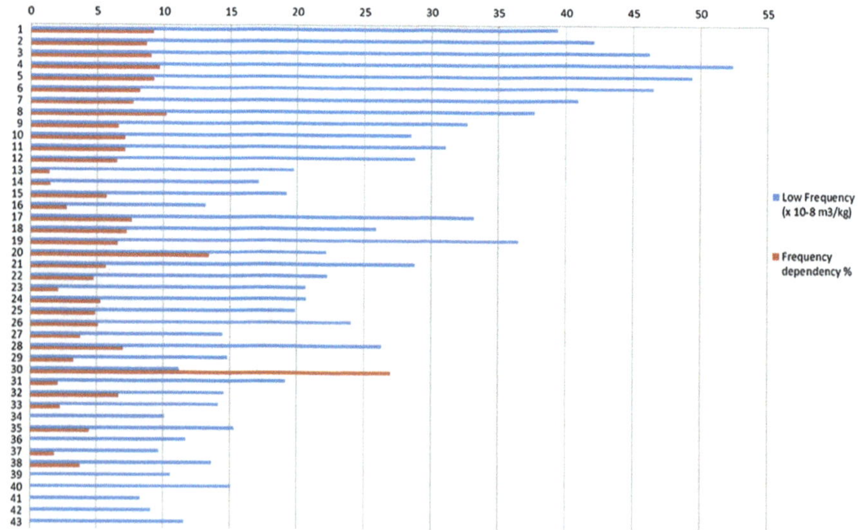

Fig. 17 X_{LF} (blue) and X_{FD} (burgundy) measurements of inner ditch samples according to depth

contamination. In this case the very high reading relates to the weak magnetic susceptibility reading of the sample (Dearing 1999).

The wide variety of both the X_{LF} and $X_{FD\%}$ can be explained as an indication of the diverse formation processes of the different samples and therefore the stratigraphic layers these are from. This can also be seen in the X_{LF} and $X_{FD\%}$ results when shown according to depth. Generally the magnetic susceptibility decreases with depth however at intervals there are higher susceptibility peaks, ID_4, ID_19, ID_28, ID_31, ID_35, ID_40 and ID_43 (Fig. 18). These can reflect formation processes and therefore infilling processes of a different kind then the layers with lower susceptibility. As the $X_{FD\%}$ is an indicator for the presence of SP grains in a sample and specific processes from primary and secondary sources, as background geology or/and soil formation processes, influence the quantity, concentration and size of these. Therefore one can assume that these difference in values can be related to different phases and kinds of infilling of the ditch. For example samples showing low $X_{FD\%}$ can be interpreted as a phase of backfilling with loess deposits or stratigraphic layers with a high percentage of loess in its makeup. This can either be through natural causes, i.e. the collapsing of the ditch flanks through thawing in winter or being undercut during rainy periods, or by cultural, i.e. purposeful backfilling of the ditch. As this seems to reflect a period were either the ditch was infilled fast due to the lack of soil formation processes mirrored in the low magnetic susceptibility and $X_{FD\%}$ values.

On the other hand samples with high X_{LF} and $X_{FD\%}$ can be seen to be from layers were soil formation processes have occurred or have been affected

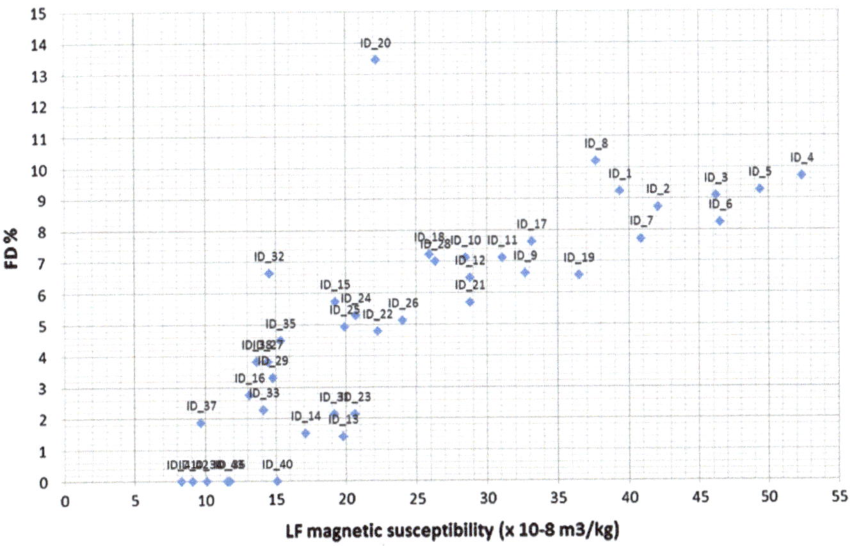

Fig. 18 X_{LF} versus X_{FD} of inner ditch samples with ID_1 the straigraphically highest and ID_43 the lowest

anthropogenically. This can either relate to periods were the ditch was open but not cleaned out therefore allowing soil to form or due to the infilling of the ditch with anthropogenically enriched soil. These peaks can be seen at various depths of the ditches infill and can equate to periods of weather favourable to soil formation whereas the layers with low values could be interpreted as periods of natural infilling, due to periods of rain or hot and cold periods, during which parts of the ditches flanks have collapsed into the ditch.

UTEMIS II Multi-frequency Measurements

The UTEMIS samples (Kainz and Cotter forthcoming) largely reflect the results of the frequency dependency measurements of the MS2 and MS3, although more refined (Fig. 19). The samples seem to form different clusters which suggest different magnetic properties for the samples and can be seen as a reflection of varying domain boundaries and grain sizes amongst the samples. This has been shown in experiments for variable magnetite grain sizes for X_{LF} (Heider et al. 1996). Samples ID_1–4 shows a higher frequency dependency and possibly this indicates the presence of higher concentrations of superparamagnetic crystals on the boundary to stable single domains in these samples. This is also indicated in the $X_{FD\%}$ laboratory results. This would be expected as these samples come from the higher magnetic susceptibility layers, which are responsible for the magnetometry response of the ditch.

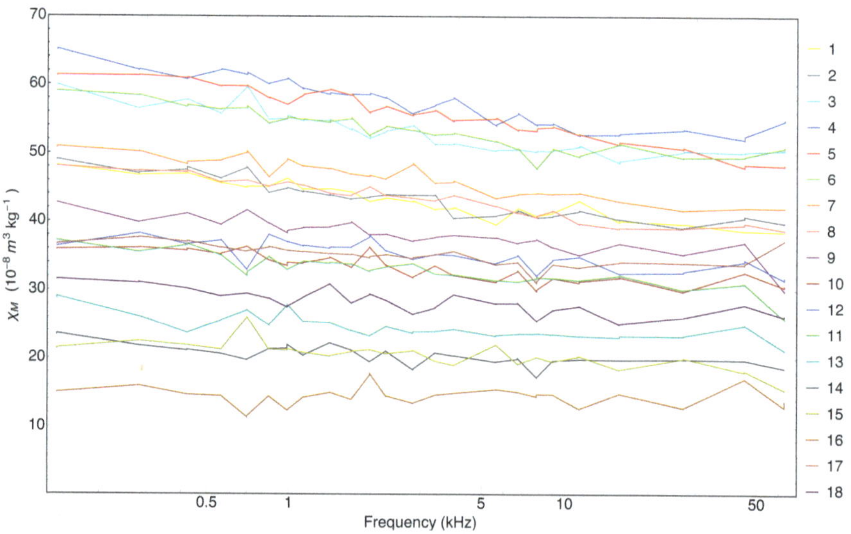

Fig. 19 Broad band magnetic susceptibility measurements over large frequency spectrum with the UTEMIS II (University of Toronto)

Another cluster is formed by samples ID_5–8, which are just above and below the high susceptibility layer (samples 1–4). Another group is made up by ID_ 9, 10 and 12 which the $X_{FD\%}$ results indicate to be of a mixed nature whereas ID_13–18 show the least frequency dependence and apart from ID_18 this is also seen in the $X_{FD\%}$ results. The layers of the ditch, from which ID_13–18 are from, consist of lighter coloured soil possibly due to the inclusion of a higher percentage of loess. Therefore the frequency dependence and magnetic susceptibility is expected to be lower than in the other samples.

Results from Hrouda et al. (2013) for different volume sizes of magnetite and maghaemite measured at different frequencies indicate that the susceptibility drops of at different volume sizes and then are blocked. This can possibly be reflected in the UTEMIS II measurements, although one has to take into account that a sample will consist of a variety of grain volumes and that these may cluster together therefore skewing the results.

ID_14 for example shows very little frequency dependence. The question is whether it is fair to assume that the sample consists of more homogenous domain states and grain sizes and therefore does not vary much when measured at different frequencies. Whereas ID_6 shows a greater variation with frequency and therefore this could indicate a higher concentration of grains at the boundary between superparamagnetic and stable single domain state. Thus suggesting different grain sizes in the sample which affect the different frequency measurements and is reflected in the lower susceptibility with increasing frequency. Therefore a question in future work will be whether the gradient can be seen as the drop of points when there is a shift in the domain state/grain sizes of the magnetic crystals.

Discussion

The field magnetic susceptibility measurements aided the confirmation as well as detection of additional features from the aerial photographs, GPR and excavation results. The magnetic susceptibility data also enabled the identification of different layers in the measured profiles, which provided useful additional information for the excavation. The different sampling strategies showed that the choice of these is especially important for the project aims, and therefore these need to be carefully selected. Within the ditches, this helped indicate when different layers were open for longer periods as well as suggesting events of faster backfilling, related to either natural or cultural activity. Especially the measurements undertaken in the palisade trench helped identify the posthole settings, which are near to invisible to the naked eye. This has helped appreciate the architecture of the KGA, as a continuous palisade has a different significance to the central area, which is presumed to have been shielded from exterior view, as one that is not continuous and allows sight into and out of the interior. However, one should not exclude the possibility of thatching between the posts, which nonetheless would still change the perception of the KGA if the palisade would be continuous and constructed out of large wood posts.

The laboratory measurements, at a X_{LF} and X_{HF} enhanced the understanding of the in situ measurements by additionally analysing the frequency dependency of the samples from the different layers. This enabled an assessment, although more qualitative than quantitative in nature, of the makeup of the magnetic properties of the samples, by giving a general indication of the different magnetic domains and grain sizes.

The UTEMIS II magnetic spectroscopy data gave further insight and allowed for a more quantitative assessment of such, although further work has to be undertaken to understand the full potential of these measurements. Future work with the data should demonstrate whether the multi-frequency spectroscopy measurements give a more detailed indication of the different magnetic domain states and/or grain sizes in the samples. Magnetic particles with certain volumes at different frequencies either are unblocked or blocked and therefore should affect the magnetic susceptibility of the samples. Whether or not these differences are reflected in the multi-frequency measurements and can be used to further help to differentiate the various layers and help the archaeological and magnetic measurements interpretation will be seen in the ongoing work. Nonetheless, certain trends can be deduced from the various clusters of layers in the data suggesting these to have similar properties with each other and implying certain processes in their creation and deposition therefore helping the archaeological interpretation.

Although not all of the data collected during the excavation has been fully evaluated a positive trend can be attributed to the application of a multi-method approach. The magnetometer and magnetic susceptibility data correlate the most, due to the properties these measure, but these datasets also corroborate the aerial photograph, orthophotos and GPR data from the site. These in turn will be helped by the future evaluation of the pXRF data, which will provide an insight into the physical properties of the anomalies by analysing their chemical contents. The magnetic properties of a soil, sediment or material are largely dependent on the inclusion and quantity of iron oxides, which in turn can alter the colour of a soil or sediment and therefore affect its visibility. This not only affects the excavator's view but also a feature's perceptibility in photographs. Furthermore, the visibility of such soil marks is largely dependent on the grain and pore size and organic content of the feature fill. Hence, these marks can vanish once they are exposed to air and sunlight, as the moisture evaporates. However, since magnetic visibility is not affected by such properties, magnetic surveys allow for corroborate information to be collected. This not only enhances the certainty of these features but can also add detail to these, as seen with the in situ magnetic susceptibility measurements of the palisade. Soil marks, especially in loess environments, sometimes only differ slightly in colour and can hamper the identification of these features purely by eyesight. Therefore additional surveys measuring other properties, i.e. magnetic and GPR as well as aerial, ortho-, ultraviolet and infrared photographs and pXRF measurements, will allow for the better identification and understanding of the properties of the features and therefore of their archaeological significance.

Acknowledgments I would like to thank the University of Vienna and the LBI Arch Pro for the resources provided during the excavation. Further gratitude goes to the help provided by Gugliemo Strappazon for helping with the magnetic susceptibility and GPR measurements as well as producing the depth slices and isosurfaces for the GPR, Matthias Kucera for the taking the photograph of the ditch profile as well as the help during the excavation and measurements, Geert Verhoeven for the aerial photograph and Roderick Salisbury for the XRF measurements as well as the help and exchanges during the excavation.

References

Aitken, M.J., 1958. Magnetic prospecting I. Archaeometry 1, 24–29.

Becker H, et al. 1996. Archäologische Prospektion. Luftbildarchäologie und Geophysik. Arbeitshefte Bayerischen Landesamtes Denkmalpflege 59:1

Becker, H. 1996. Kultplätze, Sonnentempel und Kalenderbauten aus dem 5 Jahrtausend vor Chr. – die mittelneolithischen Kreisanlagen in Niederbayern. Arbeitshefte Bayerischen Landesamtes Denkmalpflege 59:101–122

Clark, A.J. 1996. Seeing Beneath the Soil: Prospecting Methods in Archaeology. B.T. Batsford: London.

Clay, R.B. 2001. Complementary geophysical survey techniques: why two ways are always better than one. Southeastern Archaeology 20: 31–43.

Conyers, L., 2004, Ground-Penetrating Radar for Archaeology, AltaMira Press

Crowther, J. 2003. Potential magnetic susceptibility and fractional conversion studies of archaeological soils and sediments. Archaeometry 45, 685–701

Dalan, R.A. 2006. A Geophysical Approach to Buried Site Detection using Down-hole Susceptibility and Soil Magnetic Techniques. Archaeological Prospection 13 182–206

Dalan, R.A. 2008. A review of the role of magnetic susceptibility in archaeogeophysical studies in the USA: recent developments and prospects. Archaeological Prospection 15: 1–31.

Dalan, R.A., Banerjee, S.K. 1998. Solving archaeological problems using techniques of soil magnetism. Geoarchaeology 13: 3–36.

Dalan, R. A., Bevan, B. 2002. Geophysical Indicators of Culturally Emplaced Soils and Sediments. Geoarchaeology 17, 779–810.

Dearing, J.A. 1999. Environmental Magnetic Susceptibility: Using the Bartington MS2 System. Bartington Instruments Ltd.

Doneus, M. 1996. Photogrammetrical applications to aerial archaeology at the Institute for Prehistory of the University of Vienna, Austria. Int Arch Photogrammetry Remote Sensing XXXI, part B5, commission V. Vienna, pp 124–129

Dunlop, D.J., Özedemir, Ö. 1997. Rock Magnetism. Fundamentals and Frontiers. Cambridge Studies in magnetism series. Cambridge University Press.

Evans ME, Heller F. 2003. Environmental Magnetism: Principles and Applications of Environmagnetics. Academic Press: New York.

Fassbinder, J, 1994. Die magnetischen Eigenschaften und die Genese ferrimagnetischer Minerale in Böden im Hinblick auf die magnetische Prospektion archäologischer Bodendenkmäler, Buch am Erlbach

Fassbinder, J, 2015. Seeing beneath the farmland, steppe and desert soil: magnetic prospecting and soil magnetism. Journal of Archaeological Science 56:85–95

Fassbinder, J, Stanjek, H, 1993. Occurrence of bacterial magnetite in soils from archaeological sites. Archaeol Pol 31:117–128

Fassbinder, J, Stanjek H, Vali, H, 1990. Occurrence of magnetic bacteria in soil. Nature 343:161–163

Gaffney, C., 2008. Detecting Trends in the Prediction of the Buried Past: a Review of Geophysical Techniques in Arcaheology. Arcaheometry, 50 (2), 313–336

Gaffney, C., Gater, J.A., Linford, P., Gaffney, V., and White, R., 2000, Large-Scale Systematic Fluxgate Gradiometry at The Roman City of Wroxeter. Archaeological Prospection 7, 81–99.
Gaffney, C., Gaffney, V., Neubauer, W., Baldwin, E., Chapman, H., Garwood, P., Moulden, H., Sparrow, T., Bates, R., Löcker, K., Hinterleitner, A., Trinks, I., Nau, E., Zitz, T., Floery, S., Verhoeven, G., and Doneus, M., 2012, The Stonehenge Hidden Landscapes Project, Archaeological Prospection 19, 147–155
Gaffney, C, Gater, J, 2003. Revealing the Buried Past: Geophysics for Archaeologists. Tempus Publishing: Stroud.
Google Earth 2012 7.1.2.2041. Kreuttal. 48°25'58.73" N 16°26'14.92" E 48°25'58.73", elevation 282 m. Borders and Labels, Places Layer. Viewed 1 May 2015.
Heider, F., Zitzelsberger, A., Fabian, K. 1996. Magnetic susceptibility and remanent coercive force in grown magnetite crystals from 0.1 μm to 6 mm. Physics of the Earth and Planetary interiors 93, 239–256.
Hejcman, M., Karlik, P., Ondracek, J., Klir, T. 2013. Short-Term medieval Settlement Activities irreversibly changed Forest soils and Vegetation in Central Europe. Ecosystems 16, 652–663
Hrouda, F., Pokorny, J., Jezek, J., chadima, M. 2013. Out-of-phase magneitic susceptibility of rocks and soils: a rapid tool for magnetic granulometry. Geophysical Journal International 194, 170–181
Kainz, J., Cotter, J., Forthcoming. Magnetic spectroscopy of a middle Neolithic Circular Ditch Enclosure.
Keay, S., Earl, G., Hay, S., Kay, S., Ogden, J., & Strutt, K.D. (2009). The Role of Integrated Geophysical Survey Methods in the Assessment of Archaeological Landscapes: the Case of Portus. Archaeological Prospection 16, 154–166.
Kvamme, K.L., 2006. Integrating multidimensional geophysical data. Archaeological Prospection 13: 57–72.
Kvamme K.L., (2007) Integrating Multiple Geophysical Datasets, 345–374 In: Remote Sensing in Archaeology, (Eds) Wiseman, J. And El-Baz, F., Springer
Le Borgne, E, 1955. Susceptibilité magnétique anomale du sol superficiel. Ann Géophys 11, 399–419
Le Borgne, E, 1965. Les propriétés magnétique du sol: application à la prospection des sites archéologiques. Archaeo-Physika 1 (Beihefte der Bonner Jahrbücher 15):1–20
Linford, N., 2005. The application of geophysical methods to archaeological prospection. Reports on Progress in physics 69, 2205–2257
Maher, B.A. 1986. Characterisation of soils by mineral magnetic measurements. Physics of the Earth and Planetary Interiors 42, 76–92.
Maher, B.A, Taylor, R.M. 1988. Formation of ultrafine grained magnetite in soils. Nature 336, 368–371.
Melichar, P., Neubauer, W., (Eds.) 2010. Mittelneolithische Kreisgrabenanlagen in Niederösterreich. Geophysikalisch-archäologische Prospektion – ein interdisziplinäres Forschunsprojekt. Wien: Verlag der Österreichischen Akademie der Wissenschaften
Mullins, C.E. 1977. Magnetic Susceptibility of the Soil and Its Significance in Soil Science Review. Journal of Soil Science 28, 223–246.
Neubauer, W, 1990. Geophysikalische Prospektion in der Archäologie. Mitt Anthropol Ges Wien 120:1–60
Neubauer, W., 2001a. Images of the invisible-prospection methods for the documentation of threatened archaeological sites, Naturwissenschaften 88:13–24
Neubauer, W., 2001b. Magnetische Prospektion in der Archäologie, Verlag der Österreichischen Akademie der Wissenschaften
Neubauer, W. 2012. Kreisgrabenanlagen – Middle Neolithic Ritual Enclosures in Austria 4800 – 4500 BC. In Gibson A. (Ed.)Ö Enclosing the Neolithic. Recent studies in Britain and Europe. Oxford: Archaeopress (BAR International Series, 2440) 147–163
Neubauer, W., Eder-Hinterleitner, A. 1997. Resistivity and Magnetics of the Roman Town Carnuntum, Austria: An Example of Combined Interpretation of Prospection Data, Archaeological Prospection, 4: 179–189.

Petrasch, J, 1990. Mittelneolithische Kreisgrabenanlagen in Mitteleuropa. Ber Römisch-germanischen Kom 71

Podborský, V, et al (ed), 1999. Primeval socio-ritual architecture in Moravia. Brno

Powlesland, D., Lyall, J., Hopkinson, G., Donoghue, D., Beck, M., Harte, A., Stott, D., 2006. Beneath the Sand Remote Sensing - Archaeology, Aggregates and Sustainability: a Case Study from Heslerton, the Vale of Pickering, North Yorkshire, UK 2006

Ridky, J., Koncelova, M., Sumberova, R., Limbursky, P., Kvetina, P. 2014. How were Neolithic Ditches filled in? Deposition study of two Enclosures from bohemia. European Journal of Archaeology 17, 579–601.

Schmidt, A. 2001. Visualisation of multi-source archaeological geophysics data. In M. Cucarzi and P. Conti (eds) Filtering, Optimisation and Modelling of Geophysical Data in Archaeological Prospecting: 149–160. Rome: Fondazione Ing. Carlo M. Lerici.

Scollar, I, Tabbagh, A, Hesse, A, Herzog, I, 1990. Archaeological prospecting and remote sensing, Cambridge University Press, Cambridge

Simon, F-X., Koziol, A., Thiesson, J. 2012. Investigating Magnetic Ghosts on an Early Middle Age Settlement: Comparison of Data from Stripped and Non-stripped Areas. Arcaheological Prospection 19, 191–200

Tite, M.S. 1972. The influence of geology on the magnetic susceptibility of soils on archaeological sites. Archaeometry 14, 229–236.

Trinks, I., Kucera, M., Hinterleitner A., Löcker K., Nau E., Neubauer W., and Zitz T., Large-scale, high-definition Ground Penetrating Radar prospection in archaeology, Geophysical Research Abstracts, Vol. 14, EGU General Assembly 2012

Trnka, G, 1991. Studien zu mittelneolithischen Kreisgrabenanlagen. Wien: Verlag der Österreichischen Akademie der Wissenschaften

Verhoeven, G, 2009. Providing an archaeological bird's-eye view: An overall picture of ground-based means to execute low-altitude aerial photography (LAAP) in Archaeology. Archaeological Prospection 16 (4), 233–249. DOI:10.1002/arp.354

West, G., Holladay, S. 2014. UTEMIS II Operating Manual. University of Toronto Department of Physics.

Creating a Chronological Model for Historical Roads and Paths Extracted from Airborne Laser Scanning Data

Willem F. Vletter and Sandra R. Schloen

Abstract Chronological modeling is part of a methodology being developed for the use of Airborne Laser Scanning (ALS) data in reconstructing historical road and path networks in vegetated areas. It comprises four main steps. The first step tackles the (semi-) automatic visualization and extraction of linear features from ALS data. A model is presented in the second step to determine the (relative) chronology of historical roads and paths. The third step deals with the predictive modeling of unknown networks. The final step combines a 3-D environment with a time element, resulting in a temporal-spatial model of the road and path networks found. The (semi-) automatic extraction results of the first step are published in August 2014 in the proceedings of the Second International Conference on Remote Sensing and Geoinformation of the Environment (RSCy2014). Based on these extracted networks of the first step and additional manual mapping, the chronological model of the second step is created. The outcome of the model will be presented in this article. The chronological model makes use of both the Harris Matrix Composer (HMC) and the Online Cultural and Historical Research Environment (OCHRE). The latter is developed at the University of Chicago. This is a multi-project, multi-user database system that provides a comprehensive framework for diverse kinds of information at all stages of research. Both the HMC and OCHRE proved useful for creating a relative dating model for roads and paths, as resulted from the case study presented in this article.

W.F. Vletter (✉)
Vienna Institute of Archaeological Science, University of Vienna,
Franz-Klein Gasse 1, 1190 Vienna, Austria
e-mail: willem.vletter@univie.ac.at

S.R. Schloen
Research Database Specialist, OCHRE Data Service, University of Chicago,
1111 E. 60 Th St., Chicago, IL 60637, USA
e-mail: sschloen@uchicago.edu

M. Forte and S. Campana (eds.), *Digital Methods and Remote Sensing in Archaeology*, Quantitative Methods in the Humanities and Social Sciences,
DOI 10.1007/978-3-319-40658-9_18

Introduction

Historical roads and maps can provide important insights about the landscape and its past use on both local and regional scales. If we examine the literature for reference material, there is very little on the dating of roads. An exception is Wilkinson et al.'s article (2010) which combines a remote sensing technique (CORONA satellite images), field investigation (trenching) and micromorphology to investigate historic roads (hollow ways), in an approach similar to the current doctoral research of this article's author. However, besides the pottery found in the hollow ways, no absolute or relative dating techniques were applied in Wilkinson et al.'s study.

The fact that little investigation has been carried out to date roads and paths is of course not a surprise. Indeed, unmetalled roads and track ways are extremely difficult to date. They have no constructional material to aid interpretation; while artefacts are rarely present (Smith 2011). Indeed, it is often difficult to establish the age of most pathways, which preceded the introduction of paving in Europe in the late 18th century onwards (Horsten 2005). Exception being the paved roads built by the Romans.

The above-mentioned difficulties mean historical road research must be based on historical, archaeological and geographical investigation. Indeed, it should go much further than the recording of surface features. Phenomena which don't leave visible traces behind, such as the historical context of the road network, road connections and the volume of traffic itself, should be considered in the reconstruction (Denecke 1969).

Nevertheless, recent technical developments have created more possibilities to date roads and paths both relatively and absolutely. One possibility comes from Airborne Laser Scanning (ALS), a technique able to capture surface relief in very high resolution and visualize details which can't be seen by the naked eye in the field. This technique supplies the data on which the author builds a methodology for the reconstruction of historical road and path networks in his current doctoral research and is the main focus of this article. Issues concerning the extraction of roads and paths from ALS data as well as the completion of networks is discussed elsewhere (Vletter 2014). The PhD project has two study areas: the Leitha Hills in Austria and the Veluwe area in the Netherlands. The former area is used to base the methodology upon, the latter to test the applicability of the methodology in a completely different environment. Additionally, the applicability of relative and absolute dating techniques and their suitability for the case study areas is also considered below. While the results and conclusion of the dating method comparison are presented in Sects. A Case Study and Conclusion. However, before proceeding, a brief background description of the research areas and the project data will follow.

The Research Areas

The Veluwe

The Veluwe is a mainly forested area in the center of the Netherlands on a push moraine with sandy soils, which extends over an area of about 1000 km^2. The forest is a mix of deciduous and coniferous trees. The highest point here is 110 meters above sea level; the lowest point is almost at sea level. Due to melting of the ice coverage, there are presently many small (dry) side-valleys, where also most of the villages are located. The Neolithic period provide the oldest known artifacts on the Veluwe. The most visible archaeology comprises mainly Bronze Age burial mounds and, to a lesser extent, the Celtic field systems. From the Middle Ages onwards, the area has suffered from drift-sand because of de-forestation. At the beginning of the 20th Century it existed mainly out of enormous zones of heathland. Later in that century many areas have been reforested again. The Veluwe, typically for the Netherlands, has a moderate maritime climate. The ALS data covering the study area was made available by the Dutch National Board of Water management (Rijkswaterstaat) and it is completely free for the public. In a following paper, when the methodology will be applied to this area, the ALS data will be described in more detail.

The Leitha Hills

The Leitha Hills is a forested area of approximately 190 square km, 40 km south of Vienna, which rises between 200 and 300 m above the valley of the river Leitha. It is covered by a forest of mixed deciduous trees, mainly oak and beech with varying degrees of understory (Doneus and Briese 2010). Geologically, the Leiha Hills lie between the Alps and the Carpathians and link the two mountain chains. Due to fractions and subsidence, valleys were created. They also caused acidulous and sulphur containing water which came to the surface on fault lines. Also at least thirty caves have been documented. Where the hills meet the plain, the landscape is often characterized by heathland with unique vegetation. The south side of the Leitha Hills is favoured by the micro climate of the Neusiedler lake (Krizsanits and Horvath 2012).

The archaeological sites in this area, which have been detected mainly on the digital terrain model (DTM) derived from ALS data, are mostly medieval or post medieval. However, in summary, these include: four late Neolithic hill forts, several Iron Age hill forts, round barrows, building structures, stone quarries, hollow ways, medieval field systems, medieval border-markers (so called 'Hotters'), hundreds of lime-kilns, military trenches ranging from the post-medieval period to World War II and a large number of bomb craters from World War II which have been interpreted (Doneus and Briese 2010). Visible remains of human settlements in the woods are

Table 1 The meta-information of the ALS data used in this paper

ALS-project	Leitha mountains
Purpose of scan	Archaeology
Time of data acquisition	March—12th of April 2007
Point-density (pt. per sq. m)	7
Scanner type	Riegl LMS-Q680i Full-Waveform
Scan angle (whole FOV)	45°
Flying height above ground	600 m
Speed of aircraft (TAS)	36 m/s
Laser pulse rate	100,000 Hz
Scan rate	66,000 Hz
Strip adjustment	Yes
Filtering	Robust interpolation (SCOP++)
DTM-resolution	0.5 m

ruins, decayed hunting houses or monastic buildings (Krizsanits and Horvath 2012).

The Leitha hills ALS project data were captured during a dedicated ALS flight undertaken for the "LiDAR Supported Archaeological Prospection in Woodland" Project (Doneus and Briese 2010). The most important parameters of the ALS data used in this paper are stated in Table 1.

Based on this ALS data, a semi-automatic extraction was executed. The extraction was carried in two steps. First the micro-topography was visualized in grey scale using the openness module in OPALS of the Technical University of Vienna (TU Wien). The second step was creating the extraction model in the software plug-in *Feature Analyst* (Vletter 2014). In a following step the networks were manually completed and some topology was carried out to check the quality of the networks (Vletter 2015). This resulted in an extensive network in the Leitha hills. However, the question of dating this network of roads and paths remained to be answered the possibilities of which are discussed in the next paragraph.

Absolute Dating

There are several methods to date absolutely. Here will discuss their applicability for both research areas. One method for age dating, namely radiometric dating will be excluded. The reason is that this method calculates in such large time scales, that it is of no use for the research.

Archaeomatic Dating

To know the possibilities are of archaeomagnetic dating we must gain knowledge of the principles behind it. Of importance for this dating technique is the Earth's magnetic field. This magnetic field is constantly changing both in direction and intensity. It is generally agreed that the magnetic field is caused by dynamo process within the Earth's liquid outer core. Around 80% of the field can be explained by imaging a bar magnet at the centre of the Earth and inclined at an angle of 11 degrees to the axis of rotation. The remaining 20% is caused by localised turbulence in the outer core (see Fig. 1). This is called the geomagnetic secular variation which makes archaeomagnetic dating possible. For archaeological dating it is the variation in the local fields which is most applicable. In order to exploit these changes and use them for dating purposes we first need to establish a calibration curve which describes the changes in the local field over the historical period. Because of the

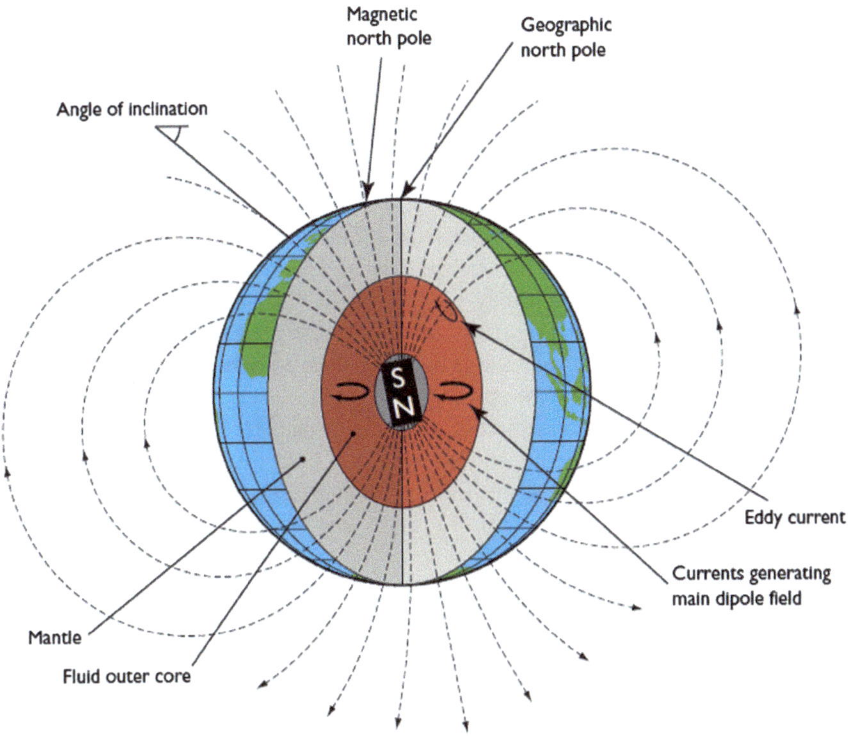

Fig. 1 The Earth's main dipolar magnetic field is depicted with dashed lines. This is generated by electric current circulation in the outer core (shown in *red*) and is similar to the field that would be produced by a bar magnetic located at the Earth's centre tilted off-vertical by about 11°, 5°. Eddy currents near the core/mantle boundary perturb this main field. The angle of dip (or inclination) is the angle that the field lines make with the horizontal plane where they cut the Earth's surface (USGS 2015)

random nature of the secular variation such a curve will follow no particular law and its shape will vary drastically from one part of the Earth to the other (McCann 2013). Hence, the archaeomagnetic dating strongly depends on the master Paleosecular Variation Curves (PSVC) used in the dating process and, accordingly, one must be careful in this respect. However, the problem of resolution and reliability in archaeomagnetic dating is difficult. The spatiotemporal distribution of the input database, the uncertainties within the database, its internal coherence and the technique used for modelling (classical PSVCs or global or regional models), define the resolution of the PSVC and, consequently, the resolution in archaeomagnetic dating. In addition to this, the relocation error introduces a bias in the dating process which is often not considered (Linford 2006).

Moreover, a feature and the conditions in which it is found, must satisfy to a lot of criteria before it can be used for archeomagnetic dating Nevertheless these obstacles, good results have been obtained. Therefore, it is worthwhile to examine it closer. As mentioned earlier archaeomagnetic dating is a physical dating technique that exploits the fact that the Earth's magnetic field changes over time. When structures such as kilns and furnaces are heated, the materials that they are built from can magnetise and record the direction and strength of the Earth's field at that time. Subsequent archaeomagnetic analysis allows these parameters to be determined and compared with calibration data to determine when the heating happened. Each time that the structure is fired, any previous magnetisation is lost, so the date obtained will be for the last firing. In some circumstances, archaeological sediments settling out from still water can also record the magnetic field at the time of their deposition and thus they can also be dated with the technique (Linford 2006). The rocks and sediments from which archaeological features are made contain trace amounts of iron oxides, which may be associated with the original rock-forming processes, or with secondary processes such as heating and weathering.

The main minerals are magnetite (Fe_3O_4), haematite (αFe_2O_3) and maghaemite (γFe_2O_3), where some of the Fe may be substituted by other cations (e.g. Ti, Al). Due to their ferromagnetic properties, they are capable of acquiring a remnant magnetisation in the presence of the geomagnetic field that is stable over archaeological (and geological) timescales (McIntosh and Catanzariti 2006).

Limestone, like in the Leitha hills, is a stone type which is usually not associated with iron minerals can often contain trace quantities of magnetic minerals capable of retaining a magnetic remanence (McIntosh and Catanzariti 2006).

Although, good results have been obtained, there are also difficulties with this method. For example wind and changed currents can have significant influence on the orientation of the magnetic field of sediment. It also indicates that palaeochannels and ox-bow lakes are ideal environments for successful archaeomagnetic dating, especially when the sediment has remained waterlogged since deposition. This means that suited sediments can meanly be found on and near lowland floodplain sites (Ellis 1998). This is not the case for the case-studies of the dissertation, which are both hilly areas. Maybe the Neusiedler lake, south of the Leiha hills, can be proper for this kind of research.

Another possibility for sediments are ditch infills. Here sediments may also may acquire a post-depositional remnant magnetisation (pDRM) at some time after their deposition, producing continuous records of directional (and occasionally relative intensity) secular variation (SV). The main drawbacks with sediments are that there is a delay in the acquisition of pDRM (related to the depth at which the pDRM is locked in and to the sedimentation rate), the amplitudes of SV may be smoothed and the values of inclination shallower than the actual field values. This can make comparison with archaeomagnetic data problematical. The key is in identifying the archaeological event associated with the remanence—such as the production of fired material or the last heating of combustion structures. In this respect, sediments are difficult to interpret as the age of magnetisation is normally older than the age of deposition and so it is hard to relate to an archaeological event (McIntosh and Catanzariti 2006). Moreover it is not clear if in the research areas proper road ditch infills can be found. Especially as ditches by preference have to be waterlogged and have had a slow accumulation of sediment (English Heritage 2006). This seems not be case for both study areas, where the inclination of the terrain is not suitable for waterlogging and slow accumulation (Fig. 2).

In the case of heated structures, the lime kilns of the Leitha hills and iron kiln of the Veluwe could be of interest. As roads and paths in hilly areas normally don't have datable structures, the age-dating should take place in an indirect way. The assumption is that archaeomagnetic dating of kilns along roads and paths provide

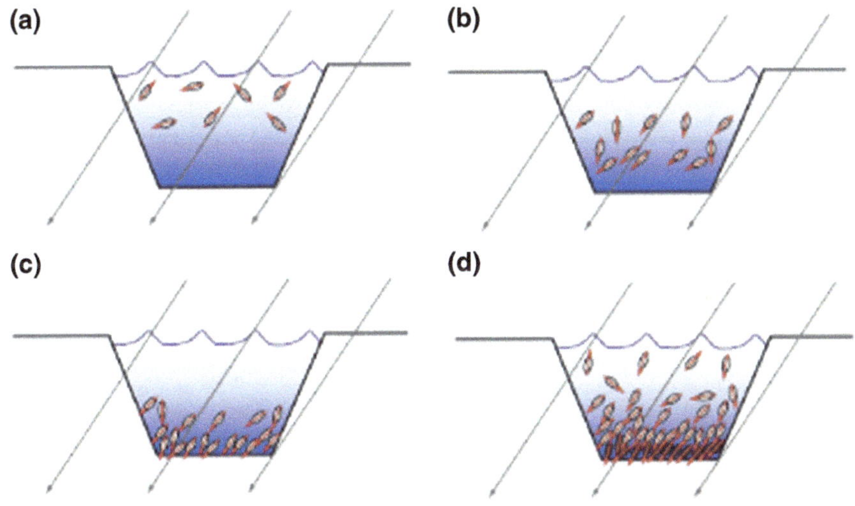

Fig. 2 Depositional remnant magnetisation. Sediment particle, each with a weak magnetisation, settle out of still water. As they fall through the water column they rotate to align their internal magnetisation directions with Earth's magnetic field (**a**, **b** and **c**). Once settled on the bed of the body of water, the weight of sediment accumulating on top of the particle locks them in place, leaving a layer of magnetised in the direction of the Earth's field (**d**) (McIntosh and Catanzariti 2006)

the age for which the roads and paths were (at least) in existence. However, it is skating on thin ice linking roads and paths to archaeological features on basis of vicinity, as there is no prove that roads and paths nearby have a relation with the kilns.

For the Veluwe (the Netherlands) there is another problem. At moment there is no good paleosecular variation curve. However there maybe the possibility to extrapolate the curve from Belgium or Germany to the Netherlands (McCann McCann 2003). In Austria Schnepp and Lanos established such a curve in 2006 (Pavón-Carrasco et al. 2011)

Optical Stimulate Luminescence (OSL) Dating

Luminescence dating is a form of geochronology that measures the energy of photons being released. In natural settings, ionizing radiation is absorbed and stored by sediments in the crystal lattice. This stored radiation dose can be evicted with stimulation and released as luminescence. The calculated age is the time since the last exposure to sunlight or intense heat. The sunlight bleaches away the luminescence signal and resets the time 'clock' (see Fig. 3). As time passes, the luminescence signal increases through exposure to the ionizing radiation and cosmic rays. Luminescence dating is based on quantifying both the radiation dose received by a sample since its zeroing event, and the dose rate which it has experienced during the accumulation period. The principal minerals used in luminescence dating are quartz and potassium feldspar (USGS 2015).

If we look at OSL we know that the presence of minerals and normally specifically quartz is needed. However, OSL can also be applied to (Volcanic) Feldspar and Potassium-Feldspar. We know that regarding the Veluwe the brown soil sand areas are relative rich of minerals. The less fertile white sands contain quartz (Berendsen 2008). These could indicate suitability for OSL dating. Of course the presence alone of quart z is not enough; also it should be preserved in excellent

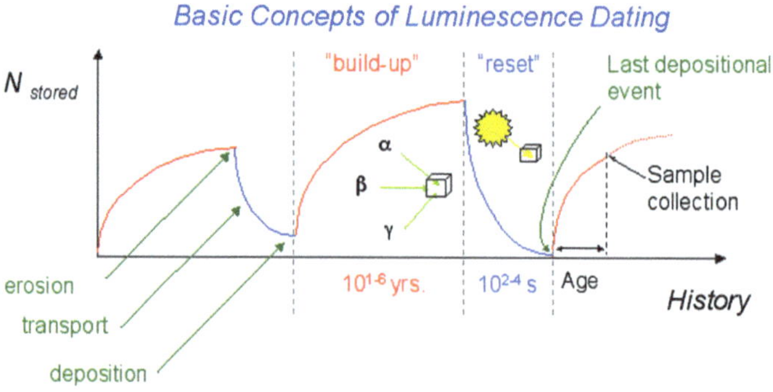

Fig. 3 Image produced by Dr. K. Lepper, North Dakota State University

conditions. Therefore, a closer look must be taken to really establish the suitability of these soils for OSL. For this reason and to gain valuable information of stratigraphy and soil data, in the spring of 2014 two trial trenches were dug across two different roads. In first case tracks were found and seven samples for OSL dating were extracted. An expert of the Wageningen University established that the samples are suitable for OSL dating. In the second trench another sample was taken from a road (late 17th century) of which we know the building date. In this way we have a check on the reliability and perhaps it could be used for calibration. The outcome of this research will be published in an upcoming paper.

The Leitha hills look at first sight not very suitable for OSL dating. Although, in some areas, like in Winden, quartz formation is surfaced, most of the soils are lime soils. The Trockenrasen ("dry lawns") may indicate the general lack of minerals in the area. However, only an investigation in the field could establish the real possibilities.

Carbon-14 Dating

The occurrence of natural radioactive carbon in the atmosphere allows archaeologists the ability to date organic materials as old as 50,000 years. Carbon-14 is continuously produced in the atmosphere and decays with a half-life of 5,730-year (+/− 40 years). Unlike most isotopic dating methods, the carbon-14 dating technique relies on the progressive decay or disappearance of the radioactive parent with time. This is now a common method for estimating the age of carbonaceous archaeological artifacts. The radioactivity of an artifact's carbon-14 content determines how long ago the specimen was separated from equilibrium with the atmosphere-plant-animal cycle. The method is based on the principle that all plants and animals, while they are alive, take in small amounts of carbon-14 and when they die, the intake ends. By measuring the loss rate of the carbon 14, the age of the object can be established (Archaeology wordsmith 2015).

Radiocarbon is the most absolute dating technique used in the Veluwe area. It has to be stressed that the samples for radiocarbon dating come from soils that were fossilized in the past. This will probably not be the case for road and paths, which normally will have been used over a long time period. An exemption could be the roads that are covered by drift-sands.

Regarding the Leitha hills no C-14 dating is known. Neither is the suitability of the area. It seems that was mentioned for OSL in Leitha hills also accounts for C-14 dating in this zone. Especially, when the acidity of the soil is considered, this is negative for the conservation of organic material. Indeed, the soil condition is very important for Carbon-14 dating possibilities. Regarding roads and paths, the only ones which are good datable are the wooden roads (Bohlwege in German) and paths in bog and peat areas. Thanks to the excellent conservation conditions the woods of these roads can often be dated by both Carbon dating and Dendrology (Denecke 2007). The latter is more accurate, but is less often applicable (Brindley and Lantink 1998). Indeed, the carbon dating of wooden roads has changed interpretation of the landscape and its use. In the Netherlands, Northwest Germany, but mostly in Ireland a lot of them have been found (Sanden 2002). However, wooden roads in bog and peat land have their own complexities and issues, which are often closely related to the hydrological situation of the bog itself (Casparie 2001). Nevertheless, the building activities of long tracks in bog and peat areas may suggest long-distance routes in prehistoric times. This is especially the case for Ireland. It doesn't make sense that people only travelled long distance on more difficult trafficable soils like bog and peat or only in Ireland. Nevertheless, in most of the cases the substantial evidence for this long distance travelling is lacking. An exemption maybe the Amber route along the Leitha hills. However, also in this case it is not clear from which northern area the amber originated.

We conclude that the possibilities for absolute dating of roads and paths in the Veluwe and in the Leitha hills are limited. Suited road ditch infills or other kinds of sediments are not likely to be available because of the inclination of the terrain which hinders not only the sediments to be waterlogged but also the slow accumulation of sediments. The only possible archaeomagnetic datable structures are kilns both for the Veluwe and the Leitha hills area. In the case of Veluwe it concerns iron kilns used to make iron out of bog ore. Limestone kilns can be found in the Leitha hills. This would be indirect dating as a link is assumed between archaeological features (the kilns) and roads which pass near and along these kilns. However, this is a hazardous enterprise as a direct relationship between the features cannot be established. A further problem regarding the Veluwe is that until now no reliable Paleosecular Variation Curves are available. An advantage for the Leitha hills is that the limestone can often contain trace quantities of magnetic minerals capable of retaining a magnetic remanence. Therefore, it is important to follow future developments in archaeomagnetics, because maybe will become possible to date the compacted (limestone) soil floors of roads and paths in forested areas. C-14 dating is possible in both areas. Although the acid soils in the Leitha hills and the white sands of the Veluwe are not the favourable conditions. Especially, if we consider road and path tracks. In the next paragraph we will look at the possibilities for relative chronology between roads and paths

Relative Chronology

Beside the possibility of absolute, there are also ways to determine the relative chronology by examining stratigraphic relations at intersections, archaeology, old maps, geomorphology and relative pollen analysis. We will start here with the latter.

Relative Pollen Analysis

The study of pollen grains in soil samples from an archaeological site which provides information on ancient human use of plants and plant resources. This technique, which is used in establishing relative chronologies as well as in environmental archaeology, was developed primarily as a technique for the relative dating of natural horizons. Pollen grains are produced in vast quantities by all plants, especially the wind-pollinated tree species. The outer skin (exine) of these grains is remarkably resistant to decay, and on wet ground or on a buried surface, it will be preserved, locked in the humus content. The Veluwe has areas that are well suited for pollen analysis and the reconstruction of the historical vegetation (Doornebosch 2014). The pollen grains of trees, shrubs, grasses, and flowers are preserved in either anaerobic conditions or in acid soils. Samples can be taken from the deposits by means of a core or from individual layers at frequent intervals in a section face on an archaeological site. The pollen is extracted and then concentrated and stained and examined under a microscope. Pollen grains are identifiable by their shape, and the percentages of the different species present in each sample are recorded on a pollen diagram. A comparison of the pollen diagrams for different levels within a deposit allows the identification of changes in the percentages of species and thus changes in the environment. As a dating technique, pollen has been used to identify different zones of arboreal vegetation which often correspond to climatic changes. The technique is invaluable for disclosing the environment of early man's sites and can even, over and series of samples, reveal man's influence on his environment by, for example, forest clearance. The sediments most frequently investigated are peat and lake deposits, but the more acid soils, such as podzols, are also analyzed (Archaeology wordsmith 2015). In peat and lake deposits are not present (anymore) in both research areas. For the Veluwe podzols are known (McIntosh and Catanzariti 2006). Indeed in the Veluwe pollen have been used to date burial mounds (Bourgeois 2013). Like for radiocarbon dating, it is important that the sample come from soils that were 'fossilized' in the past.In the earlier mentioned trial trenches across the roads investigated on the Veluwe no pollen were found. For the Leith Hills there are no data known and the geomorphological map gives.

Relative Dating Intersection

Based on the visualization of intersection linear structures, one can establish which has been latest and therefore should be considered as more recent. This has a lot in common with a stratigraphic sequence or Harris Matrix. This Matrix is the fundamental diagrammatic representation of time for an archaeological site. However it

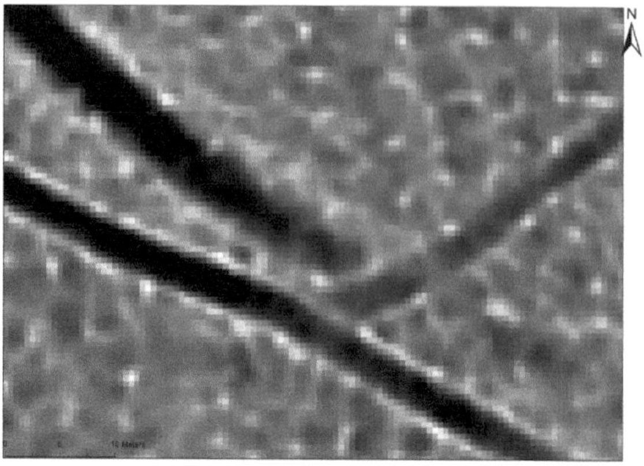

Fig. 4 It is clear from the image that the most southern Northwest-Southeast road is the youngest as it lies above the Southwest-Northeast road. The latter crosses clear the most northern Northwest-Southeast road, which is the oldest

can also be used for relative dating of roads. In the case of an archaeological site, it displays all uniquely numbered units of stratification in a sequential diagram, which represents their temporal succession. It provides the relative calendar which is the testing pattern for any further analysis. Its creation is based upon topological observations during the stratigraphic excavation process to deduce the stratigraphic relations. In historical road and path networks research, we can replace units of stratification by roads and paths and instead of excavation data we make use of elaborated ALS data to determine the 'stratigraphic relations'. The creation of a Harris Matrix does not consider any temporal relations that might be deduced from related finds in the subsequent analysis. They are used in the later grouping process of phasing or periodization by summarizing single units of stratification (roads in this case) due to structural relations or temporal relations (Schloen and Schloen 2012). In other words, the Harris Matrix can also be used to relative date roads and paths based on ALS imagery. Figure 4 gives a clear example of the visualization of an intersection in the Leitha hills of three roads.

Nevertheless, there are two considerations to make. A first consideration is the reuse or refreshment older tracks. If for example road A crosses over road B, it seems obvious that B is older A. This is true unless A is older but got out of use and came in use again after road B was established. This is theoretically possible. However the question arises in how many cases this really occurred. A second and probably major issue concerns crossings where it is no clear which road or path crosses over the other (see Fig. 7). This can have implications regarding the interpretation of the relative dating. For both the study areas good ALS are available to visualize the intersections (Fig. 5).

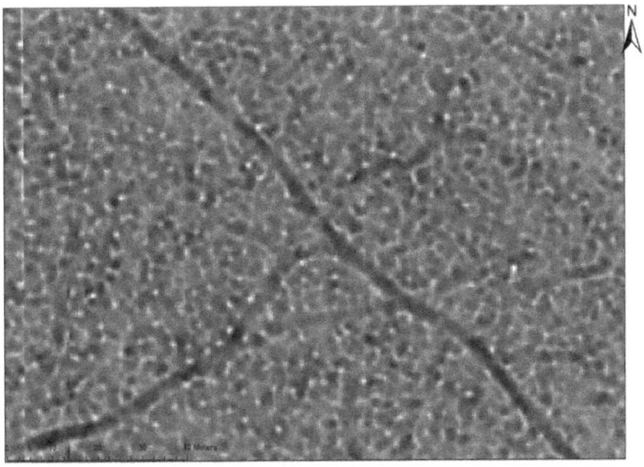

Fig. 5 The image shows the difficulty that can occur when to relative dating is based on intersection

Historical Maps

In the Leitha hills the large number of roads, tracks and hollow-ways are of special interest and some of them could be identified using old and modern maps (Doneus and Briese 2010). Indeed old maps can also provide an important indication of the era when a road or path is in used. It can be very tempting to consider a detailed historical map accurate and complete. However, there are some issues to be considered. Firstly, one has to deal with geographic accuracy of the road on the map. Especially, when more than one road lie near each other on ALS based imagery, it can be hard to discern which one is the oldest. In general military maps have a high degree of accuracy as the exact location of roads and other landscape features, like wet areas, were crucial for tactics. Further, more recent maps tend to be more precise.

Another aspect is the thrust worthy. It is quite possible that maps of certain period not reflect the real networks of roads paths that existed. When maps were based on older maps is possible, that roads were vanished over time and new ones not added. A simple reason for this would, that it was quite time and money consuming to do land survey and mapping the landscape. There are sufficient examples where roads are lacking or that towns are placed on the wrong side of a road (Aston 1985). As always with maps, it is important to know who made it and what was the purpose of it and in which circumstances it was made. In summary, one should take care when using old maps for dating for road or any historical feature. Nevertheless they can provide valuable information. For both study areas we have good historical maps from the beginning of the 19th century, which demonstrate a high accuracy. For the Leitha hills also the (historical) hiking maps are of interest, like the one from 1922 (Krizsanits and Horvath 2012). Regarding the Veluwe already lot information about main historical roads is gathered in an earlier research (Horsten 2005).

Archaeological Dating

Of course archaeological sources can help us with the dating. Nevertheless, also in this case there is an important issue to consider. If for example a road goes straight to a Neolithic hilffort, one is inclined to date this road as old (at least) as Neolithic. However, one should not date a round prehistoric because a road runs along t or even to it (Hoskins 1988). In the case this Neolithic site was also used in the medieval times, it could be that this road is 'only' medieval and that the Neolithic road vanished in, for example, the Iron Age. A counter argument could be that roads are perpetuated when a site has a more or less continued occupation and the soil is not too sensible to erosion. Also site accessibility, as measured with the help of the modern road network, turned out to have as strong effect on the longevity. Cross-roads were found to be especially important in this respect (Leusen et al. 2005). Nevertheless, archaeological sites are important to imagine possible road networks. The original lay out of a village could tell us something about the landscape and it's use (Aston 1985). For this reason one has to consider the probable different organization of the landscape. For example, some areas in England demonstrated that in the Middle Ages there were no true villages. Most of the countryside was worked from single farm isolated in the middle of their own holdings. This had of course consequence for the communication ways. Thus when we consider the study areas we have to take into account that the habitation of the landscape and it's is use could be considerable different in the past. Indeed the picture of settlement development in the landscape is a dynamic picture of great complexity, great age and constant change, but we only see it at one time. Unless, we study villages, hamlets and farmsteads as dynamic, changing, developing entities, we will miss the significance of the form and the function of them when we see them at a particular date (Aston 1985).

Traces of religious activity could also provide information of landscape and its use. Forests and mountains suited of the Hermetic tradition well. The Religious landscape, more precisely the choice of locations for medieval monasteries was influenced by classical texts in their manipulation of the landscape, as well as by the hermetic tradition of Saint Augustine. Many monasteries were sited in locations that might be described as rural retreats, and some evicted the residents of medieval villages in order to create these (Johnson 2008). Of course all the sites were connected by roads and paths to other places. Sometimes they were marked by objects along them. These "road markers" can tell us more about the age of the roads. One can think of mile stones, bridges and other buildings along a road. Archaeological research of these markers can deliver also import indications of the age of roads and paths. Looking at archaeological investigation in both study areas, we can see that lot excavations have been carried out for the Veluwe and little regarding the Leitha hills.

Geomorphology

There exist several processes that shape the morphology of the landscape. In relationship to the historical roads, only two of them are relevant for the study areas. Although, the Veluwe show clear traces of erosion of melting water, the erosion processes in Holocene have been quite limited (Bourgeois 2013). Of more importance have been the Eolic processes, the earlier mentioned drift-sands. The sands have covered up roads and also complete brooks. It caused roads to move further away, when the old track was covered by sand. This makes it possible to create a relative chronology of roads above and below the cover sand. However, it is in general very difficult to track back roads and paths covered by drift sand.

In the Leitha hills, where the slopes are steeper, a similar relative chronology can be carried out, but this time based on erosion. If for example one road goes over and landslide and another lays below it, it is possible to date them relatively. However, erosion plays also a role in the viability of the roads. It occurs often that a road or path turns into a small canal, especially after a heavy rain shower. At the moment there is no clear picture of the effects of erosion on historical roads and paths in the Leitha hills. Indeed, whilst it is easy to indicate which activities caused erosion, it is much more difficult to estimate the extent of erosion involved: the same form of erosion in one case may lead to utter destruction and in another case virtually no effect (Smith 2011). In the near future more insights into effects of erosion are expected to be gained in collaboration with the Technical University of Vienna.

A Case Study

The History of the Leitha Hills

Before to be able to date any road or path we should know about the history, and more specific, landscape history of this region. The origin of the name Leitha has different explanations. It comes from the old "Hochdeutsh" word "Litaha/ Lithaha. Probably it stems from Indo-Germanic root "loidh" that means as much mucus or slippery (Krizsanits and Horvath 2012). The Leitha hills have a history of the division. Nowadays and in the past it has been divided in the two different administrative units within Austria. The northern part is under the jurisdiction of Nieder-Östereich and the southern of Burgenland. This southern part has also been part of the Hungarian Kingdom. This history of division could be part of the explanation that little research has been carried which took the area as one entity. This explains also some of the roads as borders, like the "Grensweg" (Border road) between Mannersdorf and Hof am Leithagebirge. Sometimes boundaries followed a stream or a track way that already existed, but very often they created their own boundary lanes (Hoskins 1988).

The economic activities known, which have taken place through history in the Leitha hills are woodcutting, quarrying, cattle husbandry and communication ways for crossing the hills. The stone quarrying took place mainly on the border of the area. Probably because of logistic and costs matters. Along Leitha hills there was a cattle route from the Hungarian plane toward Vienna and further. On the maps there are still names that could refer to this period, like Groß Ochsenstand and Klein Ochsenstand. These areas are now wooded, but in former times there were open land. This could be an indication that in the past the Leitha hills was less forested in Medieval Times. For earlier periods the found Iron Age sites could be a sign that the hills were more occupied that in Modern History. However, this is not enough to suppose that during the Iron Age the area was more open.

Forestry took place in large parts of the Leitha hills and carts were used. It is interesting to see how the seasons influenced the viability in the Leitha hills. For example in autumn the falling leaves could make the roads slippery and dangerous. Besides used for economic purposes, the woods were used to hide. Indeed, people fled into the woods when Turkisch troops came to their towns. An example of a hiding place is the Teufelsloche (Devils hole) on the Sulzberg. They have in different periods protected people (Krizsanits and Horvath 2012). Interesting could be the roads or paths to these holes. Thirty holes are documented. The Hartl-Lucke is a good example. It was located near a gravel road, now it is completely overgrown. It also shows how fast places in the forest can change in only 20 or 30 years. This should make us cautious about reconstruction historical roads and paths. Moreover, we should remember that only a part of the historical roads and paths can be detected. Some holes have also served as hermitage.

The road names Brunnengasse or Quellengasse refers in a lot of places to the presence of a fountain or a well. Each town has such water well. Also other names demonstrate the former importance, like Waschgattl in Eisenstadt, where the laundry was washed or Rosschwemmen where horses were cleaned. In the Leinwandbleich the laundry was spread to bleach. Also the Ochsenbründl in Jois, named after the ones who pulled the carts through the vineyards, looked different in the past. It was like a swamp, where cows and geese came to drink from the nearby Hutweide. A Hutweide was for centuries the place were the animals of the local villages grazed. The animals stayed here during the summer and were protect by a shepherd or children from the villages. Here comes the name Hutweide from (behüten means protect). Each village had such a "Hutweide", normally between the village and the woods. The ground was too sparse to plant any crop. Often rocks and the slope made elaboration of the ground impossible. For the animals there was food enough. Drought resistance plants like hear grasses, which grow on both sandy and gravelly soils, were good enough for them. It was often an area between the village and the forest where the livestock grazed. It demonstrates again that the use and the outlook of the forest were different in the past. This is an important notion for interpretation of road features and reconstructing historical road and path networks.

One can find often traces of human live nearby the wells, which go back until the Iron Age. There was certainly a cult at these places. Often a link is made with the chapels to worship the Holy Radegundis, like in Großhoflein or Mannersdorf. Well pilgrimage took place which went back to pagan well worship. The Christian cult goes back till the 9th century in Leith hills (Krizsanits and Horvath 2012). All these holes, wells where connected to the towns with roads and paths. Therefore the archaeological record alone is not sufficient to reconstruct and understand the historical road networks. Knowledge of the geology and hydrology are also important. For this reason collaboration has been sought with the Technical University of Vienna.

In the forest near Eisenstadt there is a stone statute of Joseph and Maria. Maria looks in the direction of Eisenstadt and Joseph in the direction of Lorretto, symbolising the watershed (Krizsanits and Horvath 2012). The watershed could also be important for a ridgeway across the Leitha hills. The advantage of a ridgeway, which often date from prehistoric times, is of being dry, because of the water running down on both sides (Berendsen 2008). If we look at the Medieval border stones (Hotters) we can notice that they follow for a large part this watershed together with a road.

It can happen that in the forest that the water at some places gathers, where the ground is not permeable. These places are called locally "Wildschweinlacken", meaning wild boar pools. Indeed the boars wallow in these wet areas. Another example is the Saubründlegraben in the area of Stotzing. There are about 80 graben (trenches) in the Leitha hills. Their name are often derived from landscape features, but also to the owners. Some of the brooks are named after the crayfish that once and still live here. Like the Kroißental in St, George and Groisbachn in Sommerein (Krizsanits and Horvath 2012).

The Leitha hills is also connected with wine. For centuries it has characterized the landscape, influenced the shape of the villages and road segments. In the 18th century a lot the wine yards were turned into farmland because of increase in taxes. Indeed, in the past there were much more wine yards and fruit trees around and on the Leitha hills. Important were the cellars, mostly built against the slope of the hill and covered with earth to realise the optimal climatic circumstances. Not all the places were well suited for growing grapes. The local place names revealed often the quality. For example Goldberg for good quality and Steinweingarten for bad conditions. Fruit trees characterized the landscape as they were often built along roads.

Around the Leitha hills is during the centuries a pilgrimage culture has developed and also hunting took place. Although, it was for a long a privilege of the nobility, as they were the main owners of the forest. The forest are crossed by large "Alleen". These were built to protect the forest from fire. Mainly on the borders almost each town had limestone quarry. The stones were carried with a horse cart to Vienna. This took six to eight hours. Finally, the Leitha hills have long history of defence, from the Hallstadt culture till modern military training areas. With all this historical information a relative chronology within the path network can be established by means of existing software packages and old maps. The resulting relative chronology model can give significant insights in the development of road and path networks in the past. This will be shown in the following paragraphs.

Harris Matrix Composer

The Harris Matrix composer is based on the concept of Harris. The software Harris Matrix Composer provide the possibility to create a digital Harris Matrix. A considerable advantage of the product, is the possibility to validate the created relations. In this way erroneous relations can be avoided and thus a wrong interpretation(Harris Matrix Composer 2013).

As mentioned before we have a network for the Leitha hills area that was for the main part automatically extracted from ALS data (Denecke 1969). When needed the network was completed manually. Then in ARCGIS the whole network was dissolved and then with feature line each road segment runs from node to node. The reason for doing this is that each road segment could have part of different network and in different times. In other words, road segment do not necessarily belong to one road or one route. In the Harris Matrix Composer, this creates some extra work. However, there is also the possibility to group entities when needed. This helps to visualize the relations between roads and paths. An important aspect is that the road segment keeps the same ID number as created in ArcGIS to maintain the consistency of the investigation.

In the following a selected area 1,5 square of the Leitha hills has been used to test the applicability of the software. The specific area has been chosen because it contains both different kind of roads and paths and there are monuments in the area which give the possibility for more accurate dating. The earlier issue of re-used roads was not considered. The reason for this is that are no examples known of roads used in different periods, leading to wrong interpretation based on intersection. Therefore this possibility was not taken into account. The HMC works quite simple. Road segments that are situated above another are also above each other in the HMC diagram, connected with vertical arrows. Road segment from the same period are connected with horizontal arrows and located next to each other (Fig. 6).

OCHRE

The Online Cultural and Historical Research Environment (OCHRE) is a comprehensive database environment in which scholars record, integrate, analyze, publish, and preserve all forms of project data at all stages of research. It can be used for initial data acquisition and storage; for data querying and analysis; for data presentation and publication; and for long-term archiving and curation of data. It was developed by David Schloen (an archaeologist) and Sandra Schloen (a computer scientist) and is supported and made available to students and scholars around the world via the OCHRE Data Service at the University of Chicago. The main goal of OCHRE is to provide an environment for *integration* of all relevant data for a research project—spatial, temporal, bibliographic, image, textual, and so on—and to capture this variety of data in a way that preserves its natural format but

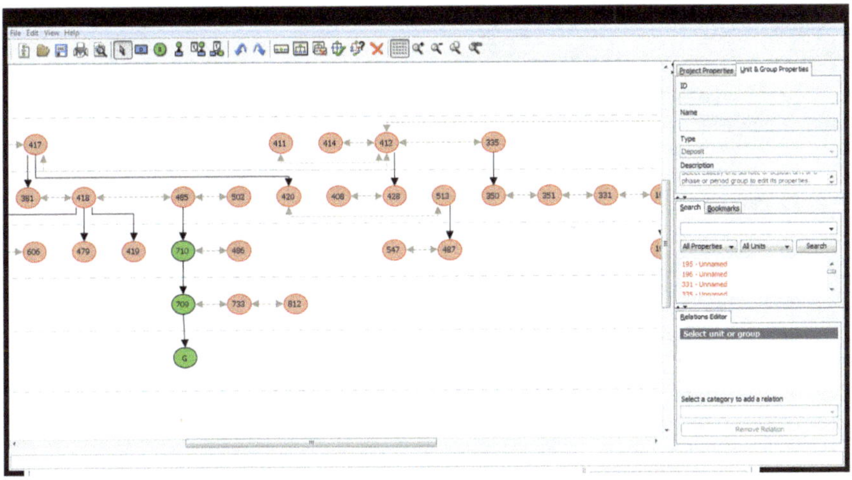

Fig. 6 Relative dating based on intersection visualised in the Harris Matrix composer. A part of the network in the study area is visible. In this part already four layers are visible. A next step in the dating process is the use of other source. This will be done within OCHRE and explained in the following paragraph

maximizes the relationships between different types of data to achieve more powerful analyses.

To accommodate a wide variety of data OCHRE uses an item-based structural model, rather than a table-based model typical of commercially available off-the-shelf software (Schloen and Schloen 2012). Any data object under consideration is uniquely identified, then described by properties, notes, links, and events. These data items are organized using hierarchical structures, a strategy which proves to be a very effective method for representing both space and time. The many categories of data managed by OCHRE are themselves structured hierarchically as shown in Fig. 7. For the scope of the *Historical Roads and Paths* project the categories **Locations & objects** and **Periods** are particularly important. Within these categories the relevant items are arranged in a tree structure.

In Fig. 7 on the right there is an example of the **Locations & objects** category in which the extracted and mapped linear features are listed within the hierarchy **All linear segments**. At this stage the linear segments are undifferentiated so they comprise, in effect, a simple list where each item is analogous to a row in a more conventional table structure. But conceptually, many of these linear segments combine to form identifiable roads and paths. One of the distinguishing features of OCHRE is that items can be arranged in more than one hierarchy, allowing the researcher to combine the items into meaningful groupings without redundancy or duplication. Here there is a second hierarchy, **Identified features**, in which identifiable roads and paths can be created from the individual linear segments in the first hierarchy, as shown in Fig. 7. In other words from segments 6, 4 and 5 one road is created (**road a**). From segments 6 and 3 **road b** is created. For this example

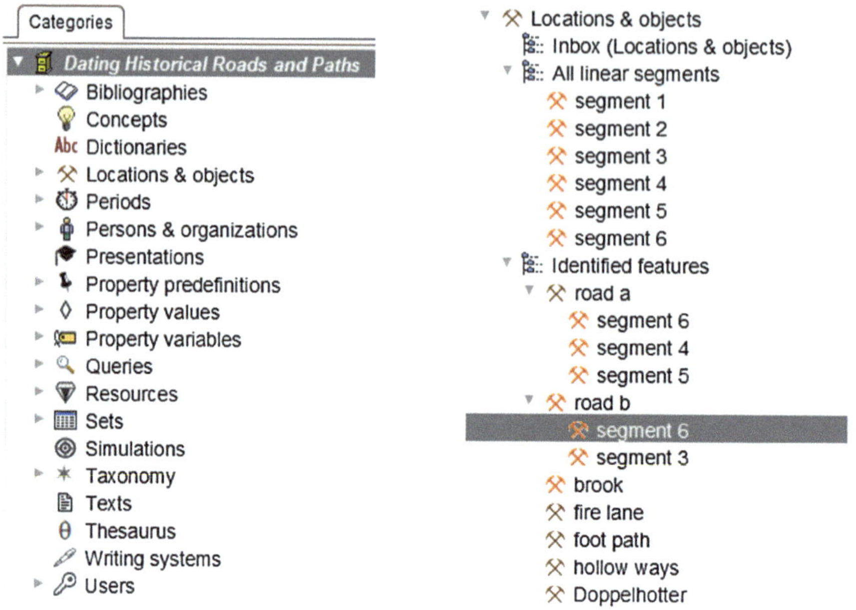

Fig. 7 On the left side, the *navigation pane* of OCHRE. On the right side, a fictitious example of the possible branching of the **Locations & objects** category

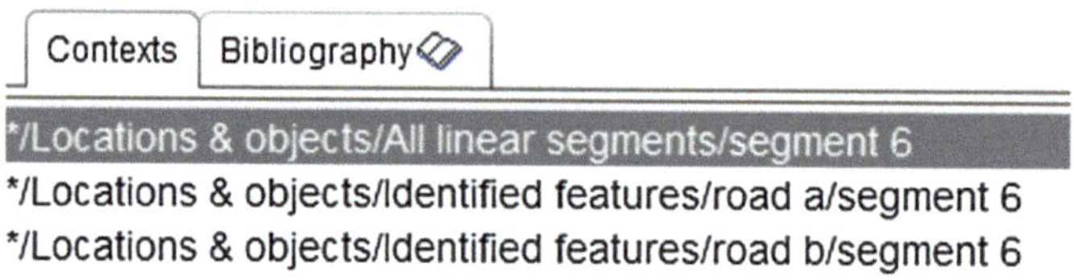

Fig. 8 Linear **segment 6** shown in each of its three organizational contexts: in the collection of all linear segments, as part of **road a**, and as part of **road b**

it becomes clear that **segment 6** is part of both **road a** and **road b**. This is of course logical when after **segment 6** the road splits up into two directions. To repeat for the sake of clarity, the creation of roads based on the individual linear segments under **All linear segments** does not mean that the segments are being copied. Instead, because of the item-based data model, the same item can exist in multiple hierarchies. In other words there no duplication of data. This is confirmed by the item's **View** which shows the item situated in three distinct contexts. Changes to *any* of these instances of the item apply directly to *all* of them because they are instances of the same database item (Fig. 8).

What applies to roads also applies to brooks. There is of course a distinction between brooks and roads or paths. A segment of a brook would also never be part of a road or path system. The case is different for the so called "fire lanes." These lanes' first objective was not as a communication way. However, from studying maps and from field walking it becomes clear that sometimes a fire lane has a double function, both as protection against fire (extension) and as a road or path. Nevertheless, this is not a problem because of the flexible, item-based model of OCHRE, in which an item (segment) can be both part of a road and part of a fire lane.

Once the segments have been organized and grouped into roads and other identifiable features, the next step is to identify the time period for those roads and features by assigning an appropriate (relative) date. Before explaining the procedure for doing this, we will first describe the **Periods** category, illustrated in **Fig.** 9.

The **Periods** category is important for the relative dating of roads, paths and fire lanes. In this category all the periods are arranged in a chronological order. Where possible a further subdivision is made. Note the use of the hierarchical structure to represent conveniently the subdivisions of the periods and the inherent relationships to the higher-level periods. If a road is dated to the **Hallstadt Culture** period, because the **Hallstadt Culture** is contained within the **Iron Age** period, the road inherits the relationship to the **Iron Age**. And so on for the other ancestors; since the **Iron Age** is contained within the **Prehistoric** period, the road will inherit the link to the **Prehistoric** period too. Because of the hierarchical inheritance, a query that searches for all roads in existence during the **Prehistoric** period will include in its matching results all roads that are assigned to the **Hallstadt Culture** period. The researcher can assign to a road the most specific period that is relevant or that can be known, but then the road inherits the relationships to all of the hierarchically related ancestor periods (Fig. 10).

Until the Roman period we can only base the evidence on archaeological finds. This is often not a very secure source. In an earlier statement, it was already

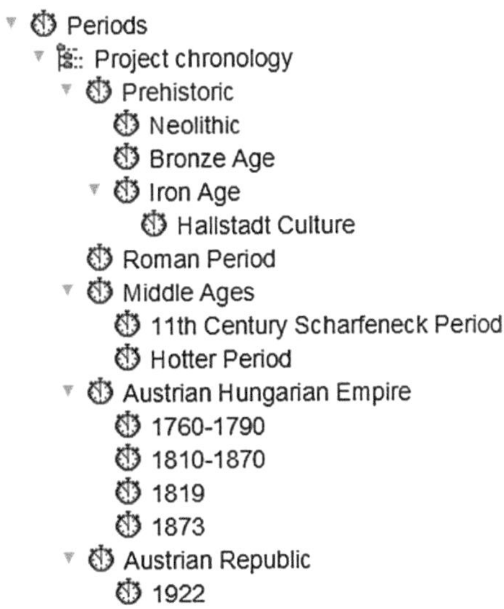

Fig. 9 The time periods distinguished for the Leitha hills

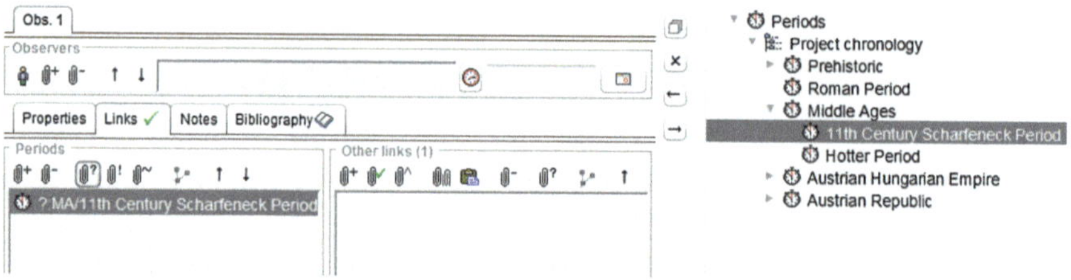

Fig. 10 An item is dated by assigning a link to an appropriate Period from the **Periods** Category using the **Linked Items** pane in the right-most OCHRE panel. Such links can be marked as uncertain

mentioned that a road running along a Neolithic monument does not mean that the road is also Neolithic. However, sometimes there are strong indications in the landscape that make dating to prehistoric times likely. OCHRE allows the researcher to tag a link to a period with a question mark, thereby indicating the uncertainty of the identification. From the Roman times onward, the written sources gain importance. We know for example that the Castle Scharfeneck was built in the 11th century. There is only one road running to the main entrance and no other options seem suitable, so we can quite confidently date this road to the same period. Other examples are the (Doppel) Hotter. These are border markers put on the ground in the Middle Ages. Often a road runs alongside such a border marker. This also gives a quite good indication of the date of the road, although less clear than the example of the Scharfeneck.

One of the strengths of the OCHRE approach is that a certain object, a group of hollow ways with a long history in this case, can be linked to different periods. If for example based on archaeological evidence, there is a strong indication that their use has mainly occurred in the Neolithic this can be assigned by clicking the paperclip link button with the exclamation mark (see Fig. 11). It is also possible that other persons have other interpretations of the predominant use. In this case a new project user can be added, who gives his or her own interpretation by adding a new *observation* to the same database item. Each observation can be tagged as to the observer and the date of the observation. In this way, scholarly discussion regarding an item of interest can be attributed and tracked.

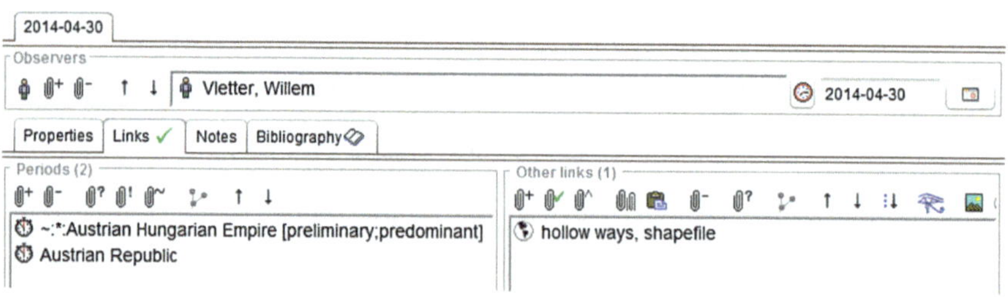

Fig. 11 An observation is made detailing the attested dates of the hollow ways

From the beginning of the 19th century accurate maps also became available. Once these are georeferenced and laid over the ALS data, the relative time depth of roads and paths becomes immediately clear. As mentioned earlier, one should consider old maps with care as they are often not complete. One can imagine for example that small paths were not mapped. A good example of this is the Koningsweg from the late 17th century near Ede on the Veluwe. With the visualization technique of Openness applied to the ALS data this road is visible. However, it is not visible on the historical map from the beginning of the 19th century. On the contrary, roads of this map are not visualized with the ALS data roads. (reference to be added).

OCHRE allows all reference works and other data sources like maps to be catalogued within its **Bibliographies** category. As with the Periods above, a simple linking mechanism using OCHRE's **Linked Items** pane allows the researcher to identify the reference for the information being assigned to the item as illustrated by Fig. 12.

Along with Periods and Bibliography, OCHRE allows the researcher to set up custom properties to describe the database items. This is done in the **Taxonomy** category which, once again, is represented as a hierarchy (see Fig. 13).

The taxonomy is an alternating arrangement of **Variables** and their permitted **Values**. A **Property** is a specific instance of a variable along with the specific value that has been assigned to a given item. For example, a road item will be identified by the variable **Type of linear feature** and the value **Road or path**. What makes hierarchies so valuable again as an organizing structure is that the variable can be repeated at the next lower level of the hierarchy to further qualify the type of road or path, in this case as a **Cart track**, a **Foot path**, or a **Modern road**. As with periods, if an item is tagged with a value deeper in the hierarchy, for example a **Cart track**, it automatically inherits its ancestor values, here the **Road or path** designation. Queries for items designated as **Road or path** will, by implication, also find those items tagged as a **Cart track**.

Also worth noting is that a variable can be assigned to an item more than once. To use the fire lanes discussed above as an example, they would be tagged with the **Function** variable assigned as a **Passage way**, and also as **Protection**. When a number of different properties are commonly assigned to items as a group, they can be collected together as a **Predefinition** in order to speed up the assignment of properties (see Fig. 14).

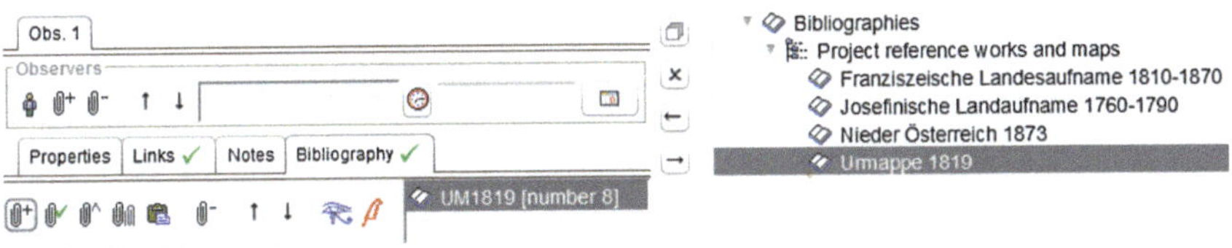

Fig. 12 A reference map is shown being linked to a fire lane item

Fig. 13 A sample user-defined taxonomy of variables and values

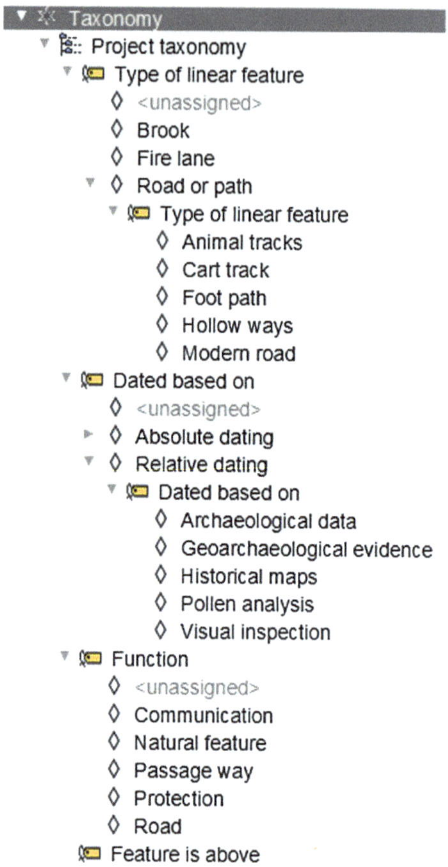

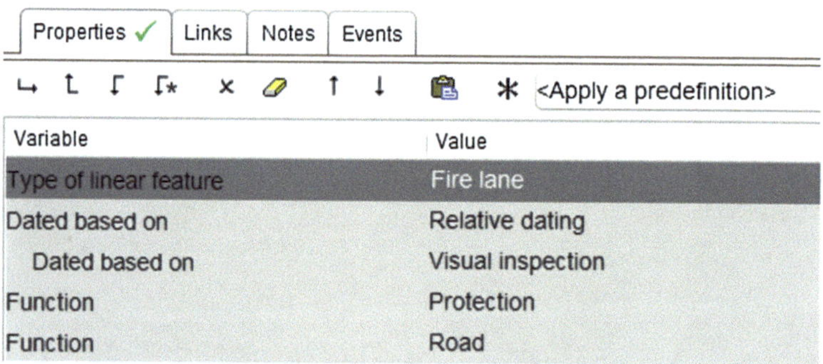

Fig. 14 The properties of a **Fire lane** collected as a **Predefinition**. Note that the **Function** variable is intentionally assigned twice

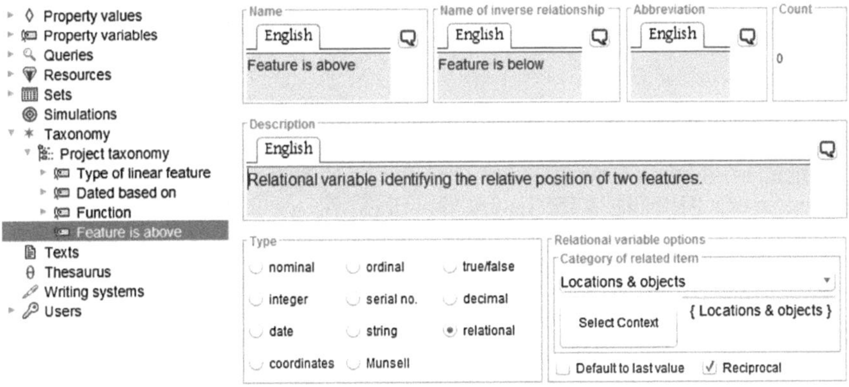

Fig. 15 A reciprocal, relational variable used to indicate the relative position of two features

One additional type of variable of particular use here is a *relational* variable. This is a special type of variable that links one item to another as a property. As shown in Fig. 15, the **Feature is above** variable will link a source item to a related target item in the **Locations & objects** category. Typically the relation is reciprocal, and OCHRE provides an alternate label to describe the inverse of the relationship when viewed from the perspective of the target item. Here the inverse name is **Feature is below** to indicate the nature of the inverse relationship.

At the moment, relative dating based on a Harris Matrix is not yet possible in OCHRE. But it is the use of relational variables such as the one described here that will make it possible to first tag the appropriate relationships between the items, and then to use those relationships to create a Harris Matrix. Plans are underway to include new tools within OCHRE to allow auto-generation, with manual adjusting, of Harris Matrix diagrams so that all of the relative dating strategies can be integrated within a single database environment.

Once all the datable periods are categorized and the item properties are assigned, the identification and (relative) dating of the roads and paths can proceed. Because of the item-based data model used by OCHRE, it is typically recommended as a best-practice strategy that geospatial data be organized in such a way so that each relevant OCHRE item knows how to draw itself independently of any others. More specifically, this means that each **location** or object within the **Locations or Objects** category may have a geographic component of some kind, either a specific latitude/longitude coordinate assignment, or a point, polyline, or polygon shape associated with it. In effect, the geospatial representation of the item serves as another property of that item.

OCHRE's built-in mapping features provide an interface for displaying and working with geospatial data including *itemizing* geospatial data; that is, assigning to each OCHRE item only the relevant shape (or shapes) for that item. This provides the high degree of flexibility required by most research projects. Items can be chosen via a query using any of the properties or links assigned to them along with any of the search criteria options available in OCHRE. Because each of the items in the set of matching results can draw just itself, a map can be generated based on the query result set. Using queries that draw on all of the data associated with the roads and paths, many different aspects of the roads networks can be visualized, providing many analytical options.

Often, however, a shapefile will represent a collection of items rather than a single item. OCHRE has a special link tool in its GIS toolkit that lets a user create a new shapefile on the fly from the current selection of shapes in a map view. This newly created subset of shapes will be linked as a new shapefile to the currently selected item (Schloen 2014). Shapes can be selected by drawing a line, circle, rectangle or polygon to constrain the selection. Multiple selections can be accumulated so that all of the parts of a feature that may not lie near each other, like brooks, can be selected and assigned together. If the selection of shapes represents a new item that does not currently exist in the database, the item itself can also be created on the fly. **Properties** and/or **Periods** can be included in a **Predefinition** that is assigned to the item when the shapes are assigned, to further describe the item.

Alternately, the researcher can make a selection from the map layer's attribute table which contains a list of all the road segments. This is of course of interest with regard to the relative dating based on intersection with the HMC. The ID numbers used to identify the road segments in the HMC are the same as those shown in the attribute table. This provides a link between the separate applications which are needed until such time as the full required functionality is available within a single environment.

In the meantime, OCHRE's mapping interface provides tools for item tagging by **Properties** or by **Periods**. Figure 16 illustrates the creation of a new item on-the-fly to represent a group of hollow ways. The predefinition for describing hollow ways is selected to be assigned to the newly created item, the name for which is provided by the researcher. The subset of currently selected line segments becomes a new shapefile that is linked to the new item.

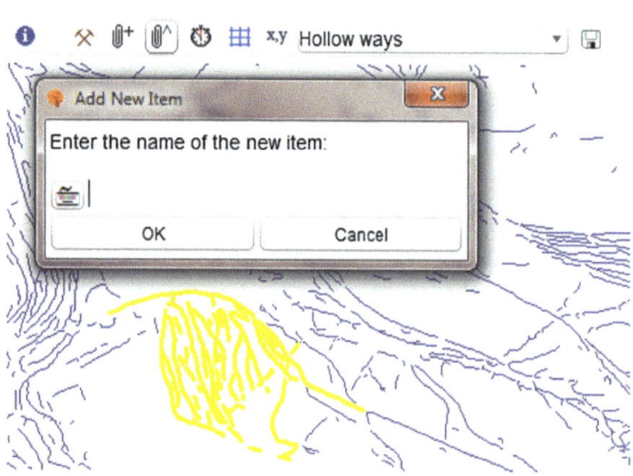

Fig. 16 A new item is created to identify a group of hollow ways; its properties are assigned by applying the **Predefinition** for **Hollow ways**. A shapefile containing the selected line segments is created and linked

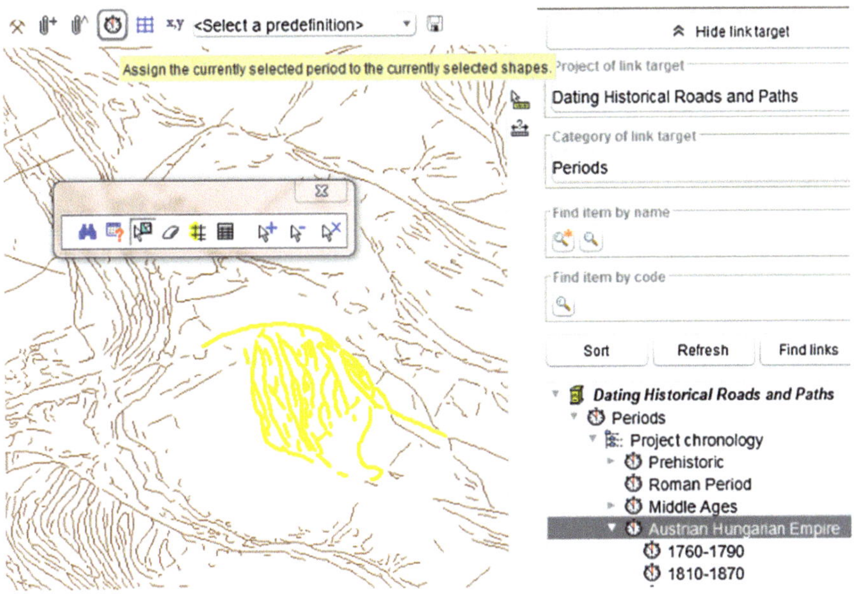

Fig. 17 An item representing a group of hollow ways is assigned to the relevant period, in this case the Austrian Hungarian Empire

As shown in Fig. 17, dating can also be assigned to selected items in the map. The currently selected **Period** in the **Linked Items** pane (in the rightmost panel) is assigned to the currently selected item (highlighted) in the map view.

Shapefiles displayed within OCHRE, or created by the itemizing process built into OCHRE, are managed by the **Resources** category. This becomes a project's catalog of the external files that are available to OCHRE and it includes additional details about where the files are stored. These external files may include shapefiles and other image types; for this example we link in the Urmappe of 1819. A useful effect is generated by using a georeferenced map as the backdrop for the OCHRE items and their individual shapefiles. This visualization helps make sense of the relationships between features (Fig. 18).

Here it becomes more evident that the hollow ways were created after the fire lane and so must be younger. The **Feature is above** property is added to the hollow ways item's properties to capture this relationship; that is, the hollow ways **Feature is above** the fire lane. Combining the visual data, along with Bibliography, Periods, Locations & objects and Properties, captures the full details of the features within the roads and paths network (Fig. 19).

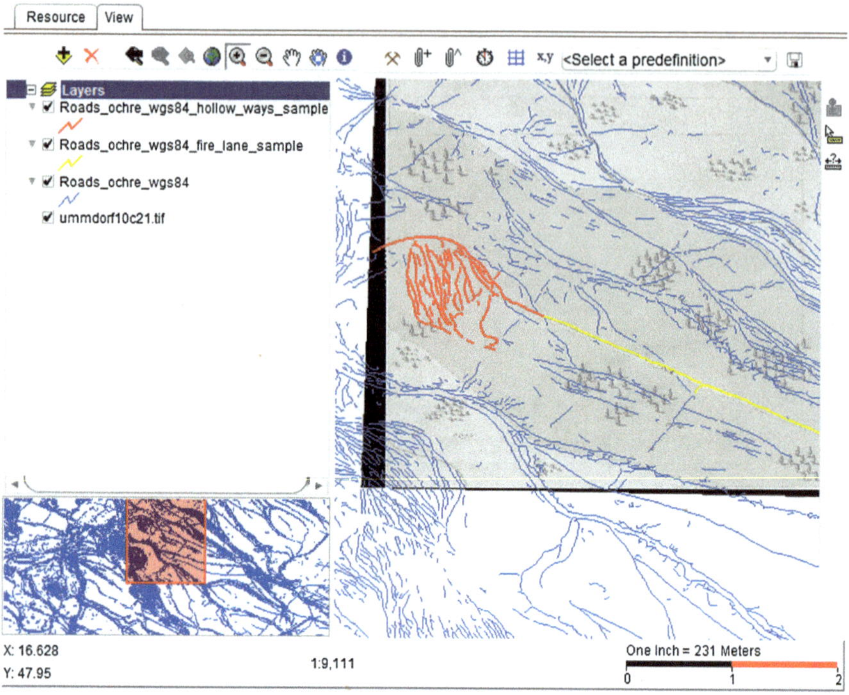

Fig. 18 The hollow ways and the fire lane overlaid together on the Urmappe from 1819

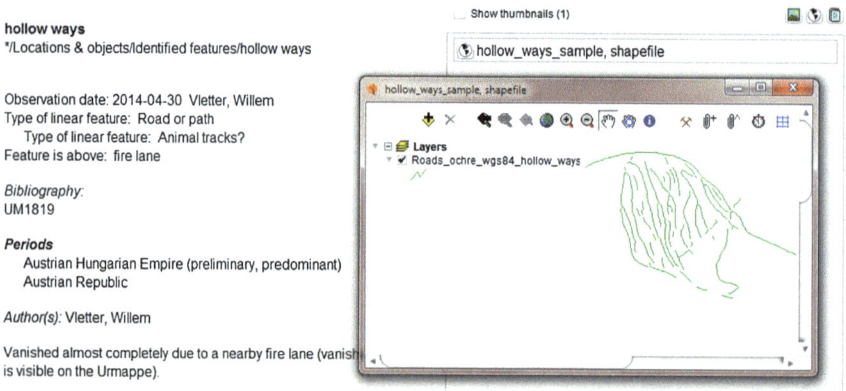

Fig. 19 The hollow ways, fully described by properties, bibliography, periods, and shape, and indicated to be above the related fire lane

Conclusion

In the end the only applicable absolute dating technique which seems worth wile for historical road research for both study areas is OSL dating in the Veluwe. Further, it can be concluded that relative dating of roads can be carried out combining different techniques and sources. An important is requisite is the availability of ALS data which can used to visualize micro relief, showing at intersections what is the relative chronology. The Harris Matrix Composer is an easy to use software package which enables to demonstrate the structural or temporal relations between roads in diagram. Important aspect of the software is that the diagram can be validated. In this way errors and thus wrong interpretation can be avoided. Extracted or mapped roads from ALS data can also be date relatively by using old maps, written sources and archaeological data. When all relative dating techniques are combined they reinforce themselves as have been shown in OCHRE. With this application a wide range of linking of items and queries are possible, which enlarge the research possibilities. One of its strengths is the tool to create subsets of the extracted or mapped network. In this way roads or brooks can be visualised separately and also be linked to certain periods. Its flexible structure allows to make distinction between road segments and to date them to different periods. The same flexibility enables multiple functions for linear features. This all contributes to a better investigation of historical roads and paths.

Acknowledgments I would like to thank Eamonn Baldwin of the University of Birmingham for his critical look at the first drafts and the staff of the OCHRE Data Service of the University of Chicago (http://ochre.uchicago.edu) for assistance and for providing complimentary student use of the OCHRE software.

References

Archaeology wordsmith. 2015. Dating and Pollen analysis. http://www.archaeologywordsmith.com/lookup.php?terms=dating, http://www.archaeologywordsmith.com/lookup.php?category=&where=headword&terms=pollen+analysis

Aston, M. 1985. *Interpreting the landscape. Landscape Archaeology in local studies*. London: B. T. Batsford

Berendsen, H. 2008. *Landschappelijk Nederland*. Assen: perspectief uitgevers.

Bourgeois, Q. 2013. *Monuments on the Horizon*. Leiden: Sidestone press.

Brindley, A.L., Lantink, J.N. 1998.Radiocarbon dates for Irish trackways. In *the Journal of Irish archaeology*, XII, 45-67.

Casparie, W. 2001. Prehistoric building disasters in Derryville Bog, Ireland: Trackways, Floodings and Erosion. In *Recent developments in wetland research*.: 115-28.

Denecke, D. 1969. *Methodische Untersuchungen zur historisch-geographischen Wegeforschung in Raum zwischen Solling und Harz*. Göttingen.

Denecke, D. 2007. Wege und Wegenetz. In *Reallexikon der Germanischen Altertumskunde* Bd.35 Berlin/New York: De Gruyter

Doneus, M., Briese, C., 2010. Airborne Laser Scanning in Forested Areas - Potential and Limitations of an Archaeological Prospection Technique. In *Remote sensing for archaeological heritage management.* Brussels: Europae Archaeologia Consilium (EAC), Association Internationale sans But Lucratif (AISBL), Siège social.

Doornebosch, M. 2014. *An environmental study of barrows.* Sidestone press. Leiden.

Ellis, C. 1998. Archaeomagnetic Dating and Palaeochannels Sediments: Data from the Mediaeval Channel Fills at Hemington, Leicestershire. In *Journal of Archaeological Science* 25: 149–63.

Harris matrix composer. 2013. http://www.harrismatrixcomposer.com/

Horsten, W.F. 2005. *Historische wegen in Nederland van 1600 tot 1850.* Amsterdam: Askant.

Hoskins, J.W. 1988. *The making of the English Landscape.* London: Guild Publishers

Johnson, M., 2008. *Ideas of landscape.* Oxford: Blackwell publishing.

Krizsanits, B., Horvath, M. 2012. *Das Leithagebirge. Grenze und Verbindung.* Weitra: Bibliothek der Provinz.

Leusen, M., van, Deeben, J., Hallewas, D., Kamermans, H., Verhagen, P., Zoetbrood, P. A. 2005 Baseline for Predictive Modelling in the Netherlands. In *Predictive modelling for archaeological heritage management: A research agenda.* (eds.) Leusen, M. van and Kamermans, H.: 28-94. Amersfoort: National Service for Archaeological Heritage.

Linford, P. 2006. *Archaeomagnetic Dating: Guidelines on Producing and Interpreting Archaeomagnetic Dates.* Portsmouth: English Heritage Publishing

McCann, W.A. 2003. The archaeomagnetic dating of two 14th-15th century hearths. In *Charter Quay, the Spirit of Change.* Salisbury: Wessex Archaeology.

McIntosh G and Catanzariti G, 2006. An introduction to archaeomagnetic dating. *Geochronometria* 25: 11-18.

Pavón-Carrasco, J., Rodríguez-González, J., Osete, M-L., Miquel Torta, J.A. 2011. Matlab tool for archaeomagnetic dating.*Journal of Archaeological Science* 38 (2):408-419.

Sanden, W. A. B. van der. 2002. Veenwegen in Drenthe: enkele nieuwe dateringe. *Nieuwe Drentse Almanak* 119: 101-112. Assen: van Gorcum

Schloen, J. D., Schloen, S.R. 2012. *OCHRE an Online Cultural and Historical Research Environment.* Winona Lake: Eisenbrauns.

Schloen, S. 2014. Itemizing shapefiles. http://ochrewiki.wikispaces.com/Itemizing+Shapefiles

Smith, N. 2011. *Pre-industrial road, track ways and canals.* English Heritage publishing.

USGS. 2015. What is Luminescence dating? http://crustal.usgs.gov/laboratories/luminescence_dating/what_is_tl.html

Vletter, W. 2014.(Semi) automatic extraction from Airborne Laser Scan data of roads and paths in forested areas. *SPIE proceedings Second International Conference on Remote Sensing and Geoinformation of the Environment.* doi:10.1117/12.2069709.

Vletter, W. 2015. Aworkflow for (Semi) automatic extraction of roads and paths in forested areas from Airborne Laser Scan data. *AARGnews* 50 March: 33-40

Part VI
Cultural Resource Management: Communication and Society

From Remote to Embodied Sensing: New Perspectives for Virtual Museums and Archaeological Landscape Communication

Eva Pietroni

Abstract Digital technologies can have a revolutionary impact in the cultural heritage domain, not only in documentation and representation but also in communication and in the emerging field of digital ecosystems. Virtual environments allow the users to learn from experience, joining sensorimotor and interpretative faculties, perceiving, acting and adapting to changing circumstances, even if in contexts that are no longer (or not yet) materially accessible today. The challenge consists in the breadth of the field and the need to bring together an exceptional spectrum of disciplines, from cognitive science and psychology to history, archaeology, art, geography, computer sciences. Despite the exciting perspectives opened in education and in terms of social and economic growth, research in the domain of Virtual Museums has not reached a sufficient level of maturity, such as Cinema or Game sectors. There is still a disconnection between the research, that develops tools with little interest in their wide application, and the industry, that builds ten-year plans addressing the market. Virtual museums, the main focus of this contribution, have reached an important advancement but still today much of the research aims at technological innovation rather than at genuine culture transmission. Given the need of enhancing the emotional and cognitive impact of virtual museums, some criteria and good practices will be discussed, dealing with engaging storytelling, embodiment, novel solutions in the interaction design and in the integration of media. The theoretical discussion will be exemplified through concrete case studies, among which the *Tiber Valley Virtual Museum*, to give more evidence and a better clarification.

E. Pietroni (✉)
National Research Council, Institute of Technologies Applied to Cultural Heritage, Rome, Italy
e-mail: eva.pietroni@itabc.cnr.it

M. Forte and S. Campana (eds.), *Digital Methods and Remote Sensing in Archaeology*, Quantitative Methods in the Humanities and Social Sciences,
DOI 10.1007/978-3-319-40658-9_19

Technologies and Cultural Heritage

Cultural artifacts, especially archaeological ones, are largely fragmented and decontextualized in museums or in urban contexts that have profoundly changed and it is often difficult to decode their meaning. Therefore they have lost the capability to convey to the public the cultural message for which they were created and displayed by their authors.

An aggravating factor is the general absence of a modern "museology": collections are often assembled without communicative criteria, following a taxonomical approach and public fruition is limited to the analyses of their pure formal aspects, instead of aiming at the comprehension of their intrinsic meaning. Communicative strategies for museums should be largely evolved, establishing meaningful narratives for collections and making the museum a place of interactive experience (Antinucci 2014; Forte and Pietroni 2006).

Through the virtual elaboration an artefact can lead us towards the knowledge of the mind and the way of living of past and present civilizations, re-activating their connections of meaning and contextualizing them in space and time. This is the essential condition for the transmission of culture.

In the last 2 decades integrated digital tools and instruments for documentation, GIS and remote sensing, computer graphics, digital libraries, real time visualization allow a wide audience to access, display, compare, analyze, interpret and transmit a huge quantity of data as never achieved before, at multi-scale level.

The possibilities promoted by remote sensing using non-destructive approach to view the buried and not directly visible evidence of past activity, has opened multiple and extraordinary challenges for digital archaeology, in the interpretation of multi-scale contexts.

Some disciplines as virtual/cyber archaeology and virtual heritage are today fields of study in their own right, studied in many Universities and based on theoretical and methodological framework as the London Charter (2006, 2009) (http://www.londoncharter.org) and the Principles of Seville (Lopez-Menchero Bendicho 2013).

Despite the awareness of the vast perspectives opened by digital technologies, research in the domain of cultural heritage communication and virtual museums has always remained distant from a logic of industry, pursuing other directions: theoretical definition of the discipline, digitization and recording, technological implementation, digital archives, taxonomies, pilot cases of immersive visualization and interactive virtual environments.

The discussion about virtual museums started at research level since the end of nineties (AA.VV. 1998) and it continues still today. Recently the V-Must network of excellence in the field of Virtual Museums (http://v-must.net) has elaborated a better definition of the theoretical framework of this promising domain, (Fig. 1). From the beginning it was clear, at least at theoretical level, that Virtual museums should not be a simple technology transfer dedicated to reproducing actual reality. Similarly, they should not reply the same structure and paradigms of real museums,

Fig. 1 *Keys2Rome, Augustus' city* exhibition, organized at the end of the *V-Must* project, Trajan Market, Sept. 2014–May 2015

nor be structured as a database or an archive (Antinucci 2007a, b). Their effective power consists in the possibility to re-organize cultural contents in order to enhance the comprehension of the artefacts, their relations and connections with the context, their narrative value (Ragghianti 1974).

What the ordinary visitor needs is simple and involving communication, that should, if possible, also intrigue and stimulate motivation.

Several evaluations of the user experience inside virtual museums realized in the past years (Pescarin et al. 2012, 2013) have shown that users' main expectation is to enter and interact inside stories, personalizing their experience, as if they would have been really there, with an active role.

Nowadays digital storytelling and the capacity to involve the public at emotional and cognitive level are still weak and not mature enough, especially in comparison with other domains like cinema and video-games.

Most of the research funds addressed to the communication of virtual heritage are driven by technology and aim at technological innovation rather than at genuine culture transmission. Cultural Heritage is often instrumental to obtain resources and visibility, it is not the real focus. Consequently there are numerous projects in which the over-deployment of technology overshadows the cultural results. Beside, technology becomes obsolete in a short time.

A good communication, both in real and virtual museums, has to face the difficult task of creating metaphors and languages able to attract, motivate and keep the attention of the public, arouse a feeling of sensing and self-identification, combing different registers, conceptual, emotional and subliminal incitements. Technology in museums has to be advanced but easy to use, sustainable and robust to survive.

It is necessary to use technologies at a deeper level in order to strengthen the relations among academy, museums and public.

Digital Ecosystems and Embodied Interaction

The interpretation of a virtual reality environment as a potential digital ecosystem has been largely examined by the scientific literature since the begin of Nineties. Many of these studies originate from Bateson's epistemology and from his cybernetic theory in the field of social, cognitive and behavioral sciences.

Concepts like 3D reconstruction, interaction, real time exploration, contextualization, inclusion, integration and connection of information in the 3D space, personalization of learning process, are well know today (Forte 2000; Forte et al. 2008; Ryan 2001).

However most part of virtual reality environments are still far from the design of a dynamic space of knowledge.

What is still missing, in several digital ecosystems for Cultural heritage, is the ability to establish a contact with the user aimed at arousing, through the emotions, a phenomenological approach and thus a cognitive interest towards the cultural object/context.

Humans open their mind to the world through intuitive experience; sensing and emotions are fundamental in the life experience and in the self-identification process, they are the engine of knowledge and development of the individuals (Kabat-Zinn 2001). Every important moment in our lives, fixed in our memory, has been marked by emotions.

"Sensing" a cultural context means also the capacity to enter in contact with those elements that let us "recognize" something and move our emotions. An artwork, a landscape, an archaeological context have the power to activate a perceptual phenomenon that allows the user to relate with the fundamental issues of life, with that kind of deep essence that generates an intimate sense of happiness and enjoyment, difficult to verbalize (Arnheim 1954; Panofsky 1996; Jodorowsky 1997).

Designers of virtual environments for cultural heritage should improve the multimodal perception of the digital space and its evocative power. As a matter of fact artists, scenographers, cognitive scientists are actually much more open to this kind of approach rather than archaeologists, art historians and museums' curators.

Through the virtual we can recreate a dynamic space that includes not only real objects but also relations, where the user can perform activities and create a "visual drama" beyond what he/she can see (Fig. 2); thus he can experience an empathic evaluation of the context. Contexts can be variously assembled, dismounted and mounted again in order to understand their deeper relations, they can be desynchronized, becoming scenarios of different simulations.

Fig. 2 *Teramo virtual city* project, example of a digital performative space. 3D reconstructions are built interactively on the archaeological site of *S. Maria Aprutientis*, the ancient cathedral of the city; CNR ITABC, 2010

Thus the concept of "affordance" is fundamental because it leads us towards the attribution of expressivity to real and digital artefacts.

The psychologist Gibson (1977, 1979) defined "affordances" all the "action possibilities" latent in the environment, objectively measurable and independent from the individual's ability to recognize them, but always in relation to agents and therefore dependent on their capabilities.

Thus the concept of affordance is referred to a relation between an object or an environment and an organism, that affords the opportunity for that organism to perform an action. What we perceive when we look at objects are mainly their affordances, not their dimensions and properties because the physical aspect of objects allows the users to understand the principles of their functionalities.

As a consequence the concept of affordance is related to the concepts of perception, usability, design, interface, shape, colour, interaction, interpretation, embodiment.

To receive, elaborate information means to establish and modify relations in the space-time (Bateson 1972). Through the experimentation and interpretation of new relations/differences we can process information and develop an appropriate epistemological approach.

Thus a digital ecosystem should integrate different levels of visualization, multi-layered models, metadata, storytelling, tools of visualization and interaction, behaviours, in order to explore, "reconstruct" and communicate a "*mindscape*", a performative space where all the information are integrated.

Beside, the emergence of new social paradigms and media enhancing cultural interactions among people induce the creation of specific social platforms for Cultural Heritage that encourage an active participation of a large number of stakeholders. There is an increasing request of digital frameworks open to the communities for the accessibility, study, participatory and sustainable management of cultural resources and assets (Forte et al. 2008).

Today digital social platforms are mainly in 2D because they are conceived for fast and immediate exchange among users. However the potential of 3D web communities is much higher in terms of embodiment and cognitive impact. The embodied communities live in a 3D space; users can perceive, meet, dialogue, touch objects, transform the surrounding virtual environment, exchange information, make and simulate hypothesis, perform actions following a common purpose. Knowledge comes from "*enaction*" (Mellet-D'Huart 2006) and it is built essentially on sensory motor skills and direct experience in the 3D cyberspace. Cognitive activity is "embodied", not separated from the body perception (Varela et al. 1991).

Of course the level of the engagement inside the cyberspace is an essential condition to create the embodiment.

The *Integrated Technologies of robotics and virtual environment in archaeology* project, financed by the Italian Ministry of the University and Scientific Research, was a pioneer multi-user VR environment in the web, realized by CNR ITABC in 2006–2009, specifically for sharing and exchanging cultural contents among the scientific community of archaeologists.

3D models, metadata, interpretative layers, multimedia contents converged in a collaborative virtual environment in the web where the scientific community could meet, chat and interact in real time, exchange and test hypothesis, share data and make dynamic simulations in the 3D space. This virtual space was an editable and dynamic environment in continuous evolution. 3D scenarios were not closed and no longer editable by the users, as in most part of VR applications; on the contrary, they were open to continuous possible re-elaboration; they could be disassembled and recomposed according to different combinations and solutions. Every expert/student could edit the reconstruction (move, delete, add objects, change materials and textures, add link to metadata, move the lights, take measures, save his hypothesis as new version of the scene, share it and discuss within the community (Fig. 3). This project enabled an experience of embodied "sensing" because the collaborative environment was conceived as a performative space, addressed to academic research. The user becomes a creator of ideas and contents, he uses tools to propose interpretations establishing a strong relationship with other users (Forte and Pietroni 2009; Pietroni et al. 2009).

The framework was developed in Virtools DEV and it required a plug into be accessed in the web. This implementation can still be considered a powerful starting point for, hopefully, new developments on open platforms (WebGL).

In a virtual museum we usually communicate a cultural context or an artwork by mean of another work that uses specific languages and aims at conceptual and aesthetic impact. The emotional involvement is an essential part of the experience since it partly states the sense of presence, or the feeling of being within the virtual environment. For this purpose the aesthetic "shell" added by the communication action has to respect the deep essence of the content that needs to be transmitted, establishing a harmonic relation with its symbolic context and the concept we want to highlight.

The beauty of the graphic, of the soundscape, the evocative style of the script breaks through the incessant succession of our thoughts, in constant motion, offering us the opportunity to feel alive, partakers of the magic dimension that is on going. When this contact is, something acts within us.

An example of sensorial cultural landscape is the *Tiber Valley Virtual Museum*, described in the last part of this chapter. Starting from the acquisition of topographical data through integrated technologies, photo interpretation, GIS systems, multi-scale representations (landscape, site, intra-site), several multi-sensory scenarios have been created, inside which visitors can feel embodied and involved, able to acquire cultural contents in a pleasant and not frustrating way (Fig. 4).

The observable archaeological landscape continuously alternates with the multiple projections of its potential reconstructions of the past (reconstructed landscape), or imaginary dimensions (mind-scape). Non linear interactive storytelling combines different languages: virtual reality paradigms, natural interaction interfaces, cinematographic and theatrical techniques, virtual set practices, augmented reality. Soundscapes are fundamental elements able to create an experience, stimulating the user's curiosity and motivations, his perceptive and interpretative faculties.

In conclusion, following Jonassen's constructivist approach to learning, (Jonassen 1991), we are aware that our learning process is not linear but reticular, information is reorganized in an unpredictable way, which is unique for everybody and always dynamic and ever-changing.

Learning is an ecosystemic process of transforming information into knowledge. Relationships are embedded or situated in a context where complex interaction, conceptual and emotional, influences the quality of learning outcomes (Bateson 1972, 1979; Forte 2007).

It is crucial to understand the important of such an approach, both for interpretative/academic study and for communication purposes.

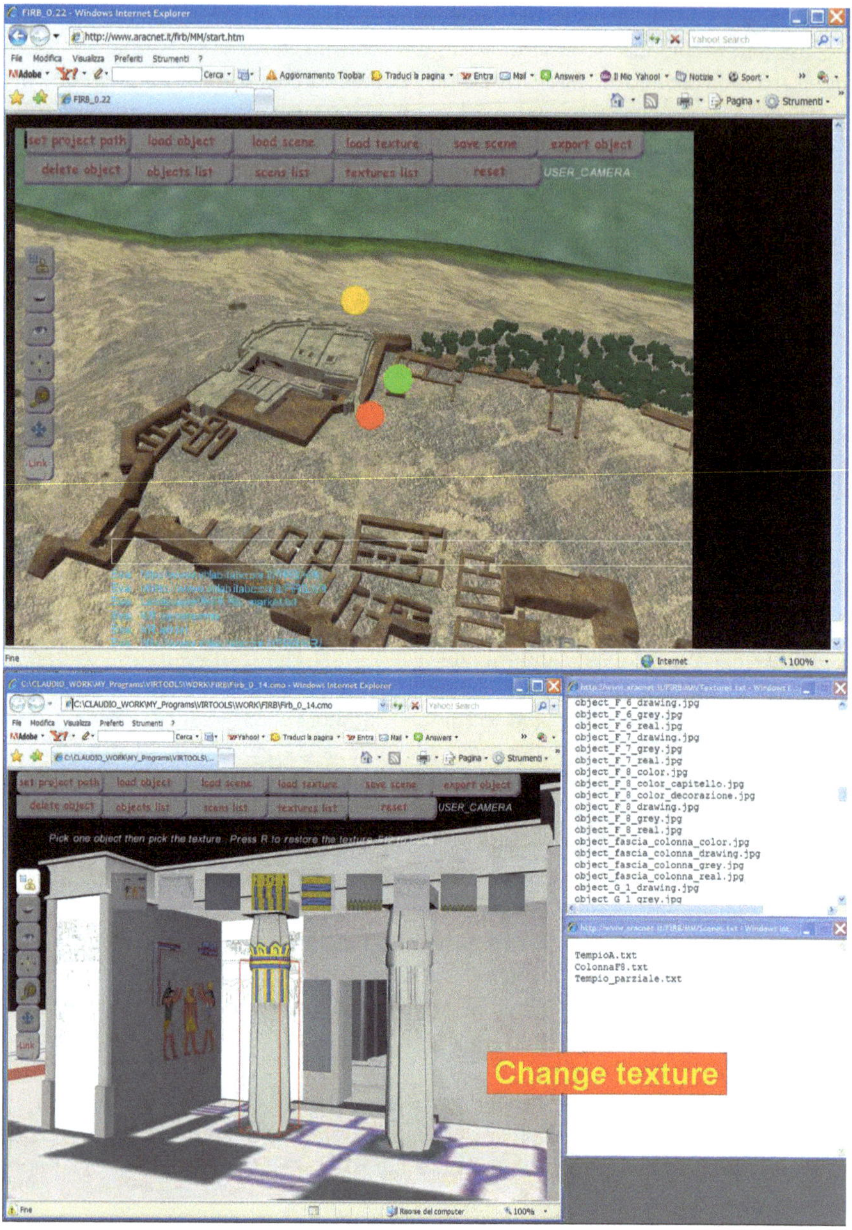

Fig. 3 Example of VR Multiuser domain in the web developed in the context of the *Integrated technologies of robotics and virtual reality in archaeology* project, realized in 2008 by CNR ITABC in collaboration with the Department of Archaeology of the University of Pisa and with Scuola S. Anna of Pisa. In the *upper image* the spheres represent the users' avatars performing actions in the 3d space, the chat is visualized in the lower part of the screen. The *lower image* is an example of real time editing of the scene, directly in the browser

Fig. 4 *Mind-scape* of the Middle Tiber Valley, realized by CNR ITABC in collaboration with E. V.O.CA..srl, 2014

Sensing Cultural Heritage: Looking for a Deeper Union Between Contents and Technologies

Starting from this premise, in the following paragraphs some suggestions for virtual museums will be proposed, in order to improve their communicative and educational impact on the public. There are not "standardized" rules and ready-made solutions in this domain, there are good practices. Communication requires a multidisciplinary approach and a continuous creative process that poses always new challenges and a broad spectrum of possibilities. Described criteria will lead not only to successful Virtual Museums applications, but also to their successful integration with the physical museums and their collections.

The integration of digital contents and technologies in the museum space is not a trivial matter. The close cooperation between museums curators and researchers in digital technologies for cultural heritage domain is fundamental for the achievement of a good and common result. We need to start from the analyses of the visitor's profiles, behaviours and needs, as necessary premise to design a communication project. The same cooperation is essential in the final step, once the digital applications have been installed in the museum: the project starts a new life that needs to be controlled, analysed, evaluated in order to improve and update its impact and usability. By observing the public reaction, we can understand the efficacy of our creation and we can improve our approach.

Such an investigation makes developers aware of how difficult communication is, especially when it is aimed at the heterogeneous public of museums. It requires the maximum effort in terms of multidisciplinary integration and a methodology combining science, art and technology.

Preliminary Steps: Choice of Key Concepts, Target, Environment, Communicative Registers, Technologies

A virtual museum should reintegrate what is illegible, contextualize what is fragmented, isolated, put back together cultural ties essential to the comprehension of a cultural object. Real and imaginary, perceptual and experiential aspects should be reconnected to the interpretive and symbolic ones. This means that a 3D model becomes an interface towards a universe of information, exactly as it happens for a real artefact (Forte and Pietroni 2006).

The choice of technologies, of paradigms and metaphors of visualization and interaction should be decided on the base of some preliminary fundamental questions: what are the few main concepts that we want the audience understand and remember? Who is our audience? Which is its average cultural level? Where do visitors come from?

Secondly, inside a museum it is crucial to focus not only on digital applications in themselves, but also on their connection with the physical space, the visitors' flow, the presence or absence of artefacts nearby, the environmental conditions. This will influence the choice and duration of contents, the level of interactivity and the interaction design, the rhythm and style of communication and, thus, the ideal communicative registers and the most appropriate technologies to adopt.

A secluded, dark and silent space, where people can stop and concentrate, will favor a condition of "slow" fruition of multimedia contents (Fig. 5). Along the paths of visit, in presence of the objects, the fruition of additional contents has to be much more essential and faster (Fig. 6). Audio contents can be also problematic if they are not conveyed in very circumscribed areas. Beautiful applications, evocative, plenty of poetic sensitivity, designed for quite environments, can be greatly

Fig. 5 *Tiber Valley Virtual Museum,* Etruscan National Museum of Villa Giulia in Rome. Example of inclusion of a spectacular VR installation in a dedicated space even if not completely secluded; CNR ITABC, 2014

Fig. 6 Roman National Museum, epigraphic section in Diocletian Baths. Example of integration of simple digital visual contents inside the showcase, among the objects of the collection; Soprintendenza Speciale per i Beni Archelogici di Roma in collaboration with Dinamo Lab and CNR ISTC, 2013

penalized and not be understood at all, if presented in fairs or noisy events where public moves frantically.

Alternation and Rhythm Variations

Both in case of multiple applications and of a unique complex installation, multiplication of the communicative registers is fundamental: narrative, educational, playful and evocative styles should alternate, suggesting to the public different relations of meaning and stimulating different levels of involvement and interaction.

Similarly the alternation between different phases of passive fruition and active interaction, of single user and collective experiences, is recommended because it contributes to create rhythm variations making the fruition not repetitive and boring but always unexpected and exciting.

Thus the museums, real and virtual, become living spaces and a wider audience can be involved: young, old people, technologically skilled or not familiar with interaction inside digital environments. The attention and interest of differentiated audience will be grabbed.

At the same time it is necessary to balance and create the right harmony among all these inputs and the right time of fruition.

A Guided Learning Process

The great potentiality of virtual reality is to allow visitors moving completely freely through the space, exploring models and metadata, acting different choices in order to personalize cognitive and learning process. This potentiality can be successfully expressed towards people with a preliminary knowledge of the subject the application deals with, but most part of the general public is not able to benefit of such a freedom.

Usually the visitor without a specific pre-alphabetization is disoriented: if we place him in the midst of several hundred choices which presume some prior knowledge of the subject, we only increase his sense of disorientation. He will get lost in some remote corner of the cyberspace.

It is necessary to guide users in their process of exploration and interpretation giving them progressive objectives and incitements in order to make them not bored, confused or abandoned in such a complex world: virtual storytelling, pre-defined camera paths (alternating with free exploration) and gaming rules are very powerful instruments for this purpose.

Storytelling, Direction, New Contamination of Media

Storytelling, interactive or not interactive, explicit or symbolic, built on a scientific base, has to be the real focus of communication. Learning process, memorization, cultural transmission, motivation, affection are facilitated by an emotional experience of intimate enjoyment.

Despite narration is one of the most effective vehicles of such an involvement, today it is one of the weakest aspects in the cultural heritage communication, both in real and in virtual museums.

In the past centuries drawings and iconographies showing cultural sites and conceived to the public fruition, usually included many narrative elements, like personages, chariots, animals, attending their daily activities. Documentation of the way of living in a place was considered an important part of the representation (Fig. 7).

On the contrary in the past years, digital technologies applied to cultural heritage and aiming at education or scientific transmission have expelled these elements connected to lifestyle, following an idea of supposed academic objective description. The great potential of digital technologies, the representation of an artefact with higher and higher level of resolution and metric accuracy, translates in cold transmission were the narrative aspect is ancillary or even completely lost. In virtual museums storytelling is often added to virtual representations as an accessory

Fig. 7 The *upper image* shows an hypothetical reconstruction of the ancient Via Appia by Luigi Canina (1795–1856). In the *lower image* the Trevi Fountain in Rome by Giovanni Battista Piranesi (1720–1778). Both representations include narrative elements, characters, animals attending their daily activities

element, planned at an advanced stage of development of the project, after that technologies, interaction design, behaviors, have been already decided.

The logical approach should be completely reversed: in cinema and games storytelling and script are the starting point of the production.

"Pure" and cold descriptions are not sufficient to incite emotional "perception" of the cultural heritage. It is possible to access thousands of descriptions in the web with our mobile devices, in few seconds and regarding every subject (Wikipedia, Google Earth), but they cannot be considered a vehicle of "experience".

Technologies need to become "warmer", supporting the capacity of the cultural heritage (real and virtual) to generate a feeling of self-identification, happiness, contentment and thus spontaneous actions of fruition, care and protection, founded on enhanced knowledge.

Storytelling, implicit or explicit, is a powerful mean. But what do we mean exactly by "storytelling"?

What storytelling is in the digital Cultural Heritage domain is still not very clear, as a lot of scripts are wrongly defined "storytelling" without being storytelling in effect. A typical element is the existence of a story/fiction (real or invented but plausible), living in a wider cultural context. Cinema and games have developed a professional approach to storytelling, (even if not always based on a scientific and historical accuracy) in terms of format, screenplay, scripts and dialogues, languages, soundscape, photography, technical means. Their fundamental components are: (1) script/story, (2) visualization, (3) soundscape, (4) interaction (in the case of games) (Fig. 8).

In the field of CH and virtual museums the state of the art of storytelling is in general very primitive, with few exceptions.

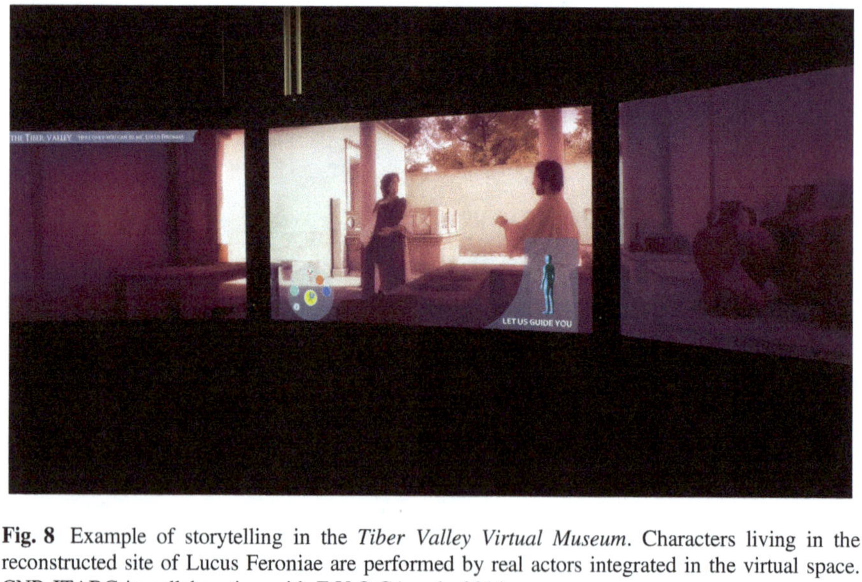

Fig. 8 Example of storytelling in the *Tiber Valley Virtual Museum*. Characters living in the reconstructed site of Lucus Feroniae are performed by real actors integrated in the virtual space. CNR ITABC in collaboration with E.V.O.CA. srl., 2014

Stories are still weak (probably because of the lack of writers and set designer supporting researchers, historians and archaeologists); music and sounds are often considered of minor importance, direction and camera movements are not so pondered and they follow quite naive approach not considering cinematographic techniques.

We need to improve our grammar to make communication more mature and competitive in this domain.

In a good practice storytelling should be one of the first elements to be planned and defined, subordinate only to the definition of the key concept, target and conditions of fruitions. Storytelling in fact determines the choices of languages, of communicative style, the light, the virtual camera behaviours, the soundscape, the rhythm, the level of interaction inside the virtual environment, the hierarchical organization of contents (Pietroni and Adami 2014).

In the case of interactive storytelling it is important to divide the story in different segments that will not be discovered always in the same predefined order (depending by the visitor's interaction). In these cases every fragment should have a complete sense and at the same moment it should encourage the visitor to go on, looking for the other fragments of the story. The ability consists in finding the right balance among the intensity of the story, its development and the need to cut it in different fragments, so that every visitor can catch some contents without losing the whole meaning.

Thus, storytelling establishes the need of a precise "direction", also in case of an interactive application or a virtual reality environment. Structuring this "direction", the designer will necessarily be induced to cross the traditional paradigms, creating novel approaches and communicative formats mixing techniques coming from virtual reality, cinema, theatre and games.

The challenge is really promising for young generations, as a possible vehicle of new competences and jobs and definitively for the development of a new industry.

Body Interaction in the Sensitive Space

From many years video-games have accustomed the public to interact within sophisticated scenarios, introducing multiuser virtual environments, embodied cyber communities, artificial intelligence and gesture-based interaction.

However most part of middle-age visitors of museums, have still problems to manage common input devices for interacting inside 3D environments and with digital artifacts: mouse, joystick, keyboard, console are not natural interfaces, they request time to become familiar.

From the point of view of human-computer interaction a thorough discussion would be impossible in this paper, however referring to digital applications for museums a novel frontier merits few short considerations: gesture-based interaction and, more specifically, mid-air gestures-based interaction where the control on the virtual environment is guided by body gestures, (Fig. 9), (Pietroni and Antinucci 2010; Pescarin et al. 2013).

Fig. 9 *The Approval of the Franciscan Rule: virtual experience among the characters of Giotto's work.* VR environment with gesture-based interaction, *Giotto's Colours* exhibition in Assisi, 2010. The original fresco, in San Francis upper Basilica, has been translated in a 3D scene as faithfully as possible, the models have been textured with Giotto's painting. Walking in the interactive area the user changes the point of view of the scene, living an immersive experience as if he would enter in the iconographic space as a leading actor. CNR in collaboration with BCAA srl., 2010

Inside a virtual reality environment the user feels spatially embodied in the system; this embodiment constitutes a new frontier of the communication and learning processes.

Gesture-based interaction allows going further in this direction, offering the ultimate possibility of the displacement of the material body from the confines of its immediate lived space (Featherstone and Burrows 1995; Fig. 10).

In fact icons, push-buttons, dialogue windows, keyboards, all disappear. What remains is the sensitive space of which the user is an integral part and, above all, an active element. We can consider this experience “body sensing”. Technology is *modeled* on the needs and the natural capabilities of a person, no particular knowledge or training is necessary to communicate with the system except that which comes from one’s natural experience. This approach represents a good opportunity for “on-site” virtual museums because contents can be accessed by every kind of users, especially young people, hopefully without efforts and in a pleasant way.

Some virtual museums introducing gesture based interaction have been developed and presented to the public in the recent years (Fig. 11; Castellano et al. 2007; Pietroni 2013; Pietroni et al. 2015). In several occasions these applications have been tested on heterogeneous audience for demographic and cultural profile.

Fig. 10 *Tiber Valley Virtual Museum,* child swimming in an underwater scenario like a fish (gesture-based interaction in the VR environment), Villa Celimontana, Library of *Società Geografica Italiana,* Rome. CNR ITABC in collaboration with E.V.O.CA. srl., 2014

Fig. 11 *Etruscanning project*, *Virtual Exploration of the Regolini-Galassi tomb,* VR environment with gesture-based interaction, Vatican Museums. Moving on interactive *hospots* on the floor and using gestures of the arms, the visitor can explore the 3D space, select and touch the virtual objects, making storytelling emerge from them. Microsoft Kinect sensor has been used for motion capture. CNR ITABC in collaboration with E.V.O.CA., 2013

Analysis were carried out on the usability, the levels of satisfaction, the ability to become familiar with the interface, the fun, the time of use, the ability to memorize and learn cultural contents, the overall cultural impact of the project. What emerges is that the possibility to dialogue and act within the digital environment through the body, generates curiosity, attraction, heightened intuition and motivation for the users, (Pescarin et al. 2013).

Most part of these projects use *Leap Motion* (Leap 2015) (Fig. 12) or *Microsoft Kinect* (Kinect 2015), motion sensing input devices able to capture the gestures performed by the user's hand (in the case of *Leap*) or by his complete skeleton *(*in the case of *Kinect),* transforming them in input.

Despite the positive response of the public, there are still some open problems: actual technical limitations of sensors in terms of resolution and noise; constraints due to environmental conditions; lack of a standardized grammar of gestures, especially at multi-cultural level; merging and alternation of natural and symbolic gestures to perform actions (for instance symbolic gestures are used to perform actions like "go forwards", "go backwards", "turn right", "turn left" and similar), that make the experience not always completely intuitive (Fig. 13).

Such experimentation will be accompanied and facilitated by the availability on the market of more precise and sophisticated sensors (new generation of sensors are

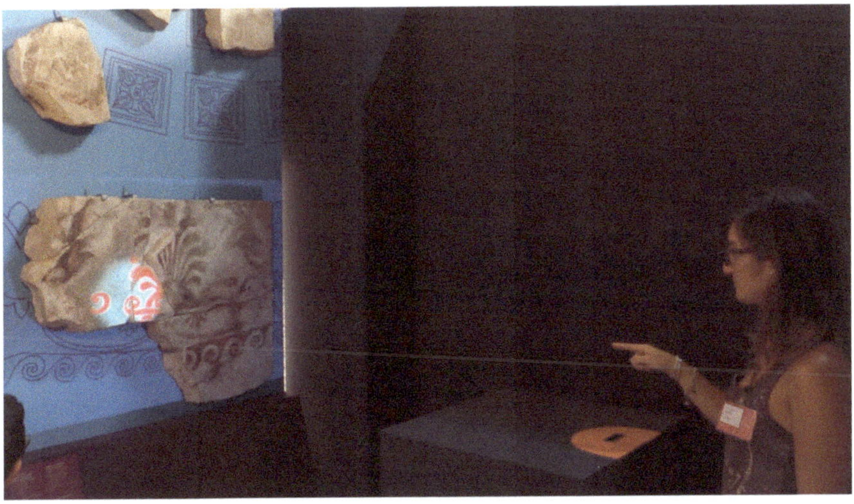

Fig. 12 *The magic flashlight*, presented in the *Keys2Rome* exhibition. The user moves his finger upon the Leap Sensor, this movement is tracked in order to project the original colour on the archaeological fragment. INRIA, French National Institute for Computer Science and Applied Mathematics, 2014

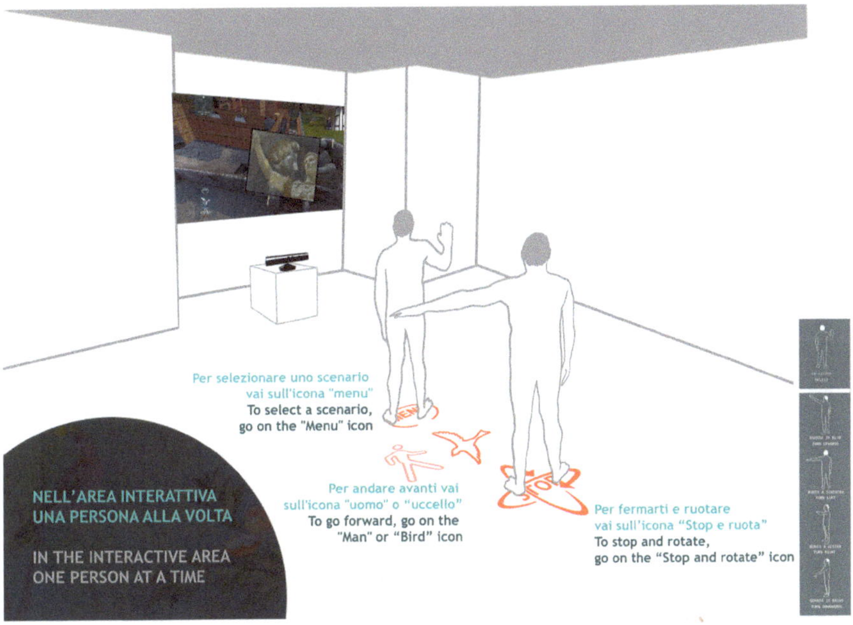

Fig. 13 *Livia's Villa Reloaded,* VR application using gesture-based interaction, National Roman Museum, Diocletian's Baths, Rome. The scheme shows the symbolic gestures that are required to move in the virtual environment. Microsoft Kinect sensor has been used for motion capture. CNR ITABC in collaboration with E.V.O.CA srl, 2013

already able to interpret facial emotions or detailed finger movements). Technological innovation in this field is very fast.

Transparency of the Scientific Approach, Reusability

Scientific knowledge and communication have to be integrated in one process. Good digital pipeline are needed to preserve the original quality of the acquired data in the successive phases of the elaboration of contents.

The whole process must be demonstrable and repeatable. Beyond storytelling, the adopted interpretative sources and criteria (usually integrating bottom-up and top-down approaches) need to be declared and made available to the public (in a dedicated website, in a complementary application, or adding a new "layer" in the informative cybermap of the virtual museum), following the rules suggested by the London Charter and Seville Principles (Forte et al. 2008). This condition will allow the public to become aware about the reliability of the proposed digital contents, 3D reconstructions and stories (Fig. 14).

Reusability of the dataset for further developments is recommended. If digital data have been optimized, organized in rational repositories and associated to proper metadata, they will be available for possible up-date and re-used in future applications.

Emotional Sensing of the Landscape

Landscape tells the stories of human beings in different ways. First of all it recounts memorable or minimum events of which it has been the stage: daily stories, forgettable events or deeds of great importance that changed the course of history. A continuous and never-ending story of individuals and social groups in a territorial environment, transformed into landscape, felt and experienced as a reality with which it is constantly necessary to relate.

Beside it recounts the history of its geological evolution through the time, the history of sedimentations, across events and generations.

Thus, we can imagine the landscape as a storage of occurred stories and facts that time is able to shape, materialize, actualize in new forms, as a result of the historical process. This process continuously renew the territory (Motta 2004).

The history of events can be stored and transmitted through the narration of inhabitants, of a reporter, the work of a writer, of a poet, the painting of an artist, the image taken by a photographer. We attribute value and meanings to the landscape and to its elements, making them speak and tell. All these are subjective perspectives of interpretation, according to our personal experiences (Turri 1982, 2004).

Fig. 14 *Teramo virtual city* project, 3D reconstruction of *Domus S. Anna*. The visualization below shows the different levels of reliability of the proposed reconstruction, through the use of chromatic fields. Multimedia installation in the Archaeological Museum of Teramo, CNR ITABC, 2010

Indeed narration usually pays much more attention to the actors rather than to the scenario. At least until the occurred event does not directly involve the scenario itself, as in the case of an earthquake, a storm or a war.

Powerful GIS tools and 3D modeling techniques are commonly used today to elaborate, analyse and represent spatial data regarding landscapes. However the concept of the landscape as an ecosystem cannot be communicated using GIS systems that are mainly used within the scientific community. The epistemological approach for the virtualization and communication of landscapes has still little attention. A landscape virtual museum should integrate both holistic and monographic visions. A landscape can be compared to a living organism: it needs to be considered as a whole entity, a unique context whose identity is much more than the sum of its single parts. This holistic vision requires a diachronic approach, the inclusion of the present and past phases of life and the consideration of dynamic relations and processes.

On the other side monographic representations are useful to go deeper into specific and exemplar stories, regarding the life and the way of thinking of specific groups of people. The identity of the landscape, the "sense of place" (Norberg-Schulz 1979) arouses from the integration of the two levels. The landscape is an organism in continuous evolution.

Given this premise, in a landscape virtual museum the multidisciplinary approach, the multiplication of communicative registers, perspectives and voices, perceptions and storytelling, acquires primary importance.

This is the challenge of the *Tiber Valley Virtual Museum* that is going to be presented in the following section, as a concrete example of the subjects previously discussed.

Tiber Valley Virtual Museum

The *Tiber Valley Virtual Museum,* realized by CNR ITABC in 2011–2015 and financed by Arcus-ALES S.p.A, has been conceived in order to increment and disseminate the knowledge and the interest towards the territory north of Rome, crossed by the Tiber river and by two important roman consular roads, via Salaria and via Flaminia, an area 40 km long × 60 km wide (Fig. 15).

The project investigates the landscape in its several cultural aspects: geological, natural, historical, archaeological, anthropological, poetic, evocative and symbolic.

A digital platform has been created, including VR and multimedia installations placed in the archaeological sites and in the museums disseminated in that area and, at central level, in Rome, inside the National Etruscan Museum of Villa Giulia, to promote the project and encourage people to visit the territory.

Starting from a cross-disciplinary study and documentation of the territory and of its evolution across the time (from 3 million years ago until today), 3D representations at different scales have been realized, from the whole landscape to specific sites.

The landscape has been investigated interpreted and represented through a complex pipeline, starting from the real and concrete experience of visit, continuously renewed in different seasons and with different purposes, up to data collection, topographical survey and GIS integration, until the elaboration of a *map-scape*. This spatial contextualization of the actual features was the base for the 3D reconstruction of the potential ancient landscape in the VIII–VII century BC (Orientalizing period) and roman period (Augustan and Trajan ages) *(reconstructed landscape).*

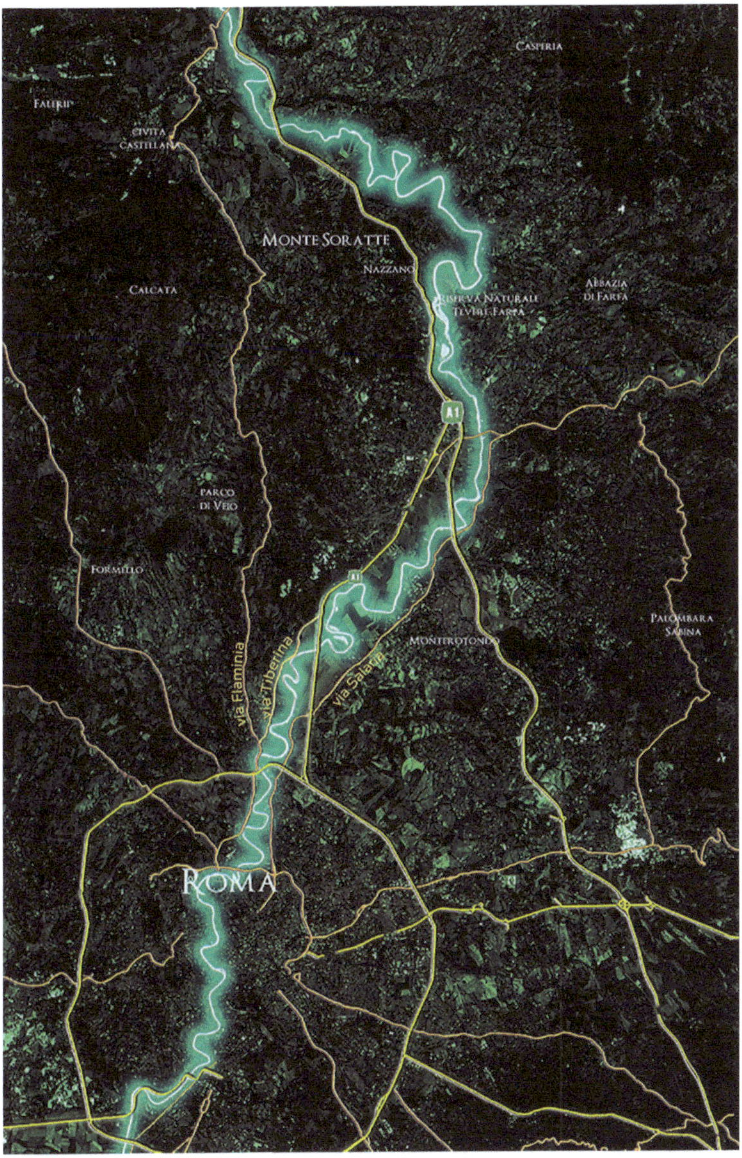

Fig. 15 *Tiber Valley Virtual Museum,* area of interest taken in consideration by the project

Beside the representation of the visionary, imaginary landscape (*mind-scape*) has been created, where it is possible to discover and experience memories, fragments of poetries, literary works, paintings, ancient legends, iconographies, traditional popular songs and music, whistles and sounds of sheep farming (Fig. 16).

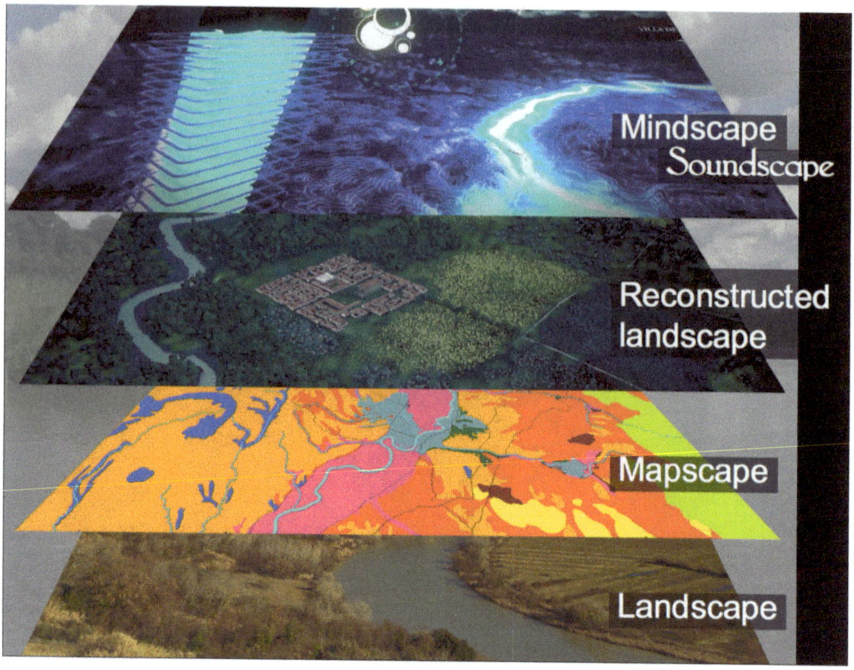

Fig. 16 *Tiber Valley Virtual Museum*, different levels of representation of the cultural landscape

Stories of individuals live on the historical background, sometimes these stories are real, in other cases they are imaginary but historically plausible. Each one looks at the landscape from a specific perspective, collective or personal, emotional or conceptual.

Map-Scape and Reconstructed Landscape: Methodological Approach

Following a multidisciplinary approach, several kinds of information have been combined: geological and ecological data, historical, archaeological and demographical studies. A complete pipeline procedure, from GIS data elaboration up to final realistic 3D renderings and communication has been developed (Pietroni et al. 2013a, b).

As a first step a preexisting digital elevation model (DEM) of the actual landscape has been re-used (coming from Istituto Nazionale di Vulcanologia, resolution 10 mt), textured with a satellite image (IRS, resolution 5 mt).

Secondly another DEM has been generated from aerial photogrammetry (resolution 5 mt). A set of 12 aerial IGM analog photos has been used, taken in 1954, before the construction of the two major modern infrastructures: the dam of Nazzano (1955) and the highway (1961). This DEM constituted the base for the 3d model of the terrain in the past ages, as we can assume that in the Orientalizing and Roman periods the geomorphology was quite similar. This DEM has been colored and characterized in a realistic way through 3D and 2D graphic libraries.

Basically, for the creation of the past ecosystems the archaeological map and the land unit map (pedo-landscape) were the starting points. The latter was useful to define the soils composition and their attitude to host specific ecosystems, both natural and cultivated by the man.

From the original 75 land units, 6 main macro-ecosystems have been obtained after a process of simplification (*eco-landscape*): (1) volcanic; (2) sedimentary (sandy-conglomeratic-clayish); (3) calcareous of low-medium height; (4) calcareous of considerable height; (5) valley (lower terraces included); (6) Tiber banks (Fig. 17).

The geomorphology and geology of the territory, the slopes, the orography, the accessibility of the river have been also analyzed, and an hypothesis of ancient courses for the Tiber has been elaborated, together with the possible presence of ancient harbours or crossing points/structures on the river. The results were compared with the archaeological remains and we found good correspondences.

Regarding the definition of the potential natural vegetation for each macro-ecosystem, the existing thematic cartography elaborated by Regione Lazio and Provincia of Rome has been taken in consideration. An ecosystem consists in a specific mixed set of plants that adapt to that soil, climate, orientation and so on. The combination and disposal of the plants in each ecosystem is natural or linked to the cultivation techniques existing in that specific age. We can assume that the potential natural vegetation has not changed in a substantial way. These data have been combined with the archaeological map and demographic analyses for the Orientalizing and the Roman periods, in order to define the areas influenced by the presence of human settlements and activities ("buffers").

A specific methodology has been followed for the buffer creation, considering several factors in relation with the geography and geomorphology of the territory, the distance of the cultivated lands from the infrastructures or the natural resources in the ladscape (roads, rivers, etc.), the economic model, the technological level and the potential movement of the population in the daily conduction of their work.

Each one of these factors has a certain level/proportion of importance in relation to the others, and this relation can change through the different historical ages. The existing bibliography about demography, settlement patterns, alimentation and food in the past ages, together with information coming from pollen analyses and archaeology, supported the characterization of the cultivations and vegetal species spread on the landscape.

The final result of this work has been the elaboration of a colour map that describes the past landscape, including natural and anthropic areas. Each colour of this map represents a specific ecosystem (Fig. 18).

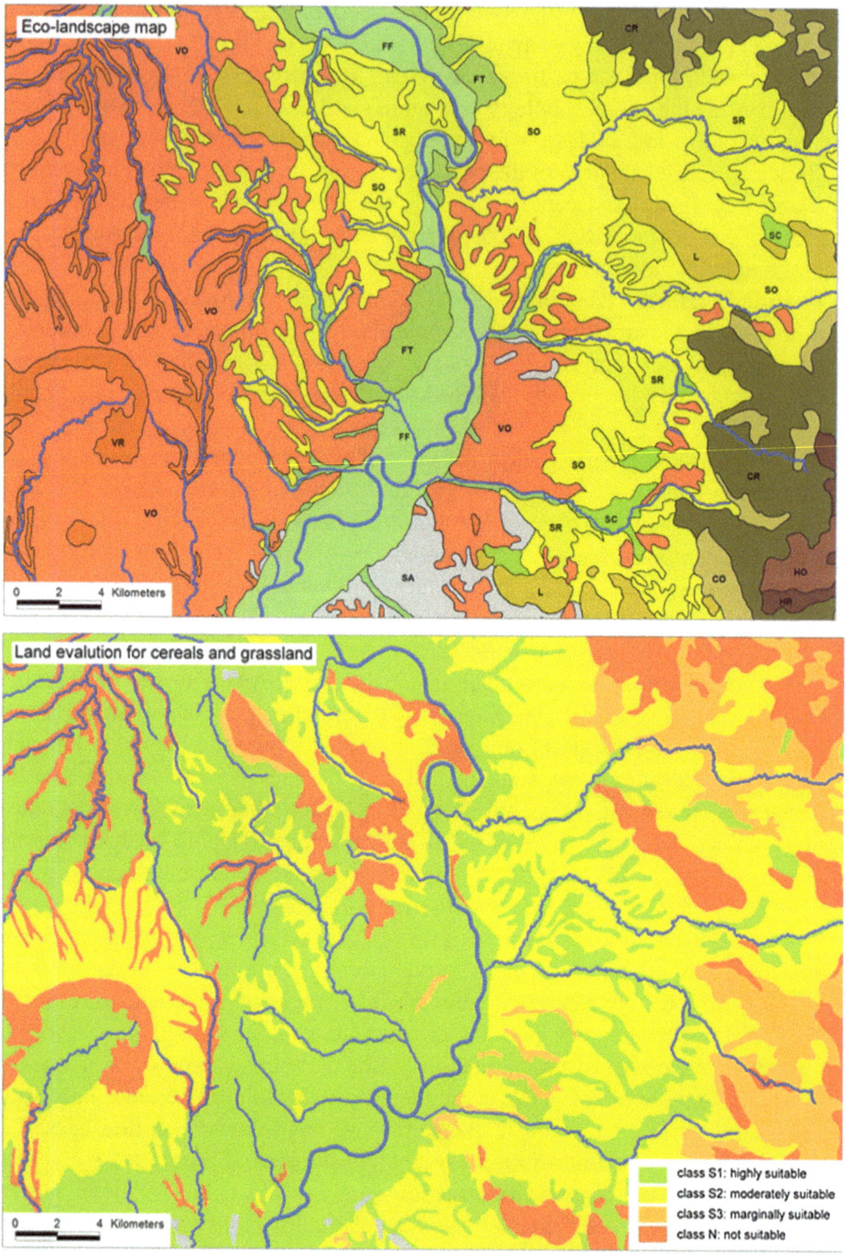

Fig. 17 *Tiber Valley Virtual Museum,* Eco-landscape map (*upper image*) and Land evaluation for cereals and grassland (*lower image*); CNR ITABC in collaboration with Digiter, srl, 2012

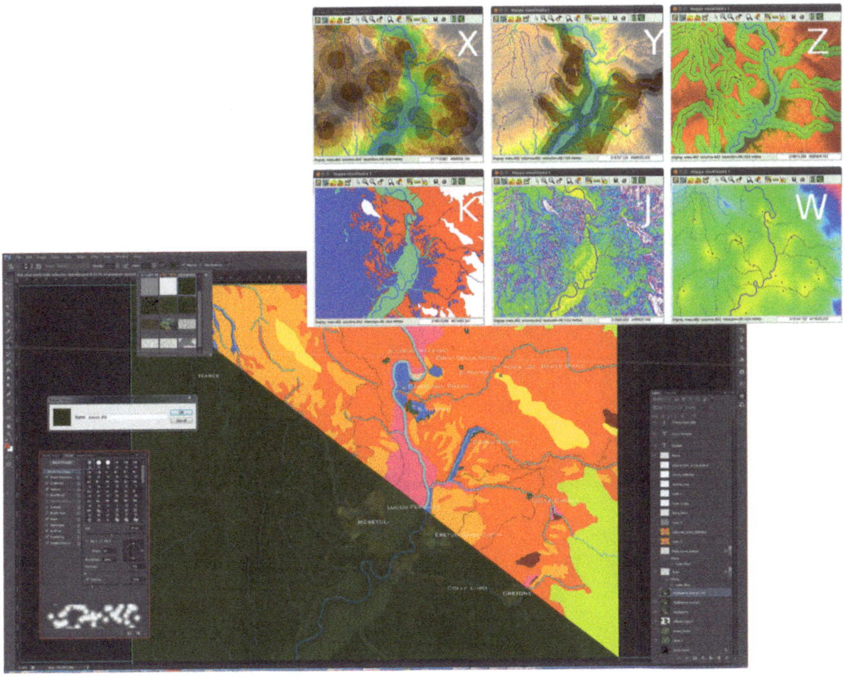

Fig. 18 *Tiber Valley Virtual Museum,* Buffers map and Colour map representing natural landscape and cultivated areas (VIII-VII sec.a.C.), Each colour refers to a specific ecosystem. CNR ITABC, 2012

The methodological process followed for the Orientalizing and Roman periods can be easily replicated for other chronological phases, changing the weight of the single parameters.

The final step has been the translation of this symbolic description of the landscape (GIS) in a realistic 3D representation, through powerful tools for landscape generation and design. The final representation, aimed also at cultural communication to the public, is of two types: a real time application developed in *Unity 3D* and movies realized in *Vue* (Fig. 19). Data have been processed and optimized differently for the two outputs.

Beside the landscape, specific archaeological sites have been documented and reconstructed in 3D. Lucus Feroniae was an important Italic sanctuary dedicated to the Goddess Feronia in the archaic age, at a later stage transformed by the Romans in a colony, under Augustus, and then restored by Trajan.

Volusii's Villa was the most important roman villa of the area, located near Lucus Feroniae, 30 km North from Rome; it belonged to one of the richest and most powerful families of Augustus' time.

These sites have been documented and represented in 3D through activities of survey, laser scanning, UAV, photogrammetry and structure from motion

Fig. 19 *Tiber Valley Virtual Museum,* 3D reconstruction of Eretum in the VIII cent. BC. Below a typical hut and a built house; CNR ITABC, 2013

(Bottom-up approach). 3D reconstructions in Augustan and Trajan periods have been realized adopting a proven methodology and a scientific approach based on archaeological evidence, literary documentation and typological comparisons (Top-down approach).

Mind-Scape: Advanced Media for Communication and Storytelling

One of the results of the project is a spectacular VR application characterized by gesture-based interaction and by a novel approach in the evocative storytelling. It has been opened as permanent installation in the National Etruscan Museum of Villa Giulia in Rome, in December 2014 (Fig. 5).

The VR application is visualized on three aligned 65" screens, arranged in a semicircle, in order to arouse a feeling of immersion and perceptive involvement. It consists of four scenarios and aims at creating an original, evocative and narrative access to the territory of the middle Tiber valley, (video demo: https://vimeo.com/129867454).

The user migrates among different "avatars" to explore the following virtual environments:

1. "*On the spirals of the Tiber: the landscape of the origins*": the user can fly, like a bird (using his arms), over an evocative 3D representation of the middle Tiber valley landscape.

The representation of the landscape is based on a accurate digital DEM and the most important places are put in evidence, but the rendering is not photorealistic; on the contrary it aims at stimulating the imagination of the visitor following a symbolic approach; 3D graphics resemble a game, sounds have been composed starting from ethno-musicological studies, redeploying traditional popular sounds and folk songs documented in the Roman countryside in this area.

Crossing magic circles in the sky the user can travel back in time and activate videos related to:

- the geological and geomorphological evolution of the territory at different stages, from 3 millions years ago until today,
- a potential reconstruction of the landscape in the VIII-VII century BC, telling about the birth of the cities from the astonished perspective of men and women attending the fast transformation of their world.

These movies present several 3D reconstructions, beside real images and video shots of the actual landscape. Despite the narrative approach, they are the result of a very complex multidisciplinary study, supported by many experts, as previously described (Fig. 20).

2. "*In the setting of the sun the Tiber gleams. The secrets of the river*: swimming underwater in the current of the Tiber, like a fish, the visitor can experience the memory of the river; he meets fluctuating images, iconographies, sounds, literary fragments taken from ancient and contemporary poets and authors. He can swim under the water in every direction, following these dynamic memories or he can awake them touching fragments of painted walls collapsed in the river. Literary quotations are spoken by a multitude of voices. The visitor uses his arms to swim and he can watch himself in a mirror fallen in the river, he also meets other fishes living in the river whose movements are controlled by artificial intelligence and swarm dynamics.

The user is involved in that flow and rhythm, his experience is multisensorial and emotional (Fig. 10 and Fig. 21).

3. "*Mena's story, Volusii's Villa*": the user is involved in an immersive experience inside this roman Villa. At the beginning he can see how the villa looks like today (in a movie), then he is brought back in Augustan time, in the 3D reconstruction of the Villa at its maximum splendor, and he can start the real time exploration. The user acts like a man walking through the space (Fig. 22).

Fig. 20 *Tiber Valley Virtual Museum,* VR installation in the Etruscan National Museum of Villa Giulia in Rome; CNR ITABC in collaboration with E.V.O.CA srl, 2014. Flight over the territory

Here he is involved in the personal story of Mena, a freed slave living his own psychic drama. This character is imaginary but plausible because his profile has been created following several literary and historical sources related to freed slaves' condition during Augustus' time and it is also historically connected to the presence of thousands of slaves working in the huge extensions of land owned by the Volusii around the villa. The archaeological and historical context is used as scientific background of this engaging story.

Through gesture-based interaction, the visitor can navigate the space: he can relax following a predefined camera path (virtual *steadicam*) while Mena's monologue develops, he can decide to stop in every moment and look around to analyze details of the architecture or decoration. In some places he finds crossroads and he can decide in which direction he wants to go on, to follow a specific part of the story (interactive storytelling). Mena's story concludes introducing the last scenario dedicated to Lucus Feroniae, where the user can find a parallel story revealing many connections and an epilogue of Mena's story itself (Fig. 23).

Fig. 21 *Tiber Valley Virtual Museum,* VR installation: underwater scenario

4. "*Here only you can see me. Lucus Feroniae*": the user walks, always in real time, through the ancient Roman settlement of Lucus Feroniae reconstructed in 3D during Tiberius' and Trajan's time.

He follows predefined camera paths (virtual *steadicam*). As in Volusii's Villa, he finds crossroads, corresponding to specific narratives, and he can choose the direction he prefers to enter the different stories and places.

Real actors have been filmed on a green screen and then integrated in the virtual scene to represent the ancient characters performing their daily activities in the reconstructed site; they dialogue and interact among them and occasionally with the user (Fig. 24). The memory of the ancient Italic goddess Feronia is also evoked:

Fig. 22 *Tiber Valley Virtual Museum,* VR installation: Volusii's Villa scenario

Fig. 23 *Tiber Valley Virtual Museum,* VR installation: real actors shot in front of a *green screen* and integrated in the 3D reconstruction; they play the role of the ancient characters of Lucus Feroniae

even if no longer worshiped in Roman times, her sanctuary was a point of attraction for many people of Center Italy in the archaic period. Only Cesia, a little child, is able to see the goddess and hear her voice. Thanks to the magic power of Feroniae we are able to see how the city of Lucus will be transformed one hundred years later.

During narratives the visualization is a single view distributed on the three screens (5760 × 1080 pixel); on the contrary, during the exploration three different views are shown on the three screens (1920 × 1080) to compare the current archaeological site to its 3D reconstruction: on the left and central screens the observed archaeological environment and the 3D reconstruction of the past structures are shown in parallel, from the same points of view, in order to offer the visitor an "augmented" perception of the space and a better comprehension of the site (Fig. 24).

On the right screen, a 3D perspective view of the whole city is shown with the evidence of the user's position during his exploration. This visualization is useful to let the user be aware of where he is. The multiplication of the points of view (first person, third person, present, past) and narrative perspectives stimulates the user's active and critical participation within the virtual environment.

Each scenario needs from 5 to 10 min to be enjoyed.

One person at a time can guide the system in the interactive area in front of the screens (4 × 4 mt). The other users (about 15 persons) can watch seating or standing in the space all around but they can alternate in every moment in the active role (Fig. 25). Microsoft Kinect has been used for motion capture.

Fig. 24 *Tiber Valley Virtual Museum,* VR installation: "augmented" experience inside Lucus Feroniae, actual and reconstructed site from the same point of view on the *left* and *central screens*; on the *right* the user's position is represented by the white disk

Fig. 25 *Tiber Valley Virtual Museum,* VR installation: a group of young students is experiencing the application, Library of Villa Celimontana, Rome. One person at a time can guide the system in the interactive area, the others can attend and alternate in the active role

The application needs to be installed in a closed and silent space/room, because under this conditions the public can concentrate on storytelling and evocation and thus can feel really immersed in the experience.

The use of a so involving storytelling led toward a new approach in the real time exploration, going beyond the traditional paradigms of virtual reality. A "direction" of cameras was needed to favor an emotional engagement (as it happens in movies), but keeping a certain degree of freedom for the user in the exploration of the 3D space. For this reason the rendering of the four scenarios is in real time, the graphic engine is Unity 3D.

The sensory and emotional experience within the virtual worlds is made possible by a number of factors:

- experimentation of a new language that marks the overcoming of the traditional paradigms of virtual reality to contaminate them with techniques derived from cinema, theatre, video games and augmented reality, in order to bring out the best expressive potentiality of each media; this strange and "metaphysic" quality of the installation is immediately perceived by the public as completely new and attractive.
- equal importance given to images, sounds, poetry and narratives, real time exploration;
- camera animations and filters in real time, the use of colour and the overall "tone" of rendering that tends to evoke rather than to coldly describe;
- use of gesture—based interaction;

- balance between active engagement of the visitor (through body interaction: flying, swimming, looking around while walking) and chance to relax, being guided along predefined paths and abandoning to storytelling;
- original soundscape and storytelling. The scientific contents are conveyed in two ways:
    1. by the characters (actors in the field or voiceovers) that tell the story and look at the landscape from their individual perspectives while, in the background of their affairs, the history of the site and of the territory is transmitted;
    2. through the quotes (audio) of ancient and modern historians, writers and poets;
- projection on three large screens that almost surrounds the visitor;
- GUI (graphic user interface) extremely simple and clear. Few colored shapes on the floor are used to activate the scenarios and the behaviors. Moreover the blue silhouette of a figure, in the central screen, suggests to the user the gestures at his disposal, to explore the virtual scenarios (Fig. 25).

Conclusions

The final goal of a virtual museum is the cognitive, perceptual and communicative enhancement of the cultural object that translates into a wider and deeper exchange with the visitors. The translation in "virtual" heritage allows to "dramatize" the cultural heritage that in the past has been often decontextualized and distanced from the general public, leaving very few possibilities of interaction.

Through the digital its value can be "recapitalized" by the activation of interactive processes, improvement of relations, expansion and diversification of the public, producing quantitative and qualitative evolution. Definitively the digital helps museums to become more and more exhaustive towards their public.

Cultural participation means capability, competitiveness, exchange and multiplication of cultural perspectives and consequently social and economic growth.

Physical and virtual fruition become part of the same process of knowledge. They strengthen and accomplish each other, according to a holistic and multi-functional approach.

The participation of the wide public to cultural processes is an essential condition to recirculate culture and stimulate people to come back and enjoy cultural sites and museums, playing an active role. Young generations should be a privileged target, as usually they are not so much attracted by museums as adults.

Of course such a progress requires improved design capabilities and a range of activities aimed at satisfying, or even evolving, the cultural expectations of the public. The museum, as a system, should be re-organized, starting from the experience design up to management and marketing strategies.

Storytelling and interaction are essential elements of the cultural experience. They are directly connected to emotions that can help the arts to get in touch with the public. Identification generates motivation, conceptual appropriation and learning.

In this context communication requires a harmonic relationship among the physical and digital contents and the surrounding context of fruition. Languages and communicative formats are crucial in the creation of such a harmony and simplification is a mean to grasp the essence of meanings.

In this chapter some fundamental rules and good practices have been suggested, together with some exemplar case studies following different approaches but landed to successful results.

Special attention has been paid to the theme of the landscape, explored in its multiple connections (real landscape, map-scape, reconstructed landscape, mind-scape) and potential attribution of diverse meanings. The landscape is the best example of co-generation and co-creation and every communication project should take care of this deep essence.

References

AA.VV. 1998. *«Sistemi intelligenti»*,- Volume: 0 - Fascicolo: 2 – pp. 159 - 306.

Antinucci F., 2007a. *Musei Virtuali*, Laterza, Roma.

Antinucci F., 2007b. *The Virtual Museum*, in "Archeologia e Calcolatori" Supplemento 1, pp. 79-86.

Antinucci F., 2014. *Comunicare nel Museo*, Laterza, Rome.

Arnheim R., 1954. *Arte e percezione visiva*. Translated by Feltrinelli in 1962.

Bateson G., 1972. *Steps To Ecology Of Mind*, San Francisco: Chandler Press.

Bateson G., 1979. *Mind And Nature: A Necessary Unit*, New York, Dutton.

Castellano G., Bresin R., Camurri A., Volpe G., 2007. *Expressive control of music and visual media by full-body movement*, in Proceedings of the Conference on New Interfaces for Musical Expression (NIME07), New York, NY, USA

Forte M., 2000. *About virtual archaeology: disorders, cognitive interactions and virtuality*, in Barcelo J., Forte M., Sanders D., (eds.),"Virtual reality in archaeology", Oxford, ArcheoPress (BAR International Series S 843), pp. 247- 263.

Forte M., 2007. *Ecological Cybernetics, Virtual Reality and Virtual Heritage*, in "Theorizing Digital Cultural Heritage. A Critical Discourse" (Edited by Fiona Cameron and Sarah Kenderdine), MIT Press, Cambridge, MA, 389-407

Forte M., (edited by) et al., 2008. *La Villa di Livia: un percorso di archeologia virtuale*, L'Erma di Bretchneider, Roma.

Forte M., Pietroni E., 2006. *The Museum's mind: a cybermap for cultural exhibition*, in "The evolution of Information Communication Technology in Cultural Heritage", VAST Proceedings edited by M. Ioannides, D. Arnold, F. Nicolucci, K. Mania, pp. 70 -73.

Forte M., Pietroni E., 2009. *Multiuser virtual environments for archaeology*, in *Archaeology and virtual environments: from excavations to virtual museums and web communities*, Alinea editrice, pp. 79 -109.

Forte M., Pietroni E., Dell'Unto N., 2008. *3D Multiuser Domain and Virtual Ecosystems for Transmission and Communication of Cultural Heritage*, in Digital Media and its Applications

in Cultural Heritage, edited by Jamal Al-Qawasmi, Michele A. Chiuni, sabry El- Hakim, pp. 179- 193.
Featherstone M., Burrows R., 1995. *Cultures of technological embodiment: An introduction,* in "Cyberspace, Cyberbodies, Cyberpunk: Cultures of Technological Embodiment", M. Featherstone, R. Burrows (eds.), Thousand Oaks, CA, Sage Publications.
Gibson J., 1977. *The Theory of Affordances*, in "Perceiving, Acting, and Knowing", edited by Robert Shaw and John Bransford, ISBN 0-470-99014-7.
Gibson J., 1979. *The Ecological Approach to Visual Perception*, ISBN 0-89859-959-8.
Jodorowsky A., 1997. *Psicomagia,Unaterapiapanica*, Eds.Feltrinelli.
Jonassen D., 1991. *Evaluating Constructivist Learning* in "Educational Technology", 36(9), pp. 28-33.
Kabat-Zinn J., 2001. *Mindfulness Meditation for Everyday Life*, Eds. Piatkus.
London Charter: http://www.londoncharter.org
Lopez-Menchero Bendicho V. M, 2013. *International Guidelines for Virtual Archaeology: the Seville Principles*, in C. Corsi et al. (eds), "Good Practices in Archaeological Diagnostic, Natural Science in Archaeology". Springer International Publishing Switzerland.
Mc Luhan M. 1967. *Gli strumenti del comunicare,* Eds. Il Saggiatore, Milano.
Mellet-d'Huart D., 2006. *A Model of (En)Action to approach Embodiment: A Cornerstone for the Design of Virtual Environments for Learning*, in WIN W. & HEDLEY N., Eds. "Journal of Virtual reality, special issue on education". Springer London. Volume 10, Numbers 3-4/ December, 2006, pp. 253-269.
Motta G., 2004. *Paesaggio, territorio, ambiente. Storie di uomini e di terre.* Eds. Franco Angeli, ISBN: 9788846464040
Norberg-Schulz C, 1979. *GeniusLoci*, Eds. Electa.
Panofsky E., 1996. *Il significatonelleartivisive*, Eds.Einaudi
Pescarin S., Pagano A., Wallergård M., Hupperetz W., Ray C., 2012. *Evaluating Virtual Museums: Archeovirtual case study*, in Proceedings "Archaeology in the Digital Era. Papers from the 40th Annual Conference of Computer Applications and Quantitative Methods in Archaeology", editors Graeme Earl, Tim Sly, Angeliki Chrysanthi, Patricia Murrieta-Flores, Constantinos Papadopoulos, Iza Romanowska and David Wheatley, Amsterdam University Press.
Pescarin S., Pietroni E., Wallegard M., Omar K., Rescic L., Rufa C., 2013. *NICH Natural Interaction in the Cultural Heritage domain: a preparatory study on cross-cultural natural gestures*, in Proceedings of Digital Heritage International Congress 2013, IEEE Advanced Technolgy of Humanity, ISBN: 978-1-4799-3169-9, Marseille France; Vol I pp. 355-362.
Pietroni E., 2013. *Natural Interaction in VR environments for Cultural Heritage: the virtual reconstruction of the Regolini Galassi Tomb in Cerveteri*, in "Archeologia e Calcolatori", n. 24/ 2013, Editore Del Giglio.
Pietroni E., Adami A., 2014. *Interacting with Virtual Reconstructions in Museums: The Etruscanning Project,* in Journal on Computing and Cultural Heritage (JOCCH), Special Issue on Interacting with the Past Volume 7, Issue 2, Article N° 9, July 2014, ACM New York, NY, USA,
DOI 10.1145/2611375
Pietroni E., Antinucci F., 2010. *The Approval of the Franciscan Rule. Virtual Experience among the Characters of Giotto's Work*, in Proceedings of the 11th International Symposium on Virtual Reality, Archaeology and Cultural Heritage VAST 2010, edited by A. Artusi, M. Joly-Parvex,G. Lucet, A. Ribes, and D. Pitzalis.
Pietroni E., Rufa C., Forte M.. 2009. *Embodied virtual communities: a new opportunity for the research in the field of cultural heritage,* in International Archives of Photogrammetry, Remote Sensing and Spatial Information Sciences Volume XXXVIII-5/W1 ISSN 1682-1777.
Pietroni E., Pagano A., Rufa C., 2013. *The Etruscanning project: Gesture-based interaction and user experience in the virtual reconstruction of the Regolini-Galassi tomb*, in Proceedings of Digital Digital Heritage International Congress 2013, IEEE Advanced Technolgy of Humanity, ISBN: 978-1-4799-3169-9, Marseille France; Vol II pp. 653-660.

Pietroni E., Palombini A., Arnoldus-Huyzendveld A., Di Ioia M., Sanna V., 2013. *Tiber Valley Virtual Museum: 3D landscape reconstruction in the Orientalising period, North of Rome. A methodological approach proposal*, in Proceedings of Digital Heritage International Congress 2013, IEEE Advanced Technolgy of Humanity, ISBN: 978-1-4799-3169-9, Marseille France; Vol II pp. 223-331.

Pietroni E., Forlani M., Rufa C., 2015. *Livia's Villa Reloaded: An Example of Re-use and Update of a Pre-existing Virtual Museum, Following a Novel Approach in Storytelling Inside Virtual Reality Environments,* in Proceedings of Digital Heritage International Congress 2015, IEEE.

Ragghianti L, 1974. *Dall'arte al museo.* Eds.Vallecchi, Firenze.

Ryan M.L., 2001. *Narrative as Virtual Reality: Immersion and Interactivity* in "Literature and Electronic Media", Baltimore and London, Johns Hopkins University press, HB.

Turri, E., 1982. *Dentro il paesaggio,* Eds. Bertani, Verona.

Turri E., 2004. *Il paesaggio e il silenzio.* Eds. Marsilio.

Varela F., Thompson E., Rosch, E. 1991. *The Embodied Mind. Cognitive Science and HumanExperience*, MIT Press, Cambridge

Leap Motion, 2015: https://www.leapmotion.com

Microsoft Kinect, 2015: https://dev.windows.com/en-us/kinect

V Must Network: www.v-must.net

Cultural Heritage and Digital Technologies

Theory, Methods and Tools for the Study and Dissemination of Knowledge in the Archaeological Practice

Riccardo Olivito, Emanuele Taccola and Niccolò Albertini

Abstract Over the years, the World of Archaeology has looked at modern technologies of 3D data acquisition and visualization in an opposite way: from an initial suspicion and refusal, it has passed to an excessive and sometimes uncritical enthusiasm. Once accepted that the use of Digital Technologies in Archaeology might not be a goal but the way by which to create knowledge, modern archaeologists must create a strong theoretical basis, and coordinate interdisciplinary teams able to cover all the different phases of the archaeological research. In order to have a both scientifically rigorous and efficient communication, it is necessary that the archaeologists are aware of the features, both positive and negative, of the devices to be used for the dissemination and sharing of very different kinds of data. Among these new tools, virtual immersive environments play a key role for the archaeological practice, since they allow for the visualization and analysis in real-time of different types of data and the interaction with them. The case study of the *agora* of Segesta represents an example of these new trends.

Sections "From a Processual- to a Cyber-Archaeological approach" and "Theoretical Issues" are by Riccardo Olivito; Sections "Museums and Virtual Museums" and "Studying and Knowing the Past Through Virtual Immersive Environments" are by Emanuele Taccola; Section "Digital tools and Virtual Reality" is by Niccolò Albertini; Section "Conclusions" is by Riccardo Olivito and Emanuele Taccola.

R. Olivito (✉) · N. Albertini
Scuola Normale Superiore, Pisa, Italy
e-mail: riccardo.olivito@sns.it

N. Albertini
e-mail: niccolo.albertini@sns.it

E. Taccola
University of Pisa, Pisa, Italy
e-mail: emanuele.taccola@unipi.it

M. Forte and S. Campana (eds.), *Digital Methods and Remote Sensing in Archaeology*, Quantitative Methods in the Humanities and Social Sciences,
DOI 10.1007/978-3-319-40658-9_20

Introduction

To trace out a synthetic picture of the dialogue between the world of Cultural Heritage, especially Archaeology, and Digital Technologies is a quite hard and complex issue. During the last decade the widespread diffusion, both in the scientific and dissemination fields, of digital tools which are increasingly more elaborated, efficient, and inexpensive, has represented one of the main elements stimulating a 'Digital Revolution' which has drastically involved the world of Cultural Heritage.

However, it would be indeed misleading to think of the digital revolution as a recent phenomenon. In order to fully understand the reasons which made this relation indissolvable (i.e. that between Archaeology/Cultural Heritage and Digital Technologies) it is necessary to quikly review the main steps that since 60s have led to think of the use of technological tools (especially the digital ones) as indispensable in the archaeological practice.

State of the Art

From a Processual- to a Cyber-Archaeological Approach

The first step in the creation of a strong relation between Archaeology and technology dates back to the 60s. It was during that period that Archaeology was involved in a deep theoretical debate concerning all the aspects of the research activity, but also those relating to data interpretation and dissemination methods.

In this regard, a triggering element was represented by the rapid spreading of scientific tools and methods to be applied to a discipline that was perceived as still strongly linked to traditional hermeneutic methodologies. With the birth of the so-called Processual Archaeology (or New Archaeology) and the criticism of the Cultural-Historical phase, L.R. Binford (Binford and Quimby 1972) and other processual archaeologists started to consider each cultural phenomenon as unique and unrepeatable. From this point of view, the widespread use of scientific methods and technological tools (quantitative, taxonomic, computational, and laboratory analysis) seemed to be the only way to bring Archaeology nearer to the so-called Hard Sciences.

At the same time, the research of the exactitude of the material data, to be investigated through 'aseptic' and objective tools and methods, led to an overestimation of the role played by scientific analysis tools to the detriment of the interpretative process.

For this reason, during the 80s, Hodder (1986) and other followers of the so-called Post-processual or Anti-Processual (Renfrew 1994) archaeological current, questioned the theoretical approach of the New Archaeology. They rejected Binford's positivism and the idea of Archaeology as a Hard Science. In this sense, the Post-processual Archaeology has correctly pointed out the variety and unpredictability of factors (of human, social, cultural, and environmental nature) which affect the historical process and cannot be considered as invariable and independent from the context in which they take place. According to Post-processual archaeologists, therefore, the historical reconstruction is not merely a sum of calculable elements. It is indeed a combination of factors and phenomena that, once investigated with the methods and scientific instruments introduced by Processual Archaeology, have to be interpreted devoting special attention to the context and the individual and psychological factors.

Since the mid-90s the debate between New Archaeology and Post-Processual Archaeology has been further enriched due to the mediation attempts by C. Renfrew and other followers of the so-called Cognitive Archaeology, whose task was: "not to set out different cognitive categories in some a priori way, to posit special 'ancient' or 'pre-modern' modes of thought, but to seek to study the way in which cognitive processes operated in specific contexts, and to investigate the interrelationship between those processes and the social context which harboured and promoted them." (Renfrew 1994).

It is in this articulated theoretical context that new techniques of computer-graphics and Virtual Reality have been recently integrated, stimulating a transition from two-dimensional to three-dimensional visualization methods. The new theoretical and technical substratum, thus, allowed for the birth and development of the so-called Virtual Archaeology and for a strong advancement in the production of archaeological reconstructions which, since the end of 90s and the beginning of the 21st century have been increasingly characterized by a high-photorealism (Forte and Siliotti 1997).

Nevertheless, as Forte has noticed (2014), since Virtual Archaeology was mainly focused on the graphic and visual issues, it did not pay too much attention to perception and behaviours and, consequently, to the capability of understanding the past and its different articulations. Quoting M. Forte, one could say that "Virtual-Archaeology was not able, substantially, to elaborate sophisticated cyber behaviours in 3D real time environment" (Forte 2010), mainly because the reconstructive process was quite independent from the interpretative one.

On the basis of these previous experiences, during the last years a new promising phase for the archaeological studies is growing up. Scholars have become increasingly interested in the necessity to emphasize the dynamic and continuously changing nature of the historical and archaeological data, that are inevitably influenced by an endless transformation, both in its formation and in the interpretation process. More than simply reflecting on 'what' to disseminate, it is now urgent to think about 'what and how' to share data and information in a way that could be as wide and complete as possible and in order to provide for different interpretative levels, not necessarily univocal and unchangeable.

In this regard, at present the word 'Cyber' represents a brand new element in the field of archaeological research. As Zubrow (2010) has noticed, Cyber involves the connotation of ‘control’ and ‘being controlled’. This connotation is central to Cyber-Archaeology, for “the core of the process is not into the model, data or environment but in the mutual relations produced by interaction/embodiment/enaction” (Forte 2010).

Virtual-Archaeology was mainly based on static ‘reconstructions’, which involved a binding and non-modifiable examination of the acquired data, whereas Cyber-Archaeology assigns a key role to the concept of ‘simulation’. The debate on the use of simulations or models in Archaeology as keys to elaborate a more dynamic knowledge of the Past is a long-standing one (Costopoulos and Lake 2010; Lake 2014).

As far as modern trends in Archaeology are concerned, ‘reconstructions’, in the way they have been used so far, involve the idea that a certain object, structure, or even context could be reconstructed as it used to be, not only in a ‘realistic’ but also in a ‘real’ manner (Clark 2010). Such a supposed truth is obviously fallacious, since it is simply impossible to have all the data and information one would need in order to re-create a certain item in a certain historical moment. On the contrary, ‘virtual simulations’, similarly to more ‘traditional simulations’ in the archaeological practice, are based on the idea that the virtual environment within which they are created is not a closed and unalterable space. It can rather increase the degree of perception and the knowledge of a given context and potential past (Forte 2010), by emphasizing affordances (Gibson 1950). Indeed, from a theoretical point of view, the open character of simulations, either traditional or the virtual ones, enables the possibility to take into consideration different variables, so that it could be possible to create an infinite number of ‘models’. Obviously, this is a very important and meaningful point in the cyber-archaeological approach. At the same time, and on behalf of the concrete presentation of the research results, cyber-archaeologists are aware of the importance to use a ‘simulation slice’, that is “a representation that reflects reasoned interpretation of how a particular place may have looked, and the kinds of things that may have happened there, at some slice in time” (Clark 2010). A ‘simulation slice’ is not a ‘reconstruction’, but rather only one of the possible interpretations of a certain context.

Cyber-Archaeology may thus represent an articulated and interdisciplinary solution, not only able to solve the dichotomy between Processual and Post-processual Archaeology, but also to overcome the limits of Virtual-Archaeology. As Zubrow (2010) notes, in fact, unlike the Virtual-Archaeology “[…] Cyber-Archaeology bridges the gap between ‘scientific’ and ‘interpretational’ Archaeology for it provides testable in the sense of adequacy material representations of either ‘interpretations’ and ‘scientific hypotheses and discoveries’”.

Nowadays, that of the Cyber-Archaeology seems to be the best approach to ensure a more thorough and dynamic investigation of the Past and, at the same time, a more detailed and effective dissemination of scientific results, even to non-experts.

Museums and Virtual Museums

The impact of these technological innovations has also involved the field of dissemination and communication, both on-site and in the museums.

The museums were progressively computerized since the 70s of the last century with the creation of virtual inventories and stock photos (Djindjian 2007). Nevertheless, since the 80s new technologies and the fundamental role of Internet have enabled the creation of real interactive virtual platforms that allow different users to immerse and interact with the 3D models. At present, many institutions and public and private foundations are increasingly relying on channels of communication, research and divulgation, which involve the use of 3D models. The main tool for the dissemination of this kind of data is the virtual museum (Antinucci 2007a).

The recent definition of virtual museum, proposed by V-MusT.net (virtual museum Transnational Network) is "a communication product made accessible by an institution to the public that is focused on tangible or intangible heritage. It typically uses interactivity and immersion for the purpose of education, research, enjoyment, and enhancement of visitor experience. Virtual museums are usually, but not exclusively delivered electronically when they are denoted as online museums, hyper-museum, digital museum, cyber-museums or web museums" (http://v-must.net/virtual-museums/glossary/virtual-museums-and-virtual-realities).

Given the complexity and wideness of the definition, the recent literature has tried to outline the typologies and the features that make an entity a virtual museum (Pescarin 2014). In addition, V-MusT.net is proposing a survey targeted at developers of virtual museums and virtual exhibitions to define the state of the art and to learn about the best practices adopted (https://docs.google.com/forms/d/1GWLZXODAvWbrsLNuBG71_cqod79ZqsFdThlK8KwJQME/viewform). This institution also takes care of the side of the user's behavior and experience, using evaluation tools and calling for more integrated approaches between cultural contents, interfaces and social and behavioral studies (Pescarin et al. 2014).

The aforementioned definition must be integrated and completed with the one proposed by Antinucci (2007b) on the opposite case, that is, what cannot be defined virtual museum. In particular:

1. A virtual museum is not the real museum transposed to the web;
2. A virtual museum is not an archive of, database of, or electronic complement to the real museum;
3. A virtual museum is not what is missing from the real museum.

Rather, as correctly noted by Antinucci, the virtual museums, with their wealth of technological tools that allow to expand the visual communication, enrich the textual apparatus to which it is traditionally given the task of explaining the contents of a museum. In this way it is possible to reach a wider audience, composed not only of scholars or experts, to encourage the creation of an effective visual communication to ordinary visitors: quoting Antinucci: "the virtual museum is the communicative projection of the real museum" (Antinucci 2007b).

However, not always the museums have been able to fully exploit the potential of new tools in order to create new knowledge, or new forms of knowledge. In more than one case, in fact, it is possible to notice that attention is still paid to what might be called 'aesthetic virtuosity', to the detriment of the quality of the transmitted information. Fortunately, the last few years have seen a growing interest in the theoretical aspects related to the use of new technologies, and thus for the possibility of increasing the divulgative impact of museums thanks to new ways of visual and interactive language. In particular, Virtual Reality, Augmented Reality and cybernetic approach, with special regard to immersion and interaction between users and data, allowed to refine the level and quality of the dissemination process, focusing on the possibility to differentiate the nature and type of information to be communicated according to the target that it is decided to reach. Similarly, the rapid spread of portable devices and techniques of augmented reality has enabled and helped to transform the experience of on-site visit, ensuring visitors to access and interactively select the contents and informations available on a website or a monument, following different narrative paths that can respond to the prior knowledge or the needs of everyone.

Despite what was said up to this point, and despite museums and exhibitions of Virtual-Archaeology are in strong increase, and with examples of excellence, still remains some suspicion in the academic world, for which the materiality of the object is essential and irreplaceable.

However, as it was pointed out, Archaeology and archaeological museums are already virtual. Without an explanation, the archaeological finds exposed in a museum are just ancient objects, and rarely masterpieces that can speak for themselves. This is even more evident for the archaeological sites, where the understanding and description of the ruins are strongly influenced by the interpretation of the archaeologist. The understanding of the objects resides in the explanation that the archaeologist gives of them, which bridges the gap between intelligence, experience and evidence. This is based on subjectivity in the process of acquisition and interpretation. In this perspective, a "virtual presentation is just the last link in a chain having more immaterial rings than material ones" (Niccolucci 2007).

Digital Tools and Virtual Reality

To draw a history of studies on the development and diffusion of technological tools in the last 50 years is a difficult task, and it becomes even more difficult if we try to focus on the last decade.

Since the 60s of the XXI century, the use of digital tools in the collection, management and presentation of data in Archaeology has progressively established itself, becoming now absolutely necessary.

But since the 90s, and even more in the last ten years, Digital Technologies have allowed Archaeology and generally the world of Cultural Heritage to make a substantial step forward. Personal computers have been gradually integrated with mobile devices, tablets, smartphones, wearable devices. Next to the widespread dissemination of more effective and rapid means of communication on a global scale (internet and social media), which have revolutionized the way we communicate and share information, the main merit of the new dimension to which Archaeology and the world of Cultural Heritage tend must surely be attributed to the diffusion of virtual reality techniques, today increasingly affordable and achievable thanks to the greater availability of tools such as 3D TV, 3D cinema, HDM, which allow us to use this technology as a tool for study and dissemination, both to an audience of experts and non-experts. Such phenomenon, in fact, besides allowing for a significant enrichment concerning the phase of data recording, represents an element equally revolutionary, since it has allowed for a progressive adjustment and a progressive habit (especially mental) to the three dimensions and the immersivity, i.e. to a perceptual dimension felt as alien and alienating until a few years ago.

In recent years, technologies for Virtual Reality (VR) have evolved continuously, the techniques have been refined, the computing capacity has increased and especially the component prices have dropped discreetly, allowing even the research centers to develop applications or tools for VR. Hardware development, and consequently software development, involved mainly Head mounted displays (HMD), CAVE-like systems, Virtual reality glasses, Data gloves, Data suits, Workbenches, Armbands gesture control and Natural User Interfaces.

The old tools for Virtual Reality were big and bulky, uncomfortable and difficult to understand by the user. The tools for virtual reality have become smaller, lighter and cheaper: the HMDs, initially large, heavy and enveloping, have been replaced with lighter models that fit neatly in the front of the face. The new HMDs allow to enjoy the best experience in immersive VR, involving the user in a total way. The main innovation in the HMDs is the Oculus Rift, which, in addition to a higher resolution than older models, is equipped with a gyroscope, an accelerometer and an external optical positional tracking. As for the CAVE-like systems, in addition to the increase in computing power and in the reduction of costs for hardware, they have substantially increased the quality and resolution of projectors, together with the spread of new more precise and functional systems of tracking and new graphics engines and development tools linked to the explosion of "consumer oriented" software (as Unity 3D, Unreal Engine, and so on). Even for what concerns the interaction, the last years have marked a radical step forward in the ability to interact actively with virtual environments. Until recently, in fact, it was essential to resort to the use of Data Gloves, i.e. gloves with sensors that allow the user to track position and gesture making him interact with the virtual environment (for instance the Peregrine e the 5DT Data Glove by LLC Realities), Data Suits (which work like data gloves but for the entire body), Joysticks and Wands.

These technologies are still used but are giving way to more natural interaction methods such as infrared and optical tracking: examples are systems like the

Optitrack, the Kinect and Leap (also developed in several versions improved significantly over the years). These systems, called NUI (Natural User Interface), make the fruition more intuitive and natural, especially for users who are approaching the VR for the first time, making the experience even more immersive. In parallel to the development and refinement of hardware and software technologies related to VR, also the VR applications relating to Archaeology have undergone a distinct shift, divisible mainly into three phases:

- Simple digitization of analogic data without really changing the workflow of the study of data;
- Creation of video or tour into pre-rendered environments with non-immersive and unnatural resources, where usually the interaction occurred with joysticks, keyboards or wands;
- Development of rigorously detailed reconstructions that include metadata, realized with devices more immersive and comfortable than older hardware.

Thanks to this last evolution of Virtual Reality Environments, a true enrichment of the workflow is obtained, giving scholars and non-expert audience the opportunity to immerse themselves into ancient archaeological contexts, with the consequent possibility of pursuing dynamic simulations rather than static and immutable reconstructions. In addition to this, it is worth recording recent acquisitions in the field of haptics (Carrozzino and Bergamasco 2010), which, albeit late compared to other lines of research in the field of Digital Technologies, constitute a sector of innovation of great interest in the perspective of an ever greater fidelity in immersive dimension and of an increase of the sense of embodiment within virtual environments.

Digital Technologies and Virtual Reality Serving Archaeology and Cultural Heritage

Other chapters in this book give a very good description of methods nowadays available for the acquisition and visualization of data and it is not necessary to sum them up here.

It is crucial not to consider them as mere tools to achieve higly impressive and photorealistic reconstructions, but as data repositories and informative vehicles allowing for a more rapid and open dissemination of knowledge.

Nowadays, this is a fundamental point in the scientific debate, since it is always necessary to remind that every representation is nothing but an interpretation or, quoting G. Bateson: “The map is not the territory, and the name is not the thing named” (Bateson 1979).

Indeed, as M. Forte has correctly pointed out, in the past years Digital Technologies have been mainly interested on the technical issues rather than looking at the “semantic level of the informative and communicational aspects. In the field of the virtual heritage, the risk was/is to enhance the amazing esthetic features despite the informative/narrative feedback and cognition within the virtual worlds" (Forte 2005).

In order to have a both scientifically rigorous and efficient communication, it is necessary that archaeologists are aware of the characteristics, both positive and negative, of the devices to be used for the dissemination and sharing of very different kinds of data. Among these new tools, Virtual Immersive Environments play a key role for the archaeological practice, since they allow for the visualization and interaction in real-time with different types of data-set, that means a new complex and articulated kind of analysis, cognition, and knowledge of the Past.

The spreading of Virtual Reality (in particular Virtual Immersive Environments) in the archaeological field and, more generally speaking, in the Cultural Heritage, allows for new considerations of both theoretical and practical nature.

Theoretical Issues

From a methodological point of view, the increasing diffusion of Virtual Immersive Reality Systems and Environments and a deeper awareness of their use fit with the most up-to-date theoretical approaches to the study of human perception and knowledge processes.

The ecological approach to human perception is certainly the one which has initially taken advantage of the use of virtual environments. In the ecological view the perception of Reality is the result of an incessant interaction between humans and external environment, and the learning process is based on the capacity to perceive differences in the ecosystem (Bateson 1972, 1979; Gibson 1979). During the cognitive process, our brain seeks constantly to interpret all the data and images, which are acquired, elaborated, and visualized, also on the basis of our past experiences (Gibson 1979). If this is true for the real world, the same can be postulated and examined while staying in a virtual environment representing, even in a different time and space, what is not virtual at all. Besides, it seems to be correct the idea that "The Virtual is mainly an Ecosystem, an Environment, so its rules are the rules of a theory of the systems. [...] The Virtual is not the opposite of the Real but of the actual, it constitutes entities" (Forte 2005).

On the way paved by the ecological thought, a higher attention for the physical interaction with the environments active during the cognitive and learning process has recently emerged. The possibility to actively operate within Virtual Immersive Environments, thus, allowed scholars to overcome a kind of interaction which was mainly visual, and to move forward towards the so-called enactivist approach (Bruner 1962; Maturana and Varela 1980, 1987; Varela et al. 1991). From the enactivist point of view, the cognitive process is indissolubly linked with the human being within the environment. Perception and knowledge are not just results of a merely sensitive action in the external world, but rather of the capability to interact in a complex way with the real world through the binomial mind-body: "The organism constituted by the brain-body partnership interacts with the environment as an ensemble, the interaction being of neither the body nor the brain alone. But complex organisms such as ours do more than just interact, more than merely generate the spontaneous or reactive external responses known collectively as behavior. They also generate internal responses, some of which constitute images (visual, auditory, somatosensory, and so on), which I postulate as the basis for mind" (Damasio 1994).

The key for understanding the transition from Virtual- to Cyber-Archaeology lies in fact in the centrality of the 'interaction' with the environment. It is not surprising, thus, that the cyber-archaeological approach recognizes a key role to the concept of 'embodiment' (Thompson and Varela 2001; Lesure 2005), to be intended as the way by which we experience the world and, due to the interaction with it, we perceive and know it. For this reason, in the cyber-archaeological practice in the archaeological pratice, together with the visual dimension it is fundamental to take into account the immersive and interactive nature of virtual environments, with the unavoidable passage from the idea of 'reconstruction' to that of 'simulation' (Clark 2010).

In this regard, the development of new techniques of natural gesture interaction in virtual environments is of the great value (Pescarin et al. 2013). These techniques do allow for reproducting and simulating, not just mentally but also physically, operations and activities we are used to during our daily life. In so doing, they can stimulate the creation of replicas of the concrete and real experience.

This consideration introduces to a further element, which is appropriate to remind in order to integrate the cognitive theories illustrated so far. If the ecological thought, and even more the Embodied Cognition Theory, contribute to the explanation of some of the principal mechanisms involved in the process of perception and knowledge of the reality, it would be a mistake not to take into account a cognitive-behaviorist approach in which the relation between body/mind and cognition/emotion, certainly fundamental for our knowledge of the world, is indeed conditioned by our previous subjective experience. The personal set of memories, together with our social and cultural background, not only influences our knowledge of the world and others organisms (human, animals, etc.), but also steers the way by which we interpret the reality, to be intended as a spatio-temporal dimension and, at the same time, as a network of personal and emotional relations.

Besides, cognitive psychologists, even those following an enactivist approach, have demonstrated that during the perception and learning process our brain does not store images considering them just like facsimiles of what we perceive by using our senses (especially the visual one). On the contrary, images "[…] are formed either under the control of sensory receptors oriented to the brain's outside (e.g., a retina), or under the control of dispositional representations (dispositions) contained inside the brain, in cortical regions and subcortical nuclei" (Damasio 1994). These representations, or dispositions, represent the core of our Knowledge and are the result of both innate and acquired knowledges. As a result, dispositional representations (i.e. our knowledge as a whole), evolve and change through time and thanks to our own experience of the world (Bartlett 1932). Quoting Damasio:

(1994) "The acquisition of new knowledge is achieved by continuous modification of such dispositional representations".

In this regard, it is not easy to accord completely with assertions of the radical constructivism by E. von Glasersfeld ("What is radical constructivism? It is an unconventional approach to the problem of knowledge and knowing. It starts from the assumption that knowledge, no matter how it is defined, is in the heads of the persons, and that the thinking subject has no alternative but to construct what he or she knows on the basis of his or her experience") (Glasersfeld 1995) or H. von Foerster ("The environment as we perceive it is our invention ") (Foerster 1984). Neither we want to sum up the differences between radical constructivism and enactivism by H. Maturana and F. Varela (Proulx 2008). While highlighting the high epistemological value of the Embodied Cognitive Theory and the key role played in the learning and knowing process by the interaction between mind/body and environments, it is however necessary to remind that "The belief that one's own view of reality is the only reality is the most dangerous of all delusions" (Watzlawick 1976).

We can move forward in this direction, referring to the very interesting research in the field of cognitive neuroscience conducted by V. Gallese who, following the revolutionary discovery of mirror neurons, could recently state that, in investigating the relation between body/mind and art, the level of description offered by cognitive neuroscience is necessary but still not enough. The starting point should be the personal experience of every single person, which has to be deconstructed and then studied through the sub-personal investigation typical of cognitive neuroscience. The obtained data could be subsequently used in order to reconsider the personal level from which the exam started, so to create a virtuous cognitive circle (Gallese 2014).

Virtual Immersive Environments, especially when they are integrated by natural gesture interaction systems allowing for a visualization and interaction with data and information as dynamic and natural as possible, constitute an extraordinarily efficient way to achieve a multimodal and multisensory learning process, in which visual, muscular, and viscero-motor neuronal circuits, together with affective and subjective stimuli, cyclically produce new knowledge (Freedberg and Gallese 2007).

From this point of view, an embodied simulation within a virtual immersive and interactive environment, either representing the unrolling of an archaeological excavation, or a monument, or even a museum or a particular historical event, will necessarily activate cognitive dynamics which will be contemporarily the results of the interaction between mind/body and external environment (in accordance with the theoretical approach of the Enactivism), and expressions of previous life experiences of the virtual environment user (in conformity with the principles of Cognitive Archaeology and of what we could define as a sociocultural constructivism).

On the basis of this considerations, the 'sense of presence', to be intended as "the subjective experience of being in one place or environment, even when one is physically situated in another" (Witmer and Singer 1998) or the result of the involvement and immersion in a virtual environment, plays a key role in the cognitive and learning process.

Virtual Immersive and Interactive Environments, where it is possible to visualize, interrogate, and interact in a dynamic and natural way with a large amount of data and information, that scholars have so far hardly used for scientific and dissemination purposes, bring us nearer to what C. Renfrew identified as the goal of a new archaeological Era, stating that "One conceives of the past as really existing in a physical world, much like the present, with human individuals living in their lives, and interacting with each other and with their environment very much as we do today. In other words, the past really happened. This clearly differs from an extreme positivist or empiricist position which might restrict our conception of the past exclusively to what which we can empirically learn about it. But this notion of a past which really happened is to be distinguished from our knowledge of the past, which has to be based upon our own observations and inferences, and is thus constructed by us using those observations" (Renfrew 1994).

By the light of these brief theoretical considerations, the increasing use of virtual interactive simulations, both in the study and dissemination of Cultural Heritage, is not only justified but worthy of further developments.

Studying and Knowing the Past Through Virtual Immersive Environments

At present, there are not so many cyber-archaeological projects allowing for a complete use of the increasing power of Virtual Immersive Environments.

With no doubt, one of the best example of this new archaeological trend and of the potentialities offered by the synergy between techniques of digital data acquisition and virtual environments is represented by the project 3D-Digging at Çatalhöyük, directed by M. Forte (Forte 2014; Berggren et al. 2015). More recently a similar cyber-archaeological project has been used in the excavations of the *agora* of Segesta, and it is presented in the following pages.

The *Agora* of Segesta Between Virtual- and Cyber-Archaeology

Introduction

The case of the *agora* of Segesta can be included fully in the discussion which has recently emphasized the evolution from Virtual-Archaeology to Cyber-Archaeology. As it will be described later, the modeling work with computer graphics of the monumental complex, begun in 2009 and still in progress, has produced a model, or, to quote Clark (2010), a *simulation slice*, initially designed for a simple passive use (rendering and movies). In 2012, following the exploit of the software of image-based modeling, the 3D documentation of the ruins already brought to light began. Finally, since 2013 the 3D documentation in real time of the excavation activity has been started. The model of the *agora* obtained with computer graphics has been integrated with the photogrammetric models within an immersive virtual environment (CAVE-like system) at the SMART of the Scuola Normale Superiore (directed by Prof. V. Barone), which allows the user to interact actively thanks to a natural interface manageable with simple hand gestures.

The Site

The *agora* of Segesta, in north-western Sicily, Italy, is a remarkable example of ancient public square of the late Hellenistic period (end of the 2nd century BC). The monumental complex consisted of a long portico (*stoa*) with two wings (82 × 20 m) disposed to delimit on three sides a large paved square (Ampolo and Parra 2012). The excavations, directed by Prof. C. Ampolo and Prof. M.C. Parra and carried out by the Laboratorio di Scienze dell'Antichità (LSA) of the Scuola Normale Superiore of Pisa in collaboration with the Department of Civilisation and Forms of Knowledge and the Laboratory of Drawing and Restoration (LADIRE) of the University of Pisa, have uncovered much of the architectural apparatus of the *agora*, frequently found in connection despite the collapse of the structures. These findings led to the formulation of a reconstructive proposal of the elevations (Abate and Cannistraci 2012). Originally, the portico was a two-story building, with an Ionic colonnade in the upper level and a Doric colonnade in the lower level. The latter was divided into two aisles by a row of octagonal pillars. The size of the site, the type of structures and the long attendance of the area (up to the Middle Age) make the *agora* of Segesta an ideal field of experimentation of representation techniques and 3D documentation.

Workflow

In 2009, the reconstructive hypothesis of the building proposed by scholars was translated into a 3D model thanks to computer graphics, using planimetries, elevations and drawings of architectural elements made using traditional 2D techniques (Taccola 2012) (Fig. 1). The model was initially employed to obtain simple rendering or movies, that is for a passive fruition by the final user. The limits of this utilization, already evident in origin, were nevertheless stimulus for an evolution of the methodological and practical approach for the acquisition, storage, integration and dissemination of successive models and related metadata, as well as for the ultimate goal, i.e. the interaction.

Indeed, in 2012 the 3D acquisition of the ruins unearthed begun, using close-range terrestrial and aerial survey techniques, thanks to the development of software for image-based modeling based on the principle of SfM. In this way, the model created with computer graphics and that one achieved by photogrammetry

Fig. 1 Image-based models of the stratigraphic sequence, realized during the fieldwork activity

have been integrated into an immersive virtual environment (CAVE), although the level of interaction was still limited to walk-metaphor interaction (Albertini et al. 2014) (https://www.youtube.com/watch?v=hpTfTCSpoD8) (Fig. 2). In the excavation campaign 2013 (Olivito and Taccola 2014), and more systematically in 2014 and 2015, the acquisition of the field activity in one of the areas of excavation with SfM techniques started. The operation, carried out during the whole duration of the fieldwork, led to the realization of ten models related to ten levels of attendance, many of them were segmented in the various stratigraphic units (Fig. 3). In addition, the models of two significant objects found within them were created. More than for an evaluation in real-time of the excavation activity (Dell'Unto 2014), the focus was to visualize ex-post the digging sequence in an immersive virtual environment, with the ability to interact with the models, query and display their metadata (Olivito et al. 2015). For this reason, all the models, properly aligned with each other in the reference system used within the excavation, were imported into the CAVE and integrated with the model of the site developed from the survey with UAV and with the simulation obtained by computer graphics. To facilitate the immersive interaction, an application has been designed, that integrates in a single device a sensor leap to 3D glasses, giving the user full mobility of the hands. The leap sensor allows to interact with the models with simple gestures, thanks to a natural interface specially created. Specifically, it is possible to select the level to analyze, extract and view individual items of particular interest, go back to metadata (available on an external device, such as a tablet) and pass to the model created with computer graphics of the entire monumental complex (for a demonstration video,

Fig. 2 The Cyber-Archaeology application within the CAVE

https://www.youtube.com/watch?v=vBJIpwSHQXA) (Fig. 4). The application, which is currently available for a single user, will be shortly multi-user (up to 4 people), thanks to the installation of a new camera system and wireless hand trackers. This will facilitate the implementation of the hands gestures and allow the introduction of new natural gestures interactions that will activate additional functions not yet available. The system is exportable and already tested on mobile wearable devices (such as the Oculus Rift DK2). This makes it possible to use the models also on-site, in order to further fill the gap between the phase of data collection and processing and the phase of interpretation during the fieldwork.

Results

The different steps in the development of the application concerning the *agora* of Segesta are a valid example of how the methodology used so far has moved from a static and 2D approach to a tridimensional, dynamic, and interactive set of tools. At the same time, these developments efficiently display some of the issues concerning our percepetion of space and, consequently, the way humans reach the knowledge of the real World (see Section "Theoretical Issues").

Although still in testing, the system has in our view a very substantial scientific, educational and divulgative potential. The immersion and interaction ensured by the application is a significant support for the study of the stratigraphic sequence, being able to validate or change the interpretations elaborated during excavation. The possibility of movement and visualization of the simulation of the entire *agora*, as well as the levels of excavation, is an effective method of data dissemination and exploitation addressed even to non-expert users and is consistent with the

Fig. 3 3D model of the agora within the CAVE

guidelines suggested in several contributions related to the relation between Digital technologies and Cultural Heritage. For instance, in our specific case the immersion in the virtual simulation of the *agora* of Segesta allows for considering issues not easily verifiable by using only 2D tools available so far, such as possible solutions adopted in the carpentry; physical and spatial relations between the *stoa*, its internal space, and the other monuments of the public square; relations between architecture, decorations, and documentary apparatus (e.g. statues and inscriptions); the role played by natural and artificial lighting in different moments of the day. Due to the possibility to acquire an increasingly wider amount of data, even during the fieldwork, and to the opportunity to make them easily available within the virtual environment, the user can both select the metadata depending on his knowledge

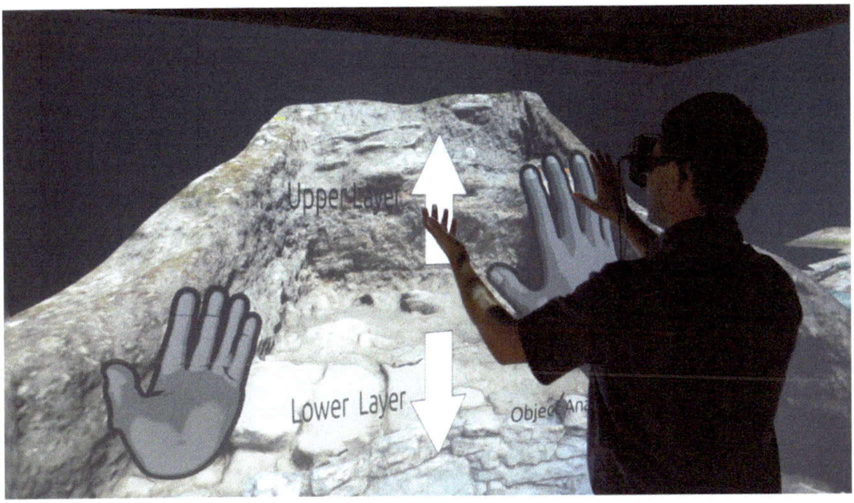

Fig. 4 3D model of the agora of Segesta

level or needs and use them to confirm or suggest new interpretations of the visualized models. In this sense, it is worth recalling J. Clark, who efficaciously stated that “Through virtual models (i.e. Virtual Simulations) one can much more readily present alternate ‘visions’ or stories of the past that can serve as valuable tools for both research and education” (Clark 2010).

Conclusions

In conclusion, it is important to remind that Digital Technologies and in particular Virtual Reality in the archaeological field might not be a goal but the way by which to create, disseminate, and share the knowledge of the Past. As scholars, we have to wonder how modern archaeologists have to deal with these new tools: rather than trying to replace the Virtual Reality and Digital Technologies experts, they must knowledgeably operate on the basis of a strong theoretical basis, and arrange and manage the work of interdisciplinary teams that could be able to cover all the different phases of the research, from the data-input to the data-output. As a result, archaeologists should ask themselves: how the fieldwork activities might be organized in order to acquire as many information as possible? And above all, how the acquired data have to be transformed and disseminated in order to create new knowledge?

Without a doubt, new Digital Technologies do allow us to pursue a collection and management of data in a much more accurate, detailed, rapid, and inexpensive way than just a few years ago. For this reason, archaeologists have to evaluate, select, and filter data and information looking at the final purposes and users they want to reach. On the one hand, the goal is to allow our scientific knowledge of the Past to be more accurate, dynamic, not limited to static and definitive reconstructions but rather open to continous investigations of a multifaceted set of data. On the other hand, the objective is to share knowledge in a wider, easier, and more rapid and accessible way to both the experts and non-experts.

The huge amount of data represents a heterogeneous database, strictly and strongly interconnected, which can be more suitably interrogated through their visual representations so to take advantage of our brain skills, which activate one-third of the neurons when processing visual information. Given the increasing computational and graphical power in modern computers, scholars have to shift the focus of their research on what kind of data it is necessary to present and how to present them, the final aim being to create new knowledges rather than higly impressive and photorealistic reconstructions with a low informative power.

References

Abate, A., and O. Cannistraci. 2012. La stoa Nord dell'agora di Segesta: alcune note preliminari sull'elevato architettonico dell'ala Ovest. In *Agorai di Sicilia, agorai d'Occidente*, ed. C. Ampolo, 305-319. Pisa: Edizioni della Normale.

Albertini, N., V. Barone, A. Brogni, S. Legnaioli, D. Licari, and E. Taccola. 2014. The agora of Segesta in immersive virtual environments. *Science and Technology for the Safeguard of Cultural Heritage in the Mediterranean Basin, Proceedings of the sixth international conference4*. Vol. I:299-304.

Ampolo, C., and M.C. Parra. 2012. L'agora di Segesta: uno sguardo d'assieme tra iscrizioni e monumenti. In *Agorai di Sicilia, agorai d'Occidente*, ed. C. Ampolo, 271-285. Pisa: Edizioni della Normale.

Antinucci, F. 2007a. *Musei virtuali : come non fare innovazione tecnologica*. 1. Aufl. Percorsi, vol. 101. Roma: Laterza.

Antinucci, F. 2007b. The Virtual Museum. *Archeologia e Calcolatori* Supplemento 1:79-86.

Bartlett, F.C. 1932. *Remembering; a study in experimental and social psychology*. The Cambridge psychological library. Cambridge Eng.: The University press.

Bateson, G. 1972. *Steps to an ecology of mind; collected essays in anthropology, psychiatry, evolution, and epistemology*. Chandler publications for health sciences. San Francisco,: Chandler Pub. Co.

Bateson, G. 1979. *Mind and nature : a necessary unity*. New York: Dutton.

Berggren, Å., N. Dell'Unto, M. Forte, S. Haddow, I. Hodder, J. Issavi, N. Lercari, C. Mazzuccato, A. Mickel, and J. Taylor. 2015. Revisiting reflexive archaeology at Çatalhöyük: integrating digital and 3D technologies at the trowel's edge. *Antiquity* 89:433-448.

Binford, Lewis R., and G.I. Quimby. 1972. *An archaeological perspective*. Studies in archaeology. New York,: Seminar Press.

Bruner, J.S. 1962. *On knowing; essays for the left hand*. Cambridge: Belknap Press of Harvard University Press.

Carrozzino, M., and M. Bergamasco. 2010. Beyond virtual museums: Experiencing immersive virtual reality in real museums. *Journal of Cultural Heritage* 11 (4):452-458.

Clark, J.T. 2010. The Fallacy of Reconstruction. In *Cyber-archaeology*, ed. M. Forte, 63-73. Oxford, England: Archaeopress.

Costopoulos, A, and M.W. Lake. 2010. *Simulating change : archaeology into the twenty-first century*. Foundations of archaeological inquiry. Salt Lake City: University of Utah Press.

Damasio, A.R. 1994. *Descartes' error : emotion, reason, and the human brain*. New York: G. P. Putnam.

Dell'Unto, N. 2014. The Use of 3D Models for Intra-Site Investigation in Archaeology. In *3D Recording and Modelling in Archaeology and Cultural Heritage. Theory and best practices*, eds. F. Remondino, and S. Campana, 151-158. BAR International Series 2598. Oxford: Archaeopress.

Djindjian, F. 2007. The virtual museum: an introduction. *Archeologia e Calcolatori* Supplemento 1:9-14.

Foerster, H. von. 1984. On constructing a Reality. In *The Invented reality : how do we know what we believe we know? : contributions to constructivism*, ed. P. Watzlawick, 41-61. New York: Norton.

Forte, M. 2005. Communicating the Virtual. In *MIDECH '05, Multimedia.Information@Design for Cultural Heritage*, eds. M. Rossi, and P. Salonia, 74-84. Roma: Aracne.

Forte, M. 2010. Introduction to Cyber-Archaeology. In *Cyber-archaeology*, ed. M. Forte, 9-13. Oxford, England: Archaeopress.

Forte, M. 2014. 3D Archaeology: New Perspectives and Challenges—The Example of Çatalhöyük. *Journal of Eastern Mediterranean Archaeology and Heritage Studies* 2 (1):1-29.

Forte, M., and A. Siliotti. 1997. *Virtual archaeology : re-creating ancient worlds*. New York: H.N. Abrams.

Freedberg, D., and V. Gallese. 2007. Motion, emotion and empathy in esthetic experience. *Trends in cognitive sciences* 11 (5):197-203.

Gallese, V. 2014. Arte, corpo, cervello : per un'estetica sperimentale. *MicroMega* (2):49-67.

Gibson, J.J. 1950. *The perception of the visual world*. Boston: Houghton Mifflin.

Gibson, J.J. 1979. *The ecological approach to visual perception*. Boston: Houghton Mifflin.

Glasersfeld, E. von. 1995. *Radical constructivism : a way of knowing and learning*. Studies in mathematics education series, vol. 6. London; Washington, D.C.: Falmer Press.

Hodder, I. 1986. *Reading the past : current approaches to interpretation in archaeology*. Cambridge Cambridgeshire; New York: Cambridge University Press.

Lake, M.W. 2014. Trends in Archaeological Simulation. *Journal of Archaeological Method and Theory* 21 (2):258-287.

Lesure, R.G. 2005. Linking theory and evidence in an Archaeology of Human Agency: iconography, style, and theories of embodiment. *Journal of Archaeological Method and Theory* 12 (3):237-255.

Maturana, H.R., and F.J. Varela. 1980. *Autopoiesis and cognition : the realization of the living*. Boston studies in the philosophy of science, vol. 42. Dordrecht, Holland; Boston: D. Reidel Pub. Co.

Maturana, H.R., and F.J. Varela. 1987. *The tree of knowledge : the biological roots of human understanding*. 1st Aufl. Boston: New Science Library : Distributed in the United State by Random House.

Niccolucci, F. 2007. Virtual museums and archaeology : an international perspective. *Archeologia e Calcolatori* Supplemento 1:15-30.

Olivito, R., and E. Taccola. 2014. 3D modelling in the agora of Segesta: techniques and data interpretation. *Archeologia e Calcolatori* 25:175-188.

Olivito, R., E. Taccola, and N. Albertini. 2015. A Hand-Free Solution for the Interaction in an Immersive Virtual Environment : the Case of the Agora of Segesta. *ISPRS-International Archives of the Photogrammetry, Remote Sensing and Spatial Information Sciences* XL-5/W4:31-36.

Pescarin, S. 2014. Virtual reality and cyberarchaeology – virtual museums. In *3D Recording and Modelling in Archaeology and Cultural Heritage. Theory and best practices*, eds. F. Remondino, and S. Campana, 129-133. BAR International Series 2598. Oxford: Archaeopress.

Pescarin, S., W. Hupperetz, A. Pagano, C. Ray, and M. Wallergård. 2014. Evaluating Virtual Museums: archaeovirtual test-case. In *Computer Applications and Quantitative Methods in*

Archaeology. Proceedings of the 40th Conference in Computer Applications and Quantitative Methods in Archaeology, Southampton, United Kingdom, 26-30 March 2012, 76-84.
Pescarin, S., E. Pietroni, L. Rescic, M. Wallergård, K. Omar, and C. Rufa. 2013. NICH: A preliminary theoretical study on Natural Interaction applied to Cultural Heritage contexts. In *1st Digital Heritage International Congress*, 355-362. IEEE.
Proulx, Jerome. 2008. Some differences between maturana and Varela's theory of cognition and constructivism. *Complicity: An International Journal of Complexity and Education* 5 (1):11-26.
Renfrew, C. 1994. Towards a cognitive archaeology. In *The ancient mind : elements of cognitive archaeology. New directions in archaeology*, eds. C. Renfrew, and E.B.W. Zubrow, 3-12. Cambridge England; New York: Cambridge University Press.
Taccola, E. 2012. Ricostruzioni 3D per l'agora di Segesta. In *Agorai di Sicilia, agorai d'Occidente*, ed. C. Ampolo, 287-289. Pisa: Edizioni della Normale.
Thompson, E., and F.J. Varela. 2001. Radical embodiment: neural dynamics and consciousness. *Trends in cognitive sciences* 5 (10):418-425.
Varela, F.J., E. Thompson, and E. Rosch. 1991. *The embodied mind : cognitive science and human experience*. Cambridge, Mass.: MIT Press.
Watzlawick, Paul. 1976. *How real is real? : Confusion, disinformation, communication*. New York: Random House.
Witmer, B.G., and M.J. Singer. 1998. Measuring presence in virtual environments: A presence questionnaire. *Presence: Teleoperators and virtual environments* 7 (3):225-240.
Zubrow, E.B.W. 2010. From Archaeology to I-archaeology : Cyberarchaeology, paradigms, and the end of twentieth century. In *Cyber-archaeology*, ed. M. Forte, 1-7. Oxford, England: Archaeopress.

Index

M. Forte and S. Campana (eds.), *Digital Methods and Remote Sensing in Archaeology*, Quantitative Methods in the Humanities and Social Sciences,
DOI 10.1007/978-3-319-40658-9

FSC
www.fsc.org
MIX
Papier | Fördert gute Waldnutzung
FSC® C083411

Kolibri 360 GmbH
Ferdinand-Jühlke-Straße 7
99095 Erfurt, Deutschland
produktsicherheit@kolibri360.de